SEMICONDUCTOR LASER DIODE TECHNOLOGY AND APPLICATIONS

Edited by **Dnyaneshwar Shaligram Patil**

Semiconductor Laser Diode Technology and Applications
Edited by Dnyaneshwar Shaligram Patil

Published by InTech
Janeza Trdine 9, 51000 Rijeka, Croatia

Publishing Process Manager Dejan Grgur

Technical Editor Teodora Smiljanic

Cover Designer InTech Design Team

Additional hard copies can be obtained from orders@intechopen.com

Semiconductor Laser Diode Technology and Applications,
Edited by Dnyaneshwar Shaligram Patil
p. cm.
ISBN 978-953-51-0549-7

For EU product safety concerns:
IN TECH d.o.o., Prolaz Marije Krucifikse Kozulić 3, 51000 Rijeka, Croatia,
info@intechopen.com or visit our website at intechopen.com.

Meet the editor

Dr D. S. Patil graduated from Poona University with honors. He received an MSc in Electronics Science with a first class, an MCM degree with an A+ grade from Poona University, and a PhD in Electronics from the North Maharashtra University, Jalgaon (Maharashtra), India. Dr Patil currently works as Associate Professor and has successfully completed a major Young scientist project. He has published and presented 157 research papers and has guided many students through their innovative research. Dr Patil worked as a reviewer for many reputed international journals, delivered many invited talks and is an active member of IEEE society.

Contents

Preface

The impact that semiconductor lasers have had on our society cannot be denied. Due to their small size, low power dissipation and reliability, they are of growing importance in this modern world. This book aims to provide an insight into the many aspects of semiconductor laser diode in the form of featured chapters. This collection of chapters will be of considerable interest to engineers, scientists, physicists and technologists working in research and development in the semiconductor laser diode field. The purpose of the book is to provide a basis for understanding the characteristics, operation, and applications of semiconductor lasers. The goal of is to bridge a gap between researchers and specialists at the leading edge of the research field, including scientists and technologists. This book presents a detailed and comprehensive treatment of semiconductor laser diode technology, which can serve a number of purposes for a number of different groups. Besides this, it provides the scientists and technologists with a detailed description of laser diode physics and its applications. The authors have extensive research experience in the field of semiconductor laser diode and their experience is reflected in the form of chapters they have contributed.

The book entitled "Semiconductor Laser diode Technology and Applications" has been organized in three main sections. The first section deals with the optimization of semiconductor laser diode parameters. Various characteristics like quantum efficiency, output power and their dependence on cavity length and composition have been discussed. There are total 5 articles included in this section. The electron transport in nitride-based semiconductors has been discussed considering use of wide band gap semiconductors for the visible light emission.

The second section is based on laser diode technology and comprises 7 chapters. It incorporates articles explaining high speed transient processes, distributed feedback laser diode, low frequency noise characteristics in single mode and multimode fibers, and advancement in laser diode arrays. There are some interesting articles like multi-wavelength light source. Fascinating applications of laser diode have been explored in the third section of this book, which contains 7 articles. Some interesting applications like use of laser diode for space, welding, printing, gas spectroscopy, terahertz wave generation, doped fiber lasers and high power laser diode packaging have been discussed.

Editing this book has been a rewarding experience for me. I have been associated with semiconductor laser diode from last 17 years and have attempted to review relevant and necessary background material for each successive article in each section of this book. One unique feature of this book is that it includes articles from optimization of semiconductor laser diode parameters to advanced applications.

Finally, I would like to thank all contributing authors for their dedicated efforts in preparing the chapters presented in this book. Their work in this emerging area is essential to the vibrancy and fast pace of progress in this modern world. I hope that this book can serve as a useful reference for engineers, scientists, technologists and physicists engaged as researchers and postdoctoral fellows.

Dr. Dnyaneshwar Shaligram Patil
Department of Electronics
North Maharashtra University, Jalgaon
India

Section 1

Optimization of Semiconductor Laser Diode Parameters

1

Effect of Cavity Length and Operating Parameters on the Optical Performance of $Al_{0.08}In_{0.08}Ga_{0.84}N/ Al_{0.1}In_{0.01}Ga_{0.89}N$ MQW Laser Diodes

Alaa J. Ghazai*, H. Abu Hassan and Z. Hassan
Nano-Optoelectronics Research and Technology Laboratory,
School of Physics, Universiti Sains Malaysia,
Malaysia

1. Introduction

In this chapter, we discuss the effects of the cavity length of the active region in quaternary $Al_{0.08}In_{0.08}Ga_{0.84}N/Al_{0.1}In_{0.01}Ga_{0.89}N$ multiquantum-well (MQW) laser diodes (LD) on its performance. Semiconductor lasers emit coherent laser light with relatively small divergence and have long operating lifetimes because their very compact sizes can be easily integrated with a solid-state structure. They have very high efficiencies and need only a few milliwatts of power because they are cold light sources that operate at temperatures much lower than the equilibrium temperature of their emission spectra. The objective of the current study is to design the smallest possible semiconductor laser diode with good performance. The effects of various values of cavity length (ranging from 400–1200 nm) for $Al_{0.08}In_{0.08}Ga_{0.84}N/Al_{0.1}In_{0.01}Ga_{0.89}N$ MQW LD on laser parameters are investigated, including internal quantum efficiency η_i, internal loss α_i, and transparency current density J_0. High characteristic temperature and low transparency current of the $Al_{0.08}In_{0.08}Ga_{0.84}N/Al_{0.1}In_{0.01}Ga_{0.89}N$ MQW LD was obtained at a cavity length of 400 µm.

2. Overview

In the last decade of the 20th century, zinc selenide (ZnSe)-based quantum-well (QW) heterostructures in the blue-green spectrum were the first laser diodes (LD) investigated by Hasse et al. (1991) and(Haase 1991) (Jeon 1991)[1, 2]. However, rapid developments in III-nitride compounds by Nakamura et al. (1993) have brought LEDs based on these materials to technological capability and commerciality (Nakamura 1993)[3]. Violet InGaN QW LD under pulsed operation was demonstrated by Nakamura and Akasaki in 1996, and major improvements have since been achieved in its performance and device durability(Akasaki 1996; Nakamura 1996) [4, 5]. High-power LD (approximately 30 mW) was launched as a commercial product in September 2000(Nagahama 2000) [6]. Recently, aluminum indium

* Corresponding Author

gallium nitride (AlInGaN) alloys have been studied as the basis for next-generation optoelectronic applications, such as optical disk technology. Moreover, quaternary AlInGaN alloy has a wide band gap energy covering IR, visible, and UV regions, and permits an extra degree of freedom by allowing independent control of the band gap energy and lattice constant. The specific properties of III-nitride, such as its wide gap, high band offset, strong polarization fields, non-ideal alloy system, and so on, need to be identified for the design and optimum performance of LDs. Quaternary alloy has the issue of simultaneous incorporation of both In and Al, but offers the further quality of a "tunable" material, in terms of both the optical emission wavelength and lattice constant. Certainly, these issues are coupled with the control of the electrical properties of the p-n junction involved. Nagahama et al. (2001) studied both GaN and AlInGaN QW LD in the near-UV region, which has led to demonstrations of continuous-wave (CW)-edge-emitting lasers at room temperature near 370 nm(Nagahama 2001) [7].

Quaternary AlInGaN LDs with emission wavelengths less than 360 nm were also developed using Al and In content between 3% and 12% (Nagahama 2000; Nagahama S. 2001; Masui S. 2003; Michael K. 2003)[6, 8–10]. Wavelength "tunability" of the lasers was achieved for different Al and In compositions in the quaternary well, and, equally important, in the corresponding variations of the threshold current density J_{th}. In particular, an increase in J_{th} with increasing Al concentrations up to approximately 12 KA/cm^2 for x(Al) = 0.08 was noted. This increase was likely mainly the result of the quality of the quaternary $Al_xIn_yGa_{1-x-y}N$, in terms of both general morphology and defects (Nagahama S. 2001)[10]. A quaternary $Al_{0.03}In_{0.03}Ga_{0.94}N$ QW device under CW operation at room temperature, with a maximum output power reaching several milliwatts lasing at 366.4 nm, was also observed (Nagahama 2001)[6]. Shingo et al. reported $Al_{0.03}In_{0.03}Ga_{0.94}N$ UV LD under CW operation with emission wavelengths of 365 nm and a lifetime of 2000 h at an output power of 3 mW. They also achieved a short lasing wavelength of 354.7 nm under pulse current injection [8]. Michael et al. demonstrated room-temperature (RT) pulsed operation of AlInGaN MQW LD emission between 362.4 and 359.9 nm. Extending toward deep UV emission wavelength seemingly involves big challenges that become increasingly complicated with decreasing lasing wavelength [9]. Y. He et al. reported an optically pumped RT pulsed laser at 340 nm based on a separate confinement AlInGaN MQW heterostructure design [11]. The improvement of lasing characteristics, such as large optical gain and reduced threshold current of the GaN/AlInGaN QW laser using quaternary AlInGaN as a barrier, was reported by Seung et al. [12].

Recently, Michael et al. demonstrated the successful injection of AlInGaN ultravoliet laser on low dislocation density bulk AlN substrates using the MOCVD technique. The lasing wavelength was 368 nm under pulsed operation [13]. Thahab et al. reported ultraviolet quaternary AlInGaN MQW LDs using ISE TCAD software. For DQW, they simulated lasing wavelength of 355.8 nm under CW operation. However, the threshold current was high [14]. Overall, these initiatives encourage more development efforts on III-nitride materials as light emitters into deeper UV.

Several attempts have been made to improve the lasing characteristic and the reliability of the laser diodes in the last few years. The small active region in the laser diodes reduce the number of threading dislocation density (TDD) in the active region, which contributes to the fabrication of reliable laser diodes [15].

This chapter focuses on the simulation of edge emitting LD, whereas, in most other lasers, incorporation of an optical gain medium in a resonant optical cavity exists. The designs of both the gain medium and the resonant cavity are critical. The gain medium consists of a material that absorbs incident radiation over a wavelength range of interest. If it is pumped with either electrical or optical energy, the electrons within the material can be excited to higher, non-equilibrium energy levels. Therefore, the incident radiation may be amplified, rather than absorbed, by stimulating the de-excitation along with the generation of additional radiation. If the resulting gain is sufficient to overcome the losses of some resonant optical mode of the cavity, this mode is said to have reached its threshold, and coherent light will be emitted.

Resonant cavity provides the necessary positive feedback for the amplified radiation; lasing oscillation can be established nonstop above threshold pumping levels. As with any other oscillator, the output power saturates at a level equal to the input, minus any internal losses.

In this chapter, the effect of cavity length parameter on the optical performance of $Al_{0.08}In_{0.08}Ga_{0.84}N/Al_{0.1}In_{0.01}Ga_{0.89}N$ MQW LD is reported. Different important operating parameters are investigated, including internal quantum efficiency η_i, the internal loss α_i, characteristic temperature T_0, and transparency current density J_0 for our structure. For the lasers, these parameters are functions of the laser structure cavity dimension.

3. Oscillation condition of Fabry-Perot Laser

The simplest LD, Fabry-Perot LD, is realized by a pair of reflector mirrors facing each other, which are built together with the active material as resonator.

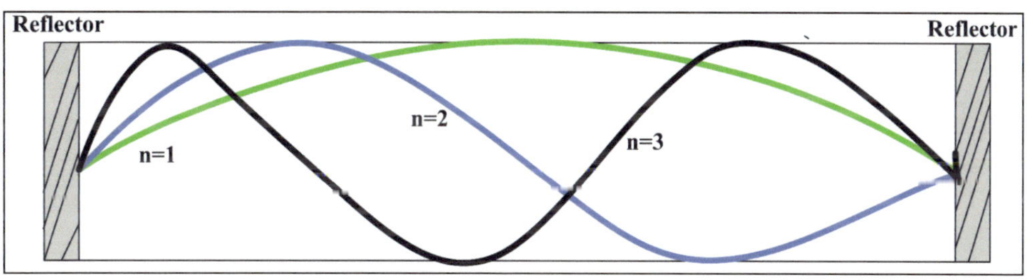

Fig. 1. Schematic description of a Fabry-Perot resonant cavity with reflecting facets on each end, the different modes supported within the cavity (n) exist in integer values.

For obtaining oscillation conditions, the plane optical waves traveling back and forth along the length of the laser are considered. These waves have optical frequencies of $\omega = 2\pi f$ with an associated propagation constant of $\beta = 2\pi/\lambda_m$, where λ_m is the wavelength in the material. Such a wave, which starts from the left-hand reflector and travels to the right, is referred to as a forward wave, and has its phase and amplitude written in complex form:

$$E_f(z) = E_0 e^{(g-\alpha_i)z} e^{-j\beta z} e^{j\omega t} \qquad (1)$$

The amplitude decays or grows with distance because the wave suffers scattering and other fixed losses α_i per unit length. However, it also experiences a material optical gain g per unit length caused by the stimulated recombination of electrons and holes. Consider that the cavity length is L, the reflectivity of the right and left facets are R_1 and R_2, respectively, and there is no phase change on the reflection from the right facets at either end. The forward wave has a reflected fraction R_1 at the right facet (z = L). This fraction then travels back from right to left. According to Eq. (1), these reverse fields are described by

$$E_r(z) = \{ E_0 e^{(g-\alpha_i)L} e^{-j\beta L} \} e^{(g-\alpha_i)(L-z)} e^{-j\beta(L-z)} \tag{2}$$

where the time variation $e^{j\omega t}$ occurring in all terms is implicitly included.

The reverse wave travels back to the left facet ($z = 0$), and fraction R_2 is reflected to form the forward wave. For stable resonance, the amplitude and phase after this single whole round trip have to be identical with the phase and amplitude of the wave when it began:

$$E_0 = E_0 R_1 R_2 e^{(g-\alpha_i)2L} e^{-j2\beta L} \tag{3}$$

This gives the amplitude condition for stable oscillation:

$$R_1 R_2 e^{(g-\alpha_i)2L} = 1 \tag{4}$$

This could be also written as

$$g = \alpha_i + \frac{1}{2L} \ln\left(\frac{1}{R_1 R_2}\right) \tag{5}$$

where the logarithmic term can be considered as a distributed reflector loss α_m. Considering that only a fraction of the photons of the guided optical wave interacts with the active region, and considering the optical confinement factor Γ, Eq. (5) should be written as

$$\Gamma g = \alpha_i + \frac{1}{2L} \ln\left(\frac{1}{R_1 R_2}\right) = \alpha_i + \alpha_m \tag{6}$$

with,

$$\alpha_i = \alpha_0(1-\Gamma) + \alpha_g \Gamma \tag{7}$$

$$\alpha_m = \frac{1}{2L} \ln\left(\frac{1}{R_1 R_2}\right) \tag{8}$$

where α_i is the loss due to absorptions inside the guide α_g and outside α_0.

The differential quantum efficiency (DQE) depends on the internal quantum efficiency η_i and photon losses η_0:

$$\eta_d = \eta_i + \eta_0 \tag{9}$$

The photon loss value η_0 can be expressed as

$$\eta_0 = \frac{\alpha_m}{(\alpha_i + \alpha_m)} \tag{10}$$

where α_i is the internal loss and α_m is the optical mirror loss, which could be expressed as in Eq. (8). The DQE is dependent on the laser length L and the reflectivity of the mirror facets of laser, R_1 and R_2, as shown in the equation below:

$$\frac{1}{\eta_d} = \frac{1}{\eta_i} + \left(\frac{L\alpha_i}{\ln(1/R_1 R_2)} + 1 \right) \tag{11}$$

The term $\eta_d^{-1}(L)$ is widely used to determine the internal quantum efficiency η_i and internal loss from (L-I) measurements with different laser lengths.

The natural logarithm of the threshold current density, ln (J_{th}), is plotted on the y-axis with temperatures on the x-axis; thus, the inverse of the slope of the linear fit to this set of data point is the characteristic temperature T_0:

$$T_0 = \frac{\Delta T}{\Delta \ln(J_{th})} \tag{12}$$

4. Laser structure and parameters used in numerical simulation

A two-dimensional (2D) ISE-TCAD laser simulation program is used in the simulation of the LDs, which is based on solving the Poisson and continuity equations of a 2D structure. The Poisson equation is given by [14, 16- 18]

$$\nabla.(\varepsilon \nabla \varphi) = q(n - p - N_D^+ + N_A^-) \tag{13}$$

where N_A is the acceptor doping density (cm^{-3}), N_D is the donor doping density (cm^{-3}), ε is the permittivity of them medium, □ is the potential energy, q is electron charge, and n and p are the number of electrons and holes, respectively. The electron and hole continuity equations are given by

$$\frac{\partial n}{\partial t} + \nabla.J_n = q(G_n - R_n) \tag{14}$$

$$\frac{\partial p}{\partial t} + \nabla.J_p = q(G_p - R_p) \tag{15}$$

where J_n and J_p are the current density of electron and hole, respectively; G_n and G_p are the electron generation rate and hole generation rate, respectively; and R_n, R_p are the electron recombination rate and hole recombination rate, respectively.

Physical models included are drift-diffusion transport with Fermi-Dirac statistic, surface recombination, Shockley-Read-Hall recombination, Auger recombination, and band gap narrowing at high doping levels. The UV LD structure was reported in our previous paper [18], which includes a 0.6 μm GaN contact layer, a cladding layer of n-$Al_{0.08}Ga_{0.92}N$/GaN modulation-doped strained superlattice (MD-SLS) that consists of eighty 2.5 nm pairs, and a

0.1 μm n-GaN wave-guiding layer. The active region consists of 3 nm $Al_{0.08}In_{0.08}Ga_{0.84}N$ MQW sandwiched between 6 nm $Al_{0.1}In_{0.01}Ga_{0.89}N$ barriers. Four other layers exists on top of the active region: a 0.02 μm p-$Al_{0.25}In_{0.08}Ga_{0.67}N$ blocking layer, a 0.1 μm p-GaN wave-guiding layer, and a cladding layer of p-$Al_{0.08}Ga_{0.92}N$/GaN MD-SLS consisting of eighty 2.5 nm pairs, and, finally, a 0.1 μm p-GaN contact layer. The doping concentrations are 5×10^{18} cm^{-3} for p-type and 1×10^{18} cm^{-3} for n-type. The LD area is 1 μm × 400 μm, and the reflectivity of the two end facets is 50% each.

5. Results and discussions

5.1 Quantum well number effective laser diodes performance

Figure 2 shows the threshold current, output power, slope efficiency, and DQE of MQW LD as a function of the QW number. Best performance is shown by LD with four QW. This is attributed to a small electron leakage current, uniform distribution of electron carriers, and enhanced optical confinement at this number of QW.

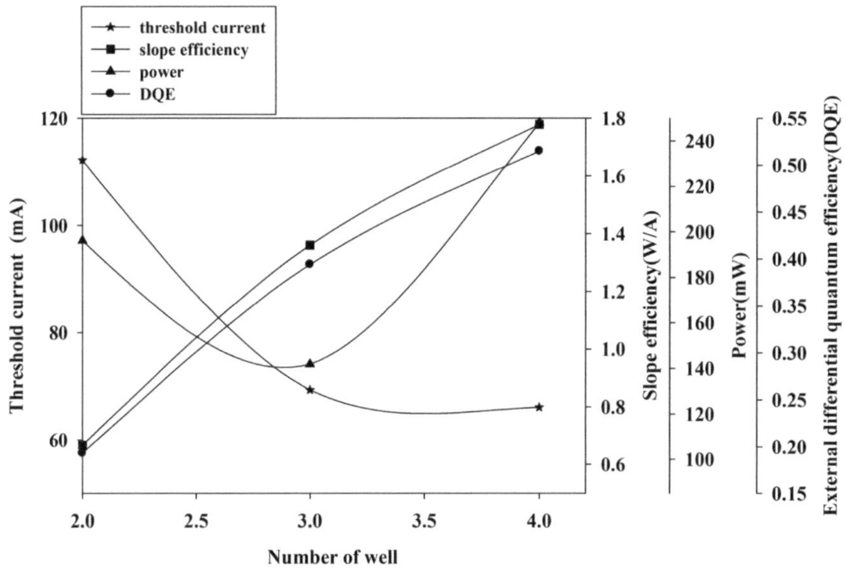

Fig. 2. The output power, threshold current, slope efficiency and DQE of $Al_{0.08}In_{0.08}Ga_{0.84}N/Al_{0.1}In_{0.01}Ga_{0.89}N$ LDs as a function of the quantum wells number.

5.2 Cavity length dependence of the threshold current and DQE

The effect of the cavity length of FQW LD on the threshold current and DQE is shown in Figure 3. The threshold current, representing by the slope efficiency increases with the decreasing in cavity length due to the increasing in mirror losses [Eq. (8)]. The external differential quantum efficiency DQE increases with increasing cavity length. The best values for threshold current, output power, slope efficiency, and DQE of four QW LD at a cavity length of 400 μm are 31.7 mA, 267 mW, 1.91 W/A, and 0.55, respectively. Longer cavity length is not recommended due to induced scattering phenomenon within the gain material of the laser structure.

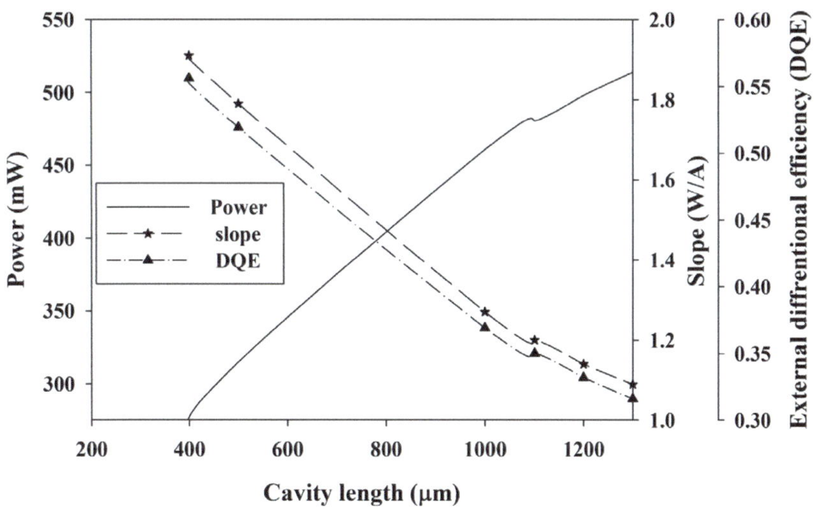

Fig. 3. The output power, slope efficiency and DQE of ($Al_{0.08}In_{0.08}Ga_{0.84}N/Al_{0.1}In_{0.01}Ga_{0.89}N$) LDs as a function of cavity length.

5.3 LD internal quantum efficiency ηi and internal loss

The internal quantum efficiency η_i and internal loss α_i values of the $Al_{0.08}In_{0.08}Ga_{0.84}N/Al_{0.1}In_{0.01}Ga_{0.89}N$ for four QW LD are calculated using Eq. (11). Figure 4 shows that at L = 0, η_i and α_i are equal to 71.4 % and 6.92 cm^{-1}, respectively, which show that good optimization of the geometrical condition of L = 400 μm considered when designing the LD.

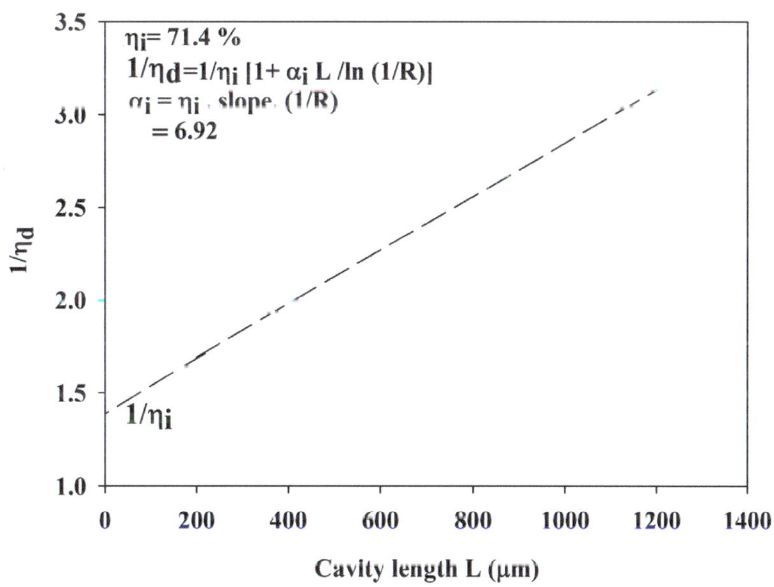

Fig. 4. The inverse DQE of ($Al_{0.08}In_{0.08}Ga_{0.84}N/Al_{0.1}In_{0.01}Ga_{0.89}N$) LDs as a function of cavity length.

5.4 Transparency threshold current density Jth

Figure 5 shows the threshold current density as a function of inverse cavity length. Hence, the intercept of the linear fit line with the vertical axis represented the transparency threshold current value J_0. The four QW LD have a J_0 value of 9.7 KA/cm², which is an acceptable value if compared with ternary LD due to the lattice match between $Al_{0.08}In_{0.08}Ga_{0.84}N$ well and $Al_{0.1}In_{0.01}Ga_{0.89}N$ barriers.

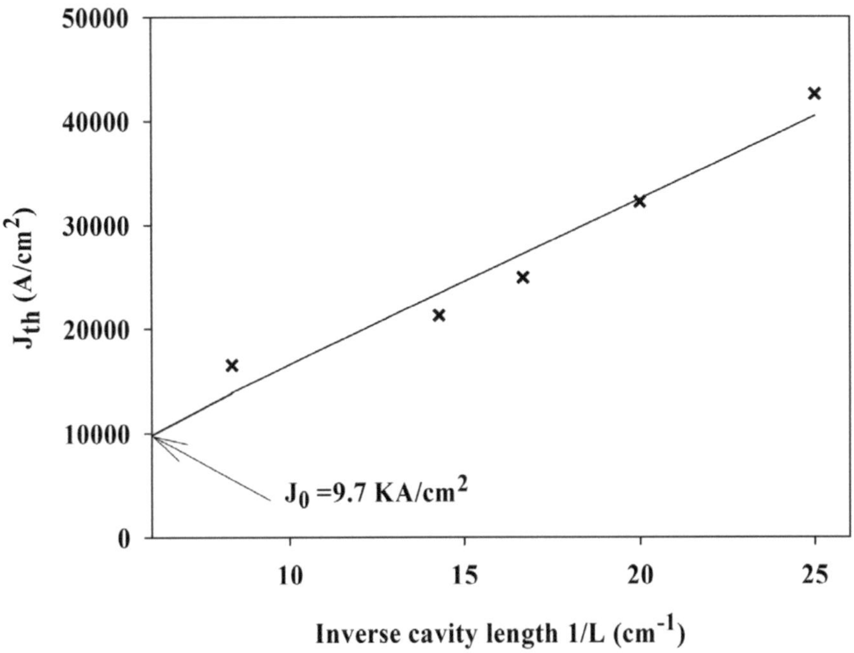

Fig. 5. Transparency threshold current density J_0 value of MQW LD.

5.5 Characteristic temperature T_0

Figure 6 shows the characteristic temperature T_0 value of four QW LD by plotting the natural logarithm of the threshold current density ln (J_{th}) on the y-axis with temperatures on the x-axis. The inverse of the slope of this plot (the linear fit to this set of data point) is the characteristic temperature T_0, which is found to be 97.5 K.

This value is somewhat lower than the characteristic temperature of ternary InGaN LD. This can be explained by the non-uniform distribution of the hole carrier density between wells due to the poor hole mobility in the InGaN layer. However, quaternary AlInGaN alloy is indeed the promising material to be used for well, barrier, and blocking layer. For a more non-uniform hole density distribution between the wells, the hole carriers require additional thermal energy to overcome the barrier potential between the wells. When the temperature increases, the hole density at the n-side of the QW increases due to the thermally enhanced hole transport from the p-side to the n-side of the QW. As a result, the gain at the n-side increases, and the thermal contribution of the hole carrier overflow is reduced with decreasing mirror loss. The characteristic temperature thus increases.

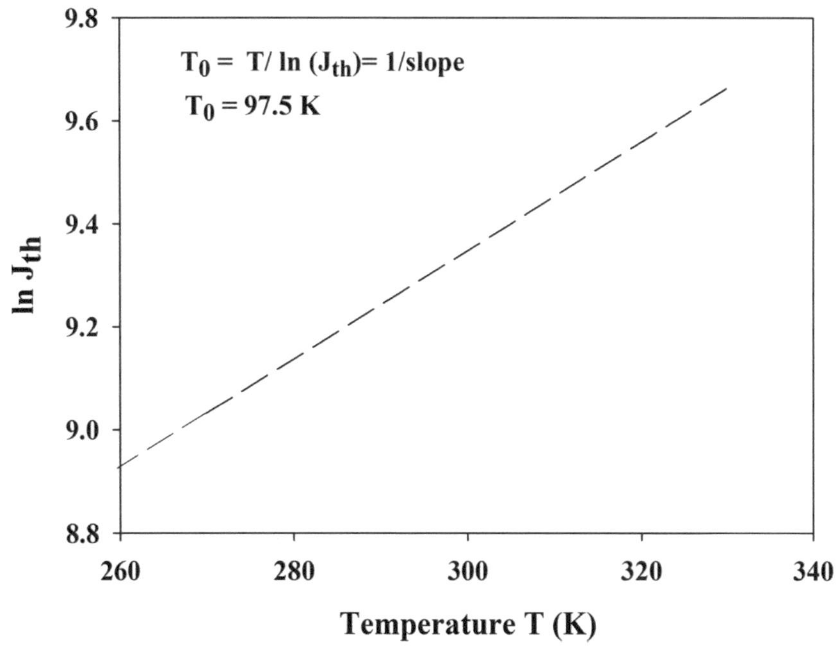

Fig. 6. The characteristic temperature T_0 value of four QWs LD.

6. Summary

The cavity length of quaternary $Al_{0.08}In_{0.08}Ga_{0.84}N/Al_{0.1}In_{0.01}Ga_{0.89}N$ MQW LD plays an important role in LD performance. The influence of cavity length on the threshold current, slope efficiency, characteristic temperature, and transparency threshold current density is studied. A higher characteristic temperature and suitable transparency current density can be obtained by decreasing the mirror loss. High characteristic temperature of 97 K, high output power of 267 mW at room temperature, and low threshold current density of 31.7 mA were achieved by applying a cavity length of 400 μm.

7. Acknowledgments

The authors would like to thank the University Science Malaysia (USM) for the financial support under the 1001/PFIZIK/843088 grant to conduct this research.

8. References

[1] Haase, M. A., Qiu, J., DePuydt, J. M., & Cheng, H. (1991). Blue-green diode lasers. *Appl. Phys. Lett.* Vol. 59, No.11, 1272-1274.

[2] Jeon, H., Ding J., Patterson W., Nurmikko A. V., Xie W., Grillo D. C., Kobayashi M., & Gunshor R. L. (1991). Blue-green injection laser diodes in (Zn,Cd)Se/ZnSe quantum wells. *Appl. Phys. Lett.* Vol. 59, No. 27, 3619-3621.

[3] Nakamura, S., Senoh M., & Mukai T. L8. (1993). p-GaN/N-InGaN/N-GaN double heterostructure blue-light-emitting diodes. Jpn. J. Appl. Phys. 32 (lA–B),

[4] Nakamura, S., Senoh, M., Nagahama, S., Iwasa, N., Yamada, T., Matsushita, T., Kiyoku, H., & Sugimoto, Y. (1996). InGaN-based multi-quantum-well-structure laser diodes. Jpn. J. Appl. Phys. 35 (IB), L74.

[5] Akasaki, I., Sola, S., Sakai, H., Tanaka, T. Koike, M., & Amano H. (1996). Shortest wavelength semiconductor laser diode, Electron. Lett. 32 (12), 1105.

[6] Nagahama S., Yanamoto T., Sano M., & Mukai T. (2001). Characteristics of Ultraviolet Laser Diodes Composed of Quaternary AlInGaN. *Jpn. J. Appl. Phys.* Vol. 40, L 788–L 791.

[7] Nagahama, S., Yanamoto T., Sano M., & Mukai T. (2001). Ultraviolet GaN single quantum well laser diodes. *Jpn. J. Appl. Phys.* Vol. 40, L785-L787

[8] Masui S., Matsuyama Y., Yanamoto T., Kozaki T., Nagahama S., & Mukai T. (2003). 365 nm ultraviolet laser diodes composed of quaternary AlInGaN alloy. *Jpn. J. Appl. Phys.* Vol. 42, L1318–L1320.

[9] Michael K., David W. T., Mark T., Naoko M., & Noble M. J. (2003). Ultraviolet InAlGaN multiple-quantum-well laser diodes. *Phys. Stat. Sol.* Vol. (a) 200, No. 1, 118–121.

[10] Nagahama, S., Iwasa N., Senoh M., Matsushita T., Sugimoto Y., Kiyoku H., Kozaki T., Sano M., Matsumura H., Umemoto H., Chocho K., & Mukai T. (2000). High-power and long lifetime InGaN multi-quantum-well laser diodes grown on low-dislocation-density GaN substrates. *Jpn. J. Appl. Phys.* Vol. 39, L647-L650.

[11] He Y., Song Y., Nurmikko A. V., Su J., Gherasimova M., Cui G., & Han J. (2004). Optically pumped ultraviolet AlGaInN quantum well laser at 340 nm wavelength. *Appl. Phys. Lett.* Vol. 84, No. 4, 463–465.

[12] Seoung-Hwan P., Hwa-Min K. & Doyeol A. (2005). Optical Gain in GaN Quantum Well Lasers with Quaternary AlInGaN Barriers. *Jpn. J. Appl. Phys.* Vol. 44, 7460–7463.

[13] Michael K., Zhihong Y., Mark T., Cliff K., Oliver S.t, Peter K., Noble M. J., Sandra S., & Leo J. S. (2007). Ultraviolet semiconductor laser diodes on bulk AlN. *J. Appl. Phys.*, Vol.101, 123103-123107

[14] Thahab S. M., Abu Hassan H., & Hassan Z. (2009). InGaN/GaN laser diode characterization and quantum well number effect. *CHINESE OPTICS LETTERS* Vol. 7, No. 3, 226-230.

[15] Iqbal Z., Egawa T., Jimbo T., & Umeno M., *IEEE PHOTONIC TECHNOLOGY LETTER.*

[16] Integrated System Engineering (ISE TCAD) AG, Switzerland, *http://www.synopsys.com*

[17] Thahab S. M., Abu Hassan H., & Hassan Z. (2007). Performance and optical, characteristic of InGaN MQWs laser diodes. *Opt. Exp.*, Vol.15, No.5, 2380-2390.

[18] Ghazai A. J., Thahab S. M., Abu Hassan H., & Hassan Z. (2011). A study of the operating parameters and barrier thickness of $Al_{0.08}In_{0.08}Ga_{0.84}N/Al_xIn_yGa_{1-x-y}N$ double quantum well laser diodes, *SCIENCE CHINA TECHNOLOGICAL SCIENCES*, Vol. 54, No. 1, 47–51.

2

Electrical Transport in Ternary Alloys: AlGaN and InGaN and Their Role in Optoelectronic

N. Bachir, A. Hamdoune and N. E. Chabane Sari
University of Abou–Baker Belkaid,
Tlemcen /Unity of Research Materials and Renewable Energies,
Algeria

1. Introduction

Since 1997, the market availability of blue, green and amber light emitting diode (LEDs), allows us to hope in time, obtaining the full range of the visible spectrum with semiconductor devices. Indeed, the development of blue emitters is very important because the blue emission is the last missing for the reconstruction of white light. Components based on gallium nitride (GaN) are the most effective in this area. In view of their low energy consumption and their high reliability, their use for the roadside (traffic lights) and domestic lighting may supersede the use of conventional incandescent or fluorescent lamps. In addition; the possibility offered by nitrides and their alloys, because of their intrinsic properties to develop blue and ultraviolet lasers, allows the production of systems that have greater storage capacity and playback of digital information (densities above the gigabit per square centimeter): the capacity is multiplied by four.

Research on GaN began in the 60s and the first blue LED based on GaN was performed in 1971 (Pankov et al., 1971). The development of GaN has been limited by the poor quality of the material obtained, and the failures in attempts to doping p. Recent research has resulted in a material of good quality, and in the development of doping p. These two achievements have developed the LEDs and lasers based on nitrides.

GaN has a direct band gap, high chemical stability, very good mechanical and physical properties, allowing it to be attractive for both optoelectronic and electronic devices operating at high temperature, high power and high frequency.

The development of InGaN alloys also allowed competing with GaP to obtain green LEDs. Table 1 shows the performance of GaN-based LEDs, compared with those obtained by other materials.

Given the performance obtained with the nitrides, the industrialization of blue and green LEDs was then very fast and has preceded the understanding of physical phenomena involved in these materials. Today, one of the major objectives of basic research on III-nitrides, is to identify key parameters that govern the emission of light in nanostructures (quantum wells or quantum dots) used as active layers of electroluminescent devices.

	Material	Wavelength emission	Light intensity	Emitted power	External quantum efficiency
Red	GaAlAs	660nm	1790mCd	4855μW	12.83%
Green	GaP	555nm	63mCd	30μW	0.07%
Green	InGaN	500nm	2000mCd	1000μW	2.01%
Blue	SiC	470nm	9mCd	11μW	0.02%
Blue	InGaN	450nm	2500mCd	3000μW	11%

Table 1. Performance of LEDs based on GaN and other materials (Agnès, 1999).

The two recurring problems concern the polarization effects related to the hexagonal structure of these materials, and also the effects of localization of carriers in the alloys.

Despite the technological and businessical advanced of these devices, some basic parameters of these materials remain little. The difficulty of developing these materials and controling their electrical properties (doping control) has long limited the determination of their parameters and their use in electronic and optoelectronic components. Gallium nitride (GaN) does not exist in nature, so it must be deposited on another material (sapphire, silicon ...) that does not have the same structural properties. This disagreement affects the optical and electronic qualities of those materials. The optimization of the devices thus requires a thorough understanding of the basic parameters and physical effects that govern the optical and electronic properties of these semiconductors.

The binary and ternary compounds based on GaN, exist in two structures: hexagonal and cubic. This second phase is far more difficult to develop and metastable, but it would present better electronic and optoelectronic performance. For this reason, we study the two ternary compounds AlGaN and InGaN in their cubic phases.

2. Band energy structures of the three nitrides

2.1 Status of the three nitrides in the family of semiconductors

The nature and bandgap energy are fundamental data in optoelectronics because direct gap materials have very large oscillator strength and the light emission is usually at energy close to that of the gap. The vast majority of semiconductor energy gap are located in the visible or near infrared. The family of nitrides stands in the UV (Fig. 1). GaN, AlN, InN and their alloys, are semiconductors with remarkable properties. The most important is undoubtedly their direct band gap ranging from 1.9eV (InN) to 3.4eV (GaN) (Nakamura S. & Fasol G., 1997), and reaches 6.2eV for AlN (fig. 1). With the concepts of the band gap engineering, developed in the context of III-V traditional semiconductors, it is possible to completely cover the visible spectral range, and to reach the ultraviolet A (320-400nm) and B (280-320nm). This is complemented by the strong stability of GaN what is responsible of the industrial production of light emitting diodes (LEDs), blue and green high brightness, and laser diodes (LDs) emitting at 0.4 microns. This makes these nitrides, the materials of choice for LEDs and laser diodes.

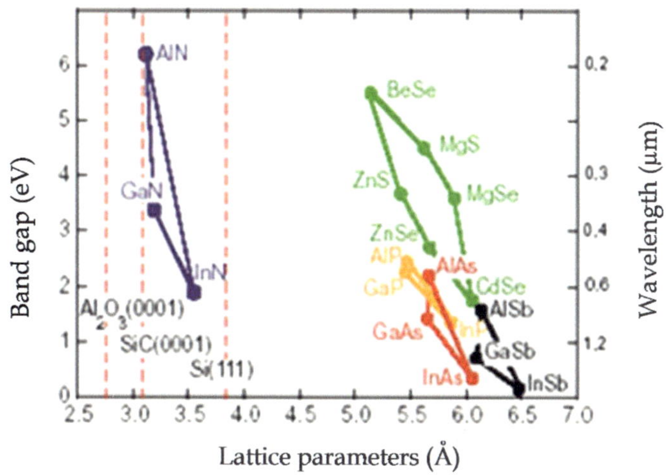

Fig. 1. Band gap and wavelength of various semiconductor compounds according to their lattice parameters (Nakamura & Fasol, 1997)

2.2 General forms of energy bands

The figure 2 shows the band structures in the cubic phase, of the gallium nitride (GaN), the aluminum nitride (AlN) and the indium nitride (InN).

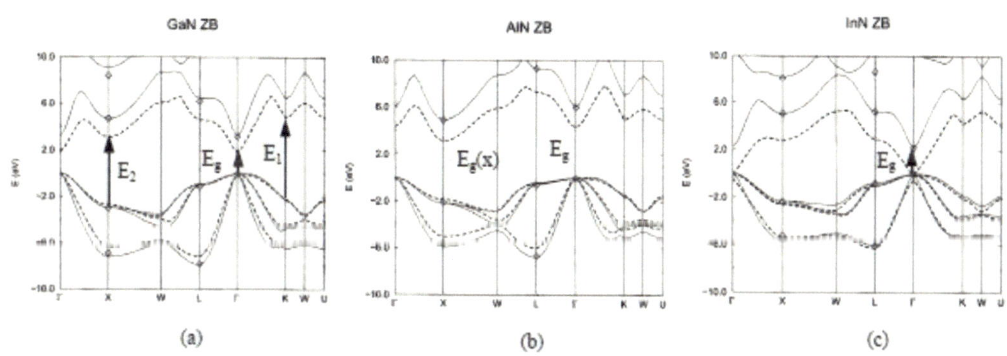

Fig. 2. Band structures of GaN, AlN, and InN in their cubic phases. The dotted lines correspond to results obtained by the "first-principles" method using VASP environment. The solid lines are those of the "semi-empirical pseudopotential" method (Martinez, 2002)

- Cubic GaN: In addition to the Γ valley, there are the L valley in the <111> direction and the X valley in the <100> direction, edge of the Brillouin zone. These last two valleys are characterized by a curvature smaller than that of the Γ valley and therefore the effective mass of electrons is higher and their mobility is lower. The minima of these valleys from the Γ band, are respectively about 2.6eV and 1.3eV (Martinez, 2002); they and are very large compared to other conventional III-V compounds.
- Cubic AlN: There are two minima between the conduction band and the valence band. However, the band structure shows a direct transition between points Γ corresponding to the gap.

- Cubic InN: The two minima between the conduction band and the valence band, have a very small difference, the gap in this case is very low.

2.3 Band-gaps (Nevou, 2008)

The first measurements of the band gap of GaN at low temperature date from the 1970s (Dingle et al., 1971). They gave a value of about 3.5eV at low temperatures. At room temperature, the band gap is about 3.39eV. Since the bandgap nitrides has been the subject of many studies (Davydov et al., 2002). The width of the band gap of AlN is of very short wavelengths that correspond to the near ultraviolet and far ultraviolet. At 300K, the energy gap is about 6.20eV, corresponding to a wavelength of 200nm. At 2K, the band gap is about 6.28eV (Vurgaftman & Meyer, 2003). The temperature dependence of the band gap is described by equation (1) of Varshni (Nakamura & Fasol, 1997):

$$Eg(T) = Eg(0) - \frac{\alpha \times T^2}{T + \beta} \tag{1}$$

Where Eg(0) is the bandgap at zero temperature, α and β are constants determined experimentally (Table 2).

material		Eg à 0K (eV)	Eg à 300K (eV)	α (eV/K)	β (K)
GaN	Cubic	3.28 (Bougrov et al., 2001)	3.2 (Bougrov et al., 2001)	0.593(Vurgaftman et al., 2001)	600(Vurgaftman et al., 2001)
	Hexagonal	3.51(Enjalbert, 2004)	3.43(Bethoux, 2004)	0,909 (Enjalbert, 2004)	830 (Enjalbert, 2004)
AlN	Cubic	6(Vurgaftman et al., 2001)	5.94(Dessene, 1998)	0.593(Vurgaftman et al., 2001)	600(Vurgaftman et al., 2001)
	Hexagonal	6.25(Enjalbert, 2004)	6.2 (Bethoux, 2004)	1.799 (Enjalbert, 2004)	1462(Enjalbert, 2004)
InN	Cubic	0.69 (Languy , 2007)	0.64 (Languy, 2007)	0.41 (Languy, 2007)	454 (Languy, 2007)
	Hexagonal	0.78(Enjalbert, 2004), 1.89(Anceau, 2004)	0.80(Bethoux, 2004), (Helman, 2004).	0.245 (Enjalbert, 2004)	624(Enjalbert, 2004)

Table 2. The energy gap at T = 0K and the parameters α and β, of the three nitrides in both phases.

The width of the band gap depends on the constraints applied to the material. For GaN, the biaxial compressive stress results in an increase in bandgap energy that is roughly linear with the applied stress. Since the lattice parameter "a" from the AlN is smaller than that of GaN (Fig. 3), the layers of the latter are in biaxial compression in the AlGaN alloy, resulting in increased energy band gap ~3,46–3,48eV (Martinez, 2002) at room temperature. As a result, the discontinuity of conduction band potential between GaN and AlN is reduced.

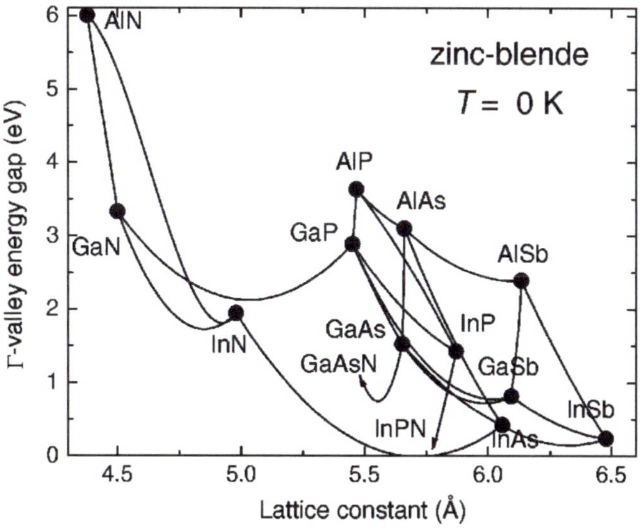

Fig. 3. The energy gap as a function of the lattice constant, for different cubic compounds at T = 0K (Vurgaftman et al., 2001).

2.3.1 The variation of AlxGa1-xN gap versus the x (Al) mole fraction

As a first approximation, the lattice parameters of AlxGa1-xN (InxGa1-xN) can be deduced from those of GaN and AlN (GaN and InN) by Vegard's law (Martinez, 2002) is a linear interpolation given by equation (2).

$$a_{Al_xGa_{1-x}N}=x\times a_{AlN}+(1-x)\times a_{GaN} \tag{2}$$

Moreover, the variation of the bandgap energy of the alloy according to the composition is not linear but quadratic; it is given by equation (3).

$$Eg(x)=x\times Eg(AlN)+(1-x)\times Eg(GaN)-bx\times(1-x) \tag{3}$$

The bowing parameter "b" is usually taken equal to 1.

Substituting Eg(AlN) and Eg(GaN) by their values at 300K in the relationship (3), we find the equations (4) and (5) which give the gaps respectively of cubic and hexagonal AlxGa1-xN, , according to x (Vurgaftman and Meyer, 2003):

$$Eg_1(x)=x^2+1.74x+3.2eV \tag{4}$$

$$Eg_2(x)=x^2-1.77x+3.43eV \tag{5}$$

By increasing the mole fraction of aluminum, the top of the valence band at Γ point, moves down and the energy gap increases.

Using MATLAB, we calculate the variation of AlxGa1-xN gap as a function of aluminum mole fraction; that is illustred by Figure 4.

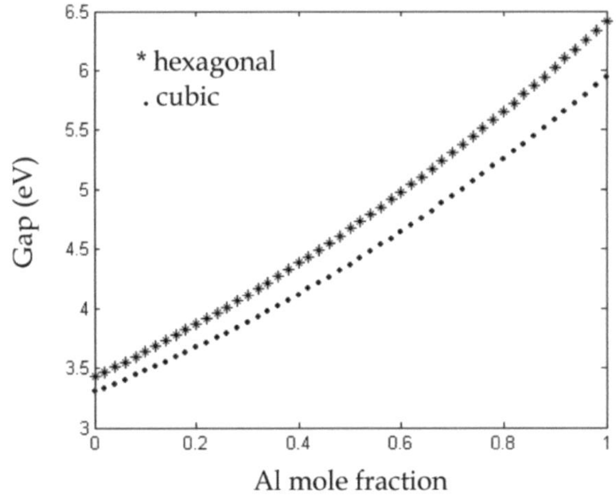

Fig. 4. Variation of AlxGa1-xN energy gap, versus the Al mole fraction (Castagné et al., 1989).

2.3.2 The variation of InxGa1-xN gap versus the x (In) mole fraction

To calculate the energy band gap, use the quadratic relationship (6) (JC Phillips):

$$Eg(InxGa1-xN)=x\times Eg(InN)+(1-x)\times Eg(GaN)-bx\times(1-x) \quad (6)$$

Substituting Eg(InN) and Eg(GaN) by their values at 300K, and taking b = 1, we find the equations (7) and (8) which give respectively gaps of cubic and hexagonal InxGa1-xN, as a function of x (Castagné et al., 1989):

$$Eg_1(x)=x^2-3.56x+3.2eV \quad (7)$$

$$Eg_2(x)=x^2-3.63x+3.43eV \quad (8)$$

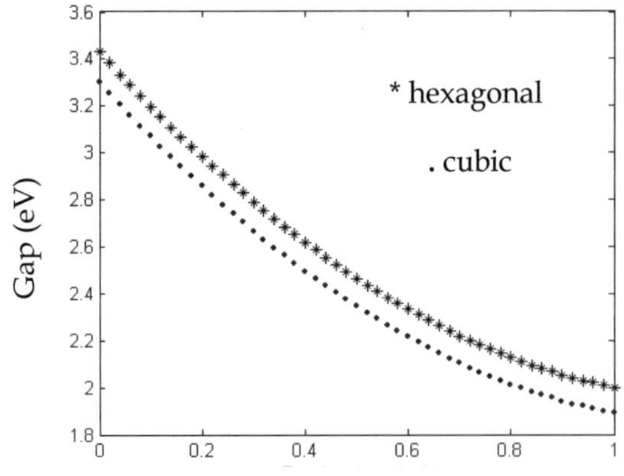

In mole fraction

Fig. 5. Variation of $In_xGa_{1-x}N$ energy gap, versus the In mole fraction (Castagné et al., 1989).

The top of the valence band at Γ point shifts up and the energy gap decreases when the indium composition increases.

3. Electrical transport in ternary alloys: AlGaN and InGaN, and their role in optoelectronic

For zinc blend AlGaN and InGaN compounds, we calculate energy and velocity of the electrons for different mole fractions and various temperatures in the steady-state. We also calculate the electrons velocity in the transient mode.

We use Monte Carlo simulation method which is a program written in Fortran 90 MSDEV, and we simulate 20,000 electrons.

This method offers the possibility of reproducing the various microscopic phenomena residing in semiconductor materials. It is very important to study the transport properties of representative particles in the various layers of material over time; it is to follow the behavior of each electron subjected to an electric field in real space and in space of wave vectors. Indeed; over time, the electrons in the conduction band will have a behavior that results from the action of external electric field applied to them and their various interactions in the crystal lattice. We consider the dispersion of acoustic phonons, optical phonons, ionized impurities, intervalley and piezoelectric in a nonparabolic band.

Consider an electron which owns energy ε(t), wave-vector k (t), and which is placed in r (t). Under action of an applied electric field E (r, t); its interaction and exchange of energy with the lattice, and the deviation of its trajectory by impurities; this will modify its energy, its wave-vector and its position. Using the mechanic and the electrodynamics laws; we determine the behavior of each electron, in time and space. To be more realistic:

1. We statistically study possible energy exchanges between electrons, modes of lattice vibration and impurities; this allows us to calculate the probability of these interactions and their action on both electron energy and wave vector.
2. We assume that these interactions are instantaneous. We can move electrons in free-flight under the only effect of electric field, between two shocks. The free-flight time is determined by the drawing of lots. When interaction takes place, we determine its nature by the drawing of lots. In this case, the electron energy and the electron wave-vector are modified. This results in a change of electrons distribution; we then calculate the electric field that results, at enough small time intervals, to assume it constant between two calculations (Enjalbert, 2004) – (Pugh et al., 1999)– (Zhang Y.et al., 2000).

We consider a simplified model of three isotropic and non parabolic bands. The wave-vector and energy of the electron are related by using the equation (9) (O'Learyet al., 2006*).

$$\frac{\hbar^2 k^2}{2m^*} = \varepsilon(1+\alpha\varepsilon) \tag{9}$$

Where m* is their effective mass in the Γ valley, ħk denotes the magnitude of the crystal momentum, ε represents the electron energy, and α is the nonparabolicity factor of the considered valley, given by equation (10) (O'Learyet al., 2006*).

$$\alpha = \frac{1}{Eg} \times \left(1 - \frac{m^*}{m_e}\right)^2 \tag{10}$$

Where m_e and Eg denote the free electron mass and the energy gap, respectively.

$m_e = 0.20m_0$ for GaN, $m_e = 0.32m_0$ for Al (Chuang et Al. 1996), and $m_e = 0.11m_0$ for InN (O'Learyet al., 2006).

The longitudinal v_l and transverse v_t, acoustic velocities (Castagné, 1989)-(Garro et Al. 2007)-(Anwar, 2001) are calculated by using equations (11) and (12) (where ρ is the density of material, and Cij are elastic constants):

$$v_l = \left(\frac{c_l}{\rho}\right)^{1/2} \tag{11}$$

$$v_t = \left(\frac{c_t}{\rho}\right)^{1/2} \tag{12}$$

The constants, c_l and c_t, are combinations of elastic constants, given by equations (13) and (14):

$$c_l = \frac{3 \times c_{11} + 2 \times c_{12} + 4 \times c_{44}}{5} \tag{13}$$

$$c_t = \frac{c_{11} - c_{12} + 3 \times c_{44}}{5} \tag{14}$$

3.1 Description of the simulation software

This software can perform two basic functions. The first is devoted to probability theory from the usual expressions, considering a model with isotropic and nonparabolic three valleys (Γ, L, X). The second role is to determine the instantaneous magnitudes defined on a set of electrons (energy, speed, position) by the "Self Scattering" procedure for which the free-flight times are distributed to each electron.

The results are highly dependent on many parameters that characterize the material and which, unfortunately, are often very poorly understood.

We developed the software by making it more friendly and simple users. The general procedure for running this software is composed of three essential steps that can be summarized as follows:

1. Reading the data file for the parameters of the used material, such as energy gap, effective masses, deformation potentials, coefficients of nonparabolicity, speed of sound, concentration of impurities, temperature of the network, applied electric fields, etc.. in a file.txt.
2. Running the software.
3. Delivery of output files: the values of interaction probabilities, speeds in different valleys, the energies...

The essence of our Monte Carlo simulation algorithm, used to simulate the electron transport within the semiconductors, AlxGa1-xN and InxGa1-xN, is given by the diagram of Fig. 6; where ∂t is the free-flight time of electrons, ε is their energy, λq(ε) is their total scattering rate, T_{total} is the simulation time.

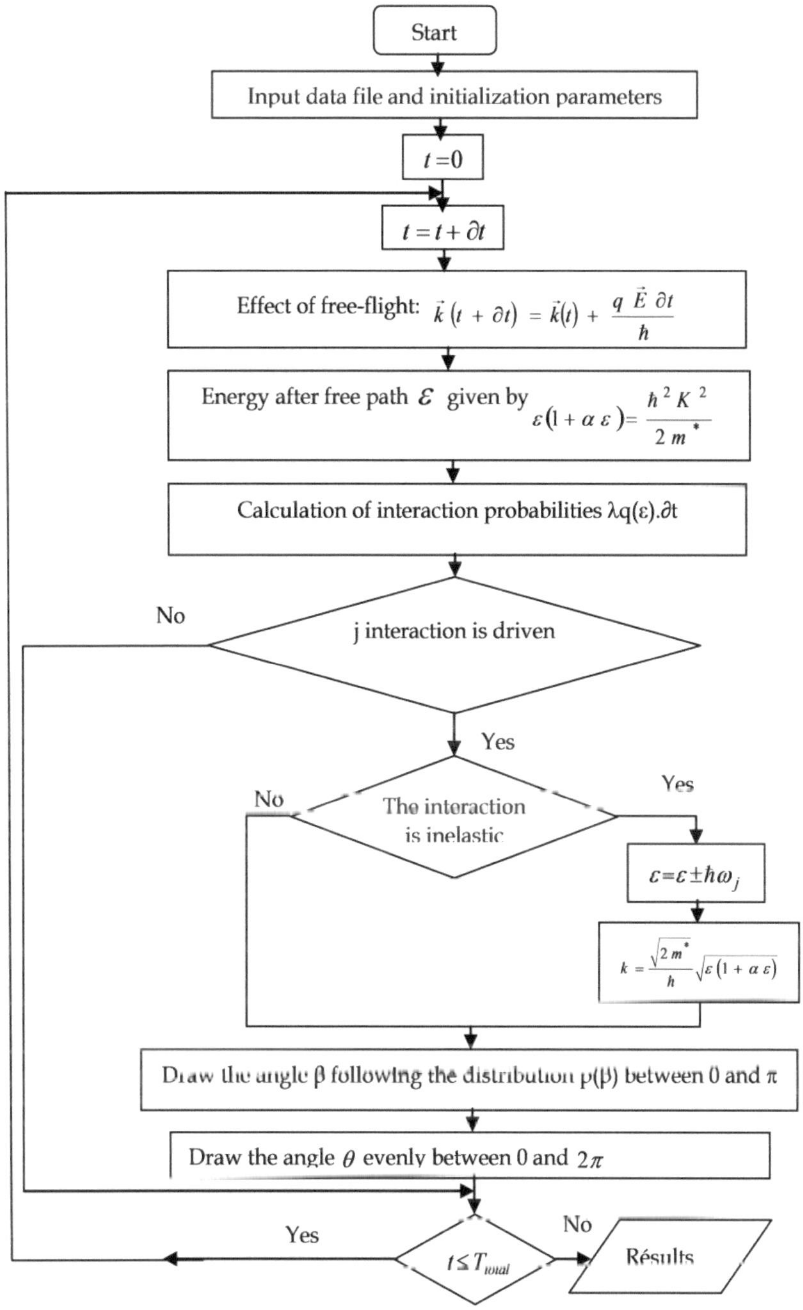

Fig. 6. Monte Carlo flowchart.

3.2 Results in the stationary regime

3.2.1 The electron energy versus the applied electric field

At room temperature and for an electron concentration of 1017cm-3, we calculate the electron energy versus the applied electric field within AlGaN and InGaN alloys, for different Al and In mole fractions. The results are illustrated by Figure 7.

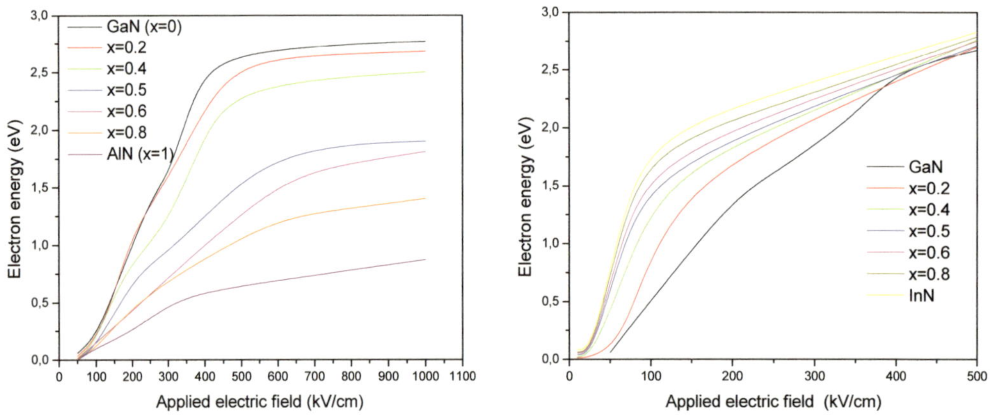

Fig. 7. The electron energy versus the applied electric field within AlGaN and InGaN alloys, for different Al and In mole fractions.

By increasing the Al mole fraction within AlGaN alloy, the energy gap increases, so it is necessary to apply an electric field harder to cross it. Therefore, the critical electric field (for which the electrons move from the valence band to the conduction band) becomes larger, it reaches 300kV/cm for AlN, and so it does not exceed 150kV/cm in GaN. In addition, the intervalley energy (E_{LX}) decreases with increasing Al mole fraction, so that the average energy decreases, it does not exceed 1eV in the case of AlN, while that of GaN is around 2.6 eV. However; within InGaN, by increasing the In mole fraction, the energy gap decreases, so the critical electric field also decreases, it is about 50kV/cm for InN. In addition, the intervalley energy becomes slightly larger and therefore the average energy increases slightly, it is around 2.85eV for InN. The energy of electrons becomes more important for electric fields relatively small compared to the AlGaN alloy.

3.2.2 Electron drift velocity versus the applied electric field

The electron drift velocity versus the applied electric field within AlGaN and InGaN alloys, at room temperature and for an electron concentration of 10^{17} cm^{-3}, is illustrated by Figure 8, for different mole fractions.

By increasing Al mole fraction within AlGaN; the energy gap, the energy between Γ and L valleys, and the electron effective mass, increase. The growth of the electron effective mass in the central valley causes decrease in its drift velocity, and displacement of the critical field to larger values of the electric field. Beyond the critical field, electrons move to the upper valley, their masses increase and they will suffer more collisions with other electrons already present in these valleys; their speed will therefore decrease.

For InGaN, increasing the In mole fraction leads to a decrease in energy gap and in effective mass of electrons in the central valley, while the effective mass of electrons in the upper valleys increases. The decrease in the effective mass of electrons in the central valley, causes an increase in their speed, and reduced energy gap results in a shift of the critical field to smaller values of the electric field. However; due to the decrease in energy between the Γ and L valleys, and the intervalley increase in collisions due to the increase of population in the satellite valleys, there is a decrease in drift velocity.

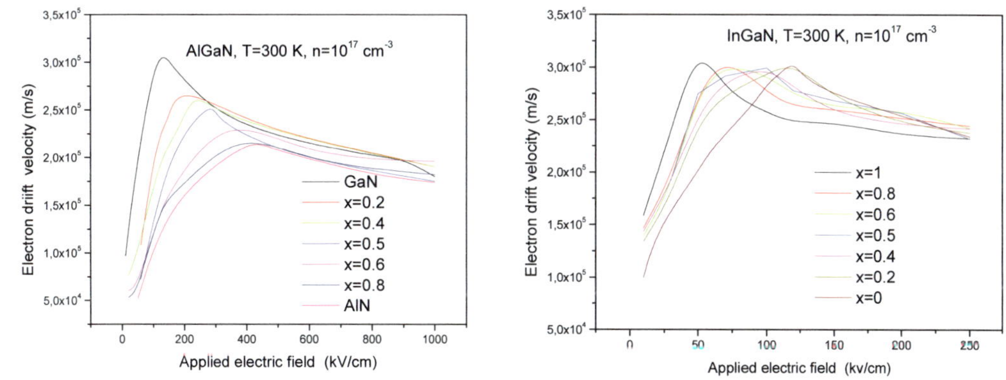

Fig. 8. The electron drift velocity within AlGaN and InGaN alloys, versus the applied electric field for different mole fractions (x = 0, 0.2, 0.4, 0.5, 0.6, 0.8, 1) at a temperature of 300K.

In conclusion, the electron velocity increases with the In mole fraction within InGaN alloy, but the electrons reach the saturation velocity for relatively small electric fields (E ≤ 500kV/cm). Within AlGaN alloy; the electron velocity is smaller, but the electrons reach the saturation velocity for electric fields much higher (E ≥ 1000kV/cm).

3.2.3 The electron drift velocity versus the applied electric field for different temperatures

Always for an electron concentration of 1017 cm-3 and for different mole fractions, we calculate the electron drift velocity versus the applied electric field for different temperatures: from 77K to 1000K within AlGaN and from 77K to 700K within InGaN. The results are respectively illustrated by Figures 9 and 10.

The electron velocity keeps almost the same pace. The higher velocities are reached for low temperatures; the best one is obtained for a temperature of 77K corresponding to the boiling point of nitrogen.

The increase in temperature allows a gain in kinetic energy to the electrons; they move more and collide with other atoms by transferring their energies; then their velocity decreases.

The scattering of electrons is dominated by collisions of acoustic phonons, ionized impurities, and polar optical phonons which are removed to very low temperatures, leading to improved mobility and improved velocity. At low electric field and for temperatures up to 300K, the impurities dispersion dominates and therefore there is an increase in their velocity. At high temperatures, the bump disappears due to the dominance of polar optical phonon collisions with a collision reduction of impurities.

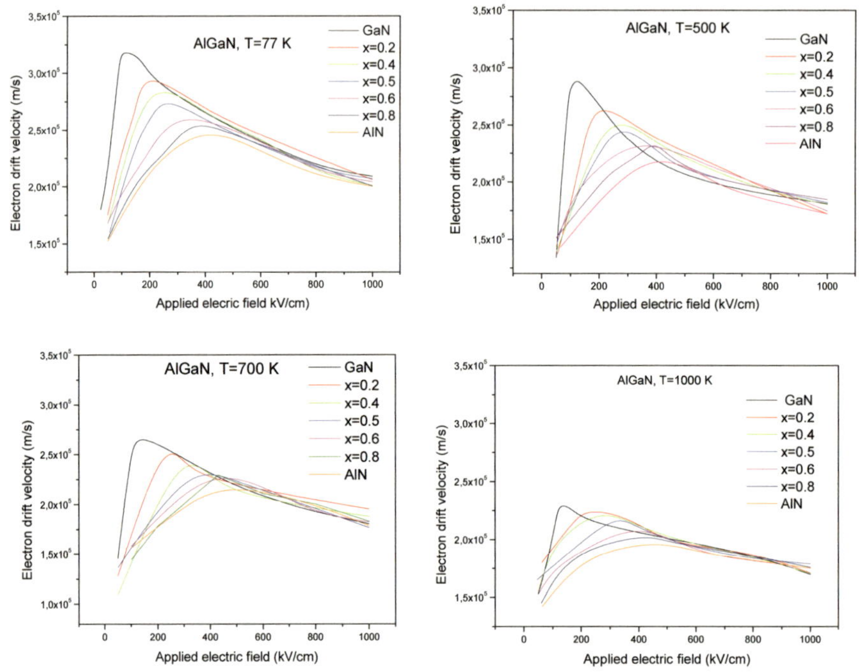

Fig. 9. The electron drift velocity versus the applied electric field within AlGaN, for different Al mole fractions at temperatures of 77K, 500K, 700K, and 1000K.

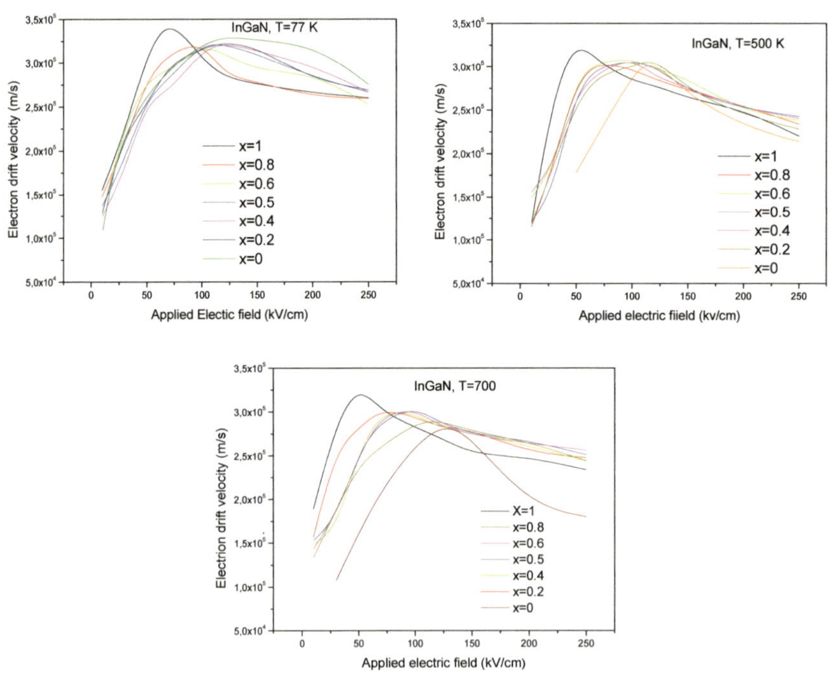

Fig. 10. The electron drift velocity versus the applied electric field within InGaN, for different In mole fractions at temperatures of 77K, 500K, and 700K.

The maximum temperature that can be applied within InGaN does not exceed 700K. Indeed, InN has a relatively small gap compared to the other nitrides, and thus even InGaN alloy has a small gap. Considering equation (1), one easily deduces that the gap becomes very small when the temperature increases.

Electron velocity is more important in InGaN for low temperatures, but the temperature can not exceed 700K. As against, the temperature can go beyond 1000K within AlGaN alloy.

3.3 The electron velocity in the transient regime

To highlight the effects of non-stationary transport that can occur in both InGaN and AlGaN alloys, we study the behavior of a bunch of electrons subject to sudden variations of the electric field, ie we apply levels of electric field.

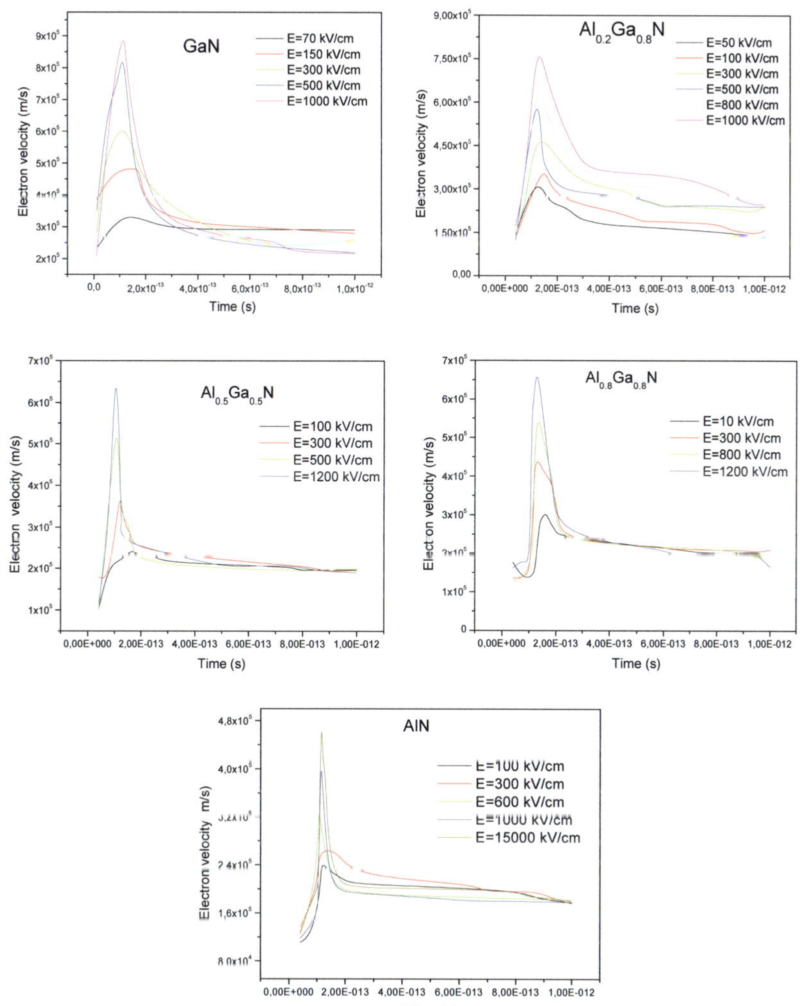

Fig. 11. The electron velocity versus the time for different levels of electric field within GaN, $Al_{0.2}Ga_{0.8}N$, $Al_{0.5}Ga_{0.5}N$, $Al_{0.8}Ga_{0.2}N$ and AlN.

To ensure a stable behavior to the electrons, we apply a field of 10kV/cm during a time equal to 1ps. Then, the electrons undergo a level of electric field. The results for AlGaN and InGaN alloys are respectively illustrated by Figures 11 and 12.

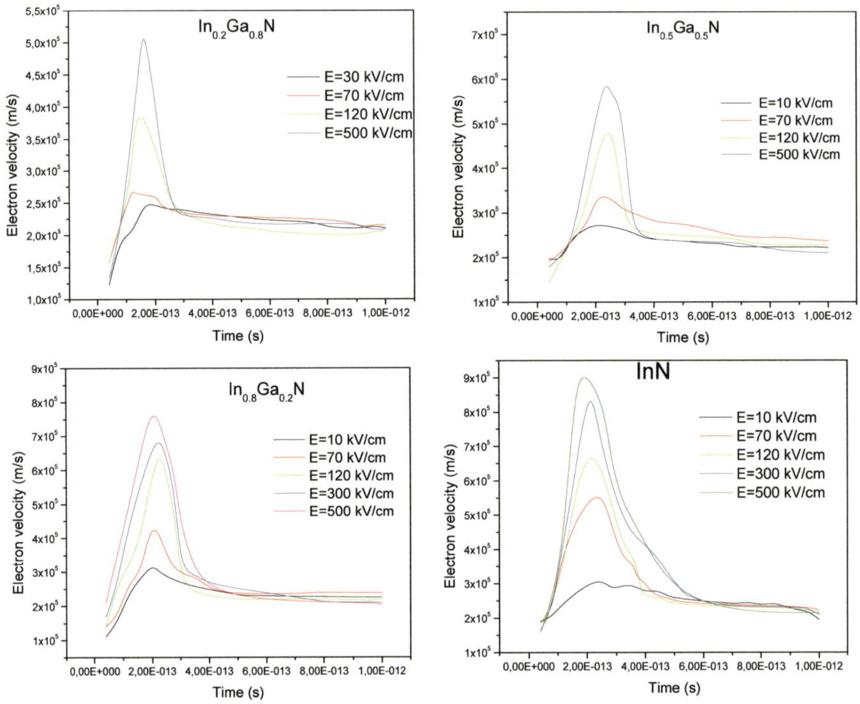

Fig. 12. The electron velocity versus the time for different levels of electric field within $In_{0.2}Ga_{0.8}N$, $In_{0.5}Ga_{0.5}N$, $In_{0.8}Ga_{0.2}N$ and InN.

Overshoots within InGaN alloy, are very large and increase with the In mole fraction. Within AlGaN alloy, they are relatively smaller and decrease with increasing the Al mole fraction.

The response time is very small in both alloys, of the order of picoseconds (see smaller); meaning that the lifetime of the electrons is very large, which is very important in optoelectronic devices.

Within InGaN, the electron energy and the electron velocity are very important for low field and low temperature. Within AlGaN, the breakdown voltage is high thanks to the large band gap, allowing high output impedance and high saturation velocity. AlGaN is more resistant to high temperatures and high pressures.

Both alloys are complementary: InGaN is more effective at low electric fields and low temperatures; AlGaN is more efficient for large electric fields and high temperatures.

LEDs based on GaN and its alloys are used in several areas: signage, automotive, display, lighting...

Among their advantages, one can mention: the low energy consumption, the high life time (100,000 hours or more), and the security (visible LEDs do not emit ultraviolet or infrared).

Laser diodes based on GaN and its alloys are used in civilian and military applications: environmental, medical, biomedical, sensing, missile guidance, etc....

4. References

Agnès P. (1999). *Caractérisation électrique et optique du nitrure de gallium hexagonal et cubique en vue de l'obtention d 'émetteurs bleus,* PhD thesis, INSA, Lyon, France.

Anceau S., (2004), *Etude des propriétés physiques des puits quantiques d'alliages quaternaires (Al,Ga,In)N pour la conception d'émetteurs ultraviolets,* PhD thesis, University of Montpellier II, France.

Anwar A.F.M., Member S., Wu S., Richard T. & Webster. (2001). Temperature dependent transport properties in GaN, AlxGa1-xN, and InxGa1-xN semiconductors, IEEE *Transactions on Electron Devices,* vol. 48, No 3, March, pp. 567–572.

Bethoux J.M., (2004), *Relaxation des contraintes dans les hétérostructures épaisses (Al,Ga)N : une piste originale pour la réalisation de diodes électroluminescentes à cavité résonante,* PhD thesis, Nice university, France.

Bougrov V., Levinshtein M.E., Rumyantsev S.L. & Zubrilov A. (2001). Properties of Advanced Semiconductor Materials GaN, AlN, InN, BN, SiC, SiGe. *Eds. Levinshtein M.E., Rumyantsev S.L., Shur M.S., John Wiley & Sons, Inc.,* New York, 1-30.

Castagné R., Duchemin J.P, Gloanie M. & Rhumelhard Ch. (1989). Circuits integers en Arsenic de Gallium, Physique, technologie et conception, *Masson edition.*

Chuang S. L. & Chang C. S. (1996). k · p method for strained wurtzite semiconductor, *Phys. Rev.* B 54, 2491.

Davydov V.Yu., Klochikhin A.A., Emtsev V.V., Kurdyukov D.A., Ivanov S.V., Vekshin V.A., Bechstedt F., Furthmüller J., Aderhold J., Graul J., Mudryi A.V., Harima H., Hashimoto A., Yamamoto A. & Haller E.E. (2002). Band Gap of Hexagonal InN and InGaN Alloys *Phys. Stat. Sol.* (b) 234, 787.

Dessene F. (1998). *Etude thermique et optimisation des transistors à effet de champ de la filière InP et de la filière GaN,* PhD thesis, University of Lille 1, France.

Dingle R., Sell D. D., Stokowski S. E. & Ilegems M. (1971). Absorption, Reflectance, and Luminescence of GaN Epitaxial Layers, *Phys. Rev.* B 4, 1211.

Enjalbert F. (2004). *Etude des hétérostructures semi-conductrices III-nitrures et application au laser UV pompé par cathode à micropointes,* PhD thesis, University of Grenoble 1, France.

Garro N., Cros A., Garcia A. & Cantarero A. (2007). Optical and vibrational properties of self-assembled GaN quantum dots, Institute of Materials Science, University of Valencia, Spain.

Helman A. (2004). *Puits et boîtes quantiques de GaN/AlN pour les applications en optoélectronique à λ≈1,55 μm,* PhD thesis, University of Orsay, Paris XI, France.

Languy F. (2007). *Caractérisation de nanocolonnes d'In(1-x)Ga(x)N en vue d'étudier l'origine de la haute conductivité de l'InN,* Faculty of University "Notre dame de la paix", Belgique.

Martinez G. (2002). *Elaboration en épitaxie par jets moléculaires des nitrures d'éléments III en phase cubique,* PhD thesis, INSA, Lyon, France.

Nakamura S. & Fasol G. (1997). The Blue Laser Diode, *Springer* - Berlin.

Nevou L., (2008), *Emission et modulation intersousbande dans les nanost ructures de nitrures,* PhD thesis, Faculty of Sciences, Orsay, France.

O'Leary Stephen K., Brian E. Foutz, Michael S. Shur & Lester F. Eastman (2006). "Steady-State and Transient Electron Transport within the III-V Nitride Semiconductors, GaN, AlN, and InN," *A Review J Mater Sci: Mater Electron,* vol. 17, pp. 87-126

Pankov J.I., Miller E.A., & Berkeyheiser J.E. (1971). *Rca Review* 32, 383.

Pugh S. K, Dugdale D. J, Brand S. & Abram R.A. (1999). Electronic structure calculation on nitride semiconductors, *Semicond. SCI, Technol,* vol 14, pp. 23-31.

Vurgaftman I. & Meyer J. R. (2003). Band parameters for nitrogen-containing semiconductors, *J. Appl. Phys.* 94, 3675.

Vurgaftman I., Meyer J. R. & Ram-Mohan L. R. (2001). Band parameters for III-V compound semiconductors and their alloys, *J. Appl. Phys.* 89, 5815.

Zhang Y.et al. (2000). Anomalous strains in the cubic phase GaN films grown on GaAs (001) by metalorganic chemical vapour deposition, *J.Appl. Phys,* vol. 88, N° 6, pp. 3762-3764.

3

Carrier Transport Phenomena in Metal Contacts to AlInGaN-Based Laser Diodes

Joon Seop Kwak
Department of Printed Electronics Engineering, Sunchon National University, Maegok, Jeonnam, Korea

1. Introduction

High power AlInGaN-based laser diodes (LDs) have attracted great attention as light sources for high-density optical storage systems and micro-display systems, as well. One of the main concerns in this area is the fabrication of high-quality ohmic contacts to p-GaN with low resistance and thermal stability, since operation power of the AlInGaN-based LDs greatly affects the life time of the LDs.[1] Achieving low resistance ohmic contacts to p-GaN has been particularly challenging, because of difficulty in obtaining a hole concentration over 10^{18} cm^{-3} and the absence of metals having a work function higher than that of p-GaN (6.5 eV).[2]

For fabrication of AlInGaN-based LDs with low input power, free-standing GaN has attracted attention as a substrate because of its low dislocation density, high thermal conductivity, and easy cleaving.[3,4] Another advantage of using the GaN substrate is to fabricate devices with a backside n-contact. This allows the fabrication process to be simple and reliable, and reduces the size of devices, which increases yield in mass-production. Free-standing GaN substrate has two faces with a different crystal polarity, Ga- and N-face polarity, which greatly influence on the electrical properties at metal/GaN interface as well as those at AlGaN/GaN heterostructure.[5,6]

In order to achieve high-quality ohmic contacts to AlInGaN-based LDs, carrier transport phenomena at the interface between metal and p-GaN as well as n-GaN should be elucidated. This chaper discusses carrier transport phenomena in metal contacts to AlInGaN-based LDs, and also introduces a design to improve the operating voltage characteistics of the AlInGaN-based LDs based upon the carrier transport mechanism in metal contacts to AlInGaN-based LDs.

2. Carrier transport in metal contacts to p-GaN contact layer of AlInGaN LDs

Many experiments on the electrical characteristics of metal contacts to p-GaN have been carried out, mainly focusing on the variation of metallization scheme, alloying condition, and surface treatment.[7-11] In order to achieve the low-resistance ohmic contacts to p-GaN, carrier transport phenomena at the interface between metal and p-GaN should be elucidated. Kim et al. reported that surface treatment of the p-GaN shifted the Fermi level to an energy level near the valence band by removing oxide layer, resulting in the reduction of the barrier height for holes.[7] In addition, Jang et al. showed the increase in carrier concentration at the regions near the surface of p-GaN for the surface-treated p-GaN using the Hall-effect measurements, and suggested that field emission dominated current flow at the interface between non-alloyed Pt contact and the surface-treated p-GaN.[12]

2.1 Dependence of contact resistivity on hole concentration

In order to understand the carrier transport phenomena at the interface between the metal contacts and p-GaN contact layer of AlInGaN LDs, the dependence of contact resistivity on the hole concentration of ohmic contacts to p-GaN should be investigated, because carrier concentration is one of the most important factor to affect carrier transport at the metal-semiconductor interface.

The hole concentration and the concentration of Mg, [Mg] for a series of p-GaN films are shown in Fig. 1 as a function of flow rate of bis-(cyclopentadienyl)-magnesium (Cp_2Mg). As shown in Fig. 1, the [Mg] increased with Cp_2Mg and reached up to $1x10^{20}$ cm^{-3} when the flow rate of Cp_2Mg was 2.8 μmole/min. The hole concentration initially increased from $7.5x10^{16}$ (sample A) to $2.2x10^{17}$ cm^{-3} (sample B) with the increase of the flow rate of Cp_2Mg from 0.70 to 1.05 μmole/min. Further increase in the flow rate of Cp_2Mg to 1.75 and 2.80 μmole/min, however, resulted in a significant decrease of the hole concentration to $1.2x10^{17}$ (sample C) and $2.0x10^{16}$ cm^{-3} (sample D), respectively. These results are well matched to the previous results, where the hole concentration initially increased with [Mg] and it decreased significantly with further increase in the [Mg].[13]

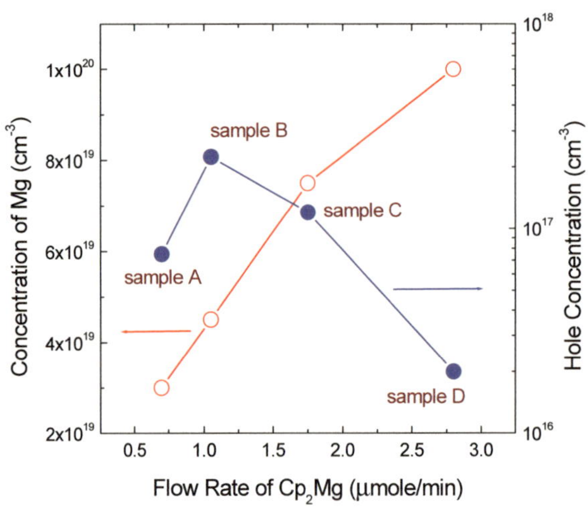

Fig. 1. The hole concentration and the [Mg] for a series of p-GaN films as a function of flow rate of Cp_2Mg.

Figure 2 shows I-V characteristics of the non-alloyed Pd/Pt/Au contacts deposited on the series of p-GaN films with a different [Mg]. For the contacts on sample A having the lowest [Mg] among the samples as shown in Fig. 1, the measured current at 0.5 V between two metal pads with the gap spacing of 5 μm was 19 μA, and it increased to 41 and 54 μA for the contacts on sample B and C, respectively. In case of the contacts on sample D with the highest [Mg] and the lowest hole concentration, the current at 0.5 V was measured as 64 μA. It is worth to notice that the non-alloyed Pd/Pt/Au contacts deposited on samples A, B, and C displayed non-linear I-V curves, while I-V curve became linear for the contacts on sample D. This is suggestive of producing the better ohmic contact on sample D having the lowest hole concentration among the samples.

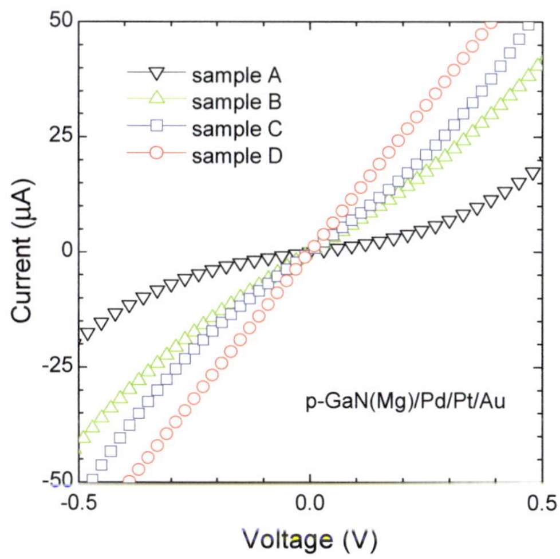

Fig. 2. I-V characteristics of the non-alloyed Pd/Pt/Au contacts deposited on the series of p-GaN films with a different [Mg].

The contact resistivity for the Pd/Pt/Au contacts on the p-GaN films having the different [Mg] is shown in Fig. 3. For comparison, the hole concentration as a function of the [Mg] is also displayed. The contact resistivity for the non-alloyed Pd/Pt/Au contacts on sample A having the [Mg] of 3x10^{19} cm^{-3} was 3.0x10^{-1} Ω-cm^2, and it was reduced to 8.9x10^{-2} Ω-cm^2 for the contacts on sample B, having the higher [Mg] than that of sample A. This can be attributed to the increase in hole concentration with the Mg until the [Mg] is 4.5x10^{19} cm^{-3}, as shown in Fig. 3. The increase in the hole concentration reduces the barrier width as well as barrier height between the contact and the p-GaN, followed by the enhancement of carrier tunneling through the barrier and the reduction of contact resistivity, as shown in Fig. 3.[14]

In case of the Pd/Pt/Au contacts on sample C, however, the contact resistivity was further reduced to 3.2x10^{-2} Ω-cm^2, although the hole concentration of sample C (1.2x10^{17} cm^{-3}) was lower than that of the sample B (2.2x10^{17} cm^{-3}), as shown in Fig. 3. Furthermore, the contact resistivity for the contacts on sample D, having the lowest hole concentration among the samples (2.0x10^{16} cm^{-3}), was significantly reduced to 5.5x10^{-4} Ω-cm^2. This value is more than two orders of magnitude lower than that of the contacts on sample B having the highest hole concentration among the samples.

Figures 2 and 3 clearly show that the non-alloyed Pd/Pt/Au contacts on sample D, having the lowest hole concentration and the highest [Mg] among the samples, had the lowest contact resistivity among the samples and displayed the linear I-V relationship. The metal-semiconductor contact theory indicates that the decrease in carrier concentration increases contact resistivity, especially in field emission which is essential for obtaining ohmic behavior for the metal-semiconductor interface with high barrier height.[14,15] This implies that the linear I-V relationship and the low contact resistivity of the non-alloyed contacts on sample D cannot be explained by the field emission in the metal-semiconductor contact theory.

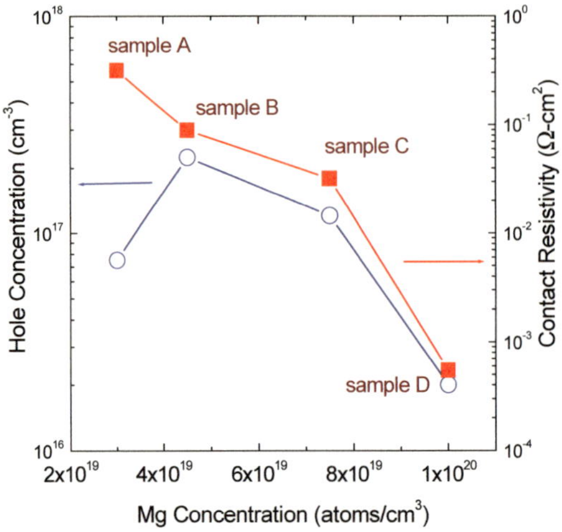

Fig. 3. The contact resistivity for the Pd/Pt/Au contacts on the p-GaN films having a different [Mg]. The hole concentration as a function of the [Mg] is also displayed.

2.2 Influence of micro-structural defects on the carrier transport

One possible explanation for the reduction of contact resistivity of Pd/Pt/Au contacts as the hole concentration of p-GaN decreased when the [Mg] is higher than $4.5x10^{19}$ cm^{-3}, as shown in Fig. 3, is the influence of microstructural defects such as dislocations on the carrier transport at the interface between the non-alloyed contact and p-GaN. High threading dislocation density, which is produced by large differences in lattice mismatch and thermal expansion between GaN and sapphire substrates, has a significant influence on the performance of InGaN light emitting devices.[16] The LEO method is a popular method used to reduce the threading dislocation density. The LEO method has been used to lower the threading dislocation density by more than two orders of magnitude.[17,18] Indeed, the use of LEO GaN has already increased the lifetime of InGaN LDs to more than 10,000 hours.[19,20]

The electrical characteristics of dislocations in n-GaN have been well studied. Weimann *et al.* suggested a model in which filled traps along dislocation lines act as Coulomb scattering centers, explaining the low transverse mobility in GaN by the scattering of electrons at charged dislocation lines.[21] Hansen *et al.* observed free surfaces of undoped GaN by scanning capacitance microscopy, and found the presence of excess negative charge in the region surrounding the dislocations, which implies the presence of deep acceptorlike trap states near the valence band associated with threading dislocations.[22]

The high dislocation density has also greatly affected electrical properties at p-n junctions. Kozodoy *et al.* examined the effect of dislocations on the reverse bias current density by comparing the reverse bias current for p-n diodes on LEO GaN with that for diodes on the dislocated GaN, and found that reverse-bias leakage current was reduced by three orders of magnitude on LEO GaN.[23] Kuksenkov *et al.* also showed that the dark current was correlated with dislocation density in GaN diodes.[24] However, scant information is available concerning the correlation between the high dislocation density and the electrical properties of ohmic contacts to p-GaN.

In addition, the effect of dislocation density on threshold current and life time of InGaN LDs were reported in several papers,[19,20,25] while, little results on the effects of dislocation density on the operating voltage of the InGaN LDs have been reported until now. Nam *et al.* reported that the lifetime of the InGaN LDs was greatly influenced by the operating voltage, although it is well known that the lifetime is affected by the operation current and the temperature of the LDs.[26]

To examine the influence of microstructure of p-GaN on carrier transport at the interface between the non-alloyed contact and p-GaN, Kwak and Park have focused on investigating the effect of dislocation on the contact resistivity of ohmic contacts on p-GaN through measurement of the I-V characteristics for non-alloyed Pd ohmic contacts on LEO GaN and on dislocated GaN. They fabricated TLM patterns with a very narrow mesa structure, which enabled the production of TLM patterns on LEO GaN and on dislocated GaN. They grew p-GaN films on *c*-plane sapphire substrates by metalorganic chemical-vapor deposition (MOCVD). Trimethylgallium (TMGa), trimethylindium (TMIn), and ammonia (NH_3) were used as precursors for Ga, In, and N, respectively. The LEO process was performed to decrease the dislocation density of the layers. First, 2-μm-thick GaN layers were grown on the sapphire substrates. Second, silicon dioxide and metal masks were deposited on the GaN layer; striped patterns with widths of 4 μm and a periodicity of 10~18 μm were formed in the <1-100> direction of GaN using standard lithography and inductively coupled plasma reactive ion etching (ICP-RIE). Third, window areas of GaN layers were etched down to the sapphire substrate by ICP-RIE. Finally, the striped-pattern GaN wafers were loaded into the MOCVD reactor for the LEO process.

After the growth of LEO GaN, both the 2-μm-thick, Mg-doped, p-GaN layer and the laser diode structures with a 405 nm wavelength were grown on LEO GaN wafers. For p-GaN, bis-(cyclopentadienyl)-magnesium (Cp_2Mg) was used as a precursor for Mg. the 2-μm-thick p-GaN layer had a Mg concentration of 2×10^{19} cm^{-3}, where the hole concentration measured by the Hall measurement system was 7×10^{17} cm^{-3}. In addition, an additional thin layer (10 nm) having a Mg concentration of 1×10^{20} cm^{-3} was grown on the 2-μm-thick p-GaN layer. Although the p-GaN having a Mg concentration of 1×10^{20} cm^{-3} had the hole concentration as low as 2×10^{16} cm^{-3}, it can reduce the contact resistivity of the non-alloyed contacts to p-GaN significantly.[27]

For the InGaN-based laser diode, a Si-doped GaN layer was first grown on the LEO-GaN layer. Doped AlGaN/GaN superlattice structures and GaN were used as cladding and waveguide layers, respectively. The AlGaN/GaN superlattice structures consisted of 25-Å-thick AlGaN and 25-Å-thick GaN layers. The active layers were InGaN/InGaN MQW structures consisting of a 40-Å-thick $In_{0.08}Ga_{0.92}N$ well and a 100-Å-thick $In_{0.02}Ga_{0.98}N$ barrier. A Mg-doped, AlGaN, electron-blocking layer was grown between the active layer and the p-type waveguide layer. Finally, an additional thin layer (10 nm) having a Mg concentration of 1×10^{20} cm^{-3} was grown as a p-contact layer. Conventional annealing was conducted below 900 °C to achieve good p-type conduction for both samples.

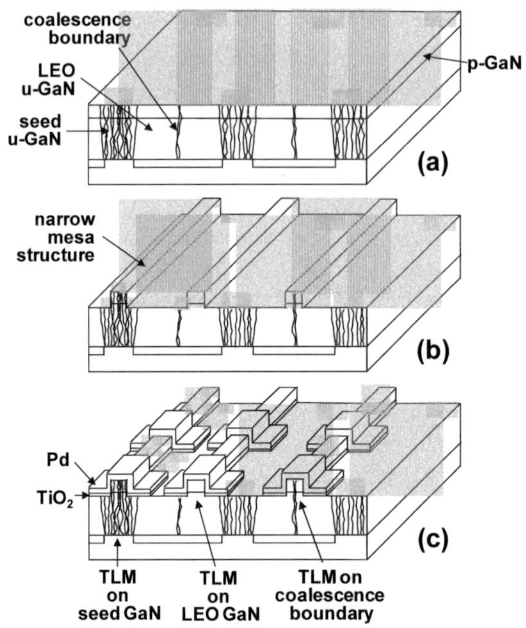

Fig. 4. Schematic diagrams showing the process flow for fabricating the narrow TLM patterns.

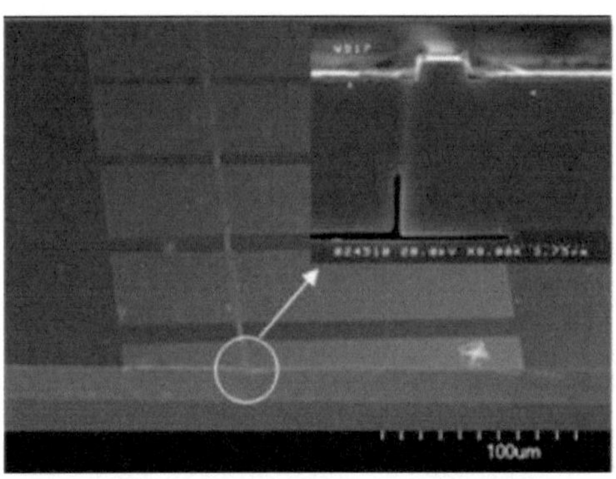

Fig. 5. SEM image of the fabricated narrow TLM patterns, having 2 μm-wide narrow mesa structures.

To investigate the effect of dislocation on the contact resistivity of ohmic contacts on p-GaN, Kwak and Park fabricated TLM patterns with a very narrow mesa structure (narrow TLM), which enabled the production of TLM patterns on LEO as well as on dislocated GaN. The narrow TLM patterns were fabricated as described in the process flow shown in Fig. 4. First, a chemically assisted ion beam method was used to etch a 2-μm-wide mesa stripe on the wing and seed regions of the LEO-GaN structures, as shown in Fig. 4(b). This was followed by deposition of the 0.2-μm-thick TiO_2 passivation layer and 0.1-μm-thick Pd contacts, as shown in Fig. 4(c). Before deposition of the Pd contacts, the TLM patterns were etched in buffered HF solution, rinsed in DI water, and dried with nitrogen before being loaded into the vacuum system for metal deposition. Pd (100 nm) films were deposited by electron beam evaporator. The Pd provides good specific contact resistance in the as-deposited state, therefore, Pd was used as an ohmic material to narrow TLM.[27,28] Figure 5 is a scanning electron microscope (SEM) image of the fabricated narrow TLM patterns with narrow (2-μm in width) mesa structures. The cross-sectional SEM image (inset) shows that the narrow TLM pattern was well fabricated on the wing region of the LEO GaN.

Figure 6 shows the I-V characteristics of the Pd contacts on the LEO GaN (low dislocation density region) as well as on the seed GaN (dislocated region) measured at a gap spacing of 10 μm. The Pd contacts, both on the LEO GaN and the seed GaN, produced linear I-V characteristics, as shown in Fig. 6. It is noteworthy that variations of the I-V characteristics of the contacts on the LEO GaN, as well as on the seed GaN, were small and identical, which implies that dislocation had little effect on the I-V characteristics of the Pd contacts.

Figure 7 shows the specific contact resistances of the Pd contacts on the LEO GaN and that of the contacts on the dislocated GaN, as measured using the narrow TLMs. The non-alloyed Pd contacts on the LEO-GaN (low dislocation density region) showed a specific contact resistance of $2\sim3\times10^{-3}$ Ω-cm^2. The contacts on the dislocated regions (both the seed and coalescence region) also showed a specific contact resistance of $2\sim3\times10^{-3}$ Ω-cm^2. In addition, the Pd contacts located between the LEO-GaN region and the dislocated regions had a specific contact resistance of $2\sim3\times10^{-3}$ Ω-cm^2. These results clearly showed that dislocation had little influence on the specific contact resistance of the non-alloyed Pd contacts on p-GaN.

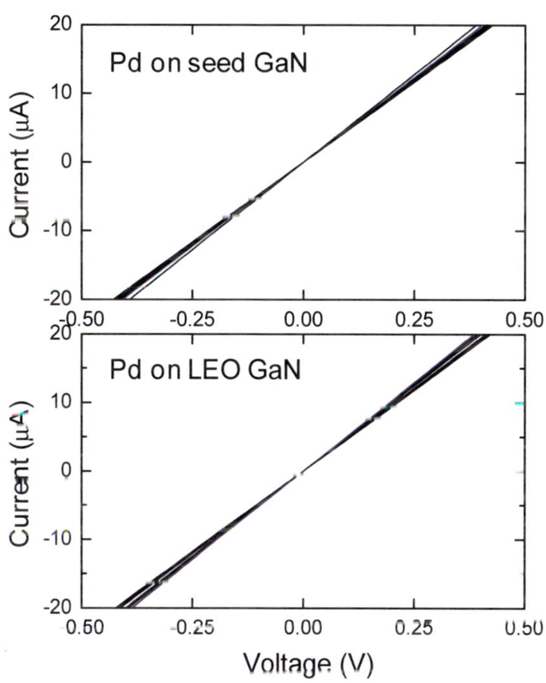

Fig. 6. I-V characteristics of the Pd contacts on the LEO GaN as well as on the seed GaN measured at 10 μm gap spacing.

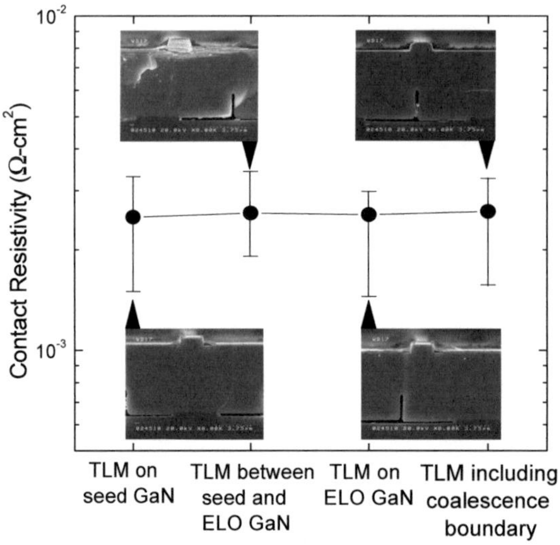

Fig. 7. The specific contact resistances of the Pd contacts on LEO GaN and that of the contacts on dislocated GaN, as measured using narrow TLMs.

It has been reported that a neutral region is formed around each dislocation due to a mid-gap state, which reduces electron mobility of n-GaN.[29] The neutral region along the dislocations likely does not contribute to carrier transport through the metal/semiconductor (M/S) interface, which implies that the actual area at the M/S interface could be reduced. In our sample, however, the carrier concentration was relatively high (p=7.0x10^{17} cm^{-3}), which is essential to InGaN based laser diodes, a smaller depleted area due to the dislocation was expected. Shiojima *et al.* evaluated the effect of dislocation on current-voltage (I-V) characteristics of Au/Ni/n-GaN Schottky contacts by using a submicrometer Schottky dot array formed by electron beam lithography, and showed that neither dislocations nor steps affected the I-V characteristics, which could be attributed to a relatively higher carrier concentration (n=5.8x10^{17} cm^{-3}).[30] To confirm this explanation, however, further studies, such as the same experiment with low hole concentrations, are needed to estimate the actual area which contributes the carrier transport at the M/S interface.

To elucidate whether the dislocation affects the voltage characteristics of the InGaN LDs, we fabricated InGaN LDs, in which non-alloyed Pd contacts were produced on a 2- μm -width ridge waveguide structure. The location of the ridge waveguide varied from seed region to coalescence region. Figure 8(a) shows the variation of output power and voltage of the InGaN LDs as a function of injected current for the LDs with the Pd contacts on the LEO GaN and on the dislocated GaN. As shown in Fig. 8(a), the output power-current curve was differed greatly for the different locations of the Pd contacts. In contrast, the voltage-current curve changed little with the location of the contacts. Figure 8(b) displays the variation of the operating voltage at 50 mA and threshold current as a function of location of the ridge waveguide of the InGaN LDs. As shown in Fig. 8(b), the InGaN LDs, which have a ridge structure on the LEO-GaN region, showed a low threshold current of 55 mA. The threshold current of the LDs where the ridge structure is located between the seed and the LEO regions, however, increased to 90 mA. Furthermore, it increased as high as 119 mA in the seed region. This implies that dislocation has a significant influence on the threshold current characteristics of InGaN LDs. These results agree well with previous reports. It was reported that an increase in dislocation density significantly deteriorated the operating current characteristics of the InGaN LDs.[26] A significant increase in nonradiative recombination in the dislocation region may cause a significant increase in the threshold current of the dislocated region of the InGaN LDs, as shown in Fig. 8.[31]

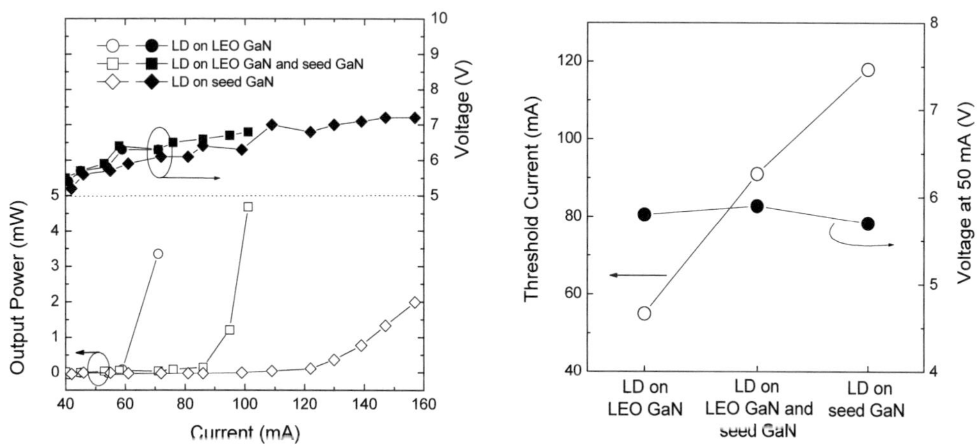

(a) (b)

Fig. 8. (a) Variation of output power and voltage of the InGaN LDs as a function of injected current for the LDs with the Pd contacts on the LEO GaN and the dislocated GaN, and (b) variation of the operating voltage at 50 mA and threshold current as a function of location of the ridge waveguide of the InGaN LDs.

The density of dislocation, however, had minimal influence on the operating voltage of the ridge waveguide InGaN LDs. As shown in Fig. 8(b), we obtained 5.8 V as the operating voltage at 50 mA for the LEO-GaN region of LDs. Similar operating voltage values were obtained even for the dislocated regions of LDs. This can be attributed to the results shown in Fig. 7, where the density of dislocation had little influence on the specific contact resistance of the non-alloyed Pd contacts. These results clearly suggest that dislocation exerts significant influence on the threshold current of InGaN LDs, but has minimal influence on the specific contact resistance of the contacts to p-GaN.

2.3 Influence of deep level defects on the carrier transport

Another possible explanation for the abnormal dependence of the contact resistivity on the hole concentration observed in this study is the influence of deep level defects in p-GaN on the carrier transport at the Pd/p-GaN interface. Götz *et al.* reported the presence of compensating deep centers in Mg-doped GaN using capacitance spectroscopy.[32] Bayerl *et al.* also detected a deep level defect named MM1 located near the mid gap using optically detected magnetic resonance and observed different defect signals for p-GaN with different [Mg].[33] Hofmann *et al.* suggested that the structure of MM1 center could be a nitrogen vacancy-Mg pair defect ($V_N Mg_{Ga}$).[34] In addition, Shiojima *et al.* found that carrier capture and emission from acceptor-like deep level defects caused depletion layer width to vary significantly in Ni contacts to p-GaN.[35]

Kwak *et al.* have examined the temperature-dependent contact resistivity of non-alloyed Pd/Pt/Au contacts to *p*-GaN films as well as the temperature-dependent sheet resistivity of the *p*-GaN films, in order to understand the anomalously low contact resistivity (~10^{-4} Ω-cm^2) considering the large work function difference between the metal and *p*-GaN and the abnormal dependence of contact resistivity on carrier concentration, as well. In their study, three different *p*-GaN samples having a different Mg concentration, [Mg], were used (samples A, B, and C). A thin buffer layer and a 1-um-thick undoped GaN layer on a sapphire substrate was grown by the MOCVD, followed by the growth of 1-um-thick *p*-GaN layers having three different [Mg]. Variation of [Mg] in *p*-GaN was conducted by changing the flow rate of bis-(cyclopentadienyl)-magnesium. Sample A, B, or C has the [Mg] of $4.5x10^{19}$, $7.5x10^{19}$, or $1.0x10^{20}$ cm^{-3}, respectively, which was measured using secondary ion mass spectroscopy. The hole concentration of the sample A, B, or C was determined as $2.2x10^{17}$, $1.2x10^{17}$, or $2.0x10^{16}$ cm^{-3}, respectively, which was obtained from Hall-effect measurements. These results indicate that the hole concentration decreased as increase in the [Mg] in the *p*-GaN films in this study. The temperature dependence of the contact resistivity and the sheet resistance were performed in a cryostat using transfer length method (TLM). Prior to the fabrication of the TLM patterns, mesa structures were patterned using chemically assisted ion beam etching with Cl_2. Then, they were etched in buffered HF solution, rinsed in DI water, and dried with nitrogen before being loaded into the vacuum system for metal deposition. Pd (20 nm)/Pt (30 nm)/Au (80 nm) films were deposited by electron beam evaporator.

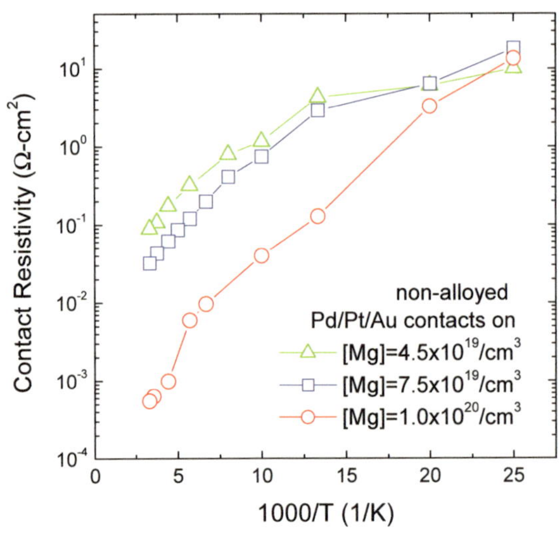

Fig. 9. The temperature dependence of contact resistivity of the non-alloyed Pd/Pt/Au contacts on the *p*-GaN samples having the different [Mg]. The [Mg] of the samples A, B, and C were $4.5x10^{19}$, $7.5x10^{19}$, and $1.0x10^{20}$ cm^{-3}, respectively, and the hole concentration of the samples A, B, and C were $2.2x10^{17}$, $1.2x10^{17}$, and $2.0x10^{16}$ cm^{-3}, respectively

The temperature dependence of contact resistivity of the non-alloyed Pd/Pt/Au contacts on the *p*-GaN samples having the different [Mg] is shown in Fig.9. As the measured temperature decreased, as shown in Fig. 9, the contact resistivity of the Pd/Pt/Au contacts on the *p*-GaN samples rapidly increased. For the contacts on sample C having the [Mg] of $1x10^{20}$ cm^{-3}, the contact resistivity increased from $5.5x10^{-4}$ to $4.0x10^{-2}$ Ω-cm^2, as the temperature decreased from 300 to 100 K. In case of the contacts on the sample A or B, the contact resistivity also increased by more than one order of magnitude as the temperature decreased from 300 to 100 K. These results clearly show that the carrier transport at the Pd/*p*-GaN interface is not dominated by pure-tunneling mechanism because it implies a temperature-independent contact resistance.[15] It is worthwhile to note that the contact resistivity of the non-alloyed Pd/Pt/Au contacts on the sample C ($5.5x10^{-4}$ Ω-cm^2) having a low hole concentration of $2.0x10^{16}$ cm^{-3} was anomalously low at room temperature, considering the large work function difference between Pd and *p*-GaN. In addition, in a temperature region higher than 75 K, the contact resistivity of the contacts on the sample C having the lowest hole concentration ($2.0x10^{16}$ cm^{-3}) was more than one order of magnitude lower than that of the contacts on the sample A or B having a higher hole concentration ($2.2x10^{17}$ or $1.2x10^{17}$ cm^{-3}, respectively). This is abnormal since the metal-semiconductor contact theory indicates that the decrease in carrier concentration increases contact resistivity.[15]

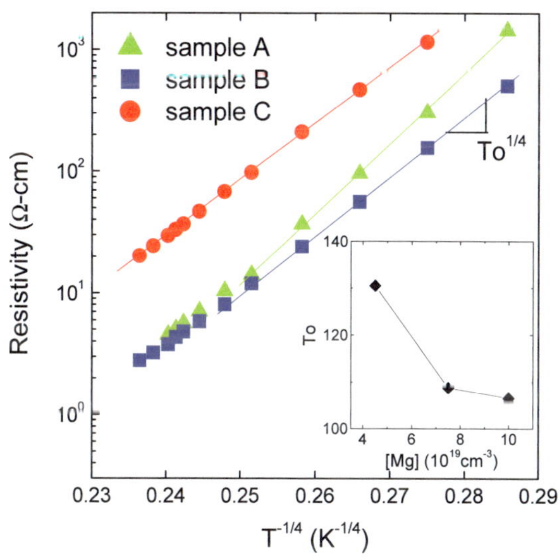

Fig. 10. The change of sheet resistivity of the *p*-GaN samples as a function of measured temperature. The inserted figure shows the dependence of To on the [Mg] in *p*-GaN.

Figure 10 shows the variation of sheet resistivity of the *p*-GaN films as a function of measured temperature. In the temperature-dependent sheet resistivity of the *p*-GaN films, as shown in Fig. 2, we obtained a linear relationship between the sheet resistivity of *p*-GaN and $\exp(To/T)^{1/4}$. The To is given by $2.1(\alpha^3/k_B N_{DE})^{1/4}$ where α^{-1} is localization length, k_B is Boltzman constant, and N_{DE} is density states at the energy level of defects. The obtained linear relationship, as shown in Fig.10, indicates that carriers may flow by variable-range hopping conduction via a deep level defect band.[36] The linear relationship was observed at temperature below 250 K for samples B and C, meanwhile it was maintained up to room temperature for sample C, which may imply that the variable-range hopping conduction via a deep level defect band still dominates current flow at room temperature for sample C. The linear relationship between sheet resistivity and $\exp(To/T)^{1/4}$ was also reported in low-temperature grown (LTG) GaAs, in which a dense EL2-like deep defect band was produced.[37]

The figure inserted in Fig. 10 shows the dependence of To on the [Mg] in *p*-GaN. As the [Mg] increased from 4.5×10^{19} to 1.0×10^{20} cm^{-3}, as shown in the inserted figure, the To was reduced from 130 to 106, which implies that the defect density increased as the [Mg] in *p*-GaN increased since To is $2.1(\alpha^3/k_B N_{DE})^{1/4}$. It is worthy of note that we can calculate the defect density, N_{DE}, from the obtained To value, if we know localization length, α^{-1}. However, since the localization length of the deep level defect band in the *p*-GaN films has not been reported yet, we could not obtain the defect density of the p-GaN. Instead, we can estimate the defect density in the *p*-GaN films having the different [Mg] using the relation between contact resistivity and defect density,[38] which is given by

$$\rho_c = \frac{k_B T}{q^2 k\Theta_D / h\exp(-\gamma / aN_{DE}^{1/3})(3/4\pi N_{DE})^{1/3} N_{DE}} \exp(\frac{q\varphi}{kT}) \quad (1)$$

where Θ_D is Debye temperature, γ is constant, a is extent of the wave function, and $q\phi$ is the barrier height between the metal and the deep level defect. Figure 11 shows the variation of contact resistivity as a function of defect density calculated from the Eq. (1). As shown in Fig. 11, contact resistivity rapidly decreases as increase in the defect density, and it also decreases when the $q\phi$ decreases. In order to obtain the defect density in the *p*-GaN films, the $q\phi$ was measured by using the devised I-V method which utilizes large-area contacts and reverse-biased I-V characteristics,[39] and the measured $q\phi$ was 0.36 eV. From the measured

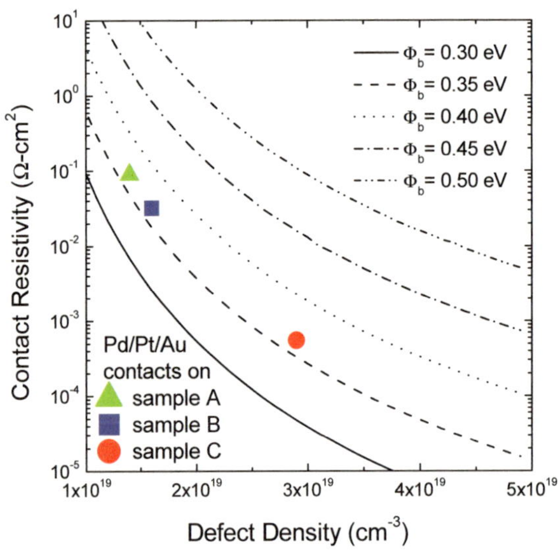

Fig. 11. The variation of contact resistivity as a function of defect density and barrier height between metal and the deep level defects, $q\phi$, which is calculated from the eq. (1). The measured $q\phi$ and the obtained contact resistivity for the Pd/Pt/Au contacts on samples A. B, and C were also depicted.

$q\phi$ and the contact resistivity of the Pd/Pt/Au contacts on the *p*-GaN having the different [Mg], as shown in Fig. 11, we obtained the defect densities in the sample A, B, and C as 1.4x10^{19}, 1.6x10^{19}, and 2.9x10^{19} cm^{-3}, respectively. It should be noted that the defect density in the *p*-GaN increased as increase in [Mg] in the *p*-GaN, which is well matched to the relation between To and [Mg], as shown in the inserted figure in Fig. 10.

Based on these results, the band diagram of a Pd/p-GaN contact was devised, as shown in Fig. 12. The deep defect level, E_{DE}, is unoccupied with electrons in most of the material, as shown in Fig. 12. However, because of the band bending near the interface between the metal and the *p*-GaN, occupied region in the deep defect level was produced by the electrons moved from the metal. The occupied region of the deep defect level, which is located below the Fermi level, E_F, as shown in Fig. 12, may allow the carriers to flow from the metal directly to the dense deep defect level in the *p*-GaN, which results in predominant current flow at the Pd/*p*-GaN interface through the deep level defect band, rather than the usual valence band. The carrier transport at the metal/*p*-GaN interface through deep level defects is also observed in the Ni/*p*-GaN interface. Shiojima *et al.* found that carrier capture and emission from acceptor-like deep level defects caused depletion layer width to vary significantly in Ni contacts to p-GaN.[40] In addition, Yu *et al.* reported a tunneling component in the Ni contacts to *p*-GaN due to the defect states located in the near surface region of the semiconductor.[41]

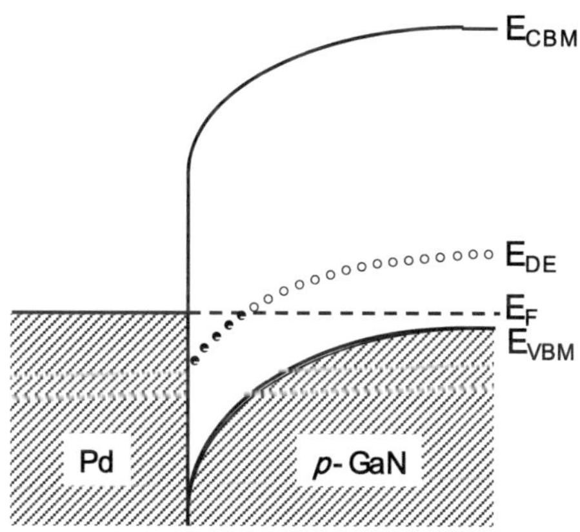

Fig. 12. Band diagram of the Pd/*p*-GaN in which the carriers flow from the metal directly to the dense deep level defects.

From the devised band diagram, as shown in Fig. 12, the abnormal dependence of the contact resistivity on the carrier concentration, as shown in Fig. 9, as well as the anomalously low contact resistivity (5.5x10^{-4} Ω-cm^2) of the Pd/Pt/Au contacts on sample C considering the large work function difference between Pd and *p*-GaN and the low hole concentration of sample C (2.0x10^{16} cm^{-3}) can be explained as follows. In the samples A, B and C, when the [Mg] in the *p*-GaN films increased from 4.5x10^{19} to 1.0x10^{20} cm^{-3}, the hole concentration decreased from 2.2x10^{17} to 2.0x10^{16} cm^{-3}, meanwhile the deep level defect density increased from 1.4x10^{19} to 2.9x10^{19} cm^{-3}, as shown in Fig. 11. The increase in the defect density allows more carriers to flow from the Pd to the *p*-GaN through the deep level defects, followed by the reduction of the contact resistivity. This can result in the abnormal dependence of the contact resistivity on the carrier concentration, that is, the decrease in the contact resistivity from 8.9×10^{-2} to 5.5×10^{-4} Ω-cm^2 even though the hole concentration was reduced from 2.2×10^{17} to 2.0×10^{16} cm^{-3}. Furthermore, the unexpectedly low contact resistivity (5.5×10^{-4} Ω-cm^2) even though the hole concentration was as low as 2.0×10^{16} cm^{-3} can be attributed to the large defect density in the sample C (2.9×10^{19} cm^{-3}), since the very large defect density can greatly lower the contact resistivity, as shown in Fig. 12. The anomalously low contact resistivity considering a low carrier concentration in semiconductors is also observed in non-alloyed ohmic contacts to LTG GaAs, which was explained by the predominant current flow at the metal/LTG GaAs interface through a very large concentration (~3×10^{19} cm^{-3}) of EL2-like deep defect level, followed by a low contact resistivity of 1.5×10^{-3} Ω-cm^2 even though the carrier concentration was less than 10^{11} cm^{-3}.[38] As for the deep level defects in the *p*-GaN films, the energy level and structure of the deep defect are not clearly understood yet. One possible acceptor-like deep defect level is an impurity band. Cheong *et al.* suggested that Mg acceptors formed an impurity band that had thermal activation energy in a range of 300-360 meV.[42] Clarifying the deep defect level seems to be the key issue to a deeper understanding of the carrier transport at the metal/*p*-GaN interface, which remains for further study.

Finally, The experimental results and theoretical calculation suggested that a deep level defect band having a large defect density over 10^{19} cm^{-3} existed in the *p*-GaN films and the density of the deep level defects was enlarged as the [Mg] increased. The unexpectedly low contact resistivity (~10^{-4} Ω-cm^2) considering the large work function difference between the metal and *p*-GaN as well as the abnormal dependence of the contact resistivity on the hole concentration could be explained by the carrier transport model in which the carriers flow from the metal directly to the dense deep level defects.

2.4 Design of p-GaN contact layer to minimize operating voltage of AlInGaN LD

Based on the understanding of carrier transport phenomena in metal contacts to p-GaN contact layer, a p-GaN:Mg contact layer can be designed to minimize the operating voltage of AlInGaN LD. As for the design of p-GaN:Mg contact layer, Kwak *et al.* reported the p-GaN contact layer having a high Mg-doped thin cap layer.[27,43,44] In this design, an additional thin layer (10 nm) having a high [Mg] (1×10^{20} cm^{-3}) was grown on p-GaN having the [Mg] of 4.5×10^{19} cm^{-3}. In x-ray diffraction, the full width at half maximum values of (0002) and ($1\bar{1}02$) for the p-GaN contact layer having a high Mg-doped thin cap layer were identical to those for the p-GaN contact layer without the cap layer, which implies that the microstructure was not changed by the addition of the thin layer. Figure 13 shows a comparison of the measured I-V characteristics of the non-alloyed Pd/Pt/Au contacts deposited on p-GaN contact layer with and without the thin layer having the high [Mg]. The I-V characteristics were measured using two metal pads with 5 μm gap spacing, and the thin layer between the two metal pads was removed by dry etching with Cl_2. The addition of the thin layer with the high [Mg], as shown in Fig. 13, significantly increased the measured current at 0.5 V from 41 μA to 670 μA and made I-V curve linear. This result indicates that the addition of a thin layer of p-GaN with high [Mg] on p-GaN with lower [Mg] greatly increased the current across the interface between the contact metal and the p-GaN films with lower [Mg].

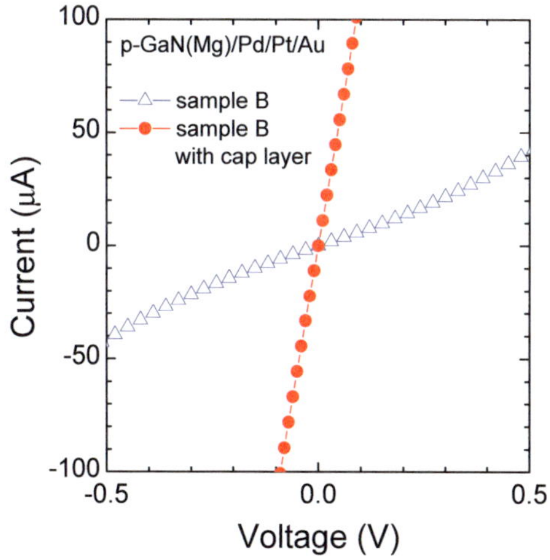

Fig. 13. A comparison of the measured I-V characteristics of the non-alloyed Pd/Pt/Au contacts deposited on sample B with and without the thin layer having the high [Mg].

The concentration of Mg, [Mg], in the thin cap layer is one of the important parameters to minimize the operating voltage of AlInGaN LD. As shown in Fig. 14, for the Pd/Pt/Au contacts on the p-GaN contact layer without the thin p-GaN layer with high [Mg], high contact resistivity of 5.2×10^{-2} Ω-cm² was obtained. When the thin p-GaN layer with the flow rate of Cp_2Mg of 1.4 μmole/min was added on the p-GaN film, however, the contact resistivity decreased by one order of magnitude (5.0×10^{-3} Ω-cm²). Further increase in the flow rate of Cp_2Mg to 2.1 and 2.8 μmole/min, as shown in Fig. 4, resulted in a reduction of the contact resistivity to 1.7×10^{-3} Ω-cm² and 9.4×10^{-4} Ω-cm², respectively.

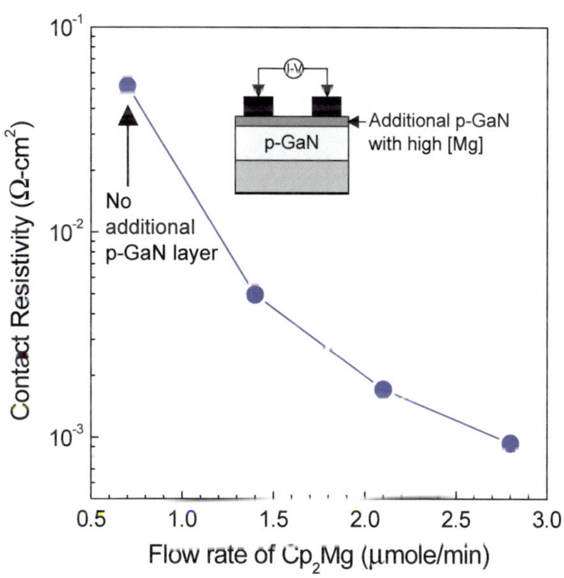

Fig. 14. The effect of the [Mg] in the additional thin p-GaN layer on the contact resistivity.

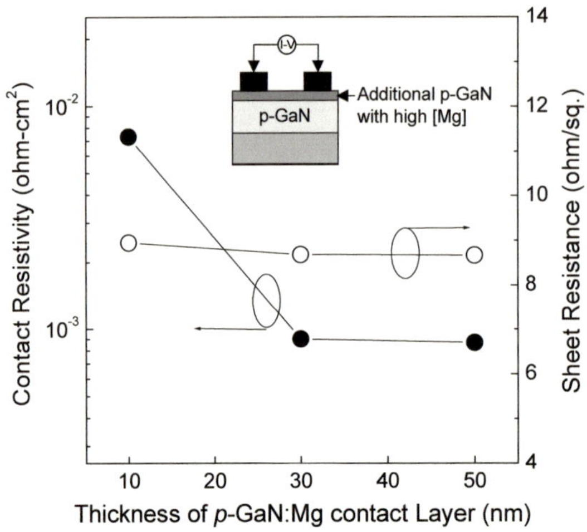

Fig. 15. Variation of the contact resistivity of the Pd/Pt/Au contacts and sheet resistance of p-GaN as a function of thickness of the highly Mg doped *p*-GaN contact layer.

The thickness of the thickness of the cap layer in the design of a *p*-GaN contact layer having a high Mg-doped thin cap layer is also important factor to minimize the operating voltage of AlInGaN LD. Figure 15 shows the variation of the contact resistivity of the Pd/Pt/Au contacts and sheet resistance of *p*-GaN as a function of thickness of the highly Mg doped *p*-GaN contact layer. As shown in Fig. 15, the sheet resistance of the *p*-GaN was not changed as increasing the thickness of the *p*-GaN contact layer. However, the contact resistivity was reduced rapidly as the thickness of the *p*-GaN contact layer increased from 10 to 30 nm. This result may be attributed to insufficient incorporation of Mg in the highly Mg doped *p*-GaN contact layer, since growth of the 10 nm-thick *p*-GaN contact layer required very rapid increase in Cp_2Mg flow rate from 1.05 to 2.80 μmol/min. Further increase in the thickness of the *p*-GaN contact layer up to 50 nm did not change the contact resistivity, as shown in Fig. 15.

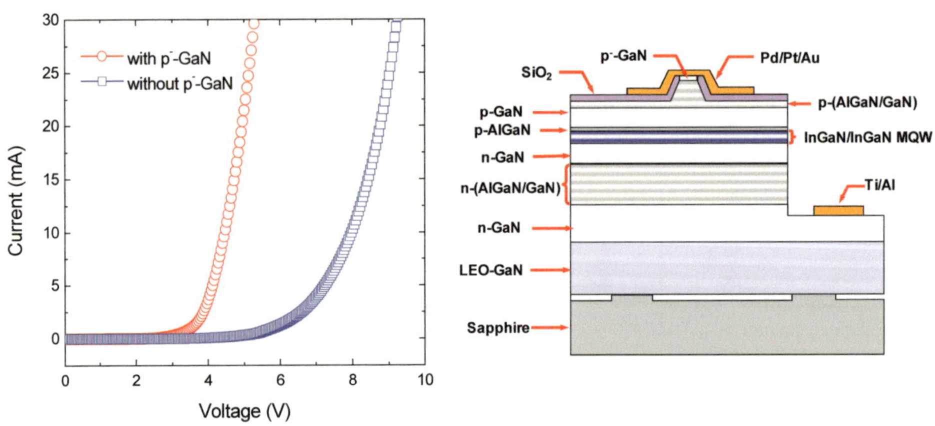

Fig. 16. The effect of a *p*⁻-GaN contact layer having a high Mg-doped thin cap layer on the lowering the operating voltage of the AlInGaN LDs.

To verify the effect of a *p*-GaN contact layer having a high Mg-doped thin cap layer on the lowering the operating voltage of the AlInGaN LDs, the thin *p*-GaN contact layer with high [Mg] at the top of the AlInGaN laser structure was introduced as a contact layer and fabricated the LDs with a ridge stripe, as shown in Fig. 16. When the thin contact layer was heavily doped with Mg, the measured voltage of the LD at 20 mA was measured as 4.9 V, meanwhile the voltage at 20 mA was as high as 8.8 V without the thin heavily doped contact layer. This clearly indicates that the thin *p*-GaN contact layer reduced the contact resistance, followed by the decrease in operating voltage of the LD.

3. Carrier transport in metal contacts to N-face n-GaN of AlInGaN LDs on free standing GaN substrates

Free-standing GaN has attracted attention as a substrate for AlInGaN LDs because of its low dislocation density, high thermal conductivity, and easy cleaving.[45,46] Another advantage of using the GaN substrate is to fabricate devices with a backside n-contact. This allows the fabrication process to be simple and reliable, and reduces the size of devices, which increases yield in mass-production.

GaN substrate has two faces with a different crystal polarity, Ga- and N-face polarity, which greatly influence on the electrical properties at metal/GaN interface as well as those at AlGaN/GaN heterostructure. Karrer *et al.* investigated the influence of crystal polarity on the properties of Pt/GaN Schottky diodes grown by plasma-induced molecular-beam epitaxy (PIMBE), and reported that different barrier heights of Pt onto the two different face were obtained to be 1.1 and 0.9 eV for Ga- and N-face GaN, respectively.[5] He suggested that this behavior could be due to the different bend bending of the conduction and valence band caused by the different spontaneous polarization in epitaxial layers with different polarity.[5] Fang *et al.* also reported that Ni/Au contacts showed higher barrier heights of 1.27 eV on Ga-face free-standing GaN than on N-face GaN (0.75 eV).[6] Since device structures including active layers are normally grown on Ga-polar GaN substrate, the backside ohmic contact should be produced on N-polar GaN side. However, little work has been performed on the comparison of electrical properties between ohmic contacts on Ga- and N-polar GaN substrate.

Kwak *et al.* investigated the influence of crystal polarity on the electrical properties of Ti/Al contacts to n-type GaN, since the Ti/Al contacts are widely used for ohmic contacts to n-GaN.[45,47,48] For this purpose, free-standing n-GaN substrates grown by hydride vapor phase epitaxy (HVPE) were used, because they have Ga-face polarity in one surface and N-face in the other surface, and a comparison between Ti/Al contacts on Ga-face and N-face n-GaN substrates is made.

The samples were grown by HVPE on sapphire substrate to a thickness of 300 μm. In order to obtain a free standing GaN substrate, the thick GaN layer was separated from the sapphire by laser induced lift-off. The GaN wafers were then mechanically polished and dry etched on both the Ga- and N-face to obtain a smooth epi-ready surface. Structural properties were investigated using double-crystal x-ray diffraction (DXRD). The full width at half maximum (FWHM) of the (0002) peak for the Ga face and the N-face were measured to be 126 and 153 arc sec, respectively. The typical dislocation density for both the Ga and N-face was lower than 10^7 cm^{-2}. The samples were doped with Si. The electron concentration and mobility obtained from Hall measurement at room temperature were 1.5×10^{17} cm^{-3} and 825 cm^2/Vs, respectively.

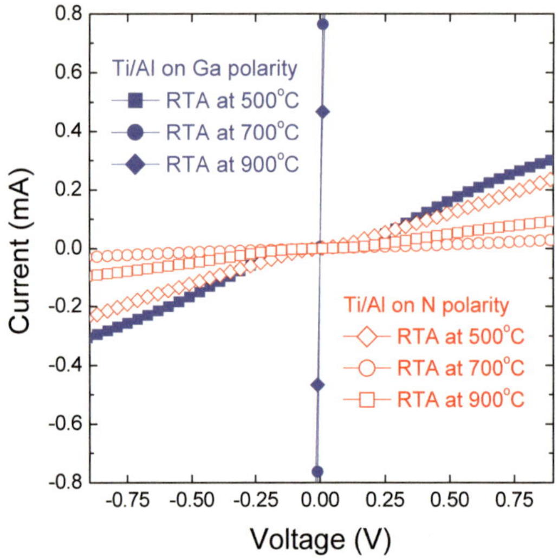

Fig. 17. Current–voltage (I-V) curves for the Ti/Al contacts on Ga- and N-face *n*-GaN annealed for 30 s at various temperatures.

Figure 17 shows I-V characteristics of the Ti/Al contacts deposited on both the Ga-face and N-face n-GaN substrate. The Ti/Al contacts displayed non-linear I-V curves after annealing at 500 °C for 30 s for both the contacts on the Ga-face and N-face n-GaN, as shown in Fig. 17. The I-V curve became linear for the Ti/Al contacts on Ga-face n-GaN after annealing at 700 and 900 °C for 30 s. These are well matched to the previous results, where the Ti/Al contacts became ohmic after annealing at temperatures higher than 600 °C.[47,48] However, the Ti/Al contacts on N-face n-GaN still exhibited non-linear I-V relations and the slope of the I-V curve deceased after annealing at 700 °C, as shown in Fig. 17.

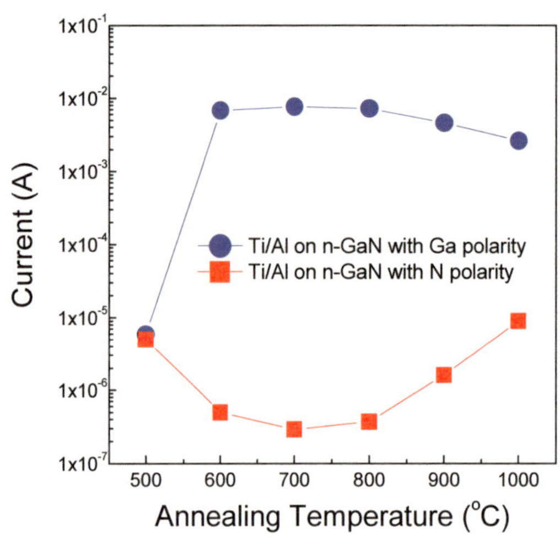

Fig. 18. Variation of currents measured at a bias voltage of 0.1 V between two Ti/Al contact pads deposited on Ga- or N-face *n*-GaN as a function of annealing temperatures.

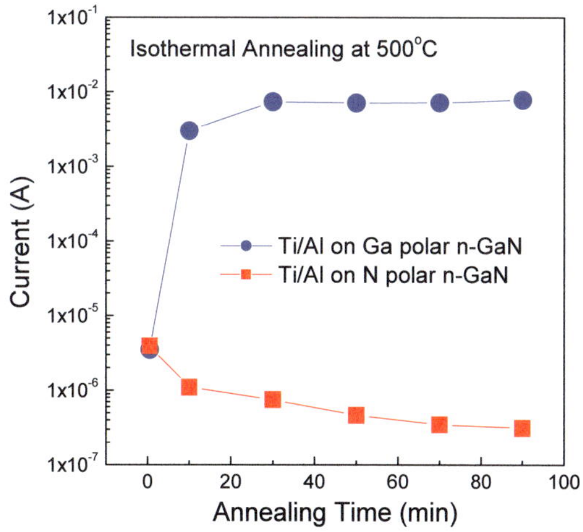

Fig. 19. Current measured at a bias voltage of 0.1 V between two Ti/Al contact pads deposited on Ga-face or N-face n-GaN as a function of the annealing time.

Figure 18 shows the variation of currents measured at a bias voltage of 0.1 V between two Ti/Al contact pads as a function of annealing temperatures. As shown in Fig. 18, the dependence of measured currents on annealing temperature is very different for the Ti/Al contacts on Ga-face n-GaN substrate compared to those on N-face n-GaN. For contacts on Ga-face n-GaN, the currents increased drastically with increasing annealing temperature from 500 to 600 °C, and contact resistivity of 2×10^{-5} Ω-cm^2 was measured from TLM after 30 s anneals at 600-800 °C. A further increase in annealing temperature resulted in a reduction in the measured current. For the Ti/Al contacts on N-face n-GaN substrate, the current at 0.1 V after annealing for 30 s at 500 °C was similar to that obtained for contacts on Ga-face n-GaN, as shown in Fig. 18. The measured current, however, dropped by one order of magnitude after 30 s anneals at 600 °C, as shown in Fig. 18. The minimum current of 300 nA at 0.1 V was obtained after 30 s anneal at 700 °C, which is four orders of magnitude lower than that obtained for the contacts on Ga-face n-GaN substrate, as shown in Fig. 18.

The most significant difference in the results shown in Fig. 18 is evident after annealing ranged from 500 to 600 °C. Contacts were further annealed at 500 °C, and currents at 0.1 V were measured after various time intervals, as shown in Fig. 19. The dependence of the measured currents on annealing time is very different for the Ti/Al contacts on the Ga-face n-GaN compared to those on the N-face n-GaN. For the contacts on the Ga-face n-GaN, the currents increased drastically after annealing for 10 min at 500°C, and the measured current at 0.1 V was 3 mA. For the Ti/Al contacts on the N-face n-GaN, however, the measured current dropped by one order of magnitude after 90 min anneals at 500 °C, as shown in Fig. 3. The measured current of the contact on the N-face n-GaN was 313 nA at 0.1 V, which is four orders of magnitude lower than that obtained for the contact on the Ga-face n-GaN. Figure 3 clearly shows that the dependences of the electrical properties of the Ti/Al contacts on annealing time are opposite, depending on the polarity of the n-GaN wafer.

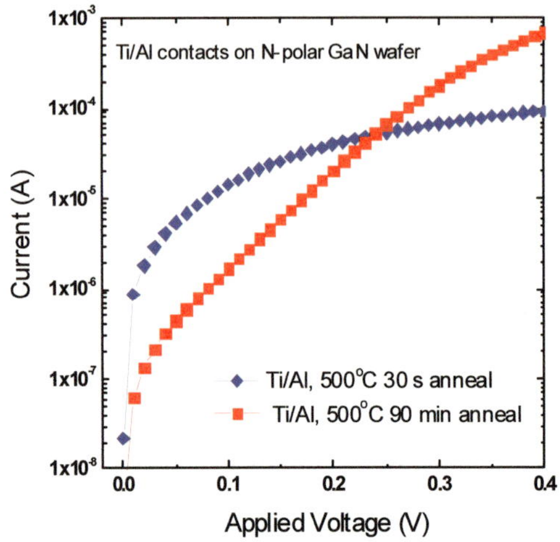

Fig. 20. I-V characteristics of the Ti/Al Schottky diodes produced on N-face n-GaN.

For further investigation of the electrical properties of the Ti/Al contacts on the N-face n-GaN, Schottky barrier heights were measured using current-voltage measurements, and the results are displayed in Fig. 20. For measuring the barrier height, we deposited Ti/Al contact on the Ga-face n-GaN and annealed it at 700 °C for 30 s; it became ohmic with a low contact resistivity. Ti/Al contacts with a 180-μm diameter were produced on the N-face n-GaN. The Ti/Al contacts on the N-face n-GaN after annealing at 500 °C for 30s, as shown in Fig. 20, did not show a linear region in lnI-V graph, although the contacts annealed at 500 °C for 90 min displayed a linear region, which is characteristic of a Schottky diode. The barrier heights and ideality factors of the Ti/Al contacts on the N-face n-GaN annealed at 500 °C for 90 min were calculated to be over 0.7 eV and 1.67, respectively.

Luther *et al.* have characterized the interface of the Ti/Al and Pd/Al contacts on GaN, and a very thin layer of AlN has been observed at the interfaces of Ti/Al contact as well as Pd/Al contact annealed at 600 °C.[49] We also analyzed the Ti/Al on N-face n-GaN using field emission Auger electron spectroscopy, and found that Al peak was detected at the surface of GaN after annealing at 600 °C for 30 s.[50] As for the role of a thin AlN layer at the interface between Ti/Al or Pd/Al contacts and GaN, Luther *et al.* suggested that the formation of thin AlN layer (2-3 nm) could affect the band lineup between the metal contact and n-GaN by eliminating any Fermi level pinning at the metal/GaN interface, or could lower the barrier height, as is the case for other metal-insulator-semiconductor structures.[47,49] This suggestion can explain the formation of ohmic contact for the Ti/Al on Ga-face n-GaN substrate, meanwhile it cannot elucidate the production of Schottky contacts for the Ti/Al on N-face n-GaN observed in this study. Another possible explanation for the role of a thin AlN layer on the formation of the ohmic or Schottky contacts of Ti/Al depending on crystal polarity is production of opposite piezoelectric field at the AlN/GaN interface resulted from different polarity of GaN. Figure 21 shows the schematic diagrams of carrier transport mechanism for Ti/Al contacts to Ga-face and N-face n-GaN. Asbeck *et al.* reported that AlGaN/GaN heterostructure with Ga polarity increased the density of two-dimensional electron gas (2-DEG) by addition of the piezoelectrically-induced donor.[51] The addition of the piezoelectrically-induced donor occurs only in strained heterostructure with Ga polarity.

Indeed, Luther *et al.* reported that the thin AlN produced by the reaction between Ti/Al or Pd/Al on GaN is under strain.[51] The increase of sheet carrier density at 2-DEG reduces Schottky barrier width and enhances tunneling of carriers through the barrier, followed by producing ohmic contacts. However, AlGaN/GaN heterostructure with opposite polarity can increase Schottky barrier height by canceling of the 2-DEG. Gaska *et al.* suggested that the opposite direction of piezoelectric field in $Al_{0.2}Ga_{0.8}N$/GaN heterostructure increased the barrier heights by more than 0.7 eV. [52] In addition, Yu *et al.* increased the barrier heights of Ni in AlGaN/GaN heterostructure by 0.37 eV using the piezoelectric effect.[53] In this study, we observed high barrier heights over 1 eV for the Ti/Al or Pd/Al contacts on N-face n-GaN substrate. Therefore, the opposite electrical properties of the Ti/Al contacts depending on crystal polarity was observed in this study, which can be explained the opposite piezoelectric field in AlN/GaN heterostructure relying on crystal polarity of the free-standing n-GaN substrate.

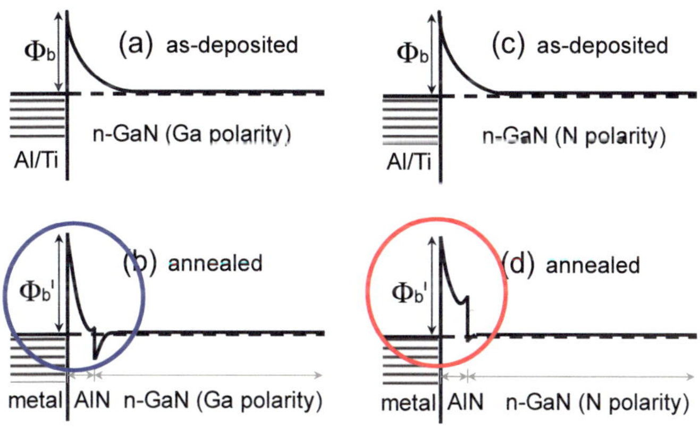

Fig. 21. Schematic diagram of carrier transport mechanism for Ti/Al contacts to Ga-face and N-face n-GaN.

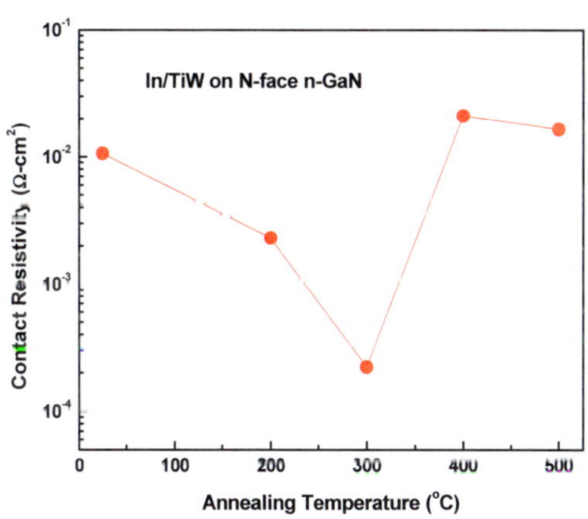

Fig. 22. Variation of the contact resistivity of the In/TiW contacts on N-face n-GaN as a function of annealing temperature.

Based upon the carrier transport phenomena of Ti/Al contacts to N-face n-GaN, we have investigated on the In-based ohmic contacts to N-face n-type GaN. The In/TiW scheme on the N-face n-GaN resulted in a low specific contact resistance of 2.2×10^{-4} Ω-cm^2 after annealing at 300 °C. These results suggest that In/TiW scheme is a promising ohmic contact scheme for N-face n-GaN for AlInGaN-based LDs.

4. Summary

In this chaper, we have discussed the carrier transport phenomena at the interface between metal and *p*-GaN as well as *n*-GaN, in order to achieve high-quality ohmic contacts to AlInGaN-based LDs. Abnormal dependence of contact resistivity on hole concentration was observed in the carrier transport at the interface between metal and *p*-GaN. This abnormal behavior was understood by predominant current flow at the metal/*p*-GaN contact layer interface through a deep level defect band, rather than the valence band, after the consideration of the influence of micro-structural defects and deep level defects on the carrier transport. Based on the understanding of carrier transport phenomena in metal contacts to *p*-GaN contact layer, a new design *p*-GaN contact layer having a thin high [Mg] cap layer was introduced to minimize operating voltage of AlInGaN LD. In use of a free standing GaN wafer for AlInGaN LDs, Ti/Al contacts to N-face n-GaN exhibited nonlinear current-voltage curve and high Schottky barrier heights, meanwhile they produced good ohmic contacts to Ga-face n-GaN. This crystal-polarity dependence of Ti/Al contacts to free-standing n-GaN substrate was understood by opposite piezoelectric field at GaN/AlN heterostructure resulted from different polarity of the GaN substrate. Further work is underway that will enhance our understanding of carrier transport phenomena at the interface between metal and *p*-GaN as well as n-GaN for future AlInGaN LD applications.

5. Acknowledgment

The author thanks the members of the Photonics Laboratory in Samsung Advanced Institute of Technology for supporting this research. This research was supported by the World Class University(WCU) program at Sunchon National University and the IT R&D program of MKE/KEIT(10040379).

6. References

[1] T. Tojyo, T. A Sano, M. Takeya, T. Hino, S. Kijima, S. Goto, S. Uchida, and M. Ikeda, Jpn. J.Appl. Phys. 40, 3206 (2001).
[2] V. M. Bermudez, J. Appl. Phys. 80, 1190 (1996).
[3] S. Nagahama, T. Yanamoto, M. Sano, and T. Mukai, Jpn. J. Appl. Phys., Part 1 40, 3075 (2001).
[4] S. S. Park, I. W. Park, and S. H. Choh, Jpn. J. Appl. Phys., Part 2 39, L1141 (2000).
[5] U. Karrer, O. Ambacher, and M. Stutzmann, Appl. Phys. Lett. 77, 2012 (2000).
[6] Z.-Q. Fang, D. C. Look, P. Visconti, D.-F. Wang, C.-Z. Lu, F. Yun, H.Morkoc¸, S. S. Park, and K. Y. Lee, Appl. Phys. Lett. 78, 2178 (2001).
[7] J. K. Kim, K.-J. Kim, B. Kim, J. N. Kim, J. S. Kwak, Y. Park, and J.-L. Lee, J. Electron. Mater. 30, 129 (2001).

[8] J. S. Kwak, J. Cho, S. Chae, O. H. Nam, C. Sone, and Y. Park, Jpn. J.Appl. Phys., Part 1 40, 6221 (2001).
[9] J.-S. Jang, S.-J. Park, and T.-Y. Seong, Appl. Phys. Lett. 76, 2898 (2000).
[10] J. Ho, C. Jong, C. C. Chiu, C. Huang, C. Chen, and K. Shih, Appl. Phys. Lett. 74, 1275 (1999).
[11] L. Zhou, W. Lanford, A. T. Ping, I. Adesida, J. W. Yang, and A. Khan, Appl. Phys. Lett. 76, 3451 (2000).
[12] J.-S. Jang and T.-Y. Seong, Appl. Phys. Lett. 76, 2743 (2000).
[13] L. T. Romano, M. Kneissl, J. E. Northrup, C. G. Van de Walle and D. W. Treat, Appl. Phys. Lett. 79, 2734 (2001).
[14] V. L. Rideout, Solid-State Electron. 18, 541 (1975).
[15] S. M. Sze, *Physics of Semiconductor Devices* (Wiley, New York, 1981) pp. 245-311.
[16] H. Morkoc, *Nitride Semiconductors and Devices* (Springer, Berlin, 1999), p. 155.
[17] T. S. Zheleva, O. -H. Nam, M. D. Bremser, and R. F. Davis, Appl. Phys. Lett. 71, 2638 (1997).
[18] A. Usui, H. Sunakawa, A. Sakai, and A. Yamaguchi, Jpn. J. Appl. Phys., Part 2 36, L899 (1997).
[19] S. Nagahama, N. Iwasa, M. Senoh, T. Matsushita, Y. Sugimoto, H. Kiyoku, T. Kozaki, M. Sano, H. Matsumura, H. Umemoto, K. Chocho, and T. Mukai, Jpn. J. Appl. Phys. 39, 647 (2000).
[20] T. Tojyo, T. Asano, M. Takeya, T. Hino, S. Goto, S. Uchida, and M. Ikeda, Jpn. J. Appl. Phys. 40, 3206 (2001).
[21] N. G. Weimann and L. F. Eastman, in Proceedings of the IEEE Cornell Conference, Cornell, NY, 1997.
[22] P. J. Hansen, Y. E. Strausser, A. N. Erickson, E. J. Tarse, P. Kozodoy, E. G. Brazel, J. P. Ibbetson, U. Mishra, V. Narayanamurti, S. P. DenBaars, and J. S. Speck, Appl. Phys. Lett. 72, 2247 (1998).
[23] P. Kozodoy, J. P. Ibbetson, H. Marchand, P. T. Fini, S. Keller, J. S. Speck, S. P. DenBaars, and U. K. Mishra, Appl. Phys. Lett. 73, 975 (1998).
[24] D. V. Kuksenkov, H. Temkin, A. Osinsky, R. Gaska, and M. A. Khan, *Proceedings of the IEEE IEDM*, 1997, p. 759.
[25] S. Tomiya, H. Nakajima, K. Funato, T. Miyajima, K. Kobayashi, T. Hino, S. Kijima, T. Asano, and M. Ikeda, Phys. Stat. Sol. (a) 188, 69 (2001)
[26] O. H. Nam, K. H. Ha, J. S. Kwak, S. N. Lee, K. K. Choi, T. H. Chang, S. H. Chae, W. S. Lee, Y. J. Sung, H. S. Paek, J. H. Chae, T. Sakong, and Y. Park, Phys. Stat. Sol. (c) 0, 2278 (2003).
[27] J. S. Kwak, O. H. Nam, and Y. Park, Appl. Phys. Lett. 80, 3554 (2002).
[28] J. S. Kwak, O. H. Nam, and Y. Park, J. Appl. Phys. 95, 5917 (2004).
[29] N. G. Weimann, L. F. Eastman, D. Doppalapudi, H. M. Ng, and T. D. Moustakas, J. Appl. Phys. 38, 3656 (1998).
[30] K. Shiojima, T. Suemitsu, and M. Ogura, Appl. Phys. Lett. 78, 3636 (2001).
[31] S. J. Rosner, E. C. Carr, M. J. Ludowise, G. Girolami, and H. I. Erikson, Appl. Phys. Lett. 70, 420 (1997).
[32] W. Götz, N. M. Johnson, and D. P. Bour, Appl. Phys. Lett. 68, 3470 (1996).

[33] M.W. Bayerl, M. S. Brandt, O. Ambacher, M. Stutzmann, E. R. Glaser, R. L. Henry, A. E. Wickenden, D. D. Koleske, T. Suski, I. Grzegory, and S. Porowski, Phys. Rev. B 63, 125203 (2001).
[34] D. M. Hofmann, B. K. Meyer, H. Alves, F. Leiter, W. Burkhard, N. Romanov, Y. Kim, J. Krüger, and E. R. Weber, Phys. State. Sol. 180, 261 (2000).
[35] K. Shiojima, T. Sugahara, and S. Sakai, Appl. Phys. Lett. 77, 4353 (2000).
[36] N. F. Mott, Phil. Mag. 19, 835 (1969).
[37] D. C. Look, D. C. Walters, M. O. Manasreh, J. R. Sizelove, C. E. Stutz, and K. R. Evans, Phys. Rev. B 42, 3578 (1990).
[38] H. Yamamoto, Z-Q. Fang, and D. C. Look, Appl. Phys. Lett. 57, 1537 (1990).
[39] T. Mori, T. Kozawa, T. Ohwaki, Y.Taga, S. Nagai, S. Yamasaki, S. Asami, N. Shibata, and M. Koike, Appl. Phys. Lett. 69, 3537 (1996).
[40] K. Shiojima, T. Sugahara, and S. Sakai, Appl. Phys. Lett. 77, 4353 (2000).
[41] L. S. Yu, D. Qiao, L. Jia, S. S. Lau, Y. Qi, and K. M. Lau, Appl. Phys. Lett. 79, 4536 (2001).
[42] M. G. Cheong, K. S. Kim, C. S. Kim, R. J. Choi, H. S. Yoon, N. W. Namgung, E. -K. Suh, and J. L. Lee, Appl. Phys. Lett. 80, 1001 (2002).
[43] J. S. Kwak, J. Cho, S. Chae, K. K. Choi, Y. J. Sung, S. N. Lee, O. H. Nam, and Y. Park, Phys. Stat. Sol. (a) 194, 587 (2002).
[44] J. S. Kwak, T. Jang, K. K. Choi, Y. J. Sung, Y. H. Kim, S. Chae, S. N. Lee, K. H. Ha, O. H. Nam, and Y. Park, Phys. Stat. Sol. (a) 201, 2649 (2004).
[45] S. Nagahama, T. Yanamoto, M. Sano, and T. Mukai, Jpn. J. Appl. Phys. 40, 3075 (2001).
[46] S. S. Park, I. W. Park, and S. H. Choh, Jpn. J. Appl. Phys. 39, L1141 (2000).
[47] B. P. Luther, S. E. Mohney, J. M. DeLucca, and R. F. Karlicek, Jr, J. Electron. Mater. 27, 196 (1998).
[48] J. S. Kwak, S. E. Mohney, J. -Y. Lin, and R. S. Kern, Semicond. Sci. Technol. 15, 756 (2000).
[49] B. P. Luther, J. M. DeLucca, S. E. Mohney, and R. F. Karlicek, Jr, Appl. Phys. Lett. 71, 3859 (1997).
[50] J. S. Kwak, J. Cho and C. Sone, J. Kor. Phys. Soc. 48, 1259 (2006).
[51] P. M. Asbeck, E. T. Yu, S. S. Lau, G. J. Sullivan, J. Van Hove, and J. Redwing, Electron. Lett. 33, 1231 (1997).
[52] R. Gaska, J. W. Yang, A. D. Bykhovski, M. S. Shur, V. V. Kaminski, and S. M. Soloviov, Appl. Phys. Lett. 72, 64 (1998).
[53] E. T. Yu, X. Z. Dang, L. S. Yu, D. Qiao, P. M. Asbeck, S. S. Lau, G. J. Sullivan, K. S. Boutros, and J. Redwing, Appl. Phys. Lett. 73, 1880 (1998).

4

Characterization Parameters of (InGaN/InGaN) and (InGaN/GaN) Quantum Well Laser Diode

Sabah M. Thahab
College of Engineering, University of Kufa, Iraq

1. Introduction

Laser characterization can facilitate improvement in laser design by allowing optical component scientists to compare different laser designs and to confirm the validity of their theories behind their designs. Various desired characteristics of a laser are discussed. All of the following characteristics will be taken into consideration seriously during laser design. The most important operating parameters in InGaN LDs are the internal quantum efficiency (η_i) , the internal loss α_i and the transparency current density J0 as these are the key parameters which control the operating characteristics of LDs. The characteristic temperature is also a very important parameter from the viewpoint of the practical application of these lasers. Well-number effect on the characteristics temperature will also be investigated through the simulation software. Experimental and theoretical work done by Domen et al. (1998) concluded that the InGaN laser diode quantum efficiency improves when the number of wells is decreased. Our work supports prior research results.

2. Construction of a blue laser diode structure

III-nitride semiconductor materials have received much attention in the past few years since they have important applications in light-emitting diodes (LEDs) and short wavelength laser diodes (LDs), due mainly to their relatively wide band gap and high emission efficiency. Especially after the development of the InGaN blue LDs that can be continuously operated for more than 10 000 h, and the InGaN LED that can emit light in the red spectral range, III-nitride semiconductor materials have become the most attractive materials among III-V and II VI semiconductors (Jiang and Lin, 2001).

However, for commercial applications, LD and LED markets in the red and yellow spectral range are still dominated by the AlGaInP semiconductor materials. This is due to the fact that InGaN devices are usually more expensive, and many aspects of the physics related to III-nitride semiconductor materials have not been well developed. When a LD is under operation, current overflow will cause a decrease in emission efficiency and an increase in the device temperature, which in turn results in the deterioration of the operation lifetime. There are many factors that might cause the current overflow.

The III-nitride LDs structure is shown in Fig. 1 the first possible cause of the overflow is the relatively small band-offset ratio between the conduction band and the valence band (Kuo et al., 2004). Under these circumstances, the potential difference between the quantum well and the barrier in the conduction band is relatively small and, hence, the electrons can easily overflow to the p side. The second factor that might cause the electronic current overflow is the high threshold currents of III-nitride LDs. Due to the lack of a lattice-matched substrate; the InGaN LDs usually have a high density of crystal defects. Moreover, high-quality p-type layers and flat reflecting mirrors are usually difficult to obtain. These problems ultimately lead to a high threshold current of the InGaN LD, which in turn increases the temperature of the LD and the possibility of current overflow (Nakamura and Fasol, 1997).

The third possible cause of electronic current overflow is the difficulty with which the holes enter the active region, due mainly to the high effective mass of the holes and the high band-offset in the valence band. Since the holes cannot easily enter the active region, more current is required to activate the laser action, thus increasing the possibility of electronic current overflow.

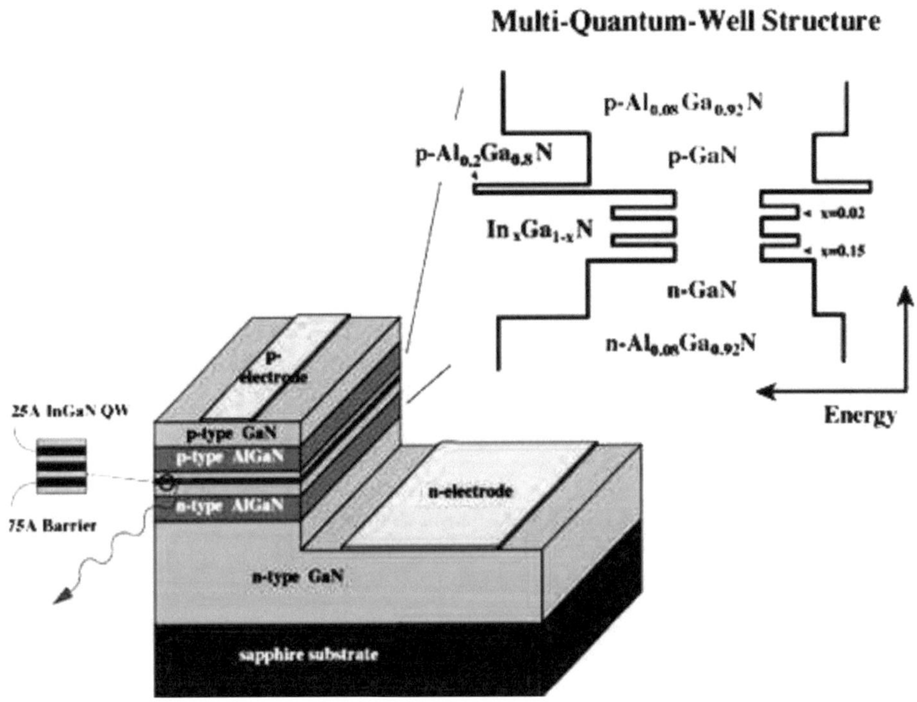

Fig. 1. The structure of InGaN multi quantum wells laser diode. (Ponce 1997)

On the other hand, Nakamura et al. studied the laser performance of several LDs with an emission wavelength of 390–420 nm as a function of the number of InGaN well layers Fig. 1. They found that the lowest threshold current density was obtained when the number of InGaN well layers was two. In addition the successes to handle the carrier overflow by using AlGaN as blocking layer as shown in Fig. 2. However, in another study, they observed that in LDs with emission wavelengths longer than 435 nm, when the number of InGaN well layers varied from one to three, the threshold current density was lowest at one, and increased with the number of InGaN well layers. This phenomenon was attributed to the dissociation of the high indium content of the InGaN well layer at a high growth temperature of 750 °C due to a high InGaN dissociation pressure

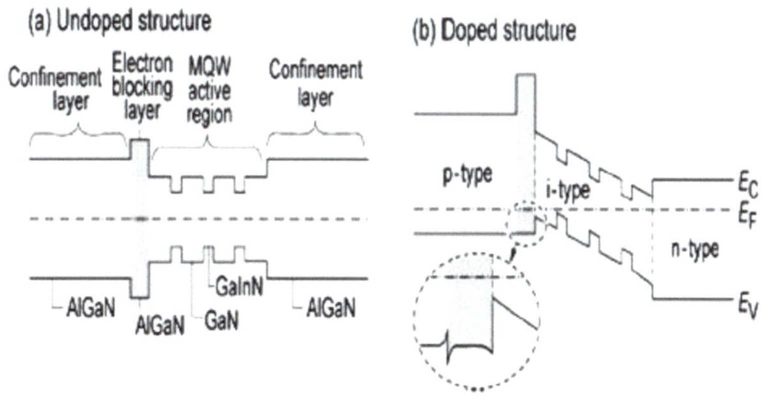

Fig. 2. The AlGaN current -blocking layer in an AlGaN/GaN/GaInN multi quantum wells. (a) Band diagram without doping. (b) Band diagram with doping. The Al content in the electron - blocking layer is higher than in the p-type confinement layer.

3. InGaN laser diode simulation procedure

The 2-D laser diode device simulations are performed using ISE-TCAD package. A conventional simulation flow would include three steps:

1. Drawing cross-sectional view of device and generating mesh in MDraw.
2. Performing 2-D device simulation using DESSIS.
3. Plotting data curves of interest and extracting parameters using INSPECT.

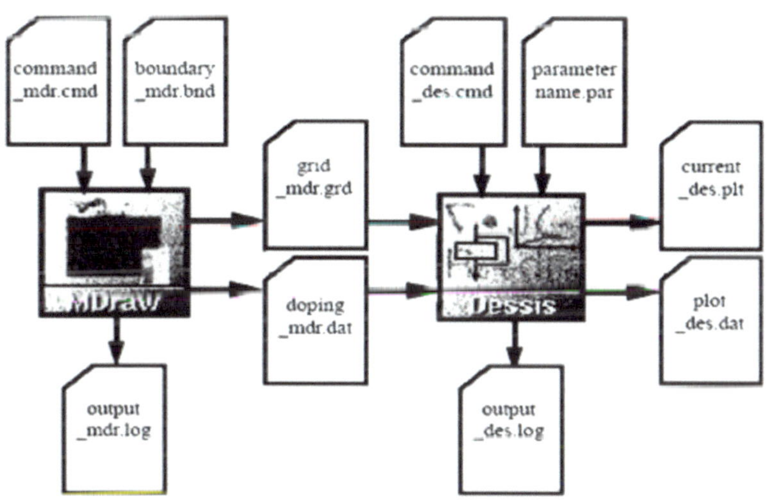

Fig. 3. Typical design flow with DESSIS device simulation. (Manual of Integrated System Engineering (ISE TCAD) AG, Switzerland)

DESSIS is used to simulate the electrical and optical characteristics of the device. Such a seamless flow through ISE TCAD tools, with the associated file types, is represented in Fig. 3. Finally, Tecplot-ISE is used to visualize the output from the simulation in 2D, and INSPECT™ is used to plot the electrical and optical characteristics.

3.1 The fundamental set of equations used in a laser simulation

Simulating a laser diode is one of the most complex problems in device simulation. The fundamental sets of equations used in a laser simulation are:

1. Poisson equation
2. Carrier continuity equations
3. Lattice temperature equation and hydrodynamic equations
4. Quantum well scattering equations (for QW carrier capture)
5. Quantum well gain calculations (Schrödinger equation)
6. Photon rate equation
7. Helmholtz equation

3.2 Coupling between optics and electronics problems

The complexity of the laser problem is apparent from the relational chart shown in Fig. 4. The key quantities exchanged between different equations are placed alongside the directional flows between the equation blocks. The optical problem must be separated from the electrical problem.

The optical problem solves the Helmholtz equation and feeds the mode and photon lifetimes back to the set of active vertices. As a result, the coupling between the optics and electronics becomes nonlocal, and this leads to convergence problems if the Newton method is used to couple these two problems. Therefore, a Gummel iteration method (instead of a coupled Newton iteration) is required to couple the electrical and optical problems self-consistently. The solution of the electrical problem provides the required refractive index changes and absorption to the optical solver, which solves for the modes. In the case of the Fabry–Perot edge-emitting laser, the wavelength is computed from the peak of the gain curve and fed into the optical problem.

Poisson equation

$$\nabla.(\varepsilon\nabla\Phi) = q(n - p - N_D{}^{+} + N_A{}^{-}) \tag{1}$$

Electron continuity equation

$$\frac{\partial n}{\partial t} - \nabla.J_n = q(G_n - R_n) \tag{2}$$

Hole continuity equation

$$\frac{\partial p}{\partial t} + \nabla.J_p = q(G_p - R_p) \tag{3}$$

$$J_n = q(-\mu_n n \nabla\Phi + D_n \nabla n) \tag{4}$$

$$J_p = q(-\mu_p p \nabla\Phi - D_p \nabla p) \tag{5}$$

In vertical cavity surface emitting laser diodes (VCSELs) and distributed Bragg reflector (DBR) lasers, the wavelength is computed inside the optical resonance problem and is an input to the electrical problem instead. The gain calculations involve the solution of the Schrödinger equation for the subband energy levels and wavefunctions. These quantities are used with the active-region carrier statistics to compute the optical matrix element in the material gain. The photon rate equation takes the material gain and mode information to compute the modal gain and, subsequently, the stimulated and spontaneous recombination rates. These optical recombinations increase the photon population but reduce the carrier population. Therefore, these recombination rates must be added to the carrier continuity equations to ensure the conservation of particles.

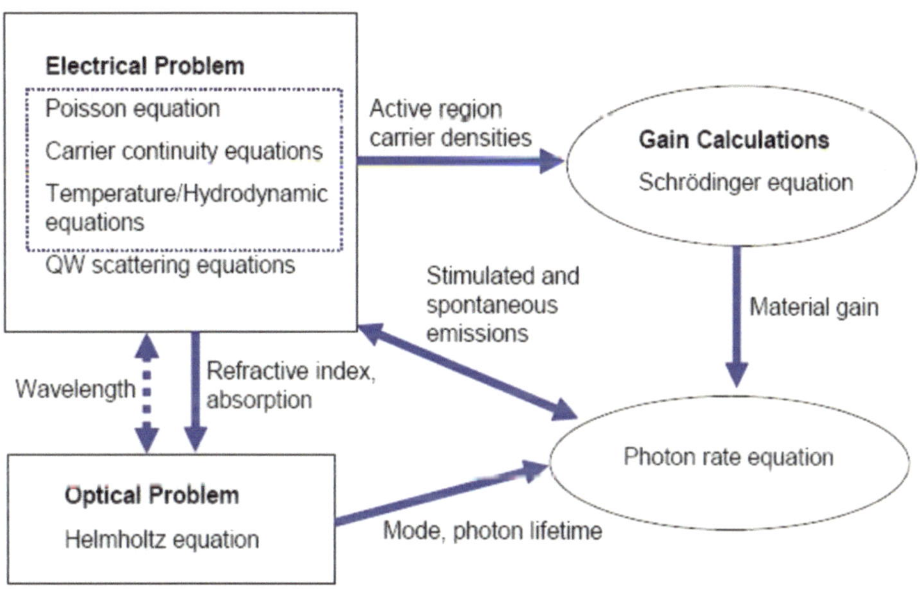

Fig. 4. The coupling between optics and electronics problems. (Manual Integrated System Engineering (ISE TCAD) AG, Switzerland)

From its bandgap dependence in other materials, a very small Auger parameter of $c - 10^{-34}$ cm^6 s^{-1} is estimated for GaN. Thus, even with large carrier densities, Auger recombination in nitride materials is negligible. A wide spectrum of bandgap bowing parameters has been obtained for ternary nitride alloys due to differences in growth and measurement conditions. For layers with low mole fraction of the alloy element (x<0.2), we employ the following room temperature relations for the direct bandgap (Stringfellow and Craford, 1997; Nakamura and Fasol, 1997).

$$E_{g\,In_xGa_{1-x}N} = xE_{g\,InN} + (1-x)E_{g\,GaN} - 1.43x(1-x) \tag{6}$$

$$E_{g\,Al_xGa_{1-x}N} = xE_{g\,AlN} + (1-x)E_{gGaN} - 1.3x(1-x) \tag{7}$$

Where Eg,InN, Eg,GaN and Eg,AlN are the bandgap energies of InN, GaN and AlN at room temperature ,respectively. The bandgap energies of InN, GaN and AlN used in our simulation are 0.77, 3.42 and 6.2 eV (Fritsch et al., 2003), respectively. A band offset ratio of $\Delta E_c / \Delta E_v = 0.7/0.3$ is assumed for InGaN/GaN as well as for AlGaN/GaN, which corresponds to an average of reported values for each case. In our strained InGaN quantum wells, binary effective mass parameters, lattice constants and elastic constants are determined from the linear interpolations to obtain InGaN values. GaN values are used for the deformation potentials (Fritsch et al., 2003).

$$m_{e\,In_xGa_{1-x}N} = m_{eGaN} + x(m_{eInN} - m_{eGaN}) \tag{8}$$

$$m_{hh\,In_xGa_{1-x}N} = m_{hhGaN} + x(m_{hhInN} - m_{hhGaN}) \tag{9}$$

$$m_{lh\,In_xGa_{1-x}N} = m_{lhGaN} + x(m_{lhInN} - m_{lhGaN}) \tag{10}$$

where $m_{eInxGa1-xN}$ is the effective mass of electrons in $In_xGa_{1-x}N$ material. $m_{hhInxGa1-xN}$ and $m_{lhInxGa1-xN}$ are the effective masses of heavy holes and light holes in $In_xGa_{1-x}N$, respectively m_{eInN} for InN is 0.1 m_0 and m_{eGaN} for GaN is 0.20m_0. While m_{hhInN} and m_{lhInN} for InN are 1.44m_0 and 0.157m_0 , respectively, and for GaN m_{hhGaN} and m_{lhGaN} are 1.595m0 , 0,261m0 respectively, and m_0 is the electron mass in free space (Kuo et al., 2004) .The optical gain mechanism in InGaN quantum wells of real lasers is still not fully understood. It may be strongly affected by a non-uniform indium distribution. Internal polarization fields tend to separate quantum confined electrons and holes, thereby reducing optical gain and spontaneous emission. However, screening by electrons and holes is expected to suppress QW polarisation fields at high current operation. The high carrier density is also assumed to eliminate exciton effects, despite the large exciton binding energy in nitrides. On the other hand, many-body models predict significant gain enhancement at high carrier densities. Considering all the uncertainties in calculating the gain of our assumed rectangular quantum wells, we start with a simple free carrier gain model, including a Lorentzian broadening function with 0.1 ps scattering time.

Optical reflection and waveguiding mainly depends on the refractive index profile inside the device. For photon energies close to the bandgap, the refractive index is a strong function of wavelength. For the design of optical waveguides, the compositional change of the refractive index is often more important than its absolute value. Reliable refractive index measurements on $In_xGa_{1-x}N$ are currently not available so that a linear interpolation of binary parameters is chosen in this study. In addition, we employ the refractive index relations given in Eqs.(11) and (12), respectively (Sink, 2000).

Spontaneous polarization as well as strain-induced polarization in nitride compounds results in polarization charges at heterointerfaces and built-in polarization fields. We calculate these charges by linear interpolation of binary material parameters. Full consideration of the calculated polarization charges results in a strong deformation of the QW potential by the built-in QW polarization field.

$$n(In_xGa_{1-x}N) = 2.5067 + 0.91x \quad (11)$$

$$n(Al_xGa_{1-x}N) = 2.5067 - 0.43x \quad (12)$$

4. InGaN laser diode simulation results and discussion

A schematic diagram of the laser diode structure is shown in Fig. 5.. n-type GaN layer that is 3μm in thickness is assumed to grow first then followed by 0.4μm n-type $Al_{0.07}Ga_{0.93}N$ cladding layer, followed by a 0.1μm n-type GaN guiding layer. The active region of the preliminary laser diode structure under study consists of a 3 nm $In_{0.13}Ga_{0.87}N$ well that is sandwiched between 5 nm $In_{0.01}Ga_{0.99}N$ barriers. 0.02 μm p-$Al_{0.15}Ga_{0.85}N$ stopper layer is assumed to be grown on the top of active region, followed by 0.1 μm p-type GaN guiding layer then 0.4 μm p-$Al_{0.07}Ga_{0.93}N$ cladding layer and 0.1 μm p-GaN contact layer. The doping concentration of n-type and p-type are 5× 10^{17} cm^{-3} and 5× 10^{18} cm^{-3}, respectively. The active region length is 800 μm and the reflectivity of the two ends (left and right) facets assumed as Fabry –Perot cavity waveguide with R= 0.3 respectively.

For the preliminary LD structure under study, the energy band diagram, and electrostatic potential of the double quantum wells (DQWs) InGaN LD are shown in Fig. 6. The right side of the diagram is the n-side and the left side is the p-side of the laser diode. The horizontal axis is the distance along the crystal growth direction. The optical material gain inside the quantum wells is shown in Fig. 7.

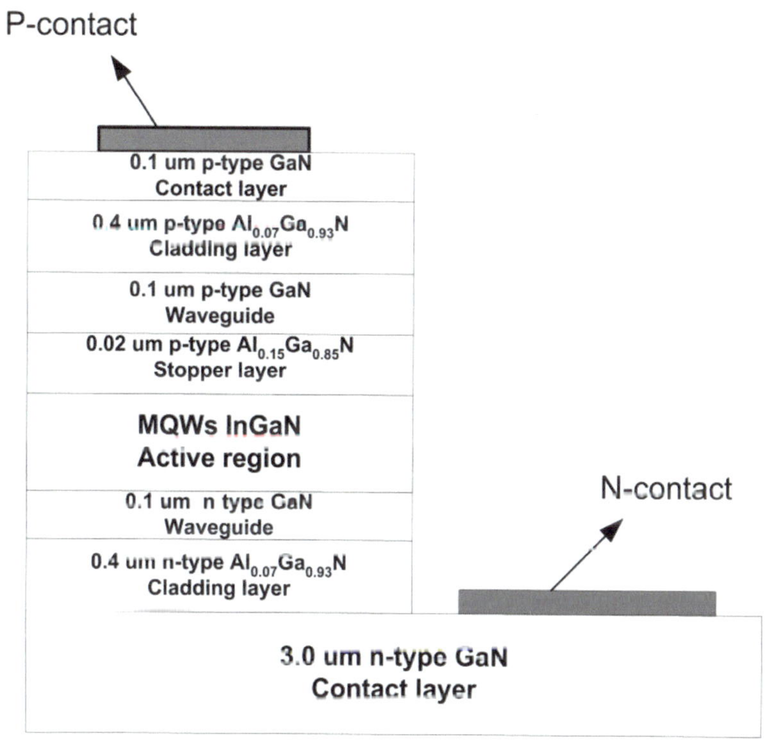

Fig. 5. Schematic diagram of the preliminary InGaN MQWs LD.

The quantum well in the left side (p-side) has a higher optical material gain due to the use of an AlGaN blocking layer in the p-side of which the electrons tend to accumulate in the left quantum well. Since holes have difficulty moving from the left quantum well to the right quantum well due to the relatively large effective mass, low mobility, and high band offset in the valence band, more holes are expected in the left quantum well.

Since the left quantum well possesses more electrons and holes as compared with the right quantum well, it has higher population inversion and hence higher stimulated recombination rate. The carriers (hole, electron) distribution inside the quantum wells determines the optical performance of the laser diode.

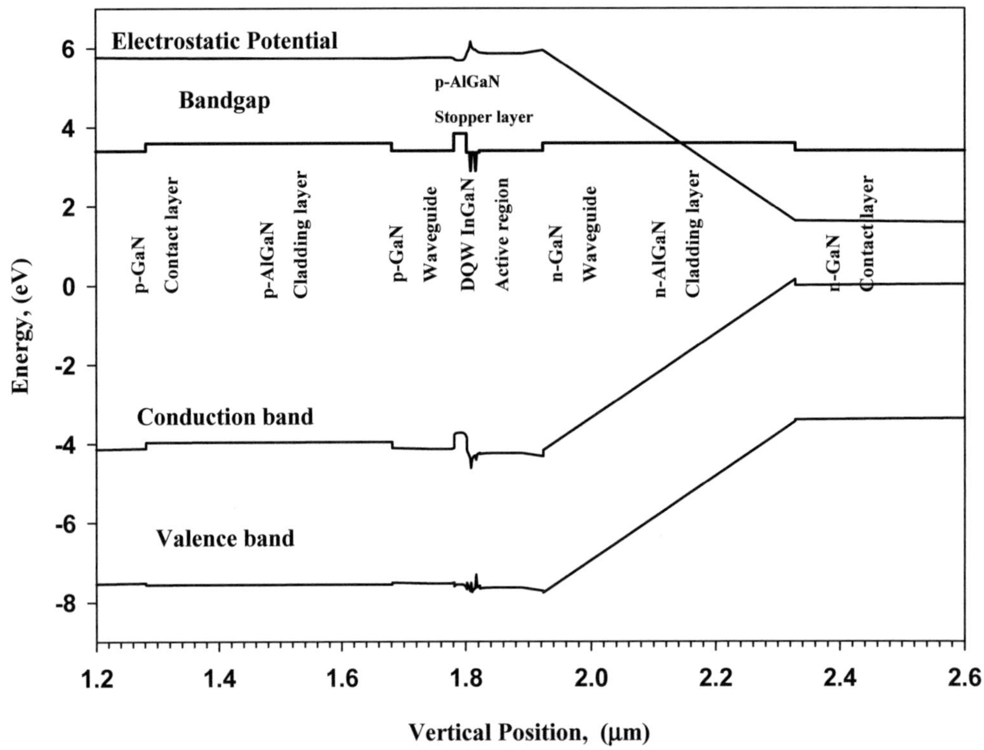

Fig. 6. Energy band diagram of the double quantum wells InGaN LD together with electrostatic potential profile.

From Fig. 8 it can be observed that the carrier distributions are inhomogeneous and are increasing towards the p-side. When the laser oscillation takes place, the hole injection becomes inhomogeneous among wells. This is ascribed to the poor hole injection due to the low mobility and thermal velocity of the hole. Thus the hole density becomes higher on the p-side and the electrons are attracted to the p-side. Optical reflection and waveguiding mainly depend on the refractive index profile inside the device. The laser threshold current, slope efficiency, output power and DQE at various well numbers are shown in Fig. 9. We observed maximum output power and lowest threshold current when the well number is two.

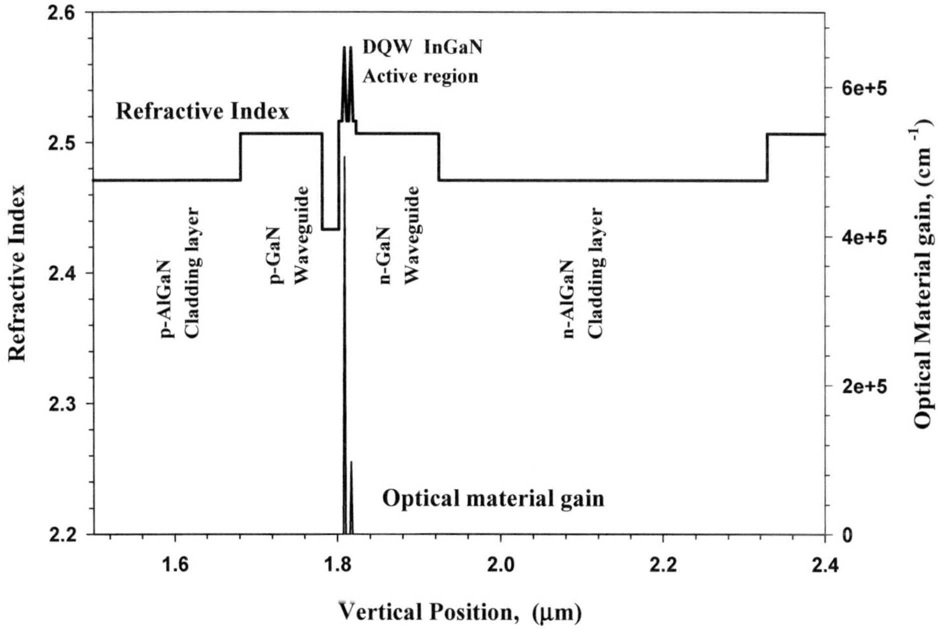

Fig. 7. Optical material gain in the InGaN double quantum wells LD.

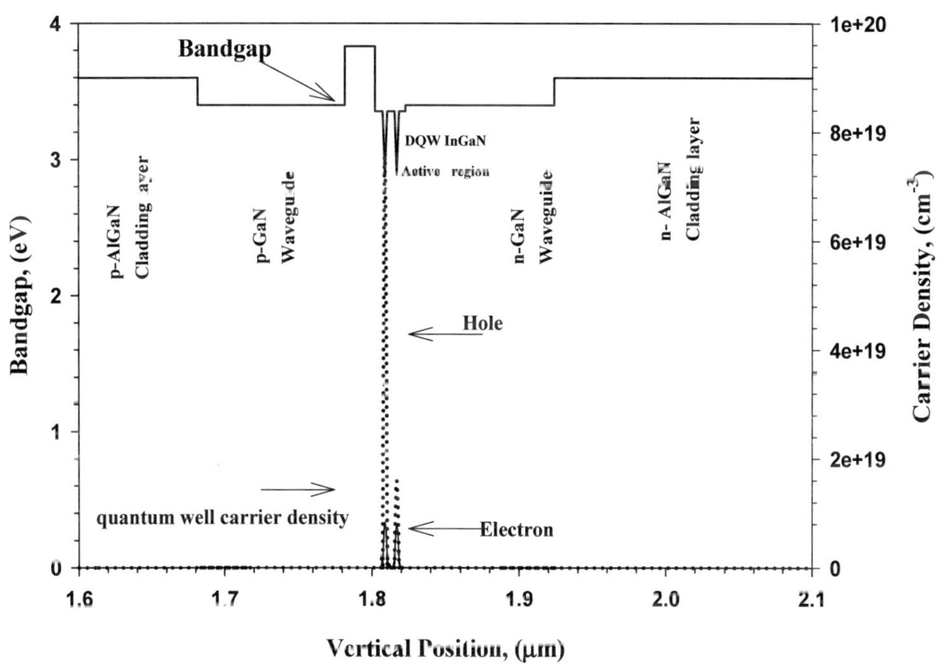

Fig. 8. Carriers density distribution profiles in the double quantum wells of InGaN LD.

These results are in agreement with the experimental results observed previously by Nakamura et al. (Nakamura et al., 1998a; Nakamura et al., 2000) in which they studied the laser performance of several laser diodes with an emission wavelength of 390-450 nm as a function of the number of InGaN well layers and found that the lowest threshold current was obtained when the number of InGaN well layers was two.Moreover, in another work they observed that when the number of InGaN well layers of the laser diodes with emission wavelengths longer than 435 nm was varied from one to three, the lowest threshold current was obtained when the number of well layers was one and the threshold current increased when the number of InGaN well layers was increased. This phenomenon was attributed to the dissociation of the high indium content InGaN well layer at a high growth temperature of 750 0C due to a high InGaN dissociation pressure. It was observed the deterioration in laser performance with increased quantum wells number in our simulation results, and attributed it to the non-uniform carrier distribution in the quantum wells. Nakamura assumption (the indium dissociation) did not take in our simulation process. The specified wavelength was selected to be the peak of the stimulated emission.

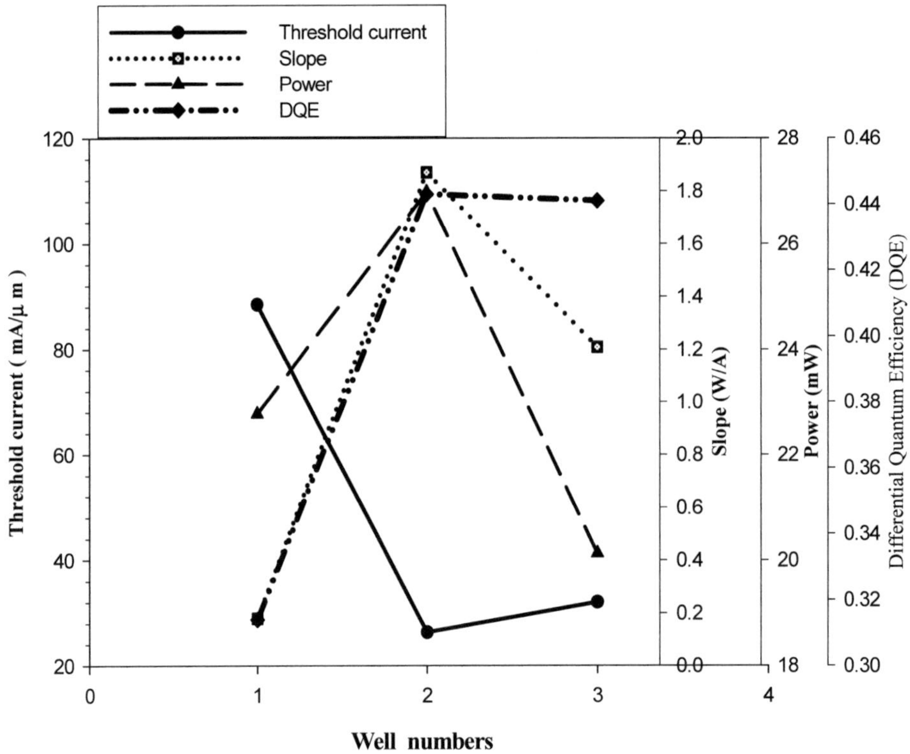

Fig. 9. Laser output power, slope efficiency, threshold current and differential quantum efficiency (DQE) as a function of well number of the MQWs InGaN LD.

The presence of the built-in electric field modifies the electronic states in QWs and lowers the optical gain of the active region of the laser as shown in Fig.10. The electrons and holes are, indeed, spatially separated by the polarization filed, but the free carrier induced field is opposite to the polarization field. The two fields tend to cancel each other out for higher sheet densities, thus re-establishing the conditions for the electron hole recombination emission. The wavelength of the light emitted from LD QW depends not only on the band gap, but also on the large internal electric field due to piezoelectric (PZ) and spontaneous polarization.

The PZ field arises from the strain due to the lattice mismatch between the InGaN well and the GaN barriers and causes redshift in the optical transition energy. The presence of the internal field in InGaN quantum wells causes separation of the electron and hole wave functions thus reducing the wave function overlap integral. As the carrier density in the quantum well increases, screening of the internal field occurs, hence a blue shift (418.5-415 nm) of the emission is observed in our laser diode.

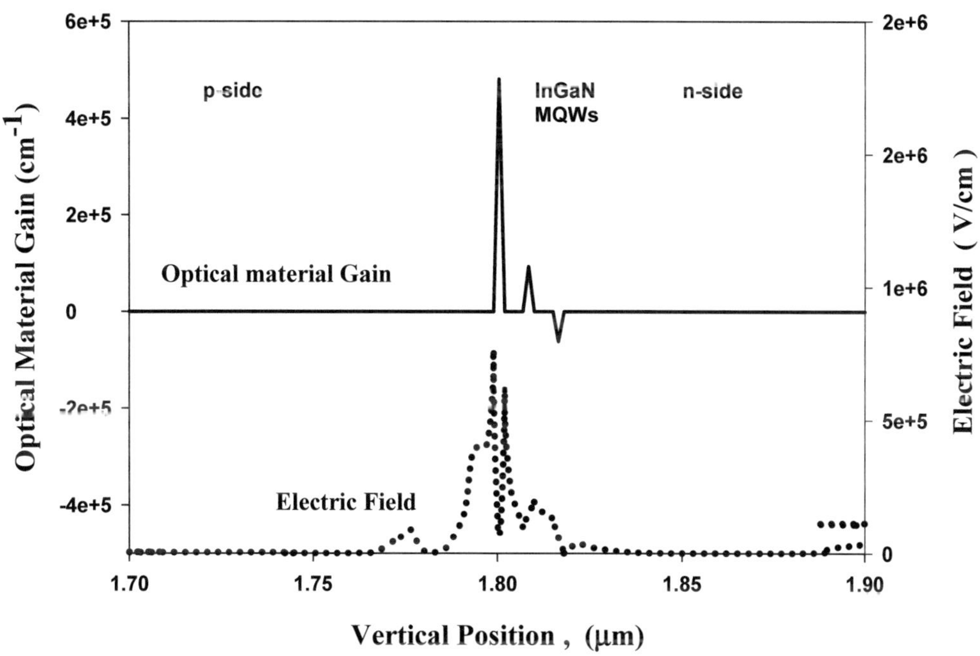

Fig. 10. Optical material gain in the InGaN TQWs LD together with internal electric field.

5. Test and characterization of InGaN/InGaN laser diode

5.1 The internal quantum efficiency η_i and internal loss α_i

The internal quantum efficiency (η_i) is determined from the vertical axis intercept point $(1/\eta_i)$ of the inverse external differential quantum efficiency (DQE) versus cavity length dependence linear fit line. Internal loss α_i is equal to the slope of the line multiplied by the $\eta_i \ln(1/R)$ parameter, where L is the laser cavity length in units of cm and R is the reflectivity of the mirror facets of the laser. Fig.11 shows the calculation method for both parameters (η_i and α_i) with respect to inverse value of DQE as a function of the cavity length. We obtained values of 95% and 9.38 cm-1 for the internal quantum efficiency (η_i) and internal loss (α_i) respectively in the case of DQWs laser diode. These internal quantum efficiency (η_i) and the internal loss α_i values are a direct indication of the efficiency of our DQWs InGaN laser diode. It can be seen that the value of the external differential quantum efficiency (η_d) is smaller than internal quantum efficiency, and this difference is related to the different internal loss mechanism in the laser diode device.

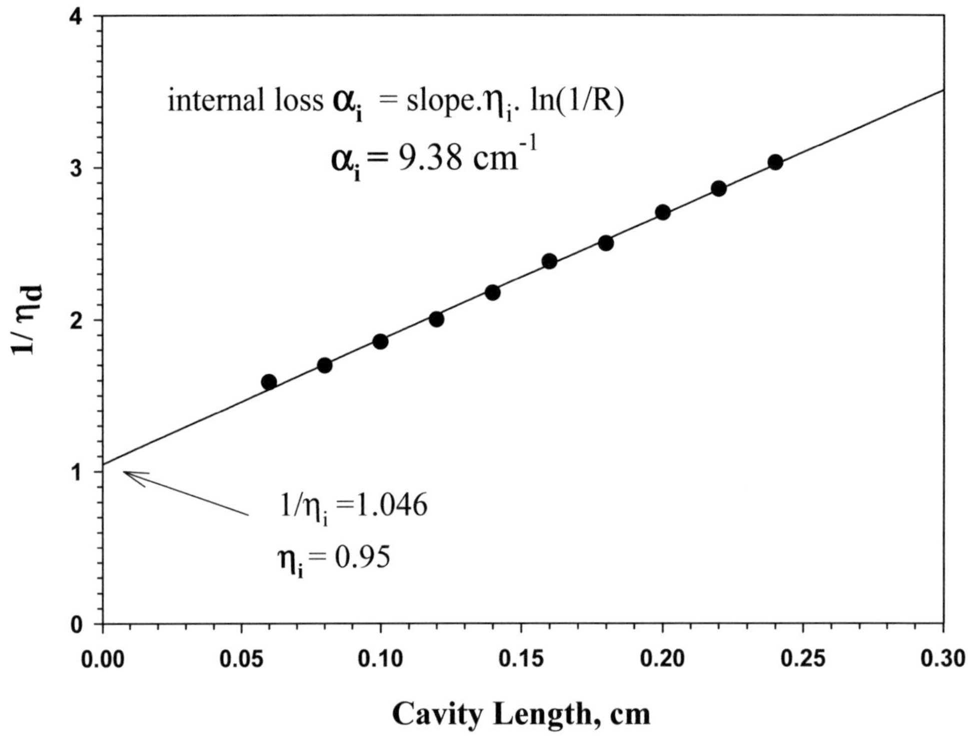

Fig. 11. Inverse of the external quantum efficiency as a function of cavity length of DQWs InGaN/ InGaN barrier LD .

5.2 Transparency threshold current density

The transparency threshold current density is denoted by the symbol J_0. Threshold current density depends upon the cavity length of the laser diode. The laser diode current that provides just sufficient injection to lead to stimulated emission just balancing absorption is called transparency current. Since there is no net photon absorption, the medium is transparent. Above transparency current there is optical gain in the medium though the optical output is not yet a continuous wave coherent radiation. Lasing oscillation occur only when the optical gain in the medium can overcome the photon losses from the cavity that is when the optical gain reaches the threshold gain g_{th} and this occurs at the threshold current density. In order to obtain J_0 value we plot the curve of threshold current density versus the inverse cavity length as shown in Fig..12. The intercept of the linear fit line of the data plotted in this curve with the vertical axis provides us with the transparency threshold current density value. Value of 397 A/cm^2 is obtained for our DQWs laser diodes structure.

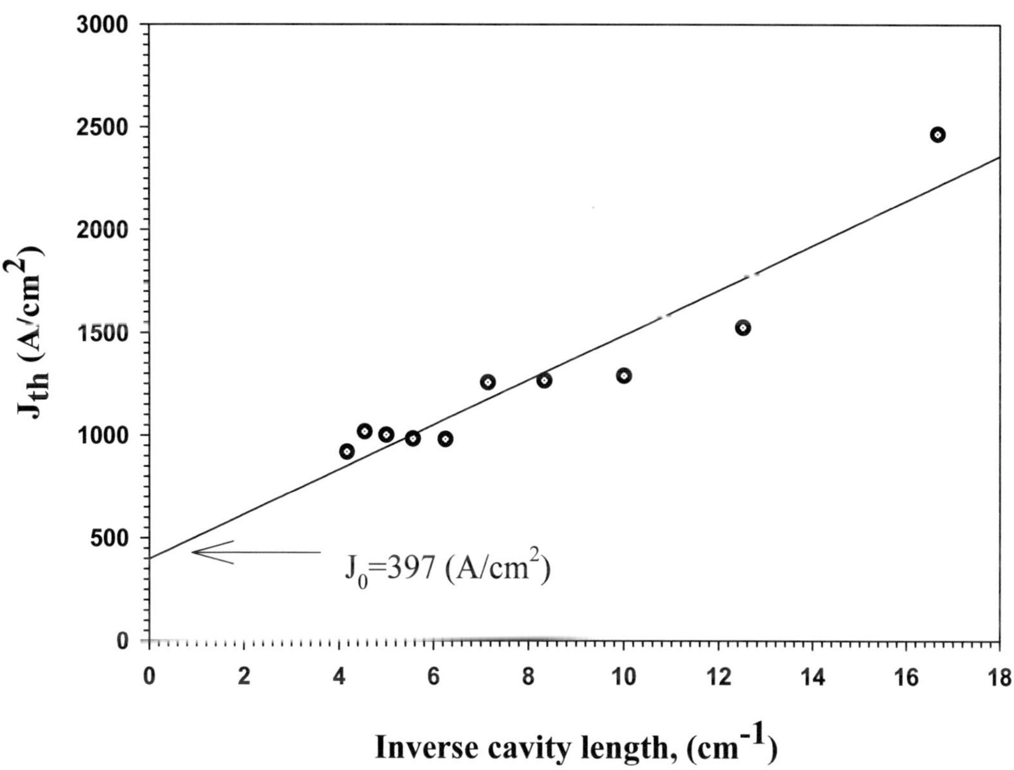

Fig. 12. The threshold current density as a function of inverse cavity length of DQWs InGaN/ InGaN barrier LD.

5.3 Characteristic temperature T_0

Higher values of characteristic temperature T_0 imply that the threshold current density and the external differential quantum efficiency of the device increase less rapidly with increasing temperatures. This translates into the laser being more thermally stable. The T_0 is determined by plotting the threshold current density (J_{th}) versus the laser device temperature on a logarithmic scale and then measuring the slope of the linear fit line as shown in Fig.13. The inverse of the slope of the linear fit to this set of data points is the characteristic temperature T_0 value. T_0 value of 181 K is obtained at temperature range from 280 to 320 K, while T_0 value drop to 29 K above 320 K due to heating effects and the sharp change in the threshold current density with temperature at this temperature value as is shown in Fig.13.

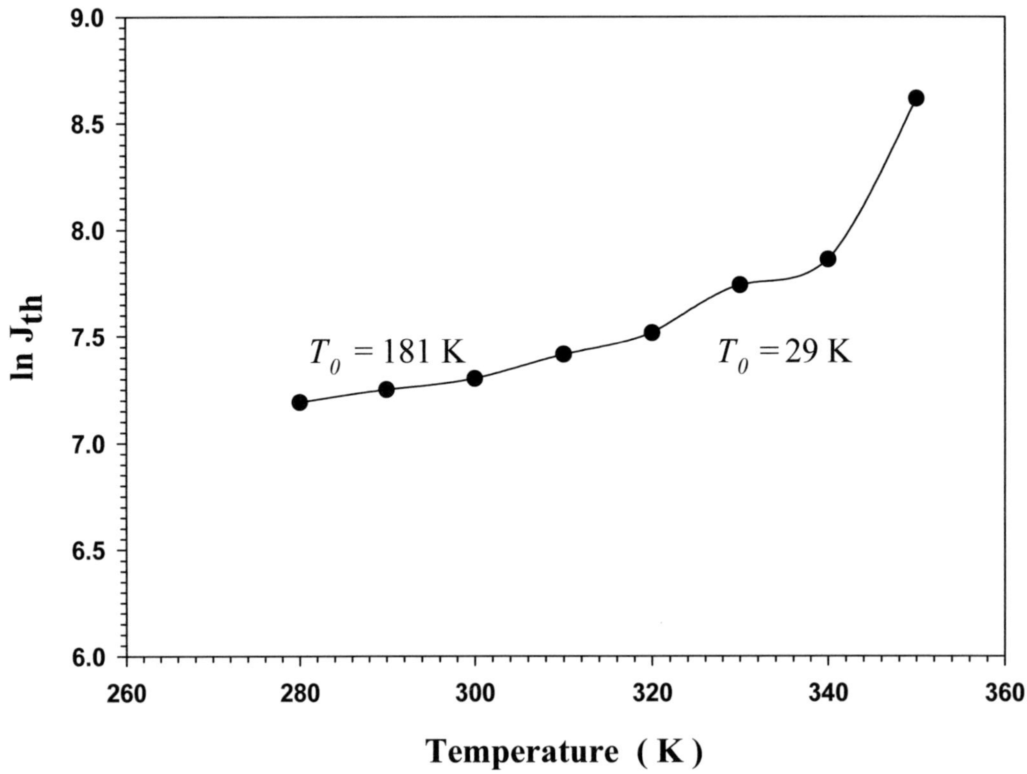

Fig. 13. The variations of the threshold current density (ln Jth) with increasing temperature of DQWs InGaN/ InGaN barrier LD.

6. InGaN/GaN multi quantum well laser diode

In this section the effect of changing the laser barrier layer in the active region from InGaN layer to GaN layer will be studied and investigated. The remaining laser diode layers are the same as in laser diode structure in section above. Figure 14 shows the output power, threshold current, slope efficiency, and external differential quantum efficiency (DQE) as a function of well numbers of InGaN/GaN laser diode. The simulation results indicate that the best laser performance is obtained when the number of QWs is two. For single quantum well (SQW), the gain begins to saturate before lasing is achieved while for triple quantum wells (TQWs), increase in current is needed to pump the additional wells above transparency. A maximum output power of 16.6 mW and lowest threshold current of 13.1

m A were obtained when the quantum well number was two. These results are in line with the other experimental results obtained by many researchers (Nakamura et al., 1998b; Nakamura et al., 2000).

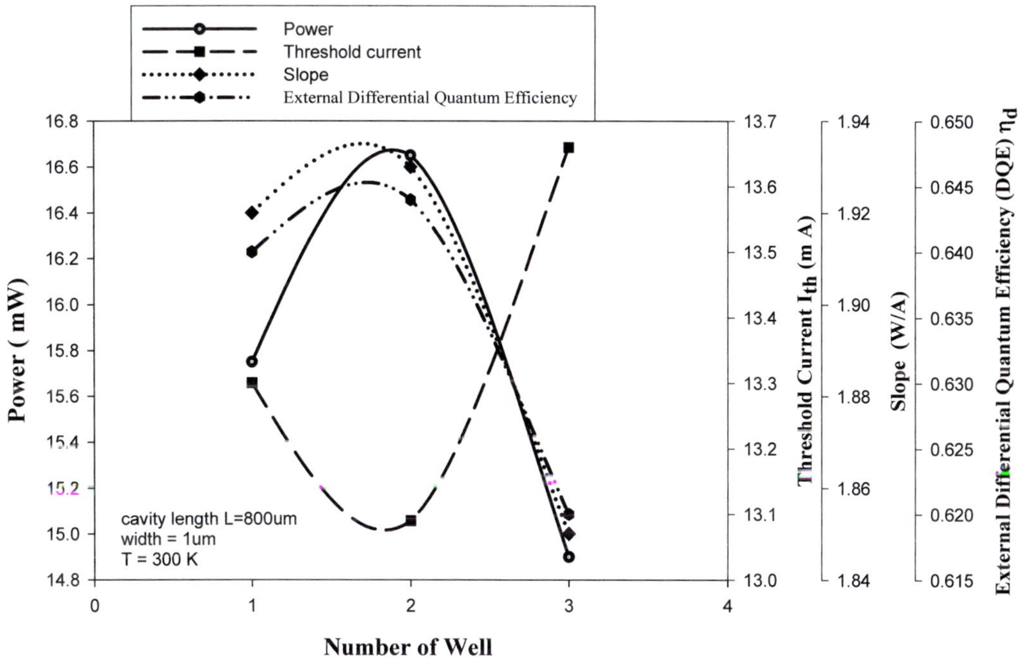

Fig. 14. The output power, threshold current density, slope efficiency and DQE as a function of quantum well number of InGaN/GaN LD.

6.1 Barrier doped and thickness effects on the performance of InGaN/InGaN laser diode

Figure 15 shows the effect of barrier doping on the output power and threshold current. We observed an increase in output power with increasing barrier doping concentration up to 30 m W for 1×10^{-19} cm^{-3} doing concentration. This may be attributed to the lowering of barrier heights in the active region with increased doping level. Low barrier heights lead to higher injection of carriers into the active region hence generating higher radiative recombinations at lower threshold currents. Also we investigated the effect of barrier thickness on the optical power and threshold current of InGaN MQWs laser. It was found that the $In_{0.01}Ga_{0.99}N$ barrier thickness also plays a key role in determining the optical characteristics of the InGaN MQWs laser diode.

Figure 16 shows that as the thickness of the $In_{0.01}Ga_{0.99}N$ barrier layer is increased, the output power decreases. Since the barrier width influences the strength of the internal fields in the wells and barriers, it also influences the absorption and gain. If the width of

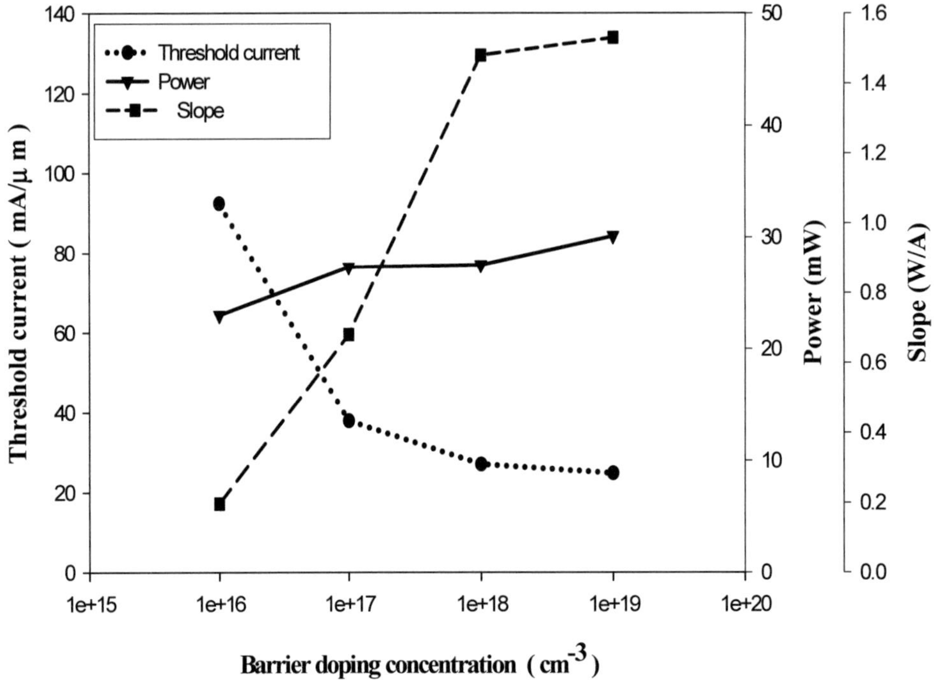

Fig. 15. Laser output power, slope efficiency, and threshold current as a function of barrier doping concentration.

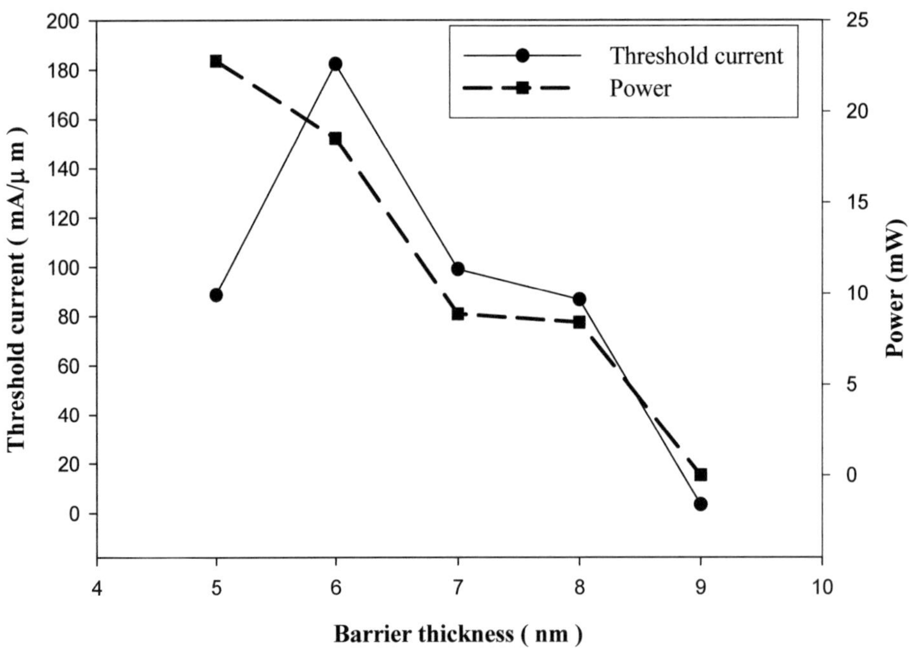

Fig. 16. Laser output power and threshold current as a function of barrier thickness.

the barriers is increased from 5nm to 9nm the internal field in the barriers is reduced. The most dominant result of this is that the electrons are no longer localized as closely to the well as before. The electrons are not confined in the well, but their confinement is provided by the tilted barriers. Thus, its localization is strongly influenced by the internal field in the barrier.

7. Summary

In this chapter we numerically investigated the performance of InGaN MQW with two different designs by changing the barrier layer type (InGaN and GaN). The effects of the well number, barrier thickness and doping on the laser performance were also studied and investigated. It was found that the problem of inhomogeneous carrier distribution in InGaN laser diode structures deteriorates with the increase of QW number for the spectral range under study. Thus, lowest threshold current is obtained when the number of InGaN well layers is two at our laser emission wavelength. We also observed that the barrier thickness and doping play important roles to determine the laser diode performance. Our simulation results in this study are in agreement with the experimental results observed by Nakamura et al. (1998a; 2000)

8. References

Domen, K., Soejima, R. Kuramata, A., Horino, K., Kubota, S., and Tanahashi, T.,(1998) "Interwell inhomogeneity of carrier injection in InGaN/GaN/AlGaN multiquantum well lasers" , *Appl. Phys. Lett.*, Vol.73, No.19.

Fritsch, D., Schmidt, H. and Grundmann, M., (2003) "Band-structure pseudopotential calculation of zinc-blende and wurtzite AlN,GaN, and InN ", *Phys. Rev.B*, Vol. 67,p. 235

Jiang. H., Lin, J., (2001) "Advances in III-nitride micro-photonic devices", Proc. 14th Annual Meeting of the IEEE Lasers and Electro-Optics Society, 12-13 Nov. 2001, LEOS 2001, Vol. 2, pp.758 - 759.

Kuo, Y., Liou, B., Chen M., Yen, S., Lin, C., (2004), "Effect of band-offset ratio on analysis of violet-blue InGaN laser characteristics", *Optics Communications*, Vol. 231,pp.395-402.

Nakamura, S. and Fasol, G. (1997) in: "The Blue Laser Diode", Springer, and Heidelberg.

Nakamura, S., Senoh, M., Nagahama, S., Iwasa, N., Matsushita, T., and Mukai, T., (2000), "Blue InGaN-based laser diodes with an emission wavelength of 450 nm", *Appl. Phys. Lett*, Vol.76, p.22.

Nakamura, S., Senoh, M., Nagahama, S., Iwasa, N., Yamada, T., Matsushita, T., Kiyoku,H. Sugimoto,Y. Kozaki,T. Umemoto,H. Sano, M. and Chocho, K. (1998b) "InGaN/GaN/AlGaN-Based Laser Diodes Grown on GaN Substrates with a Fundamental Transverse Mode", Jpn.J. Appl. Phys., Vol.37, p.L1020.

Nakamura, S., Senoh,M., Nagahama, S., Iwasa,N., Yamada,T., Matsushita,T., Kiyoku,H., Sugimoto,Y., Kozaki,T., Umemoto,H., Sano, M., and Chocho,K., (1998a) "Violet InGaN/GaN/AlGaN-Based Laser Diodes with an Output Power of 420 mW", *Jpn..J.Appl.Phys.*, Vol.37, p.L627.

Ponce. F. A. and Bour. D. P., (1997) "Nitride-based semiconductors for blue and green light-emitting devices", *Nature*, Vol.386, p.351.

Sink, R.K., (2000) "Cleaved-Facet III- Nitride Laser Diode," Ph.D. Thesis, Electrical and Computer Engineering, University of California at Santa Barbara.

Stringfellow, G. B. and Craford, M. G., (1997) "High Brightness Light-emitting Diodes", *Semiconductors and Semimetals* Vol. 48, Academic, San Diego

5

Analysis of Coherence-Collapse Regime of Semiconductor Lasers Under External Optical Feedback by Perturbation Method

Qin Zou and Shéhérazade Azouigui
Institut Telecom, Telecom SudParis, UMR 5157 CNRS, France

1. Introduction

High-performance low-cost semiconductor lasers play a crucial role in development of the future optical-fiber-based telecommunication systems. It is known that the behavior of a semiconductor laser can significantly be altered by external optical feedback, and that this effect can be used for improvement of the performance of a laser such as linewidth narrowing, threshold lowering and intensity-noise reducing. External optical feedback, on the other hand, is also responsible for undesirable phenomena such as mode instability, mode hopping and linewidth broadening, which can severely affect the bit-error rate in data transmission. For this reason, the development of lasers highly resistant to external optical feedback has been the subject of intensive investigations since the last decade.

The tolerance to external optical feedback has been investigated for quantum-well (QW) lasers emitting in the $1.3\mu m$ as well as in the $1.55\mu m$ wavelength ranges. For applications at $1.3\mu m$, 350-μm-long strained multi-quantum-well (MQW)-based antireflection (AR)/high reflectivity (HR) distributed feedback (DFB) lasers with a slope efficiency of $0.3W/A$ and a threshold of instability of $-15dB$ have been shown to be suitable for $2.5Gb/s$ transmission without optical isolators under the G.957 International Telecommunication Union recommendation, which specifies a threshold of $-24dB$ (Grillot et al., 2003). For the $1.55\mu m$ band, more sophisticated MQW-based AR/AR chirped-grating DFB lasers were fabricated, with cavities (about $500\mu m$ in length) composed of a straight section followed by a stripe section of varying width. These lasers exhibited a threshold of $-22dB$ obtained with a value of the coupling coefficient equal to $80cm^{-1}$ (Grillot et al., 2004). Considered now as a promising alternative to QW lasers for next-generation optical-fiber networks, quantum-dot (QD) lasers exhibit many interesting properties such as reduced threshold current, low chirp, and weak temperature dependence (Bimberg et al., 1999). Moreover, small values of the linewidth enhancement factor (LEF or α factor) of QD lasers are expected to reduce their sensitivity to external optical feedback. For emission at $1.3\mu m$, InAs/GaAs QD lasers have shown a high resistance to external optical feedback and a threshold of $-8dB$ has been obtained for long-cavity ($1.5mm$) devices, resulting from a low LEF value ($\alpha = 1.6$) (O'Brien et al., 2003). For emission in the $1.55\mu m$ band, a progress in the growth of InAs/InP quantum dashes (elongated dots, QDashes) (Lelarge et al., 2007) made possible a systematic investigation of the tolerance to feedback for Fabry-Perot (F-P) lasers with controlled values of the LEF and of the differential gain (Azouigui et al., 2007a, 2007b, 2008, 2009). Recently, it was shown that p-type doped InAs/InP QDash DFB lasers are in general more resistant than their F-P counterparts (Zou et al., 2010).

In parallel with experimental investigations, theoretical approaches have also been developed (see for example, Tromborg et al., 1984, Schunk & Petermann, 1988 and Binder & Cormack, 1989) in order to understand the nonlinear dynamics of a composite laser system. Good agreement between experiment and theory has been found when this latter is applied in particular to bulk and QW lasers. Most of these methods have been developed on the basis of the model proposed by Lang and Kobayashi (L-K model) (Lang & Kobayashi, 1980). More recently, asymptotic methods have been applied to modeling of the dynamic properties of a laser subject to optical feedback (Erneux, 2000, 2008). In this approach (called the asymptotic approach for simplicity), a laser under moderate optical feedback is viewed as a weakly-perturbed nonlinear dynamic system and the threshold of instability corresponds to the first Hopf bifurcation of the L-K rate equations.

This chapter investigates a preliminary interpretation of the experimental results recently obtained with InAs/InP QDash F-P lasers (Azouigui et al., 2007b, 2008). It has been found that the behaviors manifested by these lasers, when assessed for their tolerance to optical feedback, can be quite well understood using the formalism developed from the asymptotic approach. In Section 2, we will present briefly this formalism. Section 3 will be devoted to a simple stability analysis of a composite laser system. In Section 4, a qualitative interpretation of experimental results will be made. We will focus on two particular situations: the onset of coherence collapse and the fully-developed coherence collapse (Zamora-Munt et al., 2010). In Section 5, we will make a direct comparison between an analytical expression derived in Section 4 and a widely employed model for determination of the critical feedback level. A conclusion will be given in Section 6.

2. Solution of the L-K equations

In this section, the L-K rate equations will be analytically solved under the assumption of low feedback level. By means of the obtained solutions, the stability of a possible mode of a composite laser system will be completely characterized. We will review the formalism developed in (Erneux, 2000, 2008), outline the main stages of its derivation and introduce some notations that we will use in the following sections.

2.1 Dimensionless L-K equations

Consider the configuration of Fig. 1. A single-longitudinal-mode laser diode is in resonance with a F-P resonator. We assume that the mirror M is dispersionless and passive. We also consider small feedback effect (i.e. $R_2 \ll 1$), so multiple reflections are neglected.

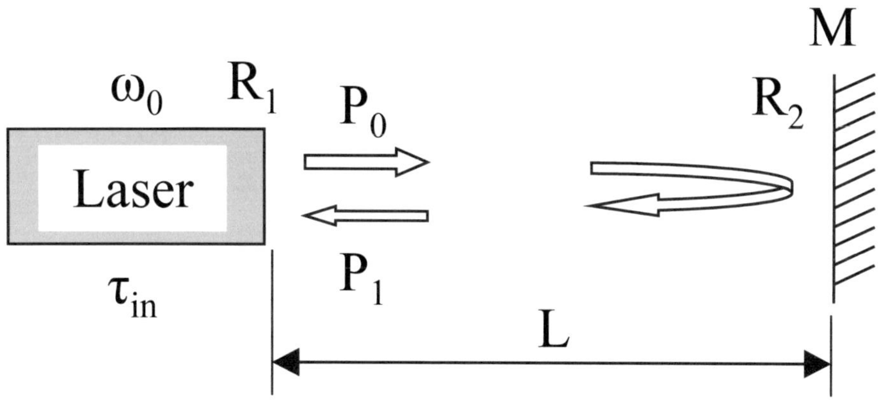

Fig. 1. Schematic drawing of a laser diode with external optical feedback. The external cavity is formed with the output facet of the laser with a reflectivity R_1 and a distant mirror M with a reflectivity R_2. ω_0: emission (angular) frequency; τ_{in}: internal round-trip time; P_0: emission power; P_1: feedback power.

The emission behavior of the above composite system is determined by the interference of the initial and time-delayed waves. Under the assumption of weak feedback level, the electric field $E(t)$ and the carrier density $n(t)$ satisfy the L-K rate equations

$$\dot{E}(t) = \frac{\xi}{2}(1 + j\alpha)n(t)E(t) + \gamma E(t-\tau)\exp(-j\omega_0\tau) \tag{1}$$

$$\dot{n}(t) = J - J_{th} - \frac{n(t)}{T_1} - \left[\Gamma_0 + \xi n(t)\right]|E(t)|^2 \tag{2}$$

where ξ (in s^{-1}) is the differential gain, α is the linewidth enhancement factor (LEF), γ (in s^{-1}) is the feedback rate, τ (in seconds) is the external delay time, $J - J_{th}$ (in s^{-1}) is the pumping current above threshold, T_1 (in seconds) is the carrier life time and Γ_0 (in s^{-1}) is the inverse photon life time. The feedback rate γ is defined as

$$\gamma = \frac{(1 - r_1^2)r_2}{r_1\tau_{in}} \tag{3}$$

In this equation, r_1 ($r_1 = \sqrt{R_1}$) and r_2 ($r_2 = \sqrt{R_2}$) are the reflection coefficients of the output facet of the laser and of the external mirror, respectively. In these notations, the relaxation oscillation (RO) frequency ω_R and the damping rate Γ_R of the solitary laser are defined by

$$\omega_R = \sqrt{\Gamma_0 \xi I} \quad and \quad \Gamma_R = \frac{1}{2}\left(\frac{1}{T_1} + \xi I\right) \tag{4}$$

where I [dimensionless, $I = (J - J_{th})/\Gamma_0$] is the photon number for the solitary laser.

It would be more convenient to use the dimensionless form of the above equations. Defining

$$N = \frac{\xi n}{2\Gamma_0}, \quad s = \Gamma_0 t \quad and \quad Y = E\sqrt{\frac{\xi T_1}{2}} \tag{5}$$

then Equations (1) and (2) become

$$\dot{Y}(s) = (1 + j\alpha)N(s)Y(s) + \eta Y(s-\theta)\exp(-j\Delta_0) \tag{6}$$

$$\dot{N}(s)T = P - N(s) - [1 + 2N(s)]|Y(s)|^2 \tag{7}$$

where

$$\eta = \frac{\gamma}{\Gamma_0}, \quad \theta = \Gamma_0\tau, \quad \Delta_0 = \omega_0\tau = \Omega_0\theta, \quad T = \Gamma_0 T_1 \quad and \quad P = \frac{\xi}{2}T_1\frac{J - J_{th}}{\Gamma_0} \tag{8}$$

In (8), η is the normalized feedback rate, Δ_0 is the initial feedback phase and P is the controlling parameter, called the pump parameter above threshold.

2.2 Stability analysis

The L-K equations (6) and (7) describe the dynamic response of a single-mode laser subject to optical feedback from a distant mirror. By using $Y = R\exp(j\Phi)$ as the expression for the electrical field, we obtain ($Z = N$)

$$\dot{R}(s) = Z(s)R(s) + \eta R(s-\theta)\cos[\Phi(s-\theta) - \Phi(s) - \Delta_0] \tag{9}$$

$$\dot{\Phi}(s) = \alpha Z(s) + \eta\frac{R(s-\theta)}{R(s)}\sin[\Phi(s-\theta) - \Phi(s) - \Delta_0] \tag{10}$$

$$\dot{Z}(s)T = P - Z(s) - [1 + 2Z(s)]R^2(s) \tag{11}$$

Equations (9)-(11) admit a basic steady-state solution

$$\tilde{Y} = \tilde{R}\exp(j\tilde{\Phi}) \quad and \quad \tilde{Z} \tag{12}$$

where $\tilde{R}$, $\tilde{\Phi}$ and $\tilde{Z}$ are given by

$$\tilde{R} = A = \sqrt{\frac{P + \eta\cos(\Delta)}{1 - 2\eta\cos(\Delta)}}, \quad \tilde{\Phi} = Q(s) = (\Omega - \Omega_0)s \quad and \quad \tilde{Z} = B = -\eta\cos(\Delta) \tag{13}$$

$\tilde{Y}$ (together with $\tilde{Z}$) is called an external-cavity mode or ECM. The feedback phase Δ ($\Delta = \omega\tau = \Omega\theta$) associated with this mode must be a solution of the following phase equation

$$\Delta - \Delta_0 = -\eta\theta[\alpha\cos(\Delta) + \sin(\Delta)] \tag{14}$$

We investigate the linear stability properties of Eqs. (9)-(11) by introducing the small perturbations u, v and w. By inserting $R = A + u$, $\Phi = Q + v$ and $Z = B + w$ into these equations, we obtain the following linearzed equations for u, v and w

$$\dot{u}(s) = -B[u(s-\theta) - u(s)] + \eta A\sin(\Delta)[v(s-\theta) - v(s)] + Aw(s) \quad (15)$$

$$\dot{v}(s) = \frac{-\eta\sin(\Delta)}{A}[u(s-\theta) - u(s)] - B[v(s-\theta) - v(s)] + \alpha w(s) \quad (16)$$

$$\dot{w}(s)T = -2A(1+2B)u(s) - (1+2A^2)w(s) \quad (17)$$

We solve these equations by looking for a solution of the form $u(s) = a\exp(\lambda s)$, $v(s) = b\exp(\lambda s)$ and $w(s) = c\exp(\lambda s)$. We then obtain the following problem for the coefficients a, b and c

$$\lambda\begin{pmatrix} a \\ b \\ c \end{pmatrix} = L_J\begin{pmatrix} a \\ b \\ c \end{pmatrix} \text{ or } \lambda V_J = L_J V_J \quad (18)$$

where V_J is the eigenvector of the Jacobian matrix L_J given by

$$L_J = \begin{pmatrix} -BF & \eta FA\sin(\Delta) & A \\ -\eta FA^{-1}\sin(\Delta) & BF & \alpha \\ -2A\varepsilon(1+2B) & 0 & -\varepsilon(1+2A^2) \end{pmatrix} \quad (19)$$

with $F = \exp(-\lambda\theta) - 1$ and $\varepsilon = T^{-1}$. A nontrivial solution is possible only if the growth rate λ satisfies the condition $\det(L_J - \lambda I_3) = 0$ (with I_3: identity matrix), leading to a characteristic equation for λ ($\lambda = \lambda_1 + j\lambda_2$). Under the approximation of low feedback rate (i.e. small values of $\gamma\tau$), there is only one ECM. By applying an asymptotic analysis to (13) and to the characteristic equation [details in (Erneux, 2000, 2008)], we obtain the normalized damping rate λ_1 and RO frequency λ_2

$$\lambda_1 = \frac{-\Gamma_R}{\Gamma_0} - \frac{\gamma}{\Gamma_0}\sin^2\left(\frac{\omega_R\tau}{2}\right)\sqrt{1+\alpha^2}\cos(\Delta_0 - \psi) \quad (20)$$

$$\lambda_2 = \pm\frac{\omega_R}{\Gamma_0} \mp \frac{\gamma}{2\Gamma_0}\sin(\omega_R\tau)\sqrt{1+\alpha^2}\cos(\Delta_0 - \psi) \quad (21)$$

where $\psi = arctg(\alpha)$. Equation (21) implies, among other things, that if $\omega_R\tau$ is an odd multiple of π, the feedback will not affect the RO frequency of the solitary laser.

3. Further discussions

3.1 First Hopf bifurcation

The stability of an equilibrium (or a fixed) point $\tilde{Y}$ of the electrical field Y is determined by the eigenvalue λ of the Jacobian matrix L_J. With increasing feedback rate, an ECM will experience a change of its stability through a Hopf bifurcation. Under the assumption of low feedback rate, this mode will become unstable if λ_1 changes from negative to positive values, corresponding to an undamped optical power proportional to $|Y|^2$. As can be seen from Eq.(20), such a situation occurs only if

$$\cos(\Delta_0 - \psi) < 0 \tag{22}$$

It follows that an ECM will lose its stability at $\lambda_1 = 0$. For a real laser system, this threshold is commonly called the onset of «coherence collapse», since the laser exhibits a drastic linewidth broadening in this state. From Eq.(20) and (8), the normalized critical feedback rate is given by

$$\eta_c = \frac{-\Gamma}{\sin^2(\frac{\omega_R \tau}{2})\sqrt{1+\alpha^2}\cos(\Delta_0 - \psi)} \tag{23}$$

where $\Gamma = \Gamma_R / \Gamma_0$. It would be worthwhile to notice here that although the term $\cos(\Delta_0 - \psi)$ should be negative $(\eta \geq 0)$, it is not necessarily equal to -1.

3.2 Stability characterization of an ECM

The stability of an ECM will be completely determined if the three components a, b and c of the eigenvector V_j are known. They can be derived from Eqs.(18) and (19) and written as functions of a. Since the value of a can arbitrarily be chosen $(a \neq 0)$, we thus take $a = 1$ and express these coefficients as follows

$$a = 1, \quad b = (k + h\alpha)/(\lambda + BF) \quad and \quad c = h \tag{24}$$

where

$$k = \frac{-\eta \sin(\Delta)}{A}, \quad f = \frac{-2A\varepsilon(1+2B)}{\lambda}, \quad g = \frac{\varepsilon(1+2A^2)}{\lambda} \quad and \quad h = \frac{f}{1+g} \tag{25}$$

In this way, the time-dependent electrical field of an ECM is determined by

$$R = A + \exp(\lambda s), \quad \Phi = Q + b\exp(\lambda s) \quad and \quad Z = B + c\exp(\lambda s) \tag{26}$$

Phase portraits in the different state spaces have been proposed to illustrate the stability properties of an ECM, as for example in the equivalent planes of (I, n) (Mørk et al., 1988) and of $(\Phi(s) - \Phi(s-\theta), -\eta\cos[\Phi(s) - \Phi(s-\theta) + \Delta_0])$ (Sano, 1994), and also in the planes of the different components of the electrical field and the carrier density (Tromborg & Mørk, 1990). Here we simply illustrate the transient trajectories (for $s = 0 - 10000$), in the complex plane, of R, Φ and Z (Figs. 2-5), for a composite laser system with the normalized feedback rate around its instability threshold. The laser parameters and their values are given in Table 1. As can be seen from these figures, equivalent to the above mentioned methods, such a representation allows a straightforward interpretation of the transient behaviors of a composite laser system.

symbol	value	description
α	5	linewidth enhancement factor
ξ	2×10^{-4} s^{-1}	differential gain
Γ_0	4×10^{11} s^{-1}	inverse photon life time
T_1	14×10^{-10} s	carrier life time
λ_0	1.55×10^{-6} m	solitary-laser wavelength
L	0.03 m	external-cavity length
Δ_0	-2.02683397 rad	initial feedback phase
Δ	-1.54276496 rad	ECM feedback phase
P	5×10^{-3}	pump parameter above threshold

Table 1. Laser parameters and their values used in most of the figures in the text.

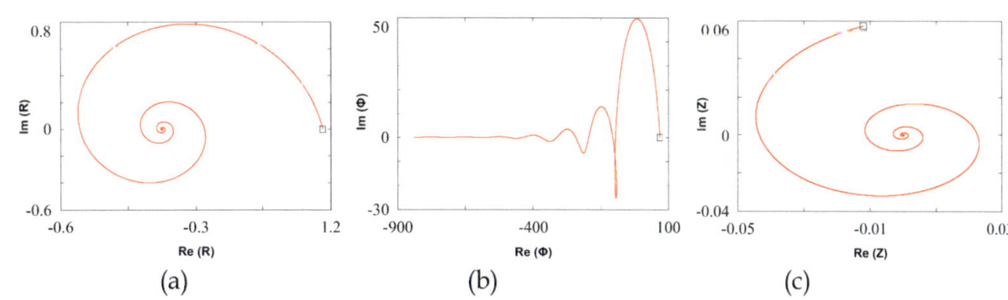

Fig. 2. Stability features of a solitary laser $(\eta = 0)$. Phase portraits in the complex plane for the electrical field Y (attractor solution). Trajectories for (a): the amplitude R, (b): the phase Φ and (c): the carrier density Z. The square symbol denotes the initial state at $s = 0$.

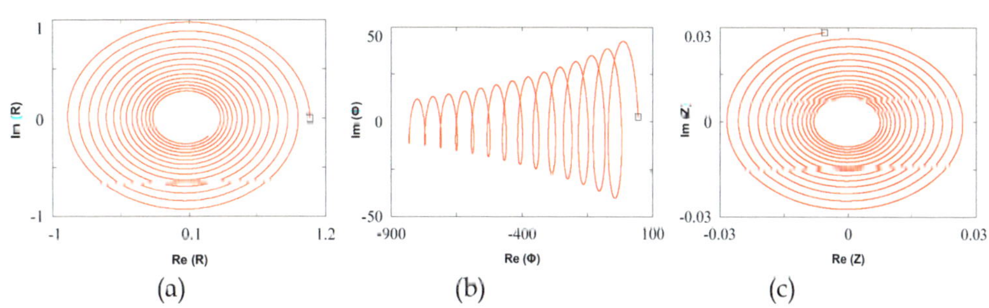

Fig. 3. Stability features of the same laser with the normalized feedback rate close to its threshold $(\eta = 0.85\eta_c)$. Phase portraits in the complex plane for the electrical field Y (attractor solution). Trajectories for (a): the amplitude R, (b): the phase Φ and (c): the carrier density Z. The square symbol denotes the initial state at $s = 0$.

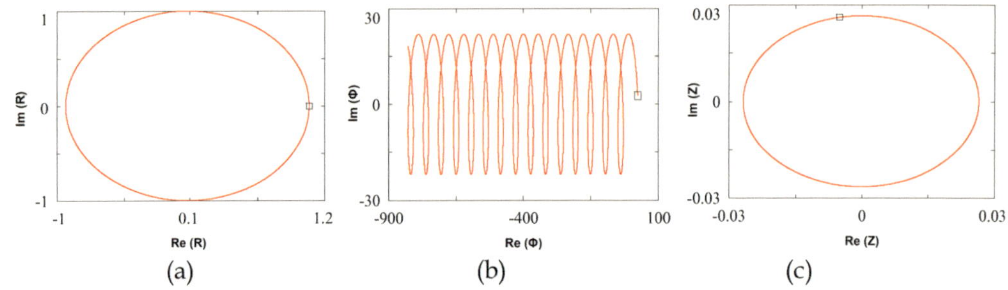

Fig. 4. Stability features of the same laser at the onset of coherence collapse $(\eta = \eta_c)$. Phase portraits in the complex plane for the electrical field Y (limit-cycle solution). Trajectories for (a): the amplitude R, (b): the phase Φ and (c): the carrier density Z. The square symbol denotes the initial state at $s = 0$.

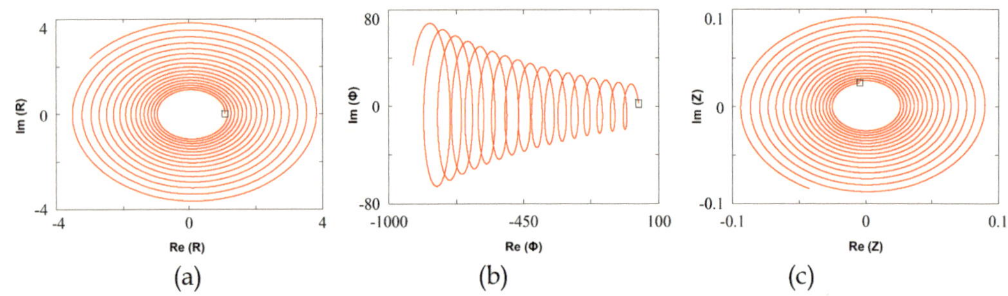

Fig. 5. Stability features of the same laser becoming now coherence-collapsed $(\eta = 1.15\eta_c)$. Phase portraits in the complex plane for the electrical field Y (repellor solution). Trajectories for (a): the amplitude R, (b): the phase Φ and (c): the carrier density Z. The square symbol denotes the initial state at $s = 0$.

4. Qualitative interpretation

As mentioned in the introduction, a preliminary interpretation of the experimental results obtained with InAs/InP QDash F-P lasers will be performed in this section using the analytical expressions derived in Section 3. We concentrate on two laser states in which some typical phenomena were observed in these experiments: the onset of coherence collapse (CC) and the regime of fully-developed coherence collapse.

4.1 Onset of coherence collapse

This state corresponds to the first Hopf bifurcation of the L-K rate equations. We hence have for the critical feedback rate

$$\gamma_c = \frac{-\Gamma_R}{\sin^2(\frac{\omega_R \tau}{2})\sqrt{1+\alpha^2}\cos(\Delta_0 - \psi)} \tag{27}$$

If the feedback level is defined as the ratio of the feedback power to the solitary-laser emission power, it can be written at the CC onset as

$$f_c = \frac{P_1}{P_0} = \frac{\left|(1-r_1^2)r_2 E(t-\tau)\exp(-j\omega_0\tau)\right|^2}{\left|E(t)\right|^2} \tag{28}$$

In steady state, we have from Eq.(3)

$$f_c = (1-r_1^2)^2 r_2^2 = \gamma_c^2 r_1^2 \tau_{in}^2 = \frac{\Gamma_R^2 \tau_{in}^2 R_1}{\sin^4(\frac{\omega_R \tau}{2})(1+\alpha^2)\cos^2(\Delta_0 - \psi)} \tag{29}$$

In general, the critical feedback level f_c is a parabolic function of the pumping current above threshold

$$f_c = \frac{\tau_{in}^2 R_1}{4T_1^2 \sin^4(\frac{\omega_R \tau}{2})(1+\alpha^2)\cos^2(\Delta_0 - \psi)}\left[1 + 2\xi T_1 \Gamma_0^{-1}(J - J_{th}) + \xi^2 T_1^2 \Gamma_0^{-2}(J - J_{th})^2\right] \tag{30}$$

It can easily be shown that the quadratic dependence of f_c on $J - J_{th}$ can be neglected if the RO frequency ω_R satisfies $\omega_R << \sqrt{2/(T_1\Gamma_0^{-1})}$. For example, if the photon lifetime Γ_0^{-1} and the carrier lifetime T_1 are taken respectively as $2.5\,ps$ and $1.4\,ns$ (Table 1), we have $\omega_R/(2\pi) << 3.8\,GHz$. In this case, Eq.(30) is simplified as

$$f_c = \frac{\tau_{in}^2 R_1}{4T_1^2 \sin^4(\frac{\omega_R \tau}{2})(1+\alpha^2)\cos^2(\Delta_0 - \psi)}\left[1 + 2\xi T_1 \Gamma_0^{-1}(J - J_{th})\right] \tag{31}$$

where the slope is proportional to the differential gain ξ and inversely proportional to the LEF α. The experimental evidence of the impact of α and ξ on f_c for InAs/InP QDash F-P lasers is reported in (Azouigui et al., 2008). In this study, the linear dependence of f_c on $J - J_{th}$ was systematically observed with these lasers, which were fabricated on the basis of three different material (dash-in-a-barrier, dash-in-a-well and tunnel-injection) structures and processed from several structures exhibiting various values of α and ξ (Fig. 6).

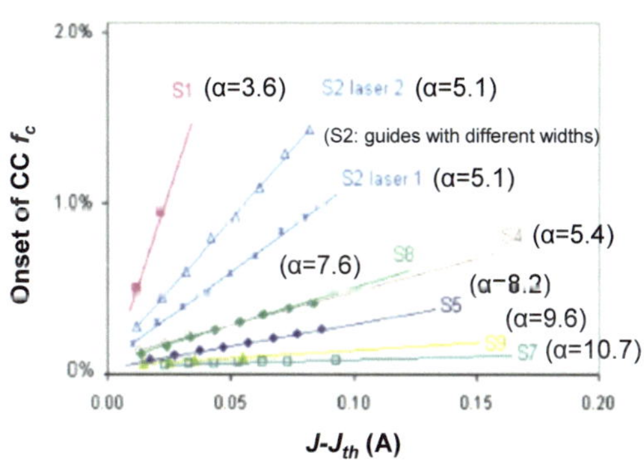

Fig. 6. Onsets of CC measured as a function of the pumping current above threshold, for seven laser structures, using an $18\,m$ -long optical fiber-based external cavity (Azouigui et al., 2008). All the devices are 600- μm -long as-cleaved F-P lasers. The most resistant lasers were fabricated from the structure S_1, since it has the lowest α value as well as the highest ξ one.

4.2 Fully-developed coherence collapse

It is widely-recognized that when a laser, subject to moderate amounts of optical feedback, operates close to its solitary threshold, it exhibits a rather complex behavior due to the so-called low-frequency fluctuations (LFF), and that for a large enough pumping current there will occur a transition from the regime of LFF to the regime of fully-developed coherence collapse (FDCC). In a recent experimental investigation, the LFF regime was observed around laser threshold ($13\,mA$) at $-1.2\,dB$ feedback, for a 600-μm-long InAs/InP QDash F-P laser emitting at $1.57\,\mu m$ and using a free-space setup with a short external cavity ($L = 0.5\,m$) (Azouigui et al., 2007b). In these experiments, the regime of FDCC was attained at about $30\,mA$. A number of feedback-induced phenomena were manifested in the frequency domain, such as RF (radio-frequency) noise enhancement, frequency shift and peak broadening (Fig. 7). In this section, we will give some explanations of all these phenomena by using the formalism developed in Section 3. An intensive discussion about the physical origin of the LFF regime and the mechanism of its transition to the FDCC regime, based on numerical simulations of the L-K rate equations, is found in (Sano, 1994) and (Zamora-Munt et al., 2010).

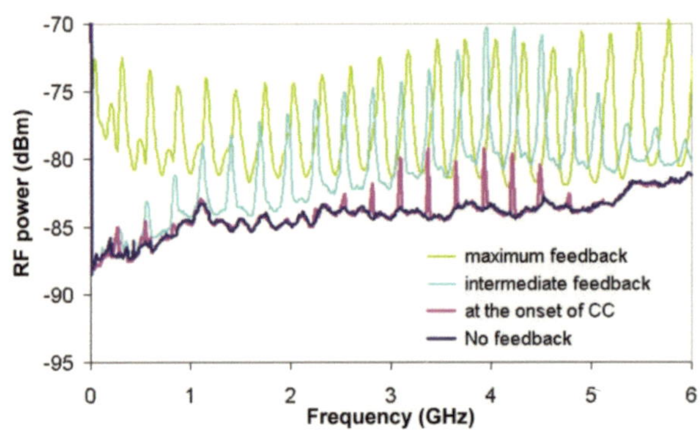

Fig. 7. Measured RF spectra with increasing feedback level, for a pumping current of $30\,mA$ (Azouigui et al., 2007b).

Let us first have a look at the expression for the intensity of an ECM. It can be derived from (26) and be written in general as

$$\begin{aligned} |Y(s)|^2 &= \exp(-2\Phi_2)|R(s)|^2 \\ &= \exp(-2\Phi_2)\left[A^2 + \exp(2\lambda_1 s) + 2A\exp(\lambda_1 s)\cos(\lambda_2 s)\right] \\ \textit{with}: \Phi_2 &= \exp(\lambda_1 s)\left[b_1\sin(\lambda_2 s) + b_2\cos(\lambda_2 s)\right] \end{aligned} \tag{32}$$

In this equation, Φ_2 is the imaginary part of Φ, which is non-null in steady state for a coherence-collapsed laser; b_1 and b_2 are respectively the real and imaginary parts of b. At the onset of CC ($\lambda_1 = 0$), this equation is reduced to

$$|Y(s)|_c^2 = \exp\left\{-2\left[b_1\sin(\lambda_2 s) + b_2\cos(\lambda_2 s)\right]\right\}\left[A^2 + 1 + 2A\cos(\lambda_2 s)\right] \tag{33}$$

In this particular case, the λ_1-dependent exponential terms disappear and the intensity of an ECM displays equal-amplitude pulses. As an example, the intensity distributions of an ECM for the normalized feedback rate around its instability threshold, calculated from Eq.(32), are illustrated in Fig. 8. In this example, the laser is pumped by a relatively strong current $(P = 5\times10^{-3})$. This, as expected, results in a rapid damping (when $\eta < \eta_c$) of its emission power to a steady-state value proportional to A^2, which is close to zero according to our notations [as can be seen in (13) for the definition of A and in (5) where $|E|^2 \propto 10^{14}|Y|^2$].

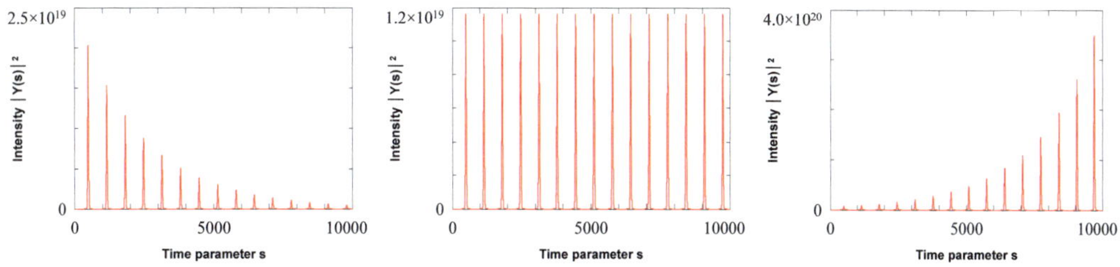

(a) (b) (c)

Fig. 8. Time traces of the transient intensity $|Y(s)|^2$ for the normalized feedback rate around its critical threshold. (a) $\eta = 0.99\eta_c$; (b) $\eta = \eta_c$; (c) $\eta = 1.01\eta_c$.

4.2.1 Feedback-enhanced RF noise

The intensity distributions of an ECM are plotted in Fig. 9 (a), for two values of the normalized feedback rate and for a great P value $(P = 5\times10^{-3})$. This result suggests that for a coherence-collapsed laser pumped by a large current, a small excess (a few percentages) in feedback level could lead to a significant increase of the RF noise, and therefore of the spectrum power as shown in Fig. 7.

4.2.2 Feedback-induced frequency shift and peak broadening

Figure 7 also shows that in the frequency domain, the increase of the RF noise with the feedback level is accompanied by a shift, to the greater frequencies (blue shift), of the peaks and their broadening. According to Eq.(32), at the CC onset as well as inside the CC regime, the undamped optical power of an ECM will oscillate at the frequency λ_2 (see also Fig. 8), which is greater than the normalized solitary-laser RO frequency ω_R / Γ_0 if $\omega_R\tau < \pi \bmod 2\pi$ [$\sin(\omega_R\tau) > 0$]. This equation also predicts a peak broadening with the feedback level. Since $\lambda_1 \geq 0$ for a coherence-collapsed laser, these peaks will have the narrowest width at the onset of CC. Figure 9 (b) depicts the calculated Fourier spectra of an ECM, showing a good qualitative agreement with the experimental findings illustrated in Fig. 7.

4.2.3 Current-induced frequency increasing and peak broadening

In the case of a given feedback rate and for a short external cavity $(\omega_R\tau < \pi/2)$, Equation (21) suggests a monotonous increasing, through the term $\sin(\omega_R\tau)$, of the RO frequency λ_2 with the pumping current J when the laser system passes from LFF to FDCC regime [Fig. 10, (a)]. The nonlinear relation between λ_2 and J can be checked by inspection of Fig. 10, (b), where the RF spectra were obtained with a $0.5\,m$-long external cavity.

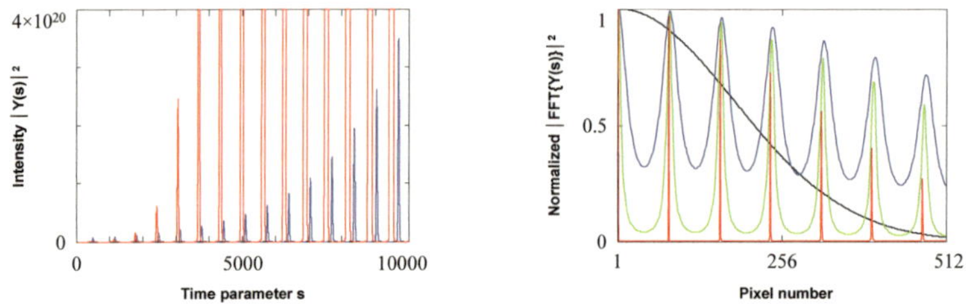

(a) (b)

Fig. 9. Feedback-induced phenomena inside the FDCC regime with $P = 5\times10^{-3}$, predicted by Eq.(32). The other parameter values are given in Table 1. (a) Time traces of the feedback-enhanced RF noise in terms of ECM intensity. Blue: $\eta = 1.01\eta_c$; Red: $\eta = 1.05\eta_c$. (b) Feedback-enhanced frequency blue shift and peak broadening (FFT: Fast Fourier Transform). Black: $\eta = 0$; Red: $\eta = \eta_c$; Green: $\eta = 1.01\eta_c$; Blue: $\eta = 1.02\eta_c$. The first pixel corresponds to the position of the normalized solitary-laser RO frequency ω_R / Γ_0 .

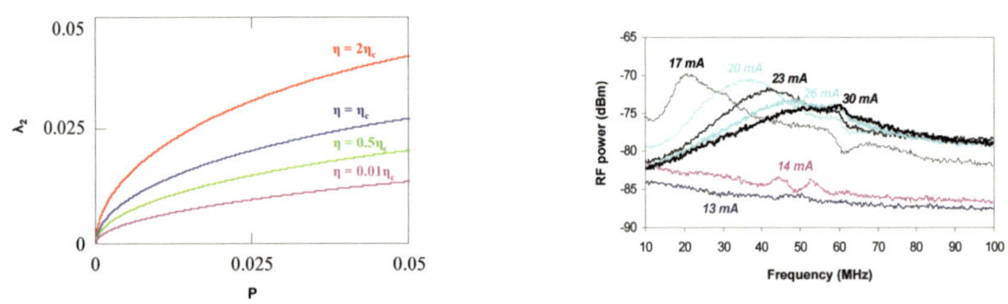

(a) (b)

Fig. 10. Current-induced frequency increasing and peak broadening during the transition from the LFF regime to the FDCC regime. (a) Variation of λ_2 as a function of P for several values of the normalized feedback rate, calculated from Eq.(21). (b) Measured RF spectra at $-1.2\ dB$ feedback (Azouigui et al., 2007b). The transition period corresponds roughly to a variation of the pumping current from $17\ mA$ to $30\ mA$.

5. Comparison with a classical model

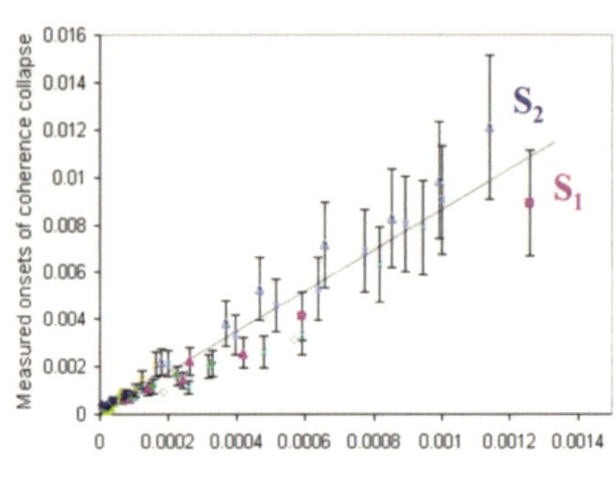

Fig. 11. Measured values of the CC onset versus predicted ones deduced from Eq.(10) in (Helms & Petermann, 1990) (Azouigui et al., 2008). Compared to the structure S_2 (see Fig. 6), the structure S_1 has a smaller slope, which means that its measured onset values are closer to the predicted ones.

In this section, the difference between Eq.(29) and the model proposed by Helms and Petermann (denoted as the H-P model, Helms & Petermann, 1990) will be discussed from a practical point of view, for the case of long external cavities (where $\omega_R\tau$ is set to be an odd multiple of π). A deeper analysis of the cause for this difference is beyond the scope of the present work. We are motivated by such an approach also because when we undertook an interpretation of Fig. 6, we found that although for all structures the threshold values predicted by the H-P model follow the same trend as the measured ones, a difference of $8-10\ dB$ is systematically obtained between the measured and predicted values (Fig. 11).

Equation (29) predicts a linear dependence of the onset of CC on the output-facet reflectivity of the laser diode. By assuming a long external cavity, it is simplified as

$$f_c = \frac{\Gamma_R^2 \tau_{in}^2 R_1}{(1+\alpha^2)\cos^2(\Delta_0 - \psi)} \tag{34}$$

Remember that $\cos(\Delta_0 - \psi)$ should take negative values. The critical feedback level predicted by the H-P model, which was established in the small-signal domain with the help of a transfer function from the modulation current to the modulated optical power, is determined by [Eq.(8) in Helms & Petermann, 1990]

$$f_c^{H-P} = \frac{\Gamma_R^2 \tau_{in}^2}{(1+\alpha^2)16C_l^2} \tag{35}$$

where C_l denotes the coupling strength from the laser cavity to the external cavity. It is given by $C_l = (1-R_1)/(2\sqrt{R_1})$ $(0 \le C_l \le \infty)$ for a F-P laser. The difference between these two models is expressed by β, defined as the ratio of f_c^{H-P} and f_c in logarithmic form

$$\beta(dB) = 10\log_{10}\frac{f_c^{H-P}}{f_c} = 10\log_{10}\frac{\cos^2(\Delta_0-\psi)}{4(1-R_1)^2} = 10\log_{10}\left[\cos^2(\Delta_0-\psi)\right] - 2.9 \tag{36}$$

where we have taken $R_1 = 0.3$ for an as-cleaved F-P laser. As indicated in this expression, β will be mainly affected by the term $\Delta_0 - \psi$. This term, varying in the range $\pi/2 \bmod 2\pi < \Delta_0 - \psi \le \pi \bmod 2\pi$, corresponds to a very large variation range of β [Fig. 12, (a)]. For example, if $\Delta_0 - \psi = \pi$, we have $\cos(\Delta_0 - \psi) = -1$ and $\beta = -2.9\ dB$. One radian less for the initial feedback phase, i.e. $\Delta_0 - \psi = \pi - 1$, will result in $\cos(\Delta_0 - \psi) = -0.54$ and $\beta = -8.3\ dB$. The absence of the cosine term in the denominator of f_c^{H-P} makes the H-P model correspond to the most unfavorable situation. For a given initial frequency ω_0, two factors can considerably influence the β ratio: the external delay time τ and the LEF α. Singularity occurs if the initial feedback phase Δ_0 approaches $\psi + (\pi/2 \bmod 2\pi)$. Moreover, β is highly sensitive to very small variations of the external-cavity length L, because of the ratio L/λ_0 in the expression of Δ_0, where L is in meter and λ_0 in micron. An example is given in Fig. 12, (b).

We can see from this specific case that for a given α value [α should be greater than 1, in order that Eq.(35) holds, according to the H-P model], the values of $|\beta|$ can be quite different due to small oscillations of the L value, and that for a positive (negative) increment of L,

$|\beta|$ increases (decreases) with α. A further analysis of this figure would give some qualitative explications of the observations stated above (Fig. 11). (1) For the curve where L is taken exactly as $18\,m$, $|\beta|$ increases with α. This means that a laser with a smaller α value would have a CC onset value closer to that predicted by the H-P model. Experimentally, we did observe such behavior. As shown in Fig. 11, the structure S_1 ($\alpha = 3.6$) has a smaller $|\beta|$ ratio than the structure S_2 ($\alpha = 5.1$); (2) The incertitude of the measurement (within hundreds of microns) of the external-cavity length can affect considerably the β value.

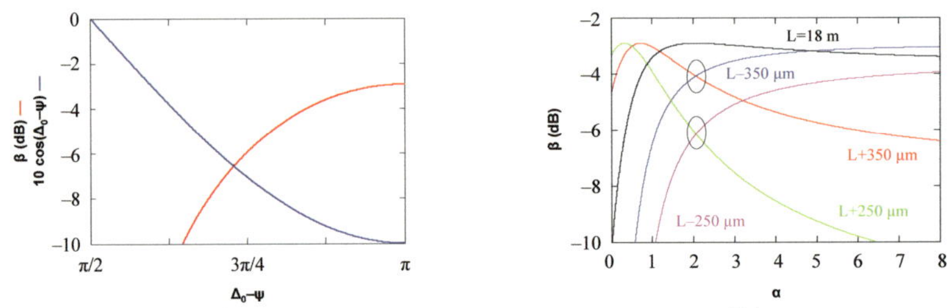

(a) (b)

Fig. 12. (a) Dependence of β on $\Delta_0 - \psi$ (red curve), in the range $\pi/2 < \Delta_0 - \psi \leq \pi$. For values of $\Delta_0 - \psi$ away from π, there exists a great difference between Eq.(34) and Eq.(35). (b) Variation of β as a function of α, for an $18\,m$-long optical-fiber-based external cavity. The refractive index of the fiber was taken as 1.5 for a solitary laser emitting at the wavelength of $1.55\,\mu m$. The small increments around L were chosen arbitrarily.

6. Conclusion

The asymptotic approach provides a useful tool for stability modeling of bulk, quantum-well as well as quantum-dash semiconductor lasers with optical feedback. The analytical solution of the Lang and Kobayashi rate equations derived using this approach enables a complete description, in the phase space and also in the time domain, of an external-cavity mode for a composite laser system close to, at the onset of as well as inside the coherence-collapse regime. More specifically, by use of this model, the temporal and spectral behaviors of the laser system in the regime of fully-developed coherence collapse, usually numerically analyzed, can be described in an analytical way. This model gives a quite good physical insight of the phenomena recently observed with 1.55-μm InAs/InP quantum-dash Fabry-Perot lasers, when these lasers were characterized for their resistivity to optical feedback originating from two different external-cavity configurations: one was based on a $0.5\,m$-long free-space setup and another was composed of an $18\,m$-long optical fiber. Compared to the H-P model, the expression for determination of the threshold of coherence collapse derived in Section 4, namely Eq.(29), is found to be in better agreement (a «gain» of at least ~ 3 dB) with the experimental data obtained with these lasers fabricated from quantum-dash-based nanostructures. However, it seems that its use would require, for each value of the pumping current, more certitude on the measurement of the solitary-laser wavelength and also on that of the external-cavity length.

The dynamics of the low-frequency fluctuations will be investigated in the near future using this approach in comparison with numerical approaches so far reported in the literature.

7. Acknowledgments

One of the authors (Q. Zou) would like to acknowledge K. Merghem, A. Martinez, and A. Ramdane (CNRS, Laboratory for Photonics and Nanostructures, France), and A. Accard, N. Chimot, J.-G. Provost, and F. Lelarge (Alcatel-Thales III-V Lab, a joint laboratory of Alcatel-Lucent Bell Labs and Thales Research & Technology, France) for fruitful collaborations. He would also like to thank T. Erneux (Université Libre de Bruxelles, Unité Optique Non-Linéaire Théorique, Belgium) for helpful discussions.

8. References

Azouigui, S.; Dagens, B.; Lelarge, F.; Provost, J.G.; Accard, A.; Grillot, F.; Martinez, A.; Zou, Q. & Ramdane, A. (2007). Tolerance to Optical Feedback of 10 Gbps Quantum-Dash-Based Lasers Emitting at 1.51 µm. *IEEE Photonics Technology Letters*, Vol.19, pp. 1181-1183

Azouigui, S.; Kelleher, B.; Hegarty, S.P.; Huyet, G.; Dagens, B.; Lelarge, F.; Accard, A.; Make, D.; Le Gouezigou, O.; Merghem, K.; Martinez, A.; Zou, Q. & Ramdane, A. (2007). Coherence Collapse and Low-Frequency Fluctuations in Quantum-Dash Based Lasers Emitting at 1.57 µm. *Optics Express*, Vol.15, pp. 14155-14162

Azouigui, S.; Dagens, B.; Lelarge, F.; Accard, A.; Make, D.; Le Gouezigou, O.; Merghem, K.; Martinez, A.; Zou, Q. & Ramdane, A. (2008). Systematic Investigation of InAs/InP Quantum-Dash Based Lasers under External Optical Feedback. *Appl. Phys. Lett.*, Vol.92, 201106

Azouigui, S.; Dagens, B.; Lelarge, F.; Provost, J.G.; Make, D.; Le Gouezigou, O.; Accard, A.; Martinez, A.; Merghem, K.; Grillot, F.; Dehaese, O.; Piron, R.; Loualiche, S.; Zou, Q. & Ramdane, A. (2009). Optical Feedback Tolerance of Quantum-Dot- and Quantum-Dash-Based Semiconductor Lasers Operating at 1.55 µm. *IEEE Journal of Selected Topics in Quantum Electronics*, Vol.15, pp. 764-773

Bimberg, D.; Grundmann, M. & Ledentsov, N.N. (1999). Quantum Dot Heterostructures (Wiley, New York).

Binder, J.O. & Cormack, G.D. (1989). Mode Selection and Stability of a Semiconductor Laser with Weak Optical Feedback. *IEEE Journal of Quantum Electronics*, Vol.35, pp. 2255-2259

Erneux, T. (2000). Asymptotic Methods Applied to Semiconductor Laser Models. *SPIE 3944, Physics & Simulations of Optoelectronic Devices VIII*

Erneux, T. (2008). First Hopf Bifurcation of the Lang-Kobayashi Equations. *Private communication*

Grillot, F.; Thedrez, B.; Gauthier-Lafaye, O.; Martineau, M.F.; Voiriot, V.; Lafragette, J.L.; Gentner, J.L. & Silvestre, L. (2003). Coherence Collapse Threshold of 1.3 µm Semiconductor DFB Lasers. *IEEE Photonics Technology Letters*, Vol.15, pp. 9-11

Grillot, F.; Thedrez, B. & Duan, G.H. (2004). Feedback Sensitivity and Coherence Collapse Threshold of Semiconductor DFB Lasers with Complex Structures. *IEEE Journal of Quantum Electronics*, Vol.40, pp. 231-240

Helms, J. & Petermann, K. (1990). A Simple Analytic Expression for the Stable Operation Range of Laser Diodes with Optical Feedback. *IEEE Journal of Quantum Electronics*, Vol.26, No.5, pp. 833-836

Lang, R. & Kobayashi, K. (1980). External Optical Feedback Effects on Semiconductor Injection Laser Properties. *IEEE Journal of Quantum Electronics*, Vol.QE-16, No.3, pp. 347-355

Lelarge, F.; Dagens, B.; Renaudier, J.; Brenot, R.; Accard, A.; Van Dijk, F.; Make, D.; Le Gouezigou, O.; Provost, J.G.; Poingt, F.; Landreau, J.; Drisse, O.; Derouin, E.; Rousseau, B.; Pommereau, F. & Duan, G.H. (2007). Recent Advances on InAs/InP Quantum Dash Based Semiconductor Lasers and Optical Amplifiers Operating at 1.55 µm. *IEEE Journal of Selected Topics in Quantum Electronics*, Vol.13, pp. 111-124

Mørk, J.; Tromborg, B. & Christiansen, P.L. (1988). Bistability and Low-Frequency Fluctuations in Semiconductor Lasers with Optical Feedback: A Theoretical Analysis. *IEEE Journal of Quantum Electronics*, Vol.24, No.2, pp. 123-133

O'Brien, D.; Hegarty, S.P.; Huyet, G.; McInerney, J.G.; Kettler, T.; Laemmlin, M.; Bimberg, D.; Ustinov, V.M.; Zhukov, A.E.; Mikhrin, S.S. & Kovsh, A.R. (2003). Feedback Sensitivity of 1.3 µm InAs/GaAs Quantum Dot Lasers. *Electronics Letters*, Vol.39, pp. 1819-1820

Sano, T. (1994). Antimode Dynamics and Chaotic Itinerancy in the Coherence Collapse of Semiconductor Lasers with Optical Feedback. *Physical Review A*, Vol.50, No.3, pp. 2719-2726

Schunk, N. & Petermann, K. (1988). Numerical Analysis of the Feedback Regimes for a Single-Mode Semiconductor Laser with External Feedback. *IEEE Journal of Quantum Electronics*, Vol.24, No.7, pp. 1242-1247

Tromborg, B.; Osmundsen, J.H. & Olesen, H. (1984). Stability Analysis for a Semiconductor Laser in an External Cavity. *IEEE Journal of Quantum Electronics*, Vol.QE-20, No.9, pp. 1023-1032

Tromborg, B. & Mørk, J. (1990). Nonlinear Injection Locking Dynamics and the Onset of Coherence Collapse in External Cavity Lasers. *IEEE Journal of Quantum Electronics*, Vol.26, No.4, pp. 642-654

Zamora-Munt, J.; Masoller, C. & García-Ojalvo, J. (2010). Transient Low-Frequency Fluctuations in Semiconductor Lasers with Optical Feedback. *Physical Review A*, Vol.81, No.033820

Zou, Q.; Merghem, K.; Azouigui, S.; Martinez, A.; Accard, A.; Chimot, N.; Lelarge, F. & Ramdane, A. (2010). Feedback-Resistant P-Type Doped InAs/InP Quantum-Dash DFB Lasers for Isolator-Free 10 Gb/s Transmission at 1.55 µm. *Appl. Phys. Lett.*, Vol.97, 231115

Section 2

Laser Diode Technology

6

DFB Laser Diode Dynamics with Optoelectronic Feedback

M. H. Shahine
Ciena Corporation, Linthicum, Maryland,
University of Maryland, Baltimore County, Maryland,
USA

1. Introduction

Semiconductor lasers have been one of the major building blocks of fiber optics based communication systems. For the past two decades, specifications of these lasers have been tailored to specific applications by defining certain performance parameters that do not necessarily overlap from one application to another. In this simulation work, we modify and enhance essential performance parameters of simple, low cost lasers and tailor them to specific applications that normally require advanced and complicated laser structures by using electronic feedback for instantaneous impairment correction, while at the same time maintain a compact size for the laser and the supporting circuitry around it. The main driver behind this work is to facilitate photonic integration by analyzing design cases with specific delays and structures for the feedback loop to deliver as a first step, a comprehensive design study for hybrid integration. Specific applications targeted are, analog transmission for wireless backhauling by reducing the laser Relative Intensity Noise and maintaining and enhancing transmitter linearity, improving laser modulation bandwidth and increasing the laser relaxation oscillation frequency, laser line-width reduction to target long haul transmission and coherent detection and finally producing self-pulsating laser for analog to digital conversion sampling application. The performance for all these applications is analyzed in both time and frequency domains. For the optical sampling source application, the feedback loop needs to be operating outside the stable regime in order for the laser to run in the self-pulsation mode, where the laser drive current can be used as a single point of control for the pulsation rep-rate.

2. Background and motivation

The work for controlling and adjusting semiconductor laser characteristics using electronic feedback was started in the early 1980's (Peterman,1991;Ohtsu,1996;Ohtsu,1988) for both controlling the laser RIN and Line-width. By the mid 1980's, it became clear to researchers that using electronic feedback control scheme would not provide the desired results due to two fundamental limitations in the feedback loops used then (Ohtsubo,2007). The first limitation was the limited bandwidth of the electronic components used in the feedback loop, which was significantly less than the modulation bandwidth of the laser itself and did not provide corrective feedback signal of the laser performance over the entire laser operational frequency range. The second limitation was large amount of delay that was introduced in the feedback loop by splitting the laser output light from the front facet and using an optical splitter and multistage amplifiers in the feedback chain.

Due to those limitations, the results achieved then were very limited in improving the laser performance and did not provide any significant breakthroughs.

In order to revive this area of research, we have proposed solutions to solve the issues introduced by the two limitations listed above. In the past few years we were able to use wideband back-facet monitors and wideband trans-impedance amplifiers that matched or exceeded the laser bandwidth, we also proposed using the back-facet to tap the laser output power hence reducing the feedback loop delay. By providing solutions to solve these issues, which proved essential to the advancement of photonic integration. Recently, DARPA has started soliciting solutions for the problem statement we specified above (DARPA, 2011). We understand that our work addresses the roadmap for delivering those solutions to DARPA which serves as a validation for reviving this area of research.

In order to address the specific issues of photonic integration, efforts were concentrated in simulations by using certain parameters that are within the acceptable ranges for such integrated solutions (i.e. loop delays between 10ps and 100ps) although, it's not difficult to expand the simulation range beyond these values, addressing this specific integration application is the goal for this chapter.

This study also provides the analysis of the laser as a pulsing source in terms of jitter and noise performance for analog to digital photonic sampling application using DFB laser, which has not been addressed previously and is also part of the DARPA proposal request.

3. Basic control theory

The theory behind this work is based on the classical control theory of negative feedback which has been well known since the pioneering work by H.S. Black who showed that noise of a classical oscillator can be suppressed by negative feedback stabilization (Black, 1934). Recent work by (Wan, 2005), has presented a rigorous, yet simple and intuitive, non-linear analysis method for understanding and predicting injection locking in LC oscillators.

A system with a negative feedback control loop is shown in figure (1).

It consists of a forward-gain element with transfer function $A(s)$, with s is the Laplace operator and can be replaced by $(j\omega)$ feedback element with transfer function $B(s)$, and a subtraction function to produce the difference between the input signal x and the output from the feedback element y.

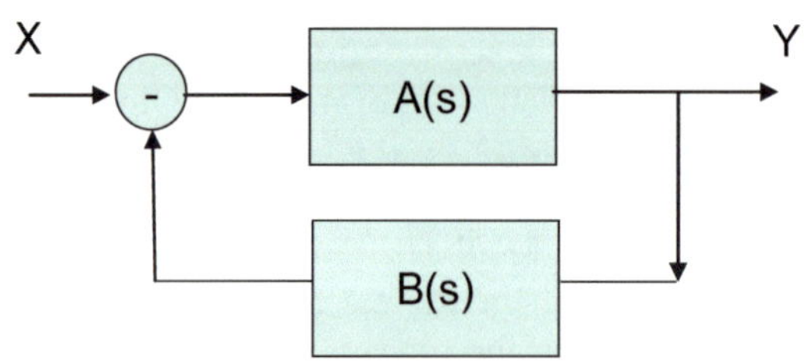

Fig. 1. Negative Feedback loop system.

Where $A(s)$ represents laser transfer function, $B(s)$ represents feedback loop transfer function, x is the injection Current and y is the Optical output power.

The closed loop transfer function of such system is:

$$T(s) = \frac{y}{x} = \frac{A(s)}{1 + A(s) * B(s)} \tag{1}$$

This system can be linearized by making the gain product $A(s)*B(s) >> 1$.

With this condition, the transfer function becomes dependent solely on the feedback gain coefficient and response of the feedback loop which can be made linear:

$$T(s) = \frac{1}{B(s)} \tag{2}$$

A feedback loop can oscillate if its open-loop gain exceeds unity and simultaneously its open-loop phase shift exceeds π. At least one of the closed-loop poles of an unstable loop will lie in the right half of the s-plane in figure (2). Analysis of stability by investigating pole location can be done by the using the Nyquist criterion which is often used instead of resolving the characteristic equation (He, 2009; Maisel, 1975; Luenberger, 1979). It is based on Nyquist plot that is a plot of real and imaginary parts of open-loop transfer function. If poles are present in the Left Half of the s-Plane, the closed-loop system is stable. If poles are shifted to the Right Hand Plane, the closed-loop system becomes unstable. In brief, the Nyquist criterion is a method for the determination of the stability of feedback systems as a function of an adjustable gain and delay parameters. It does not provide detailed information concerning the location of the closed-loop poles as a function of $B(s)$, but rather, simply determines whether or not the system is stable for any specified value of $B(s)$. On the other hand, the Nyquist criterion can be applied to system functions in which no analytical description of the forward and feedback path system functions is available.

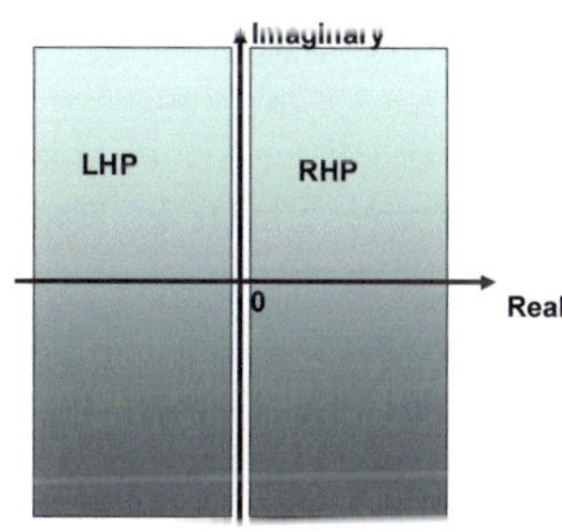

Fig. 2. S-plane plot.

For the RIN and line-width reduction systems work presented later in this chapter, the closed-loop system must to be stable.

In stabilizing an unstable system, the adjustable gain is used to move the poles into the LHP for a continuous-time system. Also the feedback can be used to broaden the bandwidth of a system by moving the pole as to decrease the time constant of the system. Furthermore, just as feedback can be used to relocate the poles to improve system performance, there is the potential that with an improper choice of feedback parameters, a stable system can be destabilized.

For the self-pulsation mode, it is well known that an active circuit with feedback can produce self-sustained oscillations only if the criterion formulated by Barkhausen is fulfilled. This criterion is based on having the denominator of the closed loop transfer function go to zero. The poles in this self-pulsation mode need to be located up and down on the imaginary axis of the s-plane plot with a zero value on the real axis and where the phase of this transfer function:

$$\angle T(j\omega) = 0 \Rightarrow \omega = \omega_0 \tag{3}$$

$$|T(j\omega_0)| = 1 \tag{4}$$

These equations (3) and (4) are referred to as the phase and gain conditions, respectively. According to Barkhausen criterion, the oscillation frequency is determined by the phase condition (3). As the poles move further to the Right Half Plane (RHP) the system becomes unstable and will enter into the chaos mode of operation.

4. Laser optimized for analog signal modulation with stable feedback settings

Transmission of analog optical amplitude modulated signals imposes stringent demands on the linearity of the system. Transmitter non-linearity causes the modulated sub carriers to mix and generate inter-modulation products, which limits the channel capacity (Helms, 1991). In order to directly modulate the laser with an analog signal and expect that the output optical power to represent that signal, the L-I (light vs. Current) relationship has to remain linear over temperature and over the laser lifetime. As stated in (Stephens, 1982; Stubkjair, 1983), that the primary cause of this non-linearity is the laser photon-electron interaction. This problem is not as critical when the laser is modulated with a low frequency signal compared to the ROF (Relaxation Oscillation Frequency) of the laser, since at such low modulation speed, even when the OMD (Optical Modulation Depth) of the signal approach 100%, the laser is virtually in a quasi-steady state as it is ramping up and down along the L-I curve, and consequently the linearity of the modulation response is basically that of a CW light-current characteristic, which is linear. However as the modulation frequency increases and start to approach the ROF of the laser, the harmonic distortions increase very rapidly. The second harmonics of the modulation signal increase roughly as the square of OMD while the IMP (Inter-Modulation Product) increase as the cube of the OMD. Previous works have been successful in producing linear transmitters for analog signals using lasers with external Mach-Zehnder modulator, and insert a linearizer circuitry after the modulator using an optical splitter (Chiu, 1999), There was also the feed-forward technique using two lasers (Ismail, 2004; Ralph, 1999), it was demonstrated recently using two lasers in every transmitter to improve linearization, however these solutions are costly and are not attractive for low cost deployment of systems for wireless backhauling.

A block diagram representing the proposed method is shown in figure (3). Based on this architecture (Shahine, 2009a), a sample of the output beam of the laser is detected through the large bandwidth back-facet monitor. The photocurrent from the back-facet is then amplified and a π phase shifted to produce the negative feedback condition. That signal is added to the input analog signal to produce the modulating signal of the laser. The overall transfer function of this system is only dependent on the transfer function of the feedback correction loop, when the gain of that loop is large enough. By having a linear transfer function of the feedback loop, the overall transfer function of the system becomes linear. This proposed solution consists of using a back-facet monitor with a wide bandwidth to accommodate the high frequency components of the modulated signal and shortens the feedback loop delay. The Feedback correction loop consists of a trans-impedance amplifier that is connected to an RF combiner, where the input signal is added to the feedback signal. The modeling of this system and the laser are done using the laser rate equations.

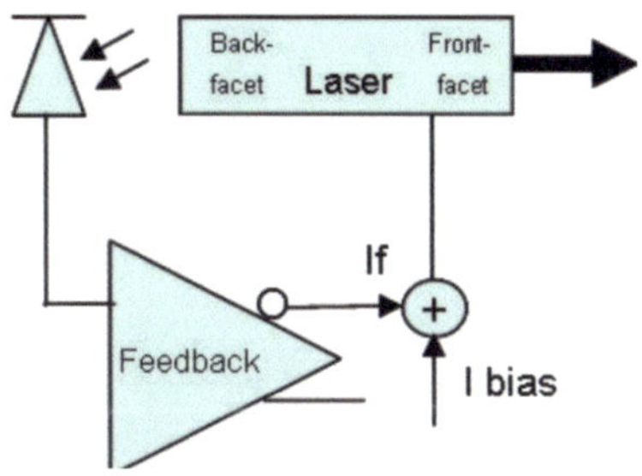

Fig. 3. The laser system being modelled.

The dynamic performance of laser diodes is usually analyzed in terms of rate equations (Tomkos, 2001) which add up all physical processes that change the densities of photons and carriers. The carrier density equation is presented in (5), the photon density rate equation is presented in (6) and the optical phase rate equation is presented in (7). These equations were modified to include the feedback loop parameters.

$$\frac{dN(t)}{dt} = \frac{I(t)}{q * V_a} - g_0 \frac{[N(t) - N_0] * S(t)}{1 + \varepsilon * S(t)} - \frac{N(t)}{\tau_n} + \left[\frac{\omega_n}{2\pi} * (\rho * S[t - \tau]) \right] + F_N(t) \tag{5}$$

$$\frac{dS(t)}{dt} = \Gamma * g_0 \frac{[N(t) - N_0] * S(t)}{1 + \varepsilon * S(t)} - \frac{S(t)}{\tau_p} + \frac{\Gamma * \beta}{\tau_n} * N(t) + F_S(t) \tag{6}$$

$$\frac{d\phi(t)}{dt} = \frac{1}{2} \alpha [\Gamma * g_0 [N(t) - N_0] - \frac{1}{\tau_p}] + F_\phi(t) \tag{7}$$

ρ represents the feedback loop gain, τ represents the feedback loop propagation delay and ω_n represents the 3dB bandwidth of the amplifier circuit.

The Langevin noise terms are the noise terms added respectively to the rate equations. These terms are present due to the carrier generation recombination process, to the spontaneous emission and the generated phase respectively. These noise terms are Gaussian random processes with zero mean value under the Markovian assumption (memory-less system) (Helms, 1991). The Markovian approximation of this correlation function is of the form:

$$\left\langle F_i(t) F_j(t') \right\rangle = 2D_{ij} \delta(t - t') \tag{8}$$

Where $i,j = S,N, or\ \phi$ is the diffusion coefficient with full derivation is presented as follows:

$$D_{SS} = \frac{\beta * V_a * N_{sd} * [(V_a * S_{sd}) + 1]^3}{\tau_n} \tag{9}$$

$$D_{NN} = \frac{V_a * N_{sd}}{\tau_n}[\beta * V_a * S_{sd} + 1] \tag{10}$$

$$D_{\varphi\varphi} = \frac{R_{sp}}{4 * S} \tag{11}$$

Where N_{sd} and S_{sd} represent the steady-state, average values of the carrier and photon populations respectively. R_{sp} is the rate of spontaneous emission.

$$N_{sd} = \frac{1}{\Gamma * \tau_p * g_0} + N_0 \tag{12}$$

$$S_{sd} = \frac{\tau_p}{\tau_n} * N_{sd} * \left(\frac{I}{I_{bias}} - 1\right) \tag{13}$$

$$R_{SP} = 2 * \Gamma * \sigma_g * (N(t) - N_0) \tag{14}$$

Where σ_g is the gain cross section, The noise power spectral density $S_{RIN}(f)$ of the laser is of the form:

$$S_{RIN}(f) = \frac{\tau_p * f_r^4 \left\langle F_N^2 \right\rangle + \psi^2 * \frac{\left\langle F_S^2 \right\rangle}{4 * \pi^2} + \tau_p * f_r^4 * \psi * \frac{\left\langle F_S F_N \right\rangle}{\pi} + \frac{\left\langle F_S^2 \right\rangle}{4 * \pi} * f^2}{\left(f^2 - f_r^2\right) + \psi^2 * f^2} \tag{15}$$

$\langle F_S F_N \rangle$ is the cross correlation and is given by:

$$\left\langle F_S F_N \right\rangle = -\frac{\beta * V_a * N_{sd} * (V_a * S_{sd} + 1)}{\tau_n} + \frac{V_a * S_{sd}}{\tau_p} \tag{16}$$

$$\psi = \frac{1}{2 * \pi}\left(\Gamma * g_0 * S_{sd} + \frac{1}{\tau_n}\right) \tag{17}$$

Where the expression for the RIN with the noise terms included is as follow:

$$RIN = 10 * \log_{10} \frac{S_{RIN}(f)^2}{S_{sd}^2} \tag{18}$$

Where f_r is the relaxation oscillation frequency of the form:

$$f_r = \frac{1}{2\pi}\sqrt{K - \frac{1}{2}(\gamma_d)^2} \tag{19}$$

γ_d is the damping factor:

$$\gamma_d = \frac{1}{\tau_n} + [\frac{\Gamma * g_0}{q * V_a}(\tau_p + \frac{\varepsilon}{g_0})(I_{Bias} - I_{th})][1 - \frac{\Gamma}{q * V_a} * \varepsilon * \tau_p * (I_{Bias} - I_{th})] \tag{20}$$

The laser output power is calculated as follows:

$$P(t) = \frac{S(t) * V_a * \eta_0 * h * \nu}{2 * \Gamma * \tau_p} \tag{21}$$

After the above description of the laser rate equations including the Langevin noise terms and RIN, Relaxation oscillation frequency and the damping factor. We will go over the description of the laser small signal transfer function which shows under the effect of the feedback on the laser modulation bandwidth, and the changes to the relaxation oscillation frequency and the damping factor.

The laser amplitude modulation response is of the form:

$$H(j\omega) = \frac{K}{[(j\omega) * (j\omega + \gamma_d)] + K} \tag{22}$$

Where

$$K = [\frac{\Gamma * g_0}{q * V_a}(I_{Bias} - I_{th})][1 - \frac{\Gamma}{q * V_a} * \varepsilon * \tau_p(I_{Bias} - I_{th})] \tag{23}$$

For the feedback loop parameters, the amplifier transfer function A is of the form:

$$A = \frac{\rho}{1 + (j\omega / \omega_n)} \tag{24}$$

Where ω_n is the 3dB bandwidth of the amplifier circuit and ρ is the feedback gain.

The Feedback loop propagation delay transfer function B is of the form

$$B = \exp(-j\omega\tau) \tag{25}$$

Where τ is the propagation time delay of the feedback loop system.

Based on the well known control theory of systems with negative feedback (Lax, 1967), the complete transfer function on this complete laser system Y is of the form:

$$Y(j\omega) = \frac{H}{1 + (H * A * B)} \tag{26}$$

In the case of the feedback effect on laser RIN, the noise spectral density under feedback is of the form:

$$S_{RIN-FB}(f) = S_{RIN}(f) * \left| \frac{1}{1 + H * A * B} \right|^2 \quad (27)$$

Based on the parameters listed in table(1) below, the laser threshold current is calculated at 9.4mA and the Bias current range is up to a maximum of 50mA for a well behaved LI curve. The slope efficiency was calculated at 0.04mW/mA.

Symbol	Value	Dimension	Description
I(t)	-	A	Laser current
S(t)	-	m^{-3}	Photon density
Γ	0.44	-	Optical confinement factor
g_0	$3x10^{-6}$	cm^{-3}/s	Gain slope
N(t)	-	m^{-3}	Carrier density
N_0	$1.2x10^{18}$	cm^{-3}	Carrier density at transparency
ε	$3.4x10^{-17}$	cm^3	Gain saturation parameter
τ_p	$1.0x10^{-12}$	s	Photon lifetime
β	$4.0x10^{-4}$	-	Spontaneous emission factor
τ_n	$3.0x10^{-9}$	s	Carrier lifetime
Va	$9.0x10^{-11}$	cm^3	Volume of the active region
Φ	-	-	Phase of the electric field from the laser
α	3.1	-	Line-width enhancement factor
P(t)	-	W	Optical power from laser
Q	$1.602x10^{-19}$	C	Electronic charge
η	0.1	-	Total quantum efficiency
h	$6.624x10^{-34}$	J.s	Plank's constant
ω_n	$75.4x10^9$	Rad/s	3dB Bandwidth of amplifier Circuit
σ_g	$2x10^{-20}$	m^2	Gain cross section

Table 1. Laser parameters used in simulations.

4.1 Effects of feedback loop on laser internal parameters

In addition to the analysis that was done in this work in the frequency domain using the feedback control theory and the effect on the total system transfer function in the previous section. We describe in this section the effect of the feedback loop on the rate equations and explain the reasons behind the results we are getting in terms of the changes of the relaxation oscillation frequency and the damping factor. We also explain the effect of this scheme on the laser line-width reduction.

By examining the rate equations presented in the previous section. Looking at equation (5) where we introduced the feedback loop effect. This feedback term in equation (5) actually reduced the carrier density in the active layer which also reduces the laser current/power slope efficiency and increases the laser threshold current. As shown in figure (4) below. We plotted the carrier density versus the feedback loop gain and as shown, as the feedback loop gain increases, the carrier density decreases.

In addition to *N(t)* decreasing with the introduction of the feedback loop. *No* has increased with the introduction of the feedback loop gain of -0.05 from 1.2e18 cm^-3 to 1.682e18 cm^-3 which translates to threshold current increase from 9.4mA in the free-running laser to 12.5mA with -0.05 loop gain.

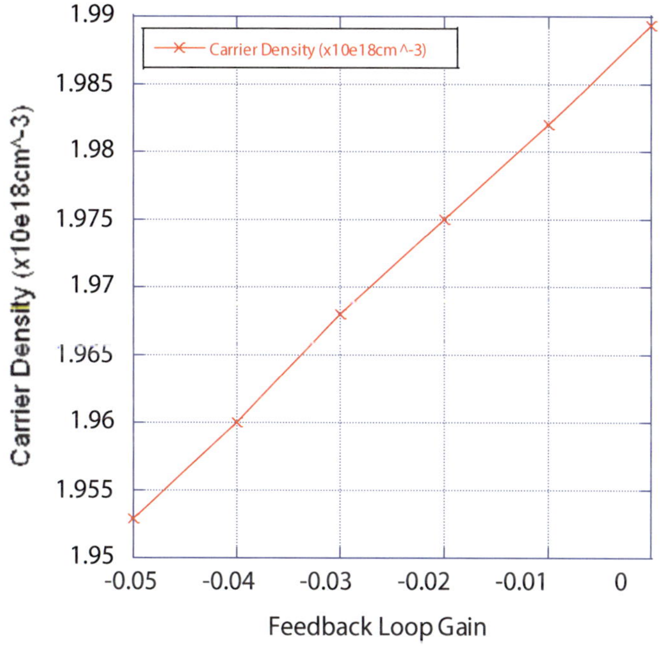

Fig. 4. Carrier density changes with the various feedback loop gain settings where the laser drive current is 50mA and the feedback loop delay is 20ps.

Where the threshold current is

$$I_{th} = \frac{q * V_a}{\tau_n}\left(N_0 + \frac{1}{\Gamma * g_0 * \tau_p} \right) \tag{28}$$

Based on the changes in the carrier density, these changes affect the differential gain (a) in the laser as follow:

$$a = \frac{g_0}{\Gamma * (N(t) - N_0)} \tag{29}$$

As the carrier density decreases with the introduction of the feedback loop, the differential gain increases. This in-turn affects the relaxation oscillation frequency and the damping factor as follow:

$$f_r = 2 * \pi * \left(\frac{\Gamma * v_g * a}{q * V_a} * \eta \; * \left(I_{Bias} - I_{th} \right) \right)^{1/2} \tag{30}$$

So as the carrier density decreases, the differential gain increases and that in-turn increases the relaxation oscillation frequency.

Also affected is the phase change of the rate equation below:

$$\frac{d\phi(t)}{dt} = \frac{1}{2}\alpha[\Gamma * g_0[N(t) - N_0] - \frac{1}{\tau_p}] + F_\phi(t) \tag{31}$$

Where the decrease in the carrier density also decreases the phase fluctuation which in-turn reduces the phase noise and the laser line-width (Agrawal, 1986).

4.2 Simulation results for the analog signal modulation laser system

We first look at the laser modulation response where the relaxation oscillation and the damping factor in the stable regime with very short loop delay will increase when negative feedback is applied due to the reduction of the carrier density. In the case of positive feedback, the ROF and the damping factor actually decrease. The ROF for the system is calculated based on (30) for various input current levels for the free running laser shown in figure (5) which also includes the feedback loop results for stable mode, this figure also shows the enhancement of the relaxation oscillation frequency with the stable feedback parameters that include a very short loopback delay. To maintain stability of the system for analog modulation application, the feedback gain was found to be 0.05 for a maximum feedback loop delay of 20ps.

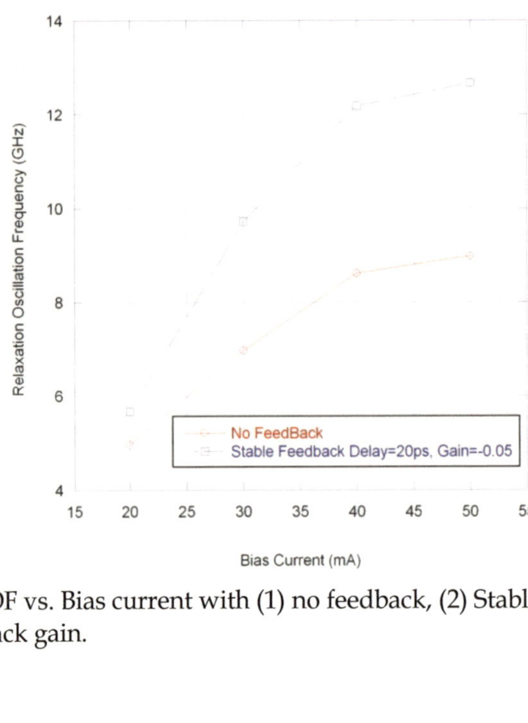

Fig. 5. Changes in ROF vs. Bias current with (1) no feedback, (2) Stable feedback with short delay and low feedback gain.

Figures (6) and (7) show the magnitude and phase transfer function plots of the system in free-running laser ($H(j\omega)$) and of the stable feedback ($Y(j\omega)$) modes respectively and they illustrate how much enhancement of the damping factor and modulation bandwidth of the laser can be attained by using negative electronic feedback loop in stable regime where there is close to 50% increase of the modulation bandwidth compared with the free-running laser condition.

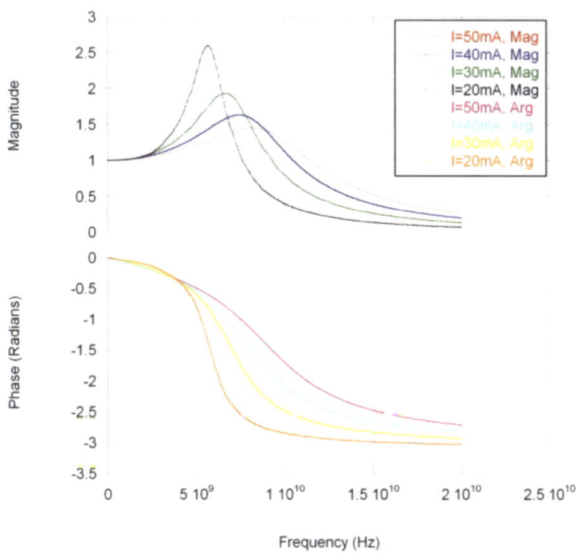

Fig. 6. Magnitude and Phase plots of the free-running laser transfer function for various current values.

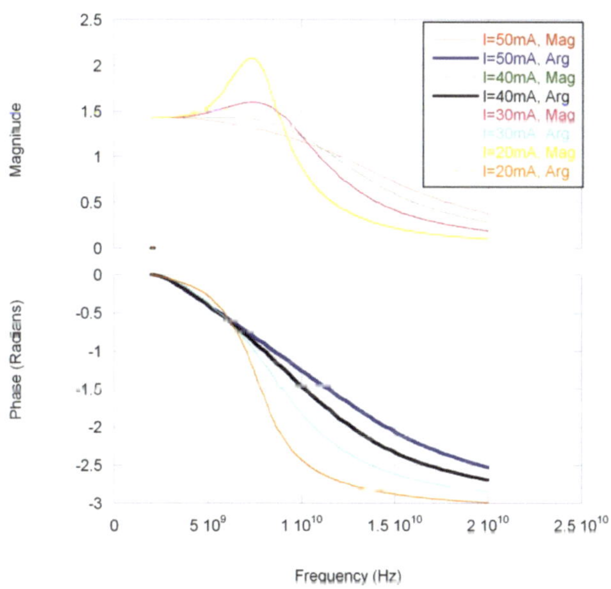

Fig. 7. Magnitude and Phase plots of the laser transfer function with feedback loop in stable regime for various current values (FB loop gain=-0.05, FB loop delay=20ps).

What was achieved with these results in terms of the shape of the laser modulation response does actually match the recent results by (Feng, 2009) where they fabricated a resonance-free frequency response laser with a fast spontaneous recombination lifetime around 29ps. This complex design was based on transistor laser design where fast recombination lifetime is obtained by a reverse-bias collector field pinning and tilting a dynamic charge population in a thin base, allowing only fast recombination. The tilted-charge laser that was demonstrated with a modulation data-rate of 10 Gb/s which was twice its 3dB bandwidth of 5.6 GHz. More details on this structure complexity in order to achieve similar results to our own results are presented in (Feng, 2009).

There are two fundamental aspects to the system we're proposing for analog signal transmission, the first fundamental goal is to reduce the laser RIN, the second goal is to modify the small signal transfer function of the system as it will become over damped and thus reducing the relaxation oscillation frequency peak which will increase the modulation bandwidth of the system. The flattening of the small-signal modulation response is achieved by introducing the negative feedback loop and maintain the operation of that loop in the stable regime where this stability condition can only be achieved by reducing the feedback loop propagation delay to less than 50ps, for the set feedback loop gain that will remain fixed through-out the operation of that system. For 50mA operating current for the laser, the stability conditions for the feedback loop will be maintained for feedback gain of 0.05 and feedback loop delay of less than 50ps. At 50ps and higher loopback delays, the system starts to self pulsate by operating as an opto-electronic oscillator as we will describe in details in section 5. Figure (8) shows the system magnitude and phase small-signal transfer function for 50mA operating current and various feedback loop delays from 10ps to 100ps where it is evident that the longer the loop delay, the less flat the response becomes and the more peaks are generated in that response.

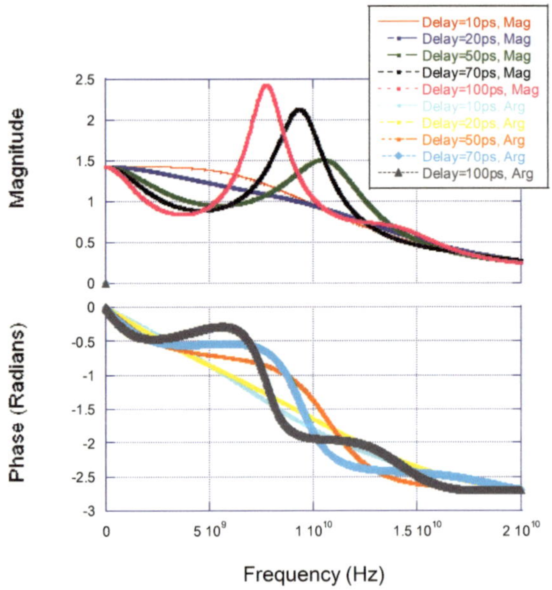

Fig. 8. Magnitude and Phase plots of the laser transfer function with feedback loop for various feedback loop delays at a fixed current value (FB loop gain=-0.05, laser current=50mA).

Upon further examination of our simulation results, and looking at feedback loop delays between 0ps and 10ps with a finer loop delay increments, the actual system response is shown in figure (9), where reducing the feedback loop delay does not produce a flatter response with the reduced delay, it actually shows that 10ps delay produces a flatter response than all lower loop delays. This behavior is affected by the feedback loop bandwidth value which was set at 12GHz.

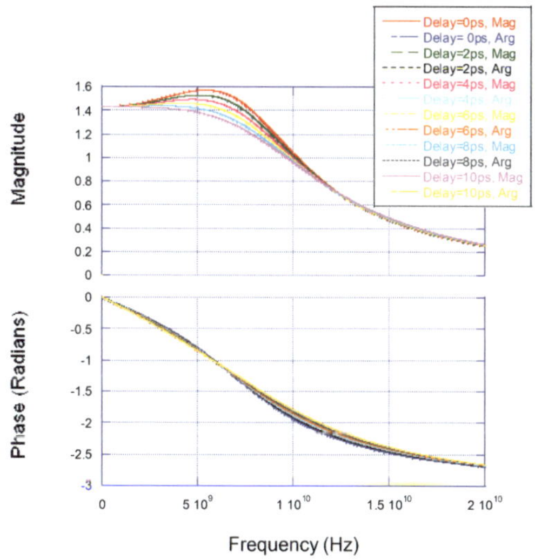

Fig. 9. Magnitude and Phase plots of the laser transfer function with feedback loop for various feedback loop delays up to 10ps, with FB loop bandwidth of 12GHz at a fixed current value (FB loop gain=-0.05, laser current=50mA).

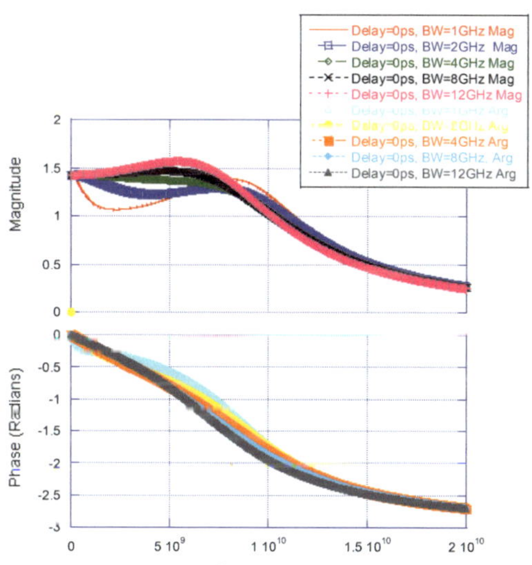

Fig. 10. Magnitude and Phase plots of the laser transfer function with feedback loop for feedback loop delay of 0ps, with FB loop bandwidths of 1,2,4,8 and 12GHz at a fixed current value (FB loop gain=-0.05, laser current=50mA).

Reducing the feedback loop bandwidth can reduce the peaking at lower loop delay values as shown in figure (10) where it shows the performance difference at 0ps feedback loop delay between 1,2,4,8 and 12 GHz feedback loop bandwidth settings. The peaking disappear with the lower bandwidth feedback loop, but the consequences of using a lower bandwidth feedback loop than 12GHz actually reduces the stability range that this system operates at on the high end of the loop delay setting, where with the lower bandwidth loop, the self-pulsation actually occurs with loop delay that is far less than the 50ps loop delay we obtain with the 12GHz feedback loop bandwidth, the stability region will be reduced with the lower feedback loop bandwidths to less than 30ps maximum loop delay for system stability and that is not practical for future experimental implementation. Note that at 1GHz bandwidth even with 0ps loop delay, it is shown that self-pulsation can actually occur.

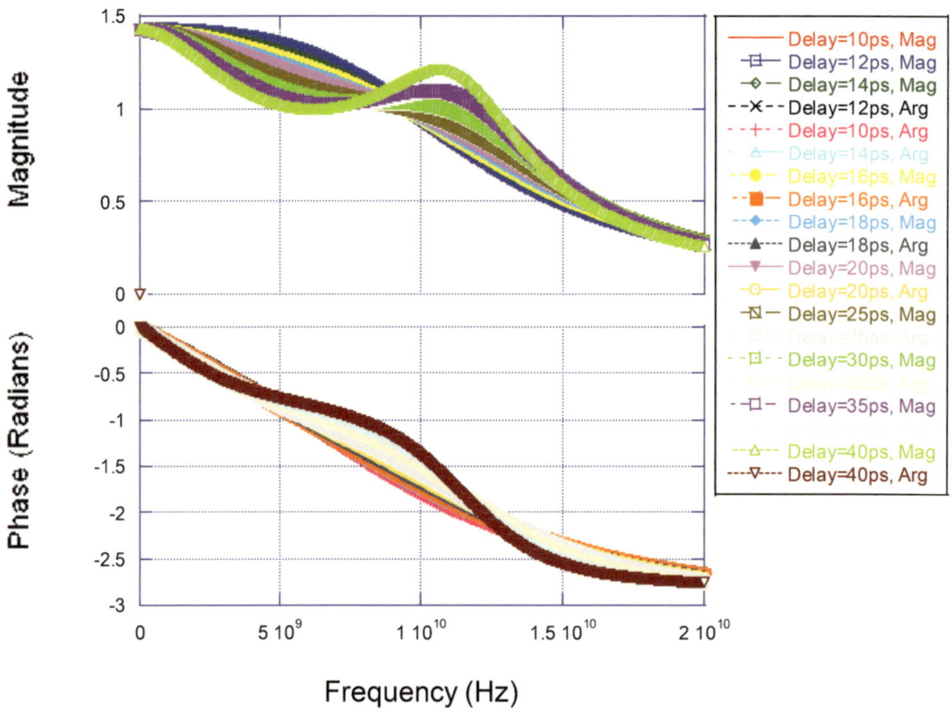

Fig. 11. Magnitude and Phase plots of the laser transfer function with feedback loop for feedback loop delays of 10ps to 100ps, with FB loop bandwidth 12GHz at a fixed current value (FB loop gain=-0.05, laser current=50mA).

Simulation results for laser RIN performance are presented as follow:

First, figure (12) show the laser free-running RIN performance for various drive current settings.

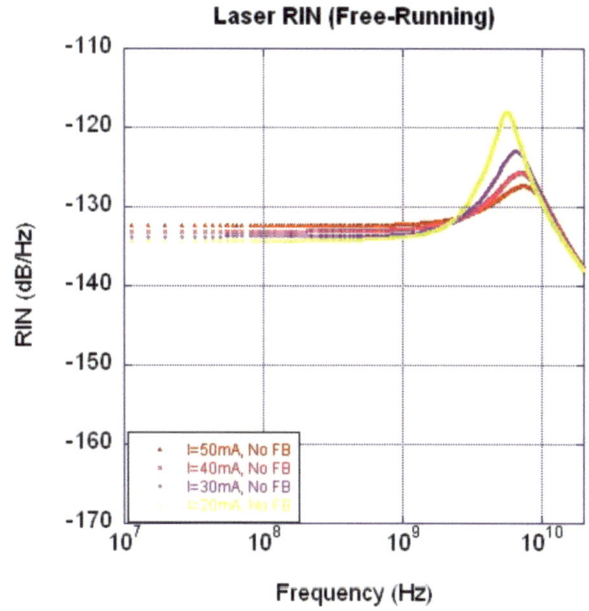

Fig. 12. Laser RIN for various drive current levels (Free-running condition).

When Feedback is applied, the RIN curve is flat and its level drops closer to the ROF in figure (12) which shows the effectiveness of the feedback loop scheme.

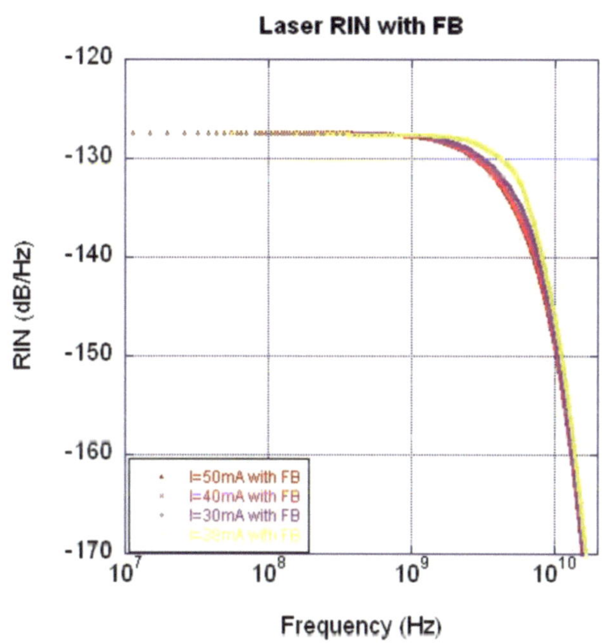

Fig. 13. Laser RIN for various drive current levels with the feedback loop applied (Gain=-0.05 and delay=20ps).

The improvement of RIN performance at 50 mA drive current level is 16dB around the ROF level as shown in figure (14).

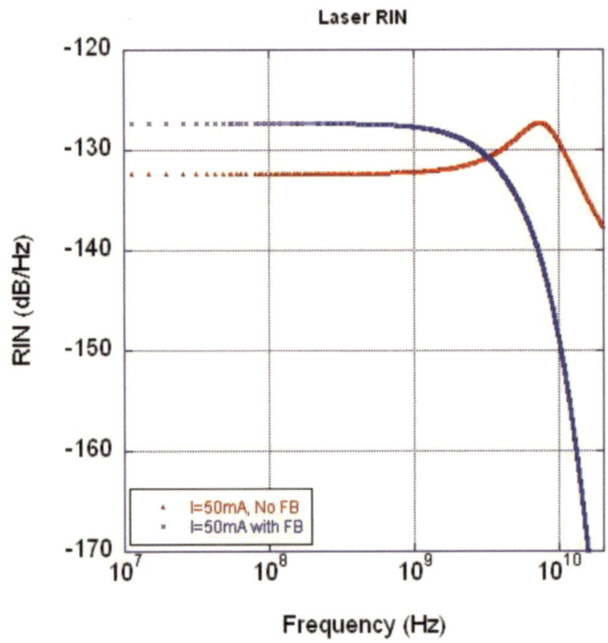

Fig. 14. Laser RIN at 50mA with and without Feedback applied (Feedback conditions, Gain=-0.05 and delay=20ps).

All the data shown so far have been generated with the laser drive condition based on fixed DC current levels. In the following, we will look at the laser performance when the laser in addition to the DC current drive, will also be modulated with sinusoidal RF signals and what that entails in terms of inter-modulation distortion and CNR.

$$CNR = 10 * \log\left(\frac{m^2}{2 * B * RIN}\right) \tag{32}$$

Where m is the modulation depth and B is the noise bandwidth.

So based on the RIN reduction results obtained above. For one dB decrease in RIN, a one dB increase in CNR, by getting 16dB decrease in RIN, we can get a 16dB CNR improvement with feedback loop applied.

Finally figure (15) shows the effect of Feedback on eliminating the turn-on ROF in the laser output power in time.

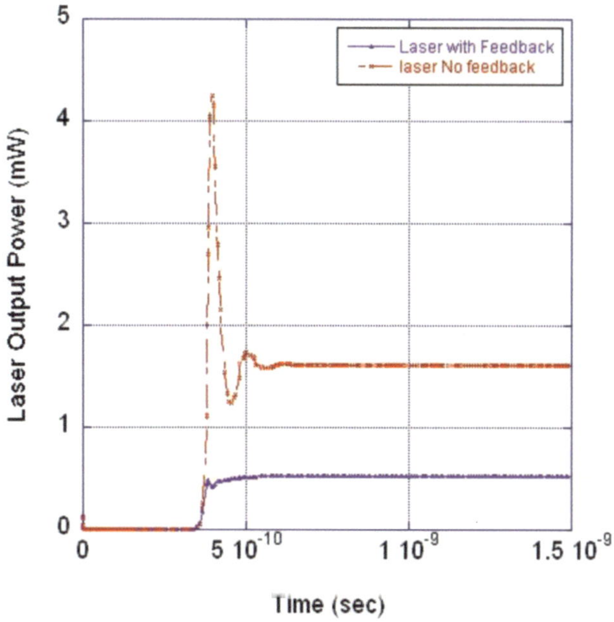

Fig. 15. Laser output power during turn-on where the effect of feedback is shown to eliminate the relaxation oscillation frequency during the power ramp-up.

5. Opto-electronic oscillator for photonic analog to digital conversion

Recent advances in coherent optical communication systems beyond 100Gb/s data rate per channel have increased the bandwidth and speed requirements for the electronic analog to digital conversion (ADC) section of the coherent receiver where the electronic ADC advances are lagging far behind the optical communication data rate growth. A detailed study of the performance of electronic analog to digital converters is presented in (Nazarathy, 1989; Walden, 2008). This review is useful to evaluate the performance requirement for photonic ADCs to achieve significant enhancement over electronic ADC performance. Typically, optics must achieve at least 10 times improvement in bandwidth improvement and/or noise reduction compared to electronics in order for it to be a viable new approach. Based on this, a 1GHz photonic ADC would have to achieve (Effective Number of Bits) ENOB >11, while at 20GHz, ENOB= 4 would be sufficient since this exceeds the comparator ambiguity limit for semiconductor circuits with transition frequency of 150GHz by at least a factor of two (Walden, 2008).

In the area of photonic analog to digital conversion, there has been recent work to address the speed of conversion (Valley, 2007). Most of the work so far has made use of Mode locked lasers (Fiber lasers or quantum dash based lasers) as the optical carrier source for the quantization circuit of the photonic ADC. This has a drawback of lacking tunability control of the pulse interval, cost and laser structure complexity (Bandelow, 2006). Previous work on identifying key photonic source specifications for the photonic ADC is found in (Clark, 1999).

We propose the use of the cost effective and widely available commercial semiconductor DFB laser optimized as a directly modulated laser for 2.5Gb/s data rate by applying electronic feedback to the laser system to make it suitable as an alternative solution to the mode-locked laser and will meet the performance criteria outlined above for the photonic ADC application. Typical directly modulated laser can be modulated with signal frequency up to its relaxation oscillation frequency (ROF), beyond the ROF, the modulation response decays rapidly (Agrawal, 1986). The laser output when modulated beyond the ROF exhibits period doubling bifurcations with increasing modulation depths.

Our proposed work eliminates the need for external high frequency signal sources, and relies only on DC bias current to generate and tune fast optical pulses using a laser with electronic feedback. The feedback loop delay variation allows us to operate the laser in stable regime with short delay, which smoothes the frequency response of the laser and extends its modulation frequency response capability while increasing the feedback loop delay beyond the stable regime forces the laser to operate in self-pulsation mode.

5.1 Conditions for self-pulsation mode

In analyzing the various configurations of this system, and by applying the FB loop gain of -0.05 and increasing the loop delay to 50ps necessary to produce the self-pulsation state, the process is explained as the sharpening and extraction of the first spike of the ROF. The feedback sharpens the falling edge of the first spike and suppresses the subsequent spikes. Hence, lasers with stronger ROF generate shorter pulses. We show the system transfer function ($Y(j\omega)$) magnitude and phase plots in figure (16) and compare those to the free-running laser transfer function ($H(j\omega)$) magnitude and phase plots for various current and delay values. What we see is in the case where the feedback loop is applied an enhanced second peak in the magnitude transfer function plot of figure (16) which indicates the generation of sharp pulsation. The inverse of the frequency peak corresponds to the pulse interval in the time domain.

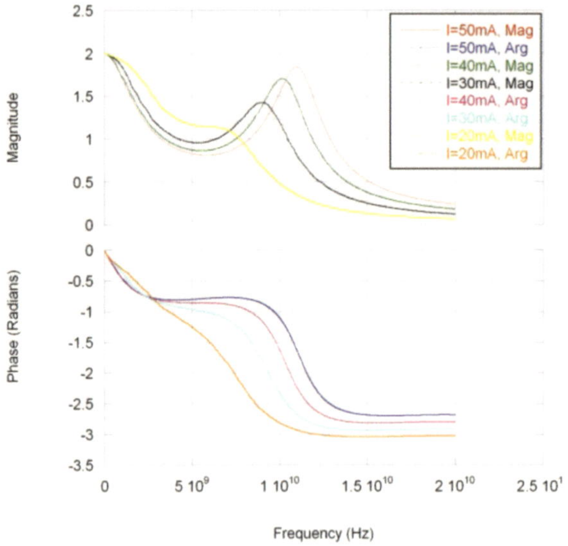

Fig. 16. Magnitude and Phase plots of the laser transfer function with feedback loop in Self-Pulsation regime for various current values (FB loop gain=-0.05, FB loop delay=50ps).

During this self-pulsation mode, figure (17) shows at 50mA bias current with feedback delay of 50ps and feedback gain of -0.05. The time domain plot of the output power of the system where the pulse interval is 147ps and the pulse width is 50ps.

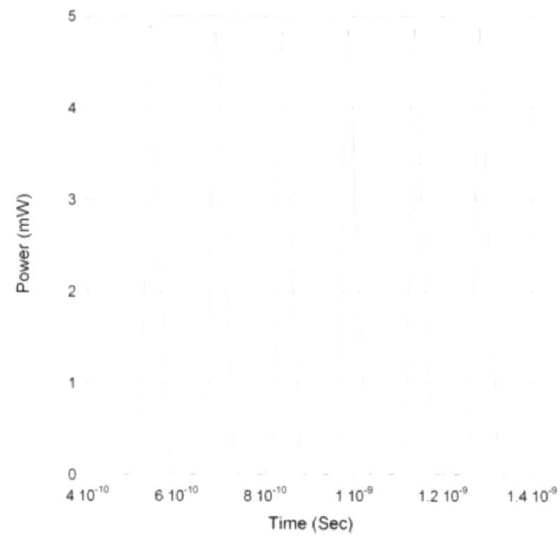

Fig. 17. Time domain plot for self-pulsation case (delay=50ps, Gain=-0.05) for 50mA bias current where the pulse interval is 147ps.

Figure (18) shows the pulse interval as a function of the bias current where the pulse interval can be fine tuned over a range > 50ps for the specified current range. The shortest pulse interval was achieved for this particular laser when setting the delay at 30ps and the gain at -0.05 with 50mA bias current was 80ps with pulse width of 30ps. These limitations on the pulse width are governed by mainly the laser carrier life-time among other physical parameters of the laser structure.

Fig. 18. Pulse interval adjustment as a function of bias current for 50ps delay and gain of -0.05.

5.2 Sampling source for photonic ADC

We propose the solution in figure (19) as an alternative to the mode-locked fiber lasers presently used in most photonic ADC applications, our proposed source has numerous advantages including lower cost, availability, tunability and most of all size and power dissipation advantage for photonic integration (Shahine, 2010). The disadvantage is that it has a larger pulse width compared to mode-locked lasers but this effect can be reduced by propagating these pulses through dispersion compensating element to match the mode-locked laser performance. This solution consists of one laser system (laser1) operating in the self-pulsating mode with feedback loop Delay=30ps and Gain= -0.05 at 50mA Bias current which would generate the lowest pulse interval at 80ps. The second laser system (laser2) is operating in the stable regime with feedback delay=15ps and Gain=-0.02 which allows the system bandwidth to increase so it would accommodate modulating the signal transferred from the first laser system feedback amplifier. Laser1 system operating in self-pulsation mode, generate the pulses which directly modulate laser2 from the non-inverting output of the feedback amplifier of laser1, modulating laser2 which is optimized with its feedback loop to accommodate fast pulse transition for direct modulation.

The SOA (semiconductor optical amplifiers) are used to match the output power levels from both systems so the combined signal would have the same amplitude when interleaving both signal. The Electrical delay adjustment is used to optimize the timing of the two interleaved signals in order to get the lowest pulse interval possible (avoiding pulse overlap).

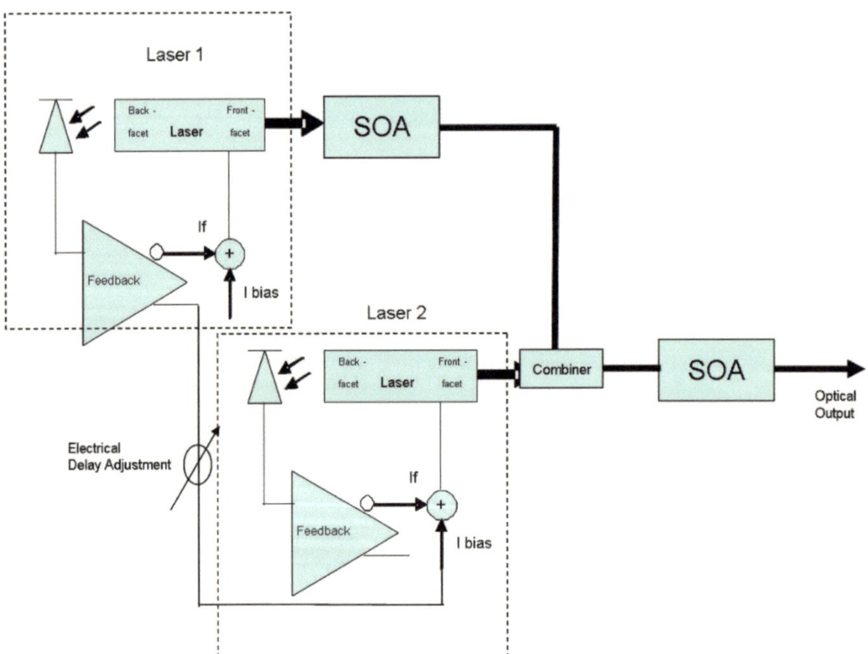

Fig. 19. Proposed Photonic ADC pulsed source as a sample and hold circuit where laser 1 is operating in the self-pulsed regime while laser 2 is operated in the stable regime and modulated by the signal from the feedback amplifier of laser 1.

Figure (20) shows the output of this system with a pulse interval of 40ps without the need of any external clock or signal sources. This solution sets the sampling rate at 25GHz which is above the 20GHz minimum requirement stated at the beginning of this section.

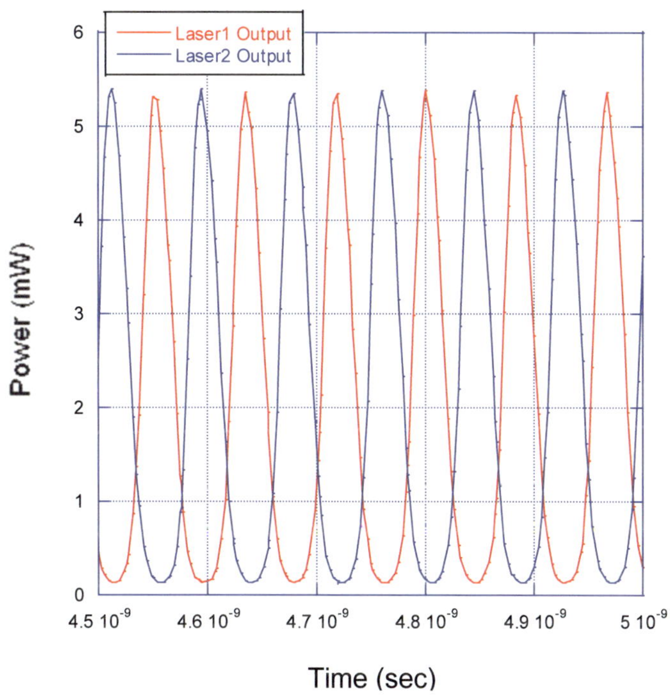

Fig. 20. Optical output of the system in figure (19) where the pulse interval is 40ps.

5.3 Sampling source noise effects analysis

In the previous sections, the effects of RIN and phase noise reduction using electronic feedback, were demonstrated by the reduction in the carrier density. Another type of noise effect analyzed is the system phase noise and the effect it has on the timing jitter performance for pulsed sources. The system phase noise $L(f)$ is produced from the effect of the laser line-width $\delta\nu$ and the power spectral density $S_\phi(f)$ of that line-width. The calculated line-width of this laser spectrum based on 50mA bias current is 2.4MHz (FWHM). The power spectral density of the laser spectrum is calculated based on (Nazarathy, 1989).

$$S_\phi(f) = \frac{A}{1 + \left(\frac{2 * f}{\delta\nu}\right)^2} \tag{33}$$

The system phase noise $L(f)$ shown in figure (21) is related to the line-width power spectral density as follow (Pozar, 2001):

$$L(f) = \frac{S_\phi(f)}{f^2} \tag{34}$$

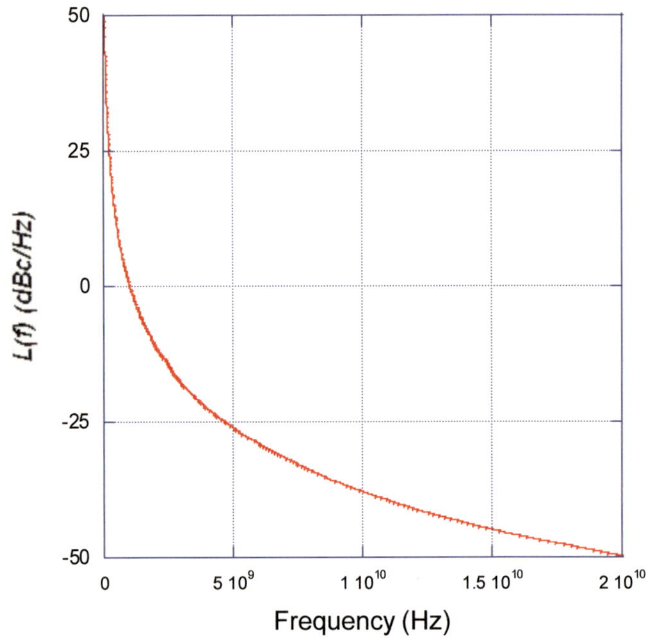

Fig. 21. Laser phase noise plot derived from the spectral density of the line-width.

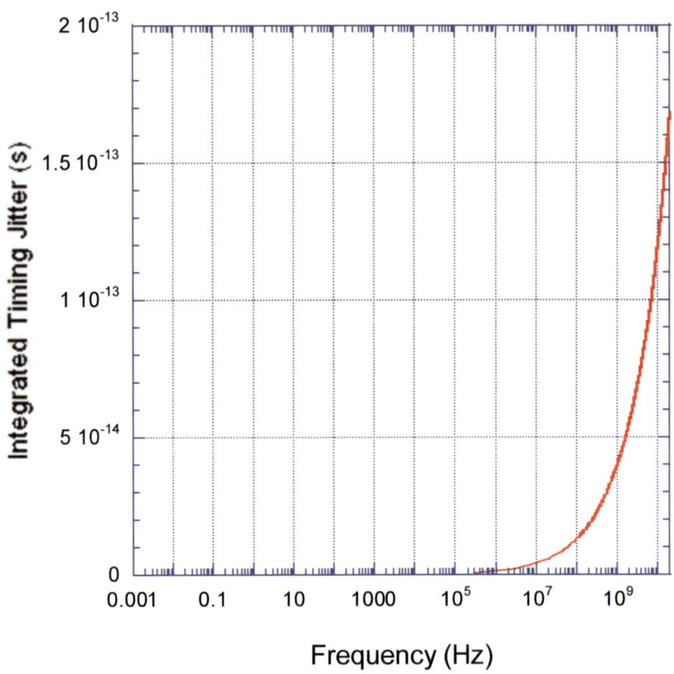

Fig. 22. rms timing jitter over the entire frequency range.

The integrated rms timing jitter σ_j which represents the upper bound of the timing jitter of the oscillator shown in figure (22) is calculated as follow (Lasri, 2003)

$$\sigma_j = \frac{PulseInterval}{2\pi}\sqrt{2 * \int_{f\,min}^{f\,max} L(f)df} \quad (35)$$

Where *f min* and *f max* are the boundary of the frequency range. For a pulsed source with a pulse interval of 80ps, the maximum tolerated rms jitter for sampling application is 120th the pulse interval according to (Jiang, 2005) which lists the requirements for such application leading to maximum tolerated rms jitter of 667fs while our calculated jitter shown in figure (22) is around 150fs which is well below the limits required for photonic ADC sampling application. In addition to the compactness of our proposed solution, this performance exceeds fiber laser performance where according to (Chen, 2007). The timing jitter for a fiber laser was 167fs for a pulse interval of 5ns with RIN of -120dB/Hz.

6. Laser line-width reduction with electronic feedback

As was described in the previous sections, when electronic feedback is introduced, the carrier density is reduced which results in phase noise reduction according to equation (31).

In order to further reduce the frequency noise in the low frequencies region (1/f noise) in the laser Frequency transfer function. A Fabry-Perot etalon is used in the feedback loop that will translate frequency changes to amplitude changes and corrects for the frequency fluctuations by modulating the feedback loop current driving the laser as shown in figure (23). Simultaneously, the frequency noise at high frequencies is also reduced when the feedback loop delay is reduced as will be described later to produce a flat laser FM response across the entire laser operating frequency range.

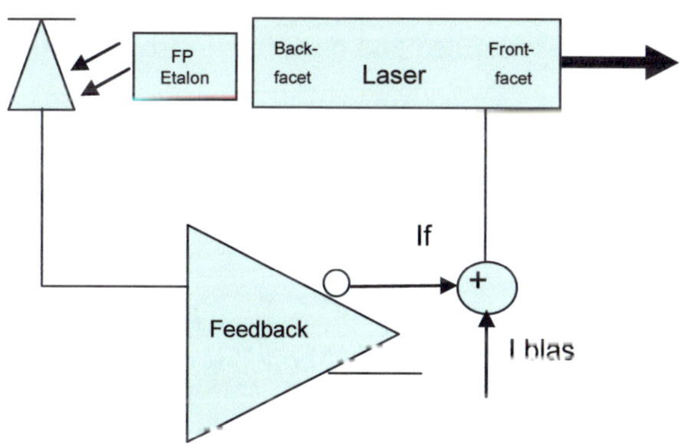

Fig. 23. System for laser line-width reduction.

This technique uses a wide-band direct frequency modulation for semiconductor laser through injection current by reducing the laser line-width, without increasing the laser cavity size (Shahine, 2009b). Furthermore, since the feedback applied to the laser is negative, a high temporal stability of the central frequency and a narrow line-width of its oscillation spectrum can be obtained simply by reducing the FM noise contributing to the line-width enhancement. By using a highly sensitive optical filter (FP Etalon) used as a frequency discriminator to translate the frequency fluctuation of the laser into a power fluctuation signal, and applying a corrective electrical signal back to the laser to tune its frequency so as to counter these optical frequency fluctuations (He, 2009; Maisel, 1975). This frequency locking system essentially transfers the frequency stability of the frequency discriminator to the laser. If the optical frequency discriminator has a low frequency noise in the locking bandwidth, the laser will inherit this low noise and will display a narrower line-width and longer coherence length. The injection current is controlled so that the laser frequency is locked at the frequency of the maximum negative slope of the transmittance curve of the FP etalon, the fluctuation of the power incident onto the detector is due to residual frequency fluctuations under the feedback condition.

The normalized residual FM noise can be defined as a ratio between the power spectral densities of the free-running laser $S_{\nu FR}(f)$and of the laser under electronic feedback $S_{\nu FB}(f)$. This is given by:

$$\frac{S_{\nu FB}(f)}{S_{\nu FR}(f)} = \left| \frac{1}{1+H(f)} \right|^2 \tag{36}$$

Where $H(f)$ represents the open loop transfer function of the feedback loop where:

$$H(f) = H_L * H_{FP} * A * B \tag{37}$$

H_L is the laser frequency modulation response.

H_{FP} is the FP Etalon transfer function.

A is the feedback loop amplifier transfer function.

B is the feedback loop delay transfer function.

A and B are the same parameters defined and used previously in section 4.

The laser frequency modulation transfer function is of the form (Tucker, 1985):

$$H_L(j\omega) = \frac{\eta * h * \nu}{2 * q} \left[\frac{1}{\left(\frac{j\omega * \tau_p}{g_0 * S_0} \right)^2 + \left(\frac{j\omega * \varepsilon}{g_0} \right) + 1} \right] \tag{38}$$

This modulation transfer function only includes the carrier effect and not the thermal effect.

The thermal contribution to the FM response of DFB lasers has been studied theoretically and experimentally in (Correc, 1994). The transfer function of the Fabry-Perot Interferometer when operating in the transmission mode is of the form (Ohtsu, 1988):

$$H_{FP}(f) = \frac{3\sqrt{3}}{4 * \Delta\nu_{FPI}} * \frac{1}{1 + (\frac{j\omega}{\omega_{FPI}})} \tag{39}$$

6.1 Simulation results for the laser line-width reduction application

The free-running laser magnitude and phase FM response for various input drive currents are shown in figures (24 and 25) respectively.

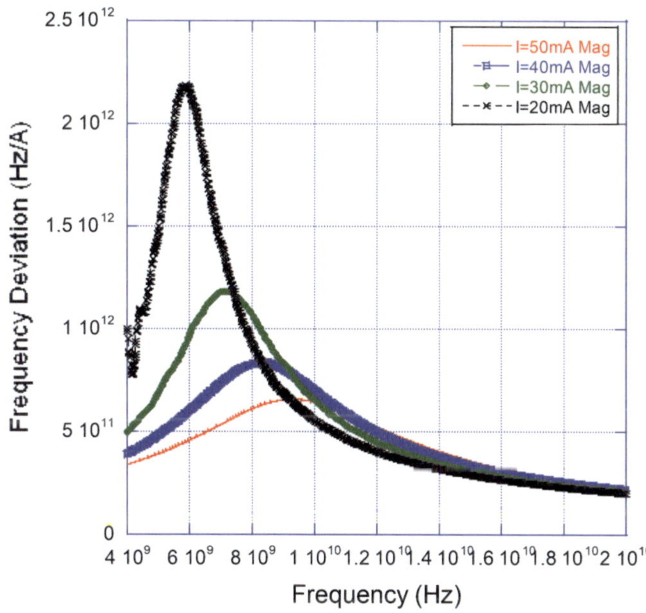

Fig. 24. Free-running laser FM response (magnitude).

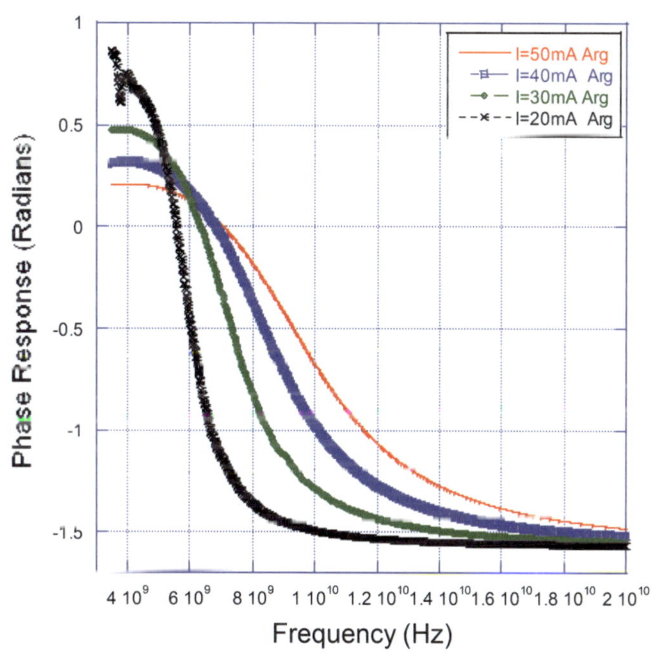

Fig. 25. Free-running laser FM response (phase).

By introducing the electronic feedback loop to the laser, the FM response for the system across the frequency range as shown in figures (26 and 27) for drive current value of 50mA and fixed loop gain of 5. The loop delay was varied from 10ps to 500ps which shows that the desired flat response is obtained with short delay of 10ps. This flat response is ideal for frequency modulation (Alexander, 1989; Gimlett, 1987; Iwashita, 1986).

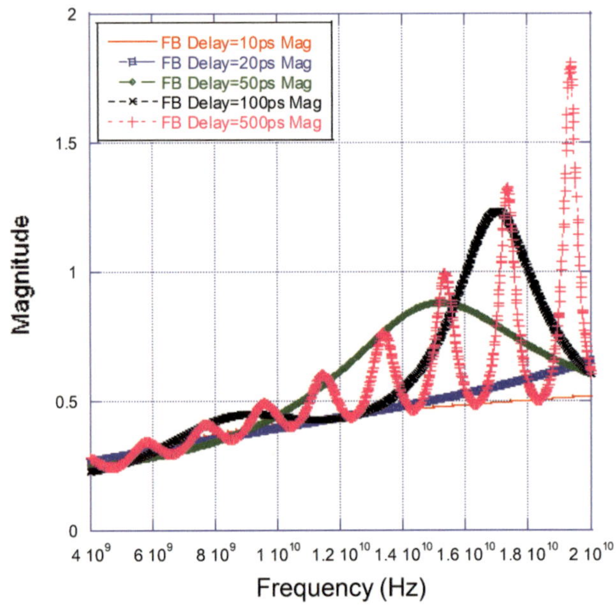

Fig. 26. Magnitude FM response with FB Gain=-5 and variable loop delay.

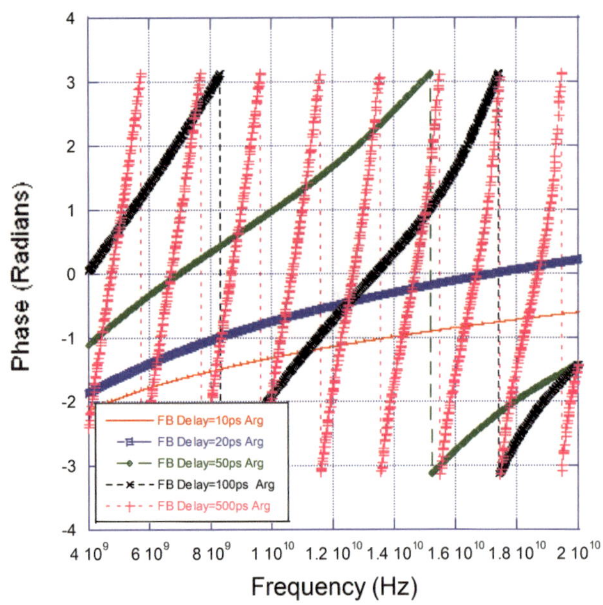

Fig. 27. Phase FM response with FB Gain=-5 and variable loop delay.

The power spectral density of the FM noise for the free-running laser can be approximated as follow (Nazarathy, 1989):

$$S_{\nu FR}(f) = \frac{A}{1+\left(\frac{2*f}{\delta\nu_{FR}}\right)^2} \tag{40}$$

Where the laser line-width, of the free-running laser with 50mA bias current is 2.4MHz.

A is a constant related to the Schawlow-Townes parameter.

By applying the feedback loop to reduce the laser line-width, with fixed loop delay of 10ps, we varied the loop gain from 5 to 1000 and shown that the PSD of the FM noise of the system with the feedback loop applied is shown in Figure (28) based on equation (41).

Where $S_{\nu FB}(f)$ is defined as:

$$S_{\nu FB}(f) = S_{\nu FR}(f) * \left|\frac{1}{1+H(f)}\right|^2 \tag{41}$$

The effectiveness of this scheme to reduce the laser line-width is calculated based on equation (41) which shows that a 1e5 reduction in PSD FM noise can be achieved with this scheme where the ratio of the FM noise power Spectral density with feedback to the free-running condition is plotted in figure (28).

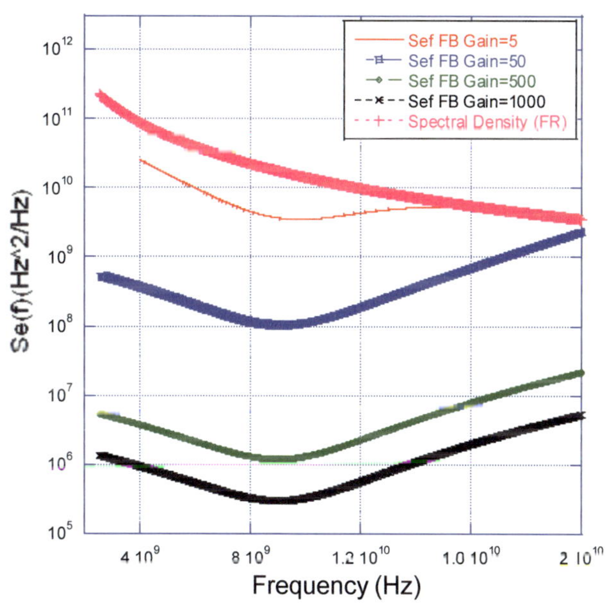

Fig. 28. PSD of FM noise for laser with FB at 50mA bias current with loop delay of 10ps and various feedback loop gain values, also shown the laser free-running PSD.

7. Conclusion

This chapter discussed the technique of electronic feedback to correct for laser impairments. By shortening the feedback loop delay and widening the feedback loop bandwidth, the laser characteristics fundamentally change and can result in far better device performance.

The simulation study that was done to optimize the design parameters of the feedback loop characteristics, in order to provide a clear roadmap for the implementation of photonic hybrid integration solutions that repair most known impairments in lasers for various applications.

This work provided detailed simulation results to improve the laser performance for analog optical signal transmission including linearization of the output power versus drive current and improve RIN performance by 15dB. It also improved the laser modulation bandwidth by 50% in increasing the relaxation oscillation frequency and increasing the damping rate.

The introduction of the FP etalon to the feedback loop as a frequency to amplitude signal translator provided a solution for reducing the laser line-width from 2.4 MHz to 24 Hz through simulation and the flattening of the laser FM response across the entire frequency range of the laser. These results showed the optimization of the laser for FM modulation and simultaneously reducing the 1/f noise at low frequencies and the carrier effect noise at high frequencies which was achieved by reducing the feedback loop delay.

We also analyzed the performance of laser with electronic feedback loop as an opto-electronic oscillator when the feedback loop was operating outside its stable regime. This work has provided a detailed analysis on how to command such oscillator by tuning the laser drive current. This analysis also covered the performance metrics of the opto-electronic oscillator including jitter performance and phase noise as compared to existing solutions.

8. Acknowledgment

The author would like to thank Dr. Yung Jui Chen for his guidance and for the stimulating discussions.

9. References

Agrawal ; Dutta. (1986). Long Wavelength Semiconductor lasers. *Book,Springer, first edition,* 1986

Alexander, S. et al. (1989). Passive Equalization of semiconductor diode laser frequency Modulation. *IEEE J. of Lightwave Technology,* Vol. 7, (1989) pp. 11-23

Black, H.S.,(1934) *Elec. Eng.,* Vol.53 , (1934) pp. 114

Chiu, Y., et al.(1999) *Photonic Technology letters,* Vol. 11 (1999), pp. 48

Correc, et. Al.,(1994), *IEEE journal of Quantum Electronics,* Vol. 30, No. 11, (1994), pp. 2485

Clark, T. R. et al.(1999), Performance of a time and wavelength-interleaved photonic sampler for analog-digital conversion. *IEEE Photonics Technology letters,* Vol. 11,(1999) , pp. 1168-1170

Chen, J. et al.(2007), High repetition rate, low jitter, low intensity noise Er-fiber Laser, *Optics letters,* Vol. 32, No. 11, (2007) pp. 1566-1568.

DARPA-BAA-11-45, (2011),Electronic-photonic heterogeneous integration (E-PHI), April 14, 2011

Feng, M. et al. (2009), Resonance-Free Frequency Response of a semiconductor laser, *Applied Physics Letters*, Vol. 95, (2009) pp. 033509.

Gimlett, J. L. et al., (1987). A 2Gb/s Optical FSK Heterodyne transmission experiment using a 1520nm DFB laser. *IEEE Journal of Lightwave Technology*. Vol. 5, (1987) pp. 1315-1324

He, et. al. (2009) *Analog Integrated Circuit Signal Processing Journal* (2009) Vol. 59, (2009), pp. 215-221

Helms, J. , et al.(2004), *IEEE Journal of Lightwave Technology*, Vol. 9 (1991), pp. 1567

Ismail, T. , et. al.(2004) *Proc of Optical Fiber Communications Conference*, (OFC 2004)- FE3

Iwashita, et al. (1986). 400Mb/s optical FSK transmission experiment over 270Km of single mode fiber. *IEEE Electronics Letters*, Vol. 22, (1986) pp. 164-165

Jiang, L. et al.(2005). Semiconductor mode-locked lasers as pulse sources for high bit rate data transmission. *Journal of Optical Fiber Commun*. Rep. 2, (2005). pp. 1-31

Lasri, J. et al. (2003) Self-starting optoelectronic oscillator for generating ultra-low-jitter high-rate optical pulses. *Optics Express*, Vol. 11, No. 12, (2003) pp. 1430-1435.

Lax, (1967). *Phys. Rev.* Vol. 160, (1967), pp. 290

Luenberger, D.,(1979). Introduction to Dynamic Systems. *Wiley Books*,(1979), Chapter 5

Maisel,(1975), *IEEE transactions on Industrial Electronics and Control Instr.*, Vol. 22, No. 2, (1975) pp. 122

Nazarathy, M. et al. (1989). Spectral analysis of optical mixing measurements. *IEEE J. Lightwave Technology*, Vol. 7, No. 7, (1989) pp. 1083-1096.

Ohtsu,(1996), Frequency control of SCL , *Book, Wiley*, (1996)

Ohtsu, (1988), *IEEE Journal of Lightwave technology*, Vol. 6, (1988), pp. 357

Ohtsubo J.(2007). Semiconductor lasers, Stability, Instability and Chaos. *second Edition, Springer Books*, (2007)

Peterman, K. (1991). Laser Diode Modulation and Noise. *KluwerAcademic Publishers* (1991)

Pozar, D. M.,(2001). Microwave and RF Design of Wireless Systems., *Book, John Wiley & Sons, Inc.*, (2001).

Ralph, T. C. et. al.(1999). *Optical and Quantum Electronics Journal*, Vol. 31,(1999), pp. 583

Shahine, M. H.,(2009a). Systems and methods for real-time compensation for non-linearity in optical sources for analog signal transmission", *United States Patent 7,505,496*, (Mar. 17, 2009)

Shahine, M. H.,(2009b). Semiconductor laser utilizing real-time line-width reduction method. *United States Patent 7,620,081*, (Nov. 17, 2009)

Shahine, M. H.; Chen, Yung Jui. (2010). Analysis for commanding the self-pulsation of DFB Laser diode with electronic feedback for photonic analog to digital conversion application. *IEEE Photonics Journal*, Vol. 2, No. 6, (2010).pp. 1013-1026

Stephens,W.E, et al. (1982) *IEEE NTC* ,(1982), pp. F4.3-1

Stubkjair, K. ,et al. (1983). IEEE Journal of Quantum Electronics, Vol.16, (1983), pp. 531

Tomkos, et al.(2001). Extraction of laser rate equations parameters for metro networks.*Optics Communications*, Vol. 194, (2001), pp. 109-129

Tucker,(1985). *IEEE transactions on Electron Devices*, Vol. 32, No.12, (1985), pp. 2572

Valley, G.C.,(2007). Photonic analog to digital converters. *Optics Express*, Vol. 15, No. 5, (2007), pp. 1955-1982

Walden, R. H.,(1999). Analog to digital converter survey and analysis. *IEEE Journal of Selected areas in Communications*, Vol.17, (1999), pp. 539-550
Walden, R.H., (2008). Analog-to-digital conversion in the early 21st century. *Book, Wiley Encyclopedia of CS and Engineering* (2008).
Wan, et al. (2005). *IEEE custom integrated circuits conference*, (2005), pp. 25-5-1

7

Ultra-Wideband Multiwavelength Light Source Utilizing Rare Earth Doped Femtosecond Fiber Oscillator

Nurul Shahrizan Shahabuddin[1],
Marinah Othman[2] and Sulaiman Wadi Harun[1]
[1]*Photonics Research Centre, University of Malaya,*
[2]*Multimedia University,*
Malaysia

1. Introduction

Multiwavelength sources are expected to play a major role in future photonic networks, where optical time-division multiplexing (OTDM) and wavelength division multiplexing (WDM) are employed (Teh et al., 2002; Ye et al., 2010; Liu et al., 2006; Parvizi et al., 2011; Harun et al., 2010; Zhong et al., 2010). Various approaches have been taken to develop this source such as by exploiting the nonlinear effects in an optical fiber as well as spectral slicing of a supercontinuum source.

Stimulated Brillouin scattering (SBS) and four wave mixing (FWM) effects are normally used to realize a multiwavelength output whereby the frequency shifts are determined by both the optical fiber structure and pump signal (Chen et al., 2010; Shahi et al., 2009a,2009b; Shahabuddin et al., 2008). A multiwavelength Brillouin fiber laser exhibits wavelength spacing of approximately 0.08 nm depending on the type of material used while FWM frequency shift is known to be dependent on the spacing of the pump signals (Johari et al., 2009; Shahi et al., 2009b; Ahmad et al., 2008). The bandwidth of operation however, is relatively narrow due to the limitation imposed by the erbium doped fiber. Previously, multiwavelength sources also have been demonstrated using superstructure Bragg grating and Fabry perot filter, both have limited range of multiwavelength region and spacing (Teh et al., 2002).

A technique known as spectral slicing can be realized using an arrayed waveguide grating or sagnac loop mirror as the wavelength selective component to slice a broad emission spectrum of amplified spontaneous emission or supercontinuum. It has been shown in many earlier works that slicing of a broadband continuum spectra is capable of generating a multiwavelength laser comb with both spacing tunability and a very wide spectral range (Nan et al., 2004). The supercontinuum light can be produced from a pulsed laser by using the interaction of multiple nonlinear effects such as self-phase modulation (SPM), four-wave mixing (FWM) and stimulated Raman scattering (SRS) in a highly nonlinear optical fiber (Buczynski et al., 2009, 2010; Kurkov et al., 2011; Genty et al., 2004; Lehtonen et al., 2003; Chen et al., 2011; Gu et al., 2010; Akozbek et al., 2006). Spectral slicing allows the channel spacing to be adjusted by changing the length of polarization maintaining fiber (PMF) used in the loop mirror (Ahmad et al., 2009). Hence, spectral slicing provides better spacing tunability as compared to other multiwavelength techniques such as stimulated brillouin scattering as the spacing of the latter highly depends on the waveguide material and structure i.e. the spacing is fixed.

In this chapter, we report a new multiwavelength source using supercontinuum slicing technique. This source use rare earth doped femtosecond fiber oscillator to generate supercontinuum, with multiwavelength operation achieved via spectral slicing of the supercontinuum. Through this technique, the multiwavelength region obtained is the broadest ever to date.

2. Development of mode-locked fiber laser

Mode-locked lasers have found widespread uses in many areas of research, medicine, and industry. The potential of making compact, rugged laser systems with low power consumption at a relatively low cost makes such fiber laser systems very promising candidates for those applications. The stability of the laser is a crucial factor for its applicability. In negative dispersion regime, a soliton pulse is maintained by the interaction of dispersion and nonlinear effect. The bandwidths of both gain fiber and mode-locked filter are neglected (Wise et al., 2008). However, in a normal dispersion fiber laser, the dissipative soliton is maintained by the effects of dispersion, nonlinearity, gain, and spectral filtering (Zhao et al., 2007; Chong et al., 2007). Accordingly, the gain-bandwidth and mode-locked filter play important roles for pulse reshaping and stability (Chong et al., 2007). A stable mode-locked fiber laser can be achieved with the use of polarization maintaining (PM) fibers and a semiconductor saturable-absorber mirror (SESAM) as the mode-locking mechanism. A setup requiring no free-space optics is highly attractive, as it increases stability, and reduces both costs, and the need for maintenance. The fiber mode-locked lasers reported to date obtains mode-locking based on the use of either SESAM or nonlinear polarization rotation (NPR) mechanism (Huan et al., 2006; Harun et al., 2011).

A saturable absorber absorbs the incoming light linearly up to a certain threshold intensity, after which it will saturate and becomes transparent. The recovery time of a semiconductor saturable absorber limits the laser repetition rate. Carbon nanotube (CNT) based saturable absorbers exhibit sub-picosecond recovery times, broadband operation, compatibility with fibers, a small footprint and is simple to fabricate besides being operable in either a transmission or a reflection mode, making them preferable over the more established SESAMs for commercial applications.

Various techniques to achieve stabilized laser sources such as by suppressing the supermode noise, starting the pulse in a laser, and compensating the small distortion caused by the gain fiber, have been reported (Li et al., 1998; Y. Li et al., 2001). In this section, an environmentally stable mode-locked fiber laser based on both NPR and a single wall carbon nanotube saturable absorber (SWCNT-SA) with zero use of free-space optics is demonstrated. The soliton pulse in the cavity can be reshaped and maintained by combining both these mode-locking mechanisms NPR and SWCNT-SA that are based on power-dependent filter. The configuration of our proposed fiber laser is shown in Figure 1. A 10 m long erbium-doped fiber (EDF), which is pumped by a 980-nm laser diode through a 980/1550 nm wavelength division multiplexer (WDM) is used as the gain medium. The EDF has a mode field diameter of 10.4 μm with maximum peak absorption of approximately -10 dB/m at 1550 nm. The group velocity dispersion (GVD) is estimated to be around -42 ps/nm.km at 1550 nm. The other part of the ring cavity uses a 20 m long standard single mode fiber (SMF-28) with a dispersion of 17 ps/nm.km at λ=1545 nm. The net cavity GVD is negative, which enables soliton shaping in the laser. A standard isolator is used to ensure unidirectional operation and to act as a polarizer. A squeezed fiber type polarisation controller (PC) is used to control the polarization states within the cavity. In this experiment, the SWCNT-SA film sandwiched between two FC/PC fiber ferrules to form the fiber-integrated saturable absorber is used for the initiation and stabilization of mode-locking at around 1561 nm region. The SWCNT-SA is fabricated by chemical vapor deposition (CVD) method with an average diameter of 0.8~0.9 nm. The output optical spectrum of the EDFL is monitored by an optical spectrum analyzer (OSA), while the output pulse train and pulse duration are measured using an oscilloscope and an autocorrelator, respectively.

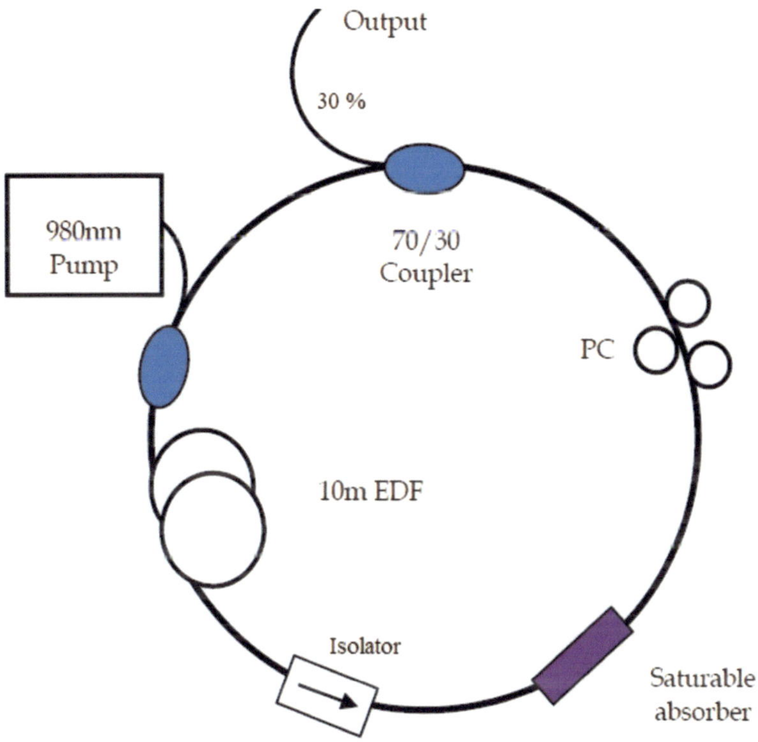

Fig. 1. Experimental setup of the mode-locked fiber laser.

To achieve mode-locking operation, the polarization state of the light within the cavity should be adjusted by using the PC. The pump power threshold for continuous wave laser operation is around 11 mW and a stable self-starting fundamental mode-locked (ML) operation commences at around 140 mW pump power. By altering the polarization state inside the cavity, the optical spectrum can be broadened to initiate Q-switching from which mode-locking operation is obtained. Figure 2 shows the pulse train observed at the output coupler after making both a careful adjustment to the imposed loss accompanied by the variation of the net birefringence of the cavity using a PC. As shown in the figure, a stable fundamental soliton pulse is obtained with a repetition rate of 142 MHz.

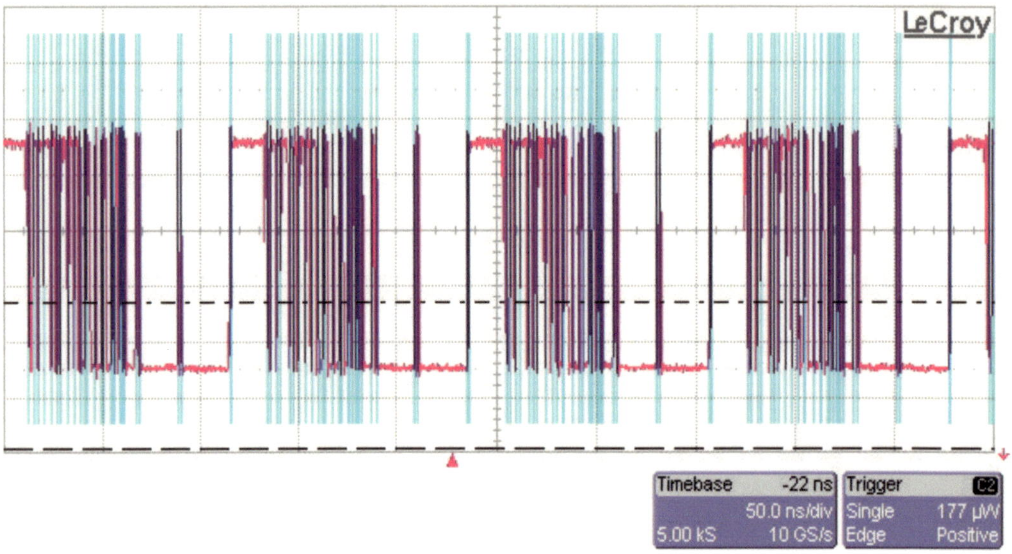

Fig. 2. Pulse train of the proposed soliton EDFL.

The spectral and temporal outputs of the proposed soliton fiber laser are also characterized using an OSA and an autocorrelator. The output spectrum of the mode-locked EDFL obtained from the output port of the coupler is shown in Figure 3. It is clearly seen that the output pulse has a 3 dB bandwidth of 4.5 nm centered at 1558 nm. Kelly sidebands are also evident indicating that the output laser is soliton pulses. These sidebands are a kind of resonant coupling, which for some optical frequencies occur when the relative phase of soliton and dispersive wave changes by an integer multiple of 2π per resonator round trip. The strong Kelly sidebands obtained suggest that the pulse duration is near the minimum possible value. Figure 4 shows an autocorrelation trace of the output pulse from the mode-locked EDFL. The pulse duration is 0.79 ps assuming that the pulse shape follows a $sech^2$ profile. For the purpose of testing the stability of this laser, we kept the laser in the fundamental mode-locked operation and pumped the power at 140 mW without disturbance for a few hours. It was observed that there was no significant change in the spectrum, central wavelength, 3 dB spectrum width and pulse width, output power and repetition rate. This indicates that a stable soliton fiber laser can be achieved using mode-locking based on the combination of both SWNT-SA and NPR.

Passively mode-locked fiber laser can be constructed based on nonlinear polarization rotation without the saturable absorber. This makes a simpler configuration.In this section, a simple mode-locked Bismuth-based Erbium-doped fiber laser (Bi-EDFL) is achieved by using a simple ring cavity structure incorporating a Bismuth-based erbium-doped fiber (EDF), an isolator and a polarisation controller. The short Bi-EDF making up the gain cavity allows the generation of both stable and clean pulses with an increase in the repetition frequency, while its high nonlinearity allows better suppression of the supermode noise. To date, this is the first demonstration of a passively mode-locked fiber laser, which uses such a short length of Bi-EDF together with the nonlinear polarisation rotation (NPR) method.

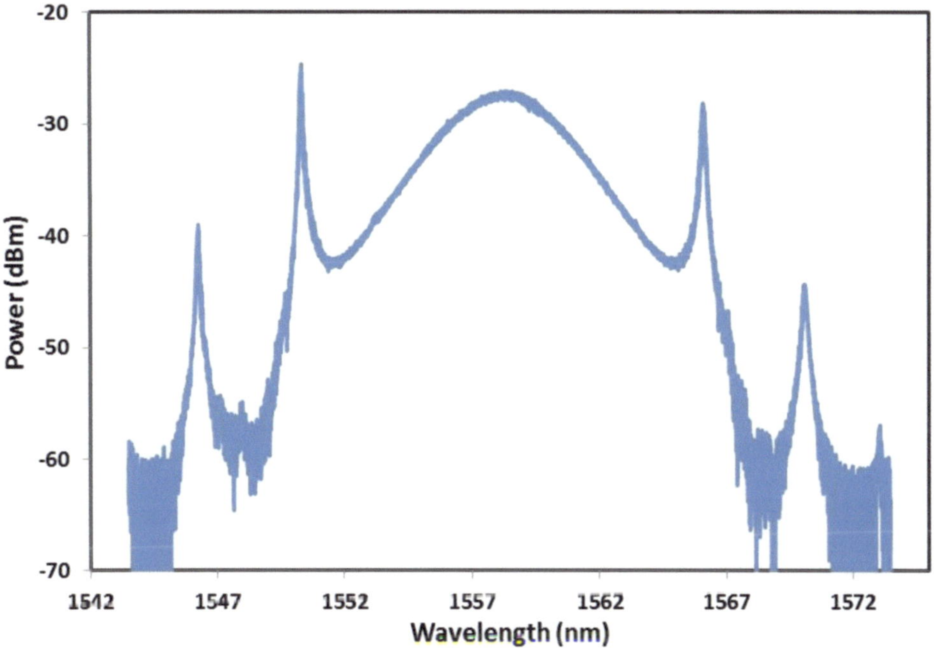

Fig. 3. Optical spectrum of the output of a soliton fiber laser, exhibiting Kelly sidebands.

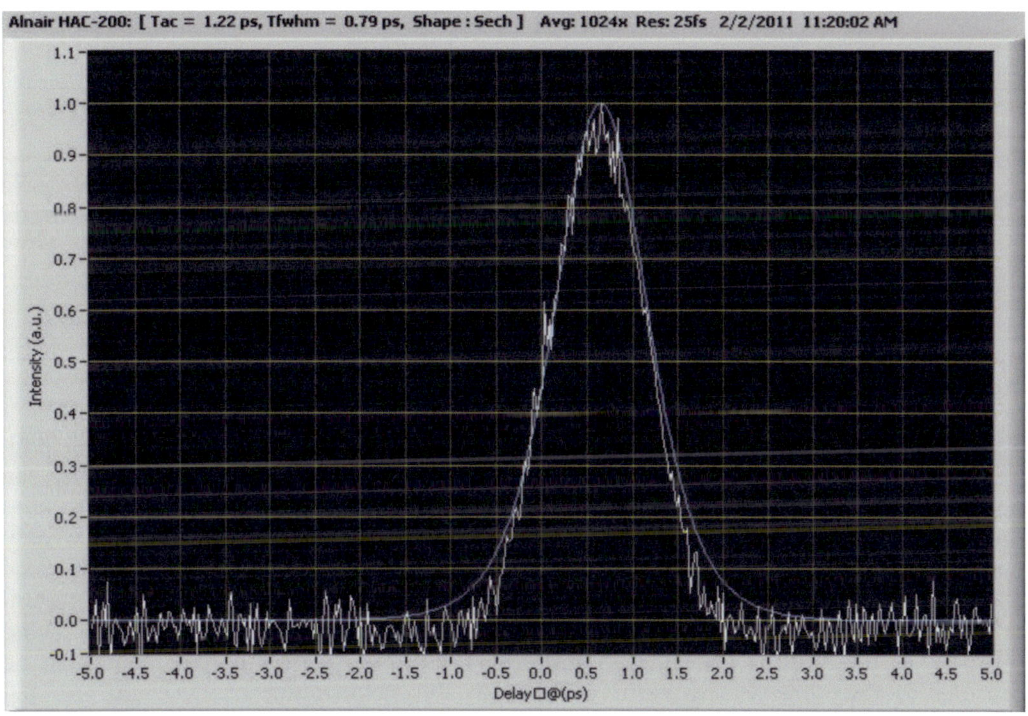

Fig. 4. Autocorrelation trace of the output pulse from the mode-locked EDFL

The experimental set-up of the proposed system for mode-locked Bi-EDFL is illustrated in Figure 5. The Bi-EDFL employs a piece of 49 cm long Bi-EDF, a wavelength division multiplexing (WDM) coupler, an isolator, a polarization controller (PC) and a 3 dB output coupler. The Bi-EDFL cavity consists of a 49 cm long Bi-EDF, and a 2.0 m long SMF-28 which is used in the cavity that is composed of a coupler, polarisation controller, isolator and 980/1550 nm WDM coupler. The Bi-EDF has a nonlinear coefficient of ~60 $(W/km)^{-1}$ at 1550 nm, an Erbium concentration of 3250 ppm, a cut-off wavelength of 1440 nm, a pump absorption rate of 130 dB/m at 1480 nm and a dispersion parameter of 130 ps/km.nm at λ=1550 nm. It is bi-directionally pumped using a 1480 nm laser diode via the WDM to provide an amplification in the C-band region. The other part of the ring cavity uses a standard single mode fiber (SMF-28) with a dispersion of 17 ps/nm.km at λ=1545 nm. A standard polarisation-independent isolator is used to ensure a unidirectional operation and acts as a polarizer. A PC is used to rotate the polarization state and to allow continuous adjustments of the birefringence within the cavity to balance both the gain and loss for the generation of the laser pulses.

A fraction of the stretched laser pulse operating at 1560 nm is extracted through the 50% output of the coupler. The pulse width and repetition rate of the laser pulse are measured to be around 131 fs and 42 MHz, respectively. The output power and spectrum are measured by a power meter and an optical spectrum analyzer (OSA), respectively. The entire experimental setup is fusion-spliced together.

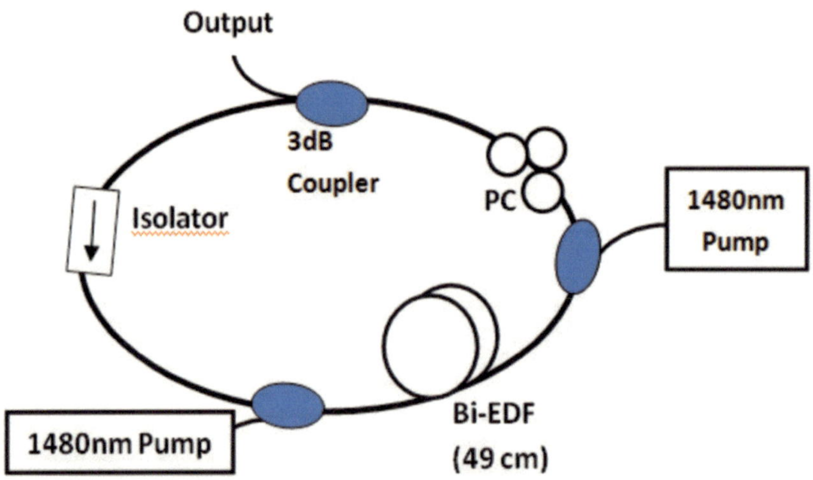

Fig. 5. Experimental set-up for mode-locked fiber laser.

To achieve mode-locking operation, the polarization state of the light within the cavity should be adjusted by using the PC. By altering the polarization state inside the cavity, the optical spectrum can be broadened to initiate Q-switching from which mode-locking operation is obtained. When a linearly polarized light is incident to a piece of weakly birefringent fiber such as a Bi-EDF, the polarization of the light will generally become elliptically polarized in the fiber. The orientation and ellipticity of the resulting polarization of the light is fully determined by the fiber length and its birefringence. However, if the intensity of the light is strong, the nonlinear optical Kerr effect in the fiber must be considered, thus introducing extra changes to the polarization of the light. As the polarization change introduced by the optical Kerr effect depends on the light intensity, if a polarizer or isolator is placed behind the fiber, the transmission of the light through the polarizer will become light intensity dependent. Upon selecting the appropriate orientation of the polarizer, an artificial saturable absorber with an ultra-fast response could then be achieved in such a system, where light of higher intensity experiences less absorption loss on the polarizer. The proposed laser makes use of this artificial saturable absorption to achieve passive mode locking. Once a mode-locked pulse is formed, the nonlinearity of the fiber further shapes the pulse into the ultrashort stretched-pulse.

In the experiment, the 1480 nm pump powers were both fixed at 125 mW. The output spectrum of the mode-locked EDFL obtained after the 3 dB coupler is shown in Figure 6. A broad spectrum with a 3dB bandwidth of 21.7 nm is obtained at the optimum polarization state, which indicates that the output laser has a stretched pulse characteristic. The spectrum has a peak wavelength at 1560 nm. Q-switching operation mode is observed by an oscilloscope in the form of an unstable pulse train with periodic variation in its pulse amplitude. Further adjustment of polarization produces a more stable mode-locked pulse train as shown in Figure 7. The mode-locked pulse train has a constant spacing of 24 ps, which translates to a repetition rate of 42 MHz. The high repetition rate pulse trains are produced from a harmonically mode-locked laser, where multiple pulses circulate within the cavity. The multiple pulses are generated passively in the laser due to the phenomenon known as the soliton energy quantization. The pulse characteristic of the mode-locked EDFL at the 3 dB coupler is also investigated by an autocorrelator. Figure 8 shows the autocorrelator trace of the pulse, which shows the $sech^2$ pulse profile with a full width half maximum (FWHM) of 131 fs. The output of the femtosecond pulses is also observed to be very stable at room temperature. The operation of the Bi-EDFL can be tuned by incorporating a tunable band-pass filter in the ring cavity. By optimizing the length of the Bi-EDF, a wideband tunable operation is expected to be achieved reaching up to the extended L-band region.

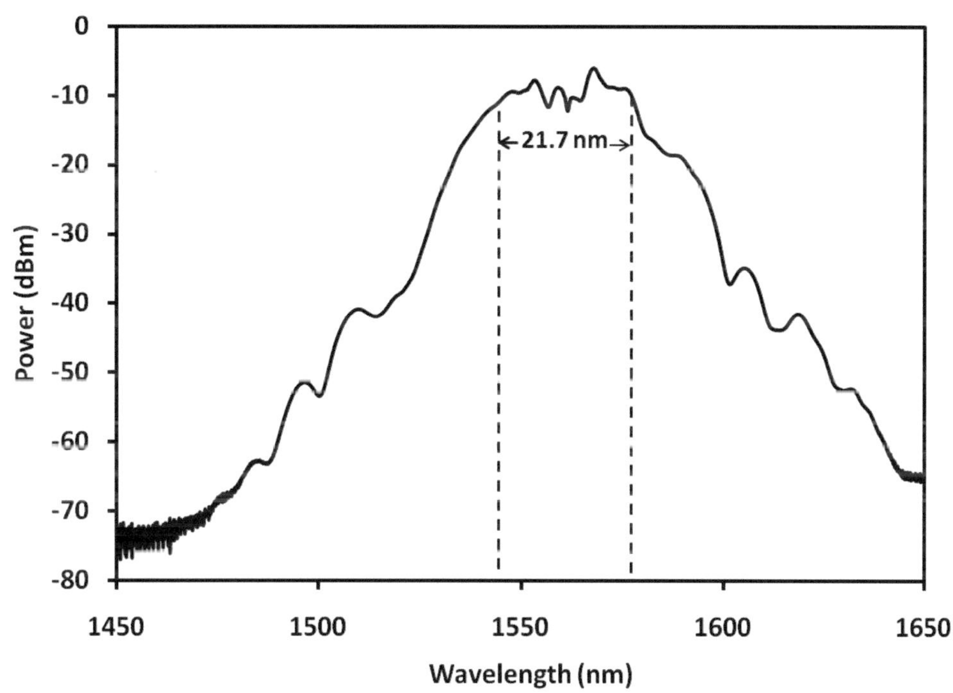

Fig. 6. Optical spectrum of the mode-locked laser.

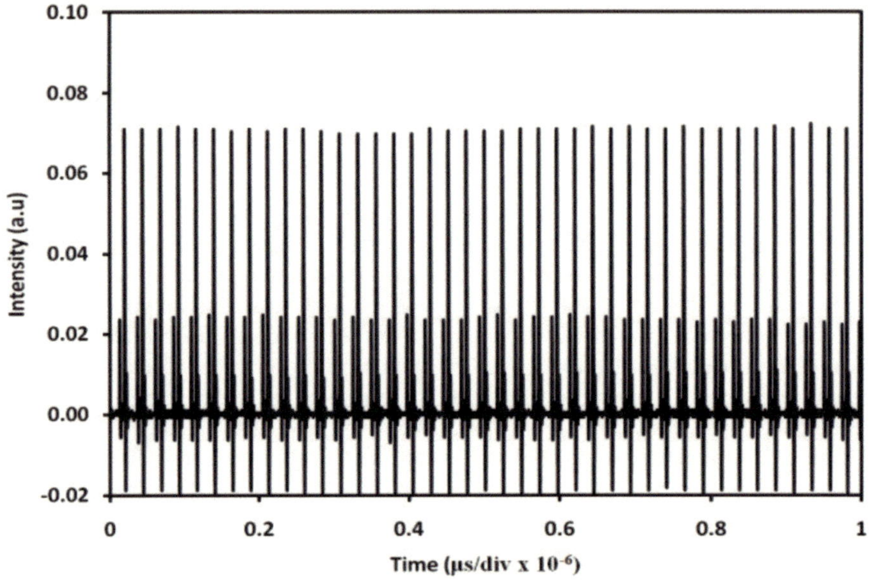

Fig. 7. Pulse train of the passive mode-locked laser with a repetition rate of 42 MHz.

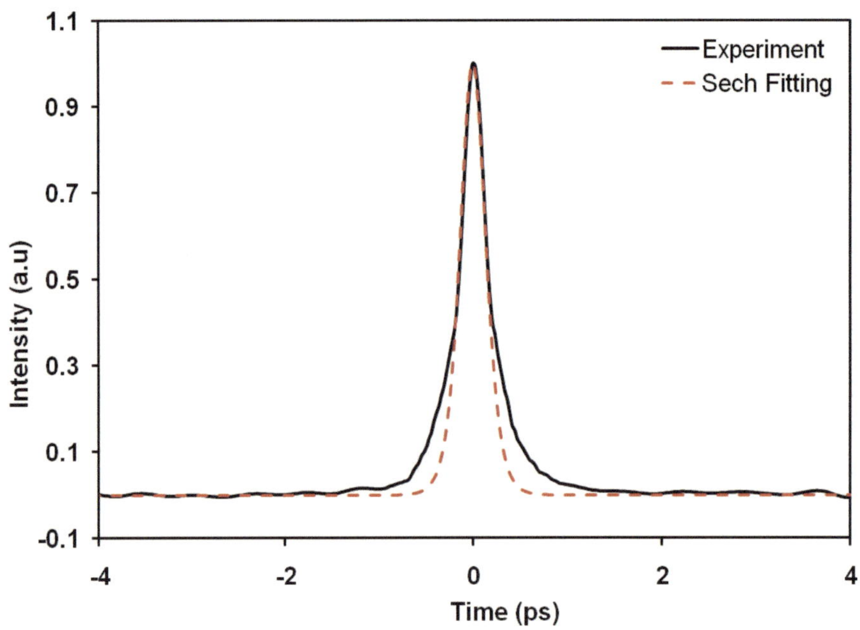

Fig. 8. Autocorrelator pulse trace with FWHM of 131 fs and $Sech^2$ pulse shape.

3. Supercontinuum generation

Supercontinuum generation is the formation of an ultrabroad spectral broadening induced by the coupling of a high peak power sub picosecond pulse laser with a nonlinear optical fiber of adequate length. The supercontinuum spectrum is determined by the sequence of events in the supercontinuum generation, and is dependent on both the pump pulse and the fiber characteristics namely the pump pulse wavelength, power, and pulse duration. The dominant nonlinear effects occurring in the generation of an SC includes self-phase modulation, stimulated Raman scattering and soliton effects. Since supercontinuum was first discovered in 1970 by Alfano (Alfano & Shapiro, 1970), many works have been performed to understand the phenomenon as well as its implementation in practical devices where it has found applications in areas of semiconductor, biology, chemistry, optical coherent tomography, sensing and optical communication, femtosecond carrier-envelope phase stabilization, ultrafast pulse compression, time and frequency metrology, and atmospheric science (Hartl et al., 2001; Kano & Hamaguchi, 2003; Mori, 2003; Alfano, 2006). These included the study of primary events in photosynthesis, nonradiative processes in photoexcited chemicals, excitation of optical phonons, carrier dynamics of semiconductor, frequency clocks and broad spectrum LIDAR.

Specialty fibers such as photonic crystal fibers (PCFs) have high nonlinearity with a managed dispersion profile, and thus, can be used to generate supercontinuum (Parvizi et al., 2010; Russell, 2003). The first supercontinuum generation in a microstructured fiber was reported in 2000 by Ranka et al. Zero dispersion and anomalous dispersion regions could have contributed in higher order soliton generation, pulse compression and ultrabroadband continuum extending from the ultraviolet to the infrared spectral regions. In addition, the pulse broadening in PCF is of great interest for its coherence, brightness and low pulse energy required to generate supercontinuum. The most widely used type of PCF consists of pure silica core surrounded by periodic arrays of air holes, where genuine photonic band-gap guidance can occur. PCFs of this type have attracted much interest because of their potential for lossless and distortion-free transmission, particle trapping, optical sensing, and for novel applications in nonlinear optics (Russell, 2000; Wiederhecker et al., 2007; Benabid et al., 2005).

The supercontinuum can be generated using a picosecond to nanosecond pulses, or even a continuous wave pump where spectral broadening is initiated in the so-called "long pulse" regime (Harun et al., 2011b; Wang et al., 2006; Gorbach et al., 2007). Research is now shifting towards supercontinuum using a more robust and cheaper mode-locked laser, achievable with some innovative designs, as well as the study of the supercontinuum process with picosecond or nanosecond pulsed lasers. As were shown in Figure 1 and Figure 5, we experimentally demonstrated two fiber-based mode-locked lasers using two methods for mode-locking; fiber-based nonlinear polarization rotation (NPR) and saturable absorber. These mode-locked fiber lasers could then be implemented in the experiment to generate supercontinuum.

Figure 9 shows the experimental setup used to generate supercontinuum. A high power amplifier is used to increase the seed pulse peak power to a maximum power of 30 dBm to allow apectral broadening. At the output of the amplifier, the pulse enters a PCF for supercontinuum generation. The supercontinuum spectrum is measured by an optical spectrum analyzer.

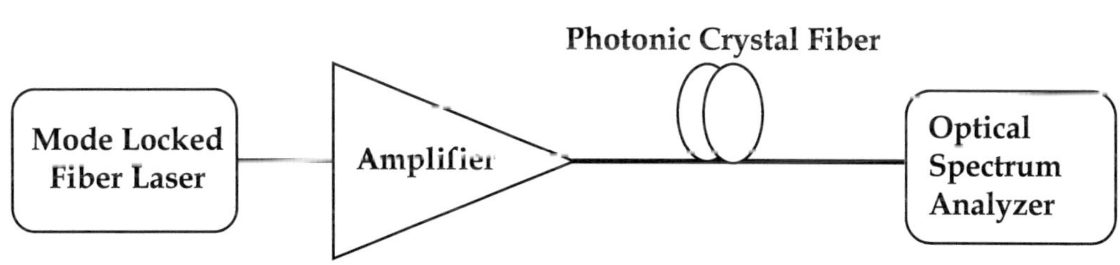

Fig. 9. Schematic diagram of the supercontinuum generation experiment.

Figure 10 shows an SC generation in a 100 m long PCF which is pumped by an amplified 1560 nm mode locked fiber laser at 30 dBm amplified pulse laser power. The mode locked fiber laser used in the experiment has a pulse width of 24 ps with a repetition rate of 42 MHz. The PCF used in the experiment has a zero and -1.5 ps/(km.nm) dispersion at wavelength of 1550 nm and 1580 nm respectively. The PCF length is fixed at the optimised length of 100 m. The nonlinearity coefficient of the PCF is around 11 $W^{-1}km^{-1}$.

As shown in Figure 10, we observe a supercontinuum starting from 600 nm up to 2100 nm at the maximum amplified pulse pump power of 30 dBm. The broad supercontinuum is obtained due to the injection of 1560 nm pulse laser in the anomalous-dispersion regime of the PCF. The pulse initially begins to self-Raman shift to longer wavelengths thus causing asymmetry in the supercontinuum spectrum. When these higher-order solitons break up, parametric four-wave mixing generates frequencies at wavelengths shorter than the zero-dispersion wavelength. With increase in the pump power, there is widening of the spectrum as well as marked improvement of the flatness characteristic.

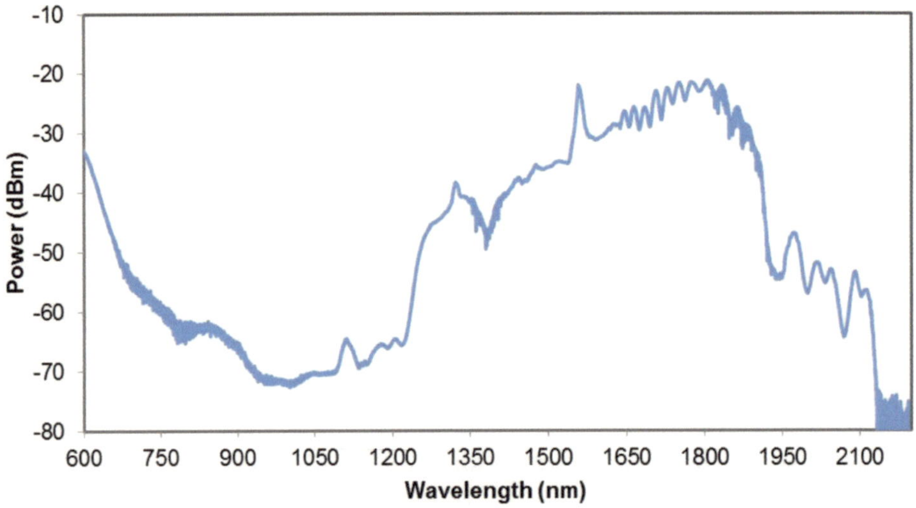

Fig. 10. Supercontinuum generation with 100 m long PCF at fixed average pump power of 30 dBm.

4. Supercontinuum slicing

The supercontinuum generated in the PCF is injected into the loop mirror as shown in Figure 11. In the loop mirror, the supercontinuum source is splitted into two by a 3-dB coupler, where one of the light beams travels in a clockwise direction and the other travels in the opposite direction of the polarization maintaining fiber (PMF). Spectral slicing occurs when the beams interfere constructively and destructively due to the phase differences encountered by the two propagating beams in the loop (Ahmad et al., 2009). Both beams are combined at the end of the fiber coupler to act as a comb filter. Figure 12 shows the multiwavelength spectrum measured by a optical spectrum analyzer through slicing the supercontinuum at pulse pump powers of 30 dBm. The spacing, $\Delta\lambda$ is related to the length, L of the PMF used by the following equation

$$\Delta\lambda = \frac{\lambda}{BL} \quad (1)$$

where B and λ are the birefringence and wavelength, respectively. It is also observed that the comb spectrum is flatter at higher pump power. The spacing of the comb increases as the operating wavelength increases, which agree well with the above equation.

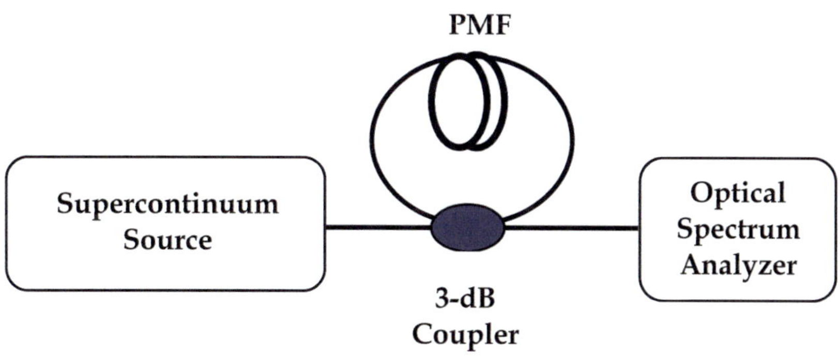

Fig. 11. Loop mirror configuration to slice the supercontinuum source.

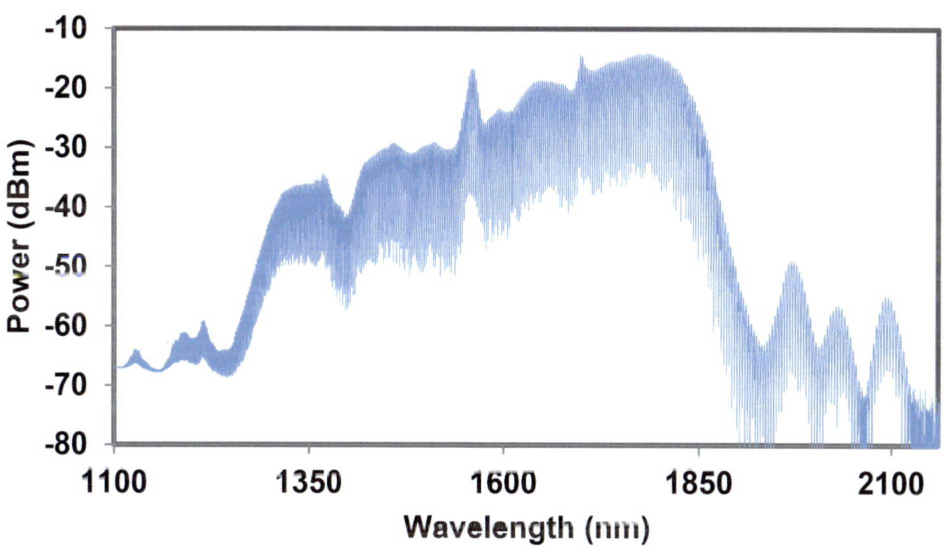

Fig. 12. Multiwavelength comb.

Figure 13 shows the superimposed multiwavelength comb spectrum at different wavelength regions for the supercontinuum with 30 dBm pump pulse power. The multiwavelength comb spectra obtained have an average channel spacings of 2.22 nm and 3.03 nm obtained at 1440 nm and 1750 nm regions, respectively. The spacing of the comb increases as the operating wavelength increases, which agree well with the above equation. In addition, the best signal to noise ratio (SNR) obtained is about 20 dB.

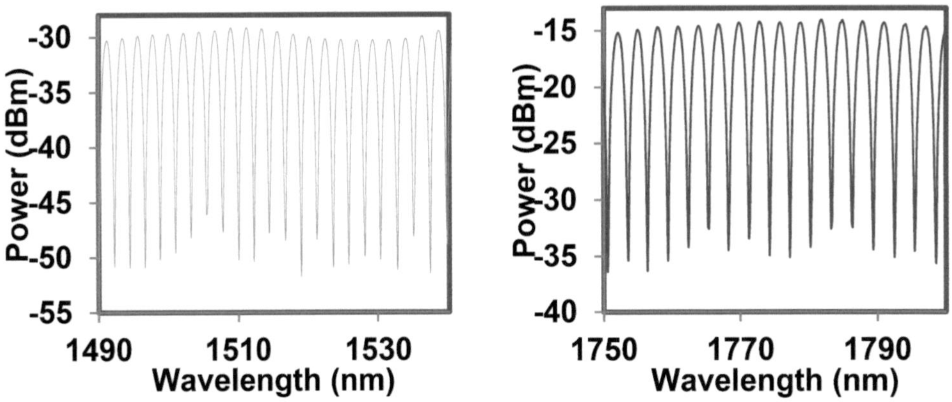

Fig. 13. Sliced spectrum at wavelength regions 1490 nm to 1540 nm and 1750 nm to 1800 nm.

5. References

F. Benabid, F. Couny, J. C. Knight, T. A. Birks and P. S. Russell, "Compact, stable and efficient all-fibre gas cells using hollow-core photonic crystal fibres", *Nature* 434, 488-491 (2005).

X. M. Liu, Y. Chung, A. Lin, W. Zhao, K. Q. Lu, Y. S. Wang and T. Y. Zhang, "Tunable and switchable multi-wavelength erbium-doped fiber laser with highly nonlinear photonic crystal fiber and polarization controllers," *Laser Phys. Lett.* 5, 904–907 (2008).

D. Chen, B. Sun and Y. Wei, "Multi-Wavelength Laser Source Based on Enhanced Four-Wave-Mixing Effect in a Highly Nonlinear Fiber," *Laser Phys.* 20, 1733-1737 (2010).

S. W. Harun, R. Parvizi, N. S. Shahabuddin, Z. Yusoff and H. Ahmad, "Semiconductor optical amplifier-based multi-wavelength ring laser utilizing photonic crystal fiber," *J. Modern Opt.* 57, 637-640 (2010).

M. I. Johari, A. Adamiat, N. S. Shahabuddin, M. N. M. Nasir, Z. Yusoff, H. A. Abdul Rashid, M. H. Al-Mansoori, and P. K. Choudhury, "On the ring cavity multiwavelength Brillouin-erbium fiber laser with partially reflective fiber Bragg grating," *JOSA B* 26(9),1675–1678 (2009).

H. Ahmad, N. S. Shahabuddin, Z. Jusoh, K. Dimyati and S. W. Harun, An enhanced Bismuth-based Brillouin/Erbium fiber laser with linear cavity, *Fiber and Integrated Optics* 27, 35-40 (2008).

N. S. Shahabuddin, S. W. Harun, M. R. Shirazi, and H. Ahmad "A Linear Cavity Brillouin/Bismuth-Based Erbium-Doped Fiber Laser with Enhanced Characteristics", *Laser Physics* 18(11) 1344 (2008).

S. Shahi, S. W. Harun and H. Ahmad, "Multi-wavelength Brillouin fiber laser using Brillouin-Rayleigh scatterings in distributed Raman amplifier," *Laser Phys. Lett.* 6, 737-739 (2009a).

S. Shahi, S. W. Harun, K. S. Lim, A. W. Naji and H. Ahmad, "Enhanced Four-Wave Mixing Efficiency of BI-EDF in a New Ring Configuration for Determination of Nonlinear Parameters ," *Journal of Electromagnetic Waves and Applications* 23, 2397-2407 (2009b)

Y. Nan, C. Lou, J. Wang, T. Wang and L. Huo, "Signal-to-noise ratio improvement of a supercontinuum continuous-wave optical source using a dispersion-imbalanced nonlinear optical loop mirror," *Applied Physics B: Laser and Optics* 76, 61-64 (2004).

R. Buczynski, D. Pysz, T. Martynkien, D. Lorenc, I. Kujawa, T. Nasilowski, F. Berghmans, H. Thienpont, and R.Stepien, "Ultra flat supercontinuum generation in silicate dual core microstructured fiber," *Laser Phys. Lett.* 6(8), 575–581 (2009).

A. S. Kurkov, E. M. Sholokhov and Y. E. Sadovnikova, "All-fiber supercontinuum source in the range of 1550–2400 nm based on telecommunication multimode fiber," *Laser Phys. Lett.* 8, 598-600 (2011).

G. Genty, M. Lehtonen, H. Ludvigsen, and M. Kaivola, "Enhanced bandwidth of supercontinuum generated in microstructured fibers," *Optics Express* 12(15) 3471-3480 (2004).

G. Genty, M. Lehtonen, and H. Ludvigsen "Effect of cross-phase modulation on supercontinuum generated in microstructured fibers with sub-30 fs pulses," *Optics Express* 12, 4614-4624 (2004)

R. Buczynski, H.T. Bookey, D. Pysz, R. Stepien, I. Kujawa, J.E. McCarthy, A.J. Waddie, A.K. Kar, M.R. Taghizadeh, " Supercontinuum generation up to 2.5 um in photonic crystal fiber made of lead-bismuth-galate glass," *Laser Phys. Lett.* 7, 666-672 (2010)

M. Lehtonen, G. Genty and H. Ludvigsen, "Supercontinuum generation in a highly birefringent microstructured fiber," *Applied Physics Letters* 82(14) 2197-2199 (2003)

S. P. Chen, J. H. Wang, H. W. Chen, Z. L. Chen, J. Hou, X. J. Xu, J. B. Chen and Z. J. Liu, "20 W all fiber supercontinuum generation from picosecond MOPA pumped photonic crystal fiber," *Laser Phys.* 21, 519-521 (2011)

Y. Gu, L. Zhan, D. D. Deng, Y. X. Wang and Y. X. Xia, "Supercontinuum generation in short dispersion-shifted fiber by a femtosecond fiber laser," *Laser Phys.* 20, 1459-1462 (2010)

N. Aközbek, S A Trushin, A Baltuˇska, W Fuß, E. Goulielmakis, K. Kosma, F. Krausz, S. Panja, M. Uiberacker, W. E. Schmid, A. Becker, M. Scalora and M. Bloemer, "Extending the supercontinuum spectrum down to 200nm with few-cycle pulses," *New Journal of Physics* 8, 177 (2006).

H. Ahmad, N. S. Shahabuddin and S. W. Harun, "Multi- wavelength source based on SOA and loop mirror," *Optoelectronics and Advanced Materials Rapid Communications* 3(1) 1-3 (2009).

F. W. Wise, A. Chong, and H. William. Renninger, "High-energy femtosecond fiber lasers based on pulse propagation at normal dispersion" *Laser & Photon. Rev.* 2, 58-72, (2008).

L. M. Zhao, D. Y. Tang, H. Zhang,T. H. Cheng,H. Y. Tam,and C. Lu, "Dynamics of gain-guided solitons in all-normal-dispersion fiber laser," *Opt., lett.* 32, 806-1808, (2007).

A. Chong, H. W. Renninger, and F. W. Wise, "Properties of normal-dispersion femtosecond fiber lasers" *J. Opt. Soc. Am. B*, 25, 2140-148, (2008).

A. Chong, H. W. Renninger, and F. W. Wise, "All-normal-dispersion femtosecond fiber laser with pulse energy above 20 nJ," *Opt. lett.* 32, 2408- 2410, (2007).

X. Huan, F. Ga, H. Y. Tama, and P. K. A. Wai, "Stable and uniform multiwavelength erbium doped fiber laser using nonlinear polarization rotation," *Optics express*, 14, 8205-8210(2006).

S. W. Harun, R. Akbari, H. Arof and H. Ahmad, "Mode-locked bismuth-based erbium-doped fiber laser with stable and clean femtosecond pulses output," *Laser Phys.* Lett. 8, pp. 449-452 (2011).

Y. Li, C. Lou, J. Wu, B. Wu, and Y. Gao, "Novel method to simultaneously compress pulses and suppress supermode noise in actively mode-locked fiber ring laser," *IEEE Photon. Technol. Lett.*, 10, 1250-1252 (1998).

Z. Li, C. Lou, K. T. Chan, Y. Li, and Y. Gao, "Theoretical and experimental study of pulse-amplitude-equalization in a rational harmonic mode-locked fiber ring laser," *IEEE J. Quantum Electron.*, 37, 33-37 (2001).

R. R. Alfano and S. L. Shapiro, "Emission in region 4000 to 7000 a via 4- photon coupling in glass," *Physical Review Letters* 24(11), 584 (1970).

I. Hartl, X. D. Li, C. Chudoba, R. K. Ghanta, T. H. Ko, J. G. Fujimoto, J. K. Ranka, and R. S. Windeler, "Ultrahigh-resolution optical coherence tomography using continuum generation in an air-silica microstructure optical fiber," *Opt. Lett.* 26, 608-610 (2001)

H. Kano and H. Hamaguchi, "Characterization of a supercontinuum generated from a photonic crystal fiber and its application to coherent Raman spectroscopy," *Opt. Lett.,* 28, 2360-2362 (2003).

K. Mori, K. Sato, H. Takara, and T. Ohara, Supercontinuum lightwave source generating 50 GHz spaced optical ITU grid seamlessly over S-, C- and L-bands," *Electron. Lett.* 39, 544-546, (2003).

R. R. Alfano, (2nd) 2006. *The Supercontinuum Laser Source: Fundamentals and Updated References*, Springer, ISBN 0-387-24504-9, New York

R. Parvizi, S. W. Harun, N. S. Shahbuddin, Z. Yusoff and H. Ahmad, "Multi-wavelength bismuth-based erbium-doped fiber laser based on four-wave mixing effect in photonic crystal fiber, *Optics and Laser Technol.* 44, 1250-1252 (2010).

P. Russell, Photonic crystal fibers, *Science* 299, 358–62 (2003)

J. K. Ranka, R. S. Windeler, and A. J. Stentz, Visible continuum generation in air-silica microstructure optical fibers with anomalous dispersion at 800 nm, *Opt. Lett.* 25, 25–27 (2000)

G. S. Wiederhecker, C. M. B.Cordeiro, F.Couny, F. Benabid, S. A. Maier, J. C. Knight, C. H. B. Cruz and H. L.Fragnito, "Field enhancement within an optical fibre with a subwavelength air core,*Nature Photonics* 1, 115-118 (2007).

S. W. Harun, R. Akbari, H. Arof and H. Ahmad, "Supercontinuum generation in photonic crystal fiber using femtosecond pulses," *Laser Phys.* 21, 1215-1218 (2011).

A. V. Gorbach & D.V. Skryabin, "Light trapping in gravity-like potentials and expansion of supercontinuum spectra in photonic-crystal fibres," *Nat. Photonics* 1, 653-657 (2007).

A. Martinez and S. Yamashita, "Carbon Nanotube-Based Photonic Devices: Applications in Nonlinear Optics, In: *Carbon Nanotubes in Electron Devices,* Jose Mauricio Marulanda, pp. 367-386, Intech, Retrieved from http://www.intechopen.com/download/pdf/pdfs_id/17299

R. Parvizi, S. W. Harun, N. M. Ali, N. S. Shahabuddin and H. Ahmad, " Photonic crystal fiber-based multi-wavelength Brillouin fiber laser with dual-pass amplification configuration," *Chinese Optics Lett.* 9, 021403 (2011).

Z. Wang, C. Y. Wang, Y. K. Han, S. Y. Cao, Z. G. Zhang and L. Lai, "Octave-spanning spectrum generation in Ti : sapphire oscillator," *Opt. & Laser Technol.* 38, 641-644 (2006).

W. Ye, W. Liu, T. Chen, D. Z. Yang and Y. H. Shen, "Erbium-ytterbium co-doped multi-wavelength double-clad fiber laser around 1612 nm," *Laser Phys.* 20, 1636-1640 (2010).

K. Zhong, J. S. Li, D. G. Xu, X. Ding, R. Zhou, W. Q. Wen, Z. Y. Li, X. Y. Xu, P. Wang and J. Q. Yao, "Multi-wavelength generation based on cascaded Raman scattering and self-frequency-doubling in KTA," *Laser Phys.* 20, 750-755 (2010).

Low Frequency Noise Characteristics of Multimode and Singlemode Laser Diodes

Sandra Pralgauskaitė, Vilius Palenskis and Jonas Matukas
Vilnius University,
Lithuania

1. Introduction

Three main fluctuating quantities are considered in the laser diode (LD) noise investigation: emitted optical power (optical noise), phase (or frequency) and LD terminal voltage (electrical noise). Both phase and amplitude fluctuations of LD affect the performance of optical communication system. Phase fluctuations determine the linewidth, which is related with radiation frequency and is very important parameter (Jacobsen, 2010; Tsuchida, 2011). Amplitude fluctuations appear in both total output power and the output levels of longitudinal modes, and may indeed contribute significantly to the error rate for externally modulated LD operating at high data rate in communication system (Jacobsen, 2010; Nilson et al., 1991). The low frequency noise can beat with the modulation signal to produce enhanced noise in the "wings" around the modulation signal and causes significant degradation in signal-to-noise (S/N) performance in optical transmission (Gray & Agrawal, 1991; Lau et al., 1993). Electrical noise governs injected carrier number in the active region and therefore emitted photon number: optical and electrical fluctuations are partly correlated.

Intensity noise arises from a variety of sources, including gain fluctuations, spontaneous emission fluctuations, and relaxation oscillations (Fronen & Vandamme, 1988). At low frequencies both the optical and electrical fluctuations of LDs usually are of $1/f^{\alpha}$-type over all operation conditions (Fronen & Vandamme, 1988; Matukas et al., 1998, 2001; Orsal et al., 1994; Smetona et al., 2001; Tsuchida, 2011), noise intensity strongly depends on the defect density and their distribution within the active region of the laser. In (Jacobsen, 2010; Tsuchida, 2011), it was shown, that linewidth of the semiconductor laser strongly depends on the level of $1/f^{\alpha}$-type noise. In (Mohammadi & Pavlidis, 2000; Palenskis, 1990; Van der Ziel, 1970), it is shown that only the superposition of generation-recombination processes through the recombination centres in macroscopic defects with a wide relaxation time distribution can explain the $1/f^{\alpha}$-type noise spectra occurring over a wide frequency range. In (Orsal et al., 1994; Simmons & Sobiestianskas, 2005), there were shown, that low-frequency terminal electrical noise is highly correlated to the optical noise and that the electrical noise measurement could be used for *in situ* noise characterization of laser diodes without any optics and accompanying elements. But this conclusion is true only in the case, when the low frequency electrical noise is caused only by the defects in the active region of the laser. Noise characteristic investigation can clear up the reasons of various effects observed in the semiconductor laser operation.

The level of the noise is a measure of an uncertainty in the system and increases if there are more individual sources of defectiveness. If a given individual source changes its condition, so, that its effect on the system is less predictable, the noise level also increases. It is normal practice to assume that a noisy device will be less reliable and to reject it for any special application, where long life is an advantage (Amstrong & Smith, 1965; Jones, 2002; Lin Ke et al., 2010; Vandamme, 1994). Long-haul high-capacity optical communication systems require high performance and reliability components. Improvement of new and modern design LD structures requires detailed investigation of their operation characteristics. However, in reviewed works the noise characterization in semiconductor lasers is often presented just as complementary information. It is difficult to find papers devoted to the detailed noise characteristic feature investigations for various design modern semiconductor lasers. Understanding the origin of noise in the device could help improving the design and fabrication methods of LD, controlling noise level, selection of reliable devices. This Chapter overviews the low frequency noise characteristics of electrical and optical fluctuations, and their cross-correlation characteristics for different modern design multiple quantum well InGaAsP/InP laser diodes (LDs): Fabry-Pérot (FP) and distributed feedback (DFB), ridge-waveguide (index-guiding) and buried-heterostructure (gain-guiding) lasers. Great attention is paid to the mode-hopping effect in Fabry-Pérot laser operation. Also laser diode quality and reliability problems, their revealing via noise characteristics investigation are discussed.

2. Measurement methods of low frequency noise characteristic of laser diode

2.1 Experimental circuit for low-frequency noise measurement

In optoelectronic device noise characteristic investigation both optical (laser output power fluctuations) and electrical (laser diode terminal voltage fluctuations) noise signals are a point of interest, also cross-correlation factor between optical and electrical fluctuations gives valuable information on processes that take place in device structure. Therefore, measurements of optical and electrical signals have to be performed simultaneously, and cross-correlation must be evaluated.

In Fig. 1, there is presented measurement circuit for two noise sources investigation: optical and electrical. Laser diode emitted light and its power fluctuations are detected by photodiode that is suitable for radiation detection around 1.5 μm and 1.3 μm wavelength. Therefore, optical noise corresponds to the photodiode voltage fluctuation due to LD output power fluctuation; electrical noise - LD terminal voltage fluctuation. Current generator mode is guaranteed for LD by choosing appropriate feeding voltage and load resistance.

Optical and electrical noise measurements are performed simultaneously, i. e. processing of both noise signals is produced using two identical channels (Fig. 1), what enables calculation of cross-correlation between two noise signals. Noise signals are amplified by low-noise amplifiers, and then enter to analogue-to-digital converter (ADC) and PC. National Instrument™ PCI 6115 board (for measurements up to 1 MHz) or standard soundboard (for measurements in 20 Hz - 20 kHz frequency range) can be used as ADC. Noise characteristic measurements are performed both over the low frequency region (10 Hz - 100 kHz) and also in one-octave frequency band by using digital filters, having the following central frequencies f_c (Hz): 15, 30, 60, 120, 240, 480, 960, 1920, 3840, 7680, 15360, 30720, 61440, and 122880.

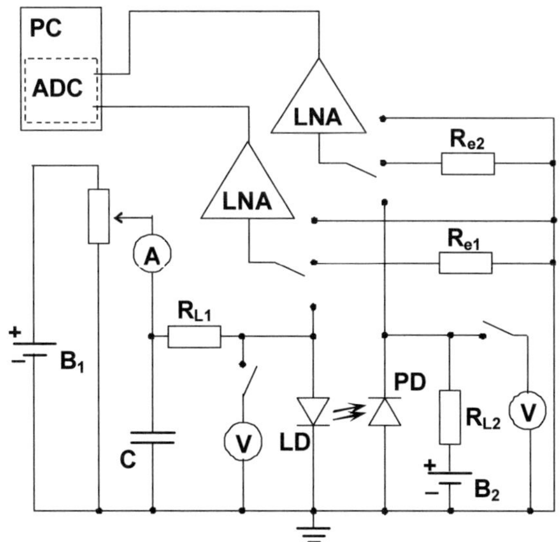

Fig. 1. Experiment circuit: LD - laser diode, PD - photodetector, R_{L1}, R_{L2} -load resistances, R_{e1}, R_{e2} - standard resistors, B_1 and B_2 - storage batteries, LNA - low-noise amplifier, ADC - analogue-to-digital converter, PC - personal computer.

Noise characteristic investigation is highly informative, quick and undestructive. However, noise signals usually are very low. Consequently, measurement system with low-noise components is needed. The most critical part of the low-frequency low-noise measurement system is the preamplifier, which usually establishes the sensitivity of the entire system. For the LD measurements special designed low-noise amplifiers (LNA in Fig. 1) with noise equivalent resistance of 90 Ω (at 1 kHz) are used. Carefully selected batteries, which exhibit low level of $1/f$ -type noise are used for measurements. The sample cell is protected from ambient electromagnetic fields by permalloy screen; the laboratory room is screened with copper and iron screens.

Laser diode radiation spectrum also can be measured together with noise characteristics. At the output the laser light beam is coupled into two fibres optical cable. One fibre is connected to the optical noise measurement equipment and the second - to the optical spectrum analyser „Advantest Q8384". Optical spectrum of the LDs is measured with 10 fm accuracy.

2.2 Evaluation of noise signal spectral density and cross-correlation factor

Noise signals acquisition was performed by PC using Cooley-Tukey Fast Furrier transformation (FFT) algorithm. The calculated FFT spectral density results were averaged by using the large number (over 400) of sets (realizations) and by using frequency weight functions. An averaging according sets gives better accuracy at low-frequencies. Averaging by using frequency weight function does away with inaccuracy at high frequencies caused by FFT algorithm.

The noise spectral density, S_V, was evaluated by comparing with the thermal noise of the reference resistor R_e:

$$S_V = \frac{\overline{V_D^2} - \overline{V_s^2}}{\overline{V_e^2} - \overline{V_s^2}} 4kT_0R_e \ [\mathrm{V^2s}]; \tag{1}$$

where $\overline{V_D^2}$, $\overline{V_s^2}$ and $\overline{V_e^2}$ are respectively the photodiode or laser, the measuring system, and the reference resistor thermal noise variances in the narrow frequency band Δf; T_0 is the absolute temperature of the reference resistor. The signal-to-noise ratio (S/N) was calculated as the ratio of output power (that is measured as photodiode load resistance voltage) and square root value from the optical fluctuation spectral density at a specified frequency. The simultaneous cross-correlation factor was expressed as

$$r = \frac{\overline{\frac{1}{N}\sum_{i=1}^{N} V_i^{\mathrm{el}} V_i^{\mathrm{opt}}}}{\sqrt{\overline{\frac{1}{N}\sum_{i=1}^{N}\left(V_i^{\mathrm{el}}\right)^2}}\sqrt{\overline{\frac{1}{N}\sum_{i=1}^{N}\left(V_i^{\mathrm{opt}}\right)^2}}} \cdot 100\%; \tag{2}$$

where $\sqrt{\overline{\frac{1}{N}\sum_{i=1}^{N}\left(V_i^{\mathrm{el}}\right)^2}}$ and $\sqrt{\overline{\frac{1}{N}\sum_{i=1}^{N}\left(V_i^{\mathrm{opt}}\right)^2}}$ are the root-mean-square values of noise signals V^{el} and V^{opt} that respectively correspond to the terminal voltage fluctuations of the laser diode and the photodetector voltage fluctuations due to laser light output power fluctuations at the same time. This formula allows obtaining not only the magnitude of cross-correlation factor between two signals, but also the correlation sign. As a positive correlation factor is taken a case, when a small increase of laser current creates an increase of both the laser terminal voltage and the light intensity; and negative correlation expresses situation, when one signal increases and the second decreases. The error of the evaluation of cross-correlation factor is less than 1 %.

3. Low frequency noise characteristics of multi-quantum-well laser diodes

In this Section, there are overviewed typical low frequency noise characteristics of singlemode distributed feedback (DFB) and multimode Fabry-Pérot (FP) laser diodes (Matukas et al., 1998, 2001; Palenskis et al., 2003, 2006; Pralgauskaite et al., 2004b). The presented results are for LDs with multiple quantum well (MQW) active region: different devices containing from 4 to 10 compressively strained quantum wells have been investigated.

Laser diodes at stable operation above the threshold current mainly distinguish by $1/f^{\alpha}$-type optical and electrical fluctuations (Figs. 2 and 3). This type of noise is characteristic for the many semiconductor devices (Jones, 1994; Mohammadi & Pavlidis, 2000; Palenskis, 1990; Vandamme, 1994; Van der Ziel, 1970;). Origin of it in semiconductor devices usually is superposition of many charge carriers generation-recombination (GR) processes through GR and capture centres with widely distributed relaxation times (Jones, 1994; Palenskis, 1990). These GR and capture centres are formed by various defects, dislocations and imperfections in the device structure. In optical fluctuations shot noise with white spectrum prevails $1/f^{\alpha}$-type fluctuations at higher frequencies (above 10 kHz). Optical shot noise is caused by the random photon emission process.

Positive cross-correlation ((10-60) %) is characteristic for $1/f^{\alpha}$-optical and electrical noise at the lasing (i. d. above threshold) operation (Fig. 3). $1/f^{\alpha}$-type noise in LDs is caused by the inherent material defects and defects created during the device formation. Resistance of the barrier (of the multiple-quantum-well structure) and adjacent to the active region layers fluctuation leads to the charge carrier number in the active region changes and therefore photon number fluctuation, and cause positively correlated optical and electrical $1/f^{\alpha}$-type noise. Thus, at the lasing operation there are active some defects, that randomly modulate the free charge carrier number in the active layer and, as a consequence, lead to the photon number fluctuations that have the same phase as the charge carrier number fluctuations. Optical and electrical noises are not 100 % correlated because part of the flowing current fluctuations is related with the regions remote from the active one (such as cap and substrate or passive layers). Remote region resistance fluctuations do not influence directly the charge carrier number in the active region, and, thus, do not affect the fluctuations of the photon number.

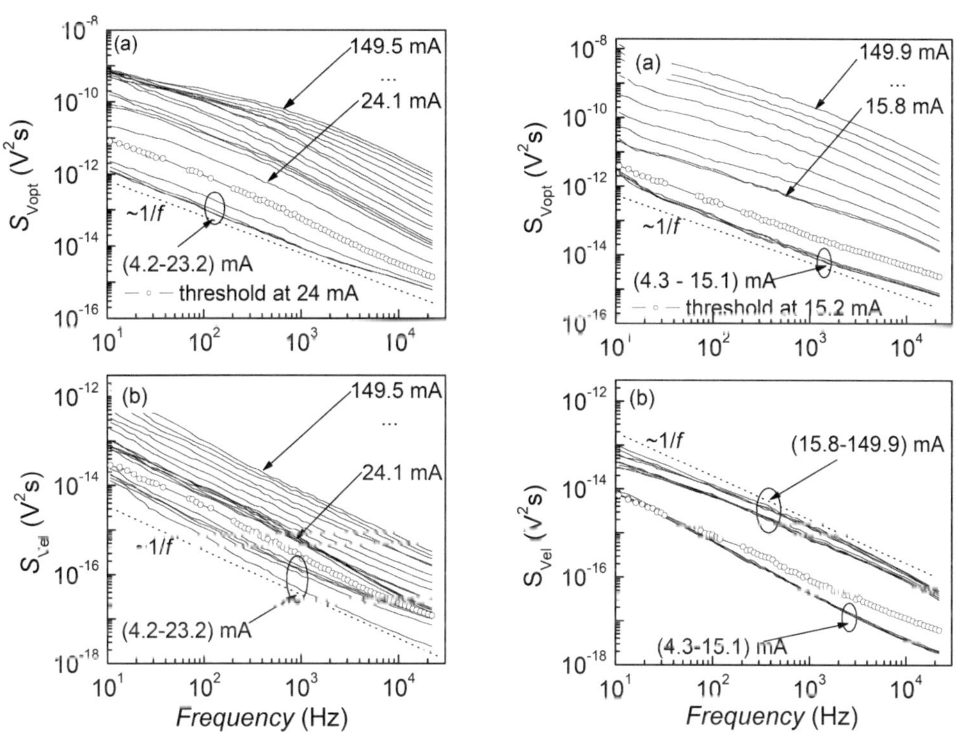

Fig. 2. Optical (a) and electrical (b) noise spectra for FP (on the left, at current ranges where there is no mode hopping) and DFB (on the right) lasers.

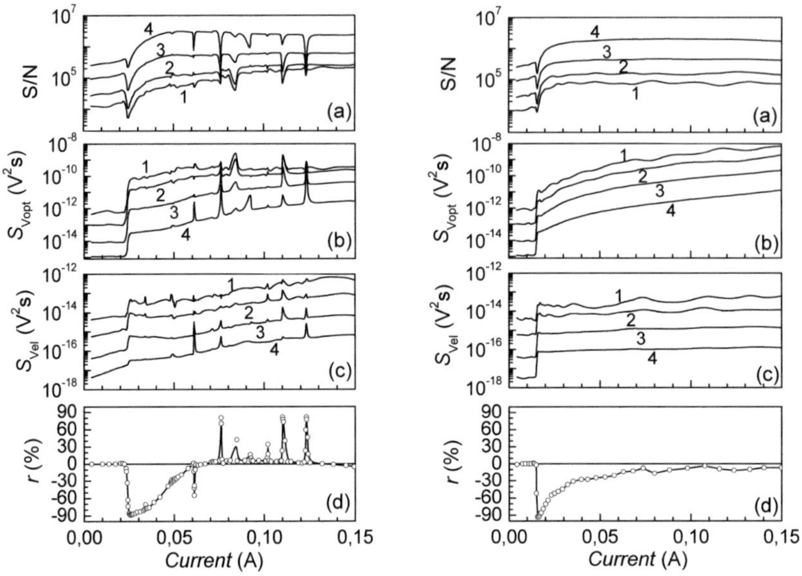

Fig. 3. Signal-to-noise ratio (a), optical (b) and electrical (c) noise spectral density (1- 22 Hz, 2- 108 Hz, 3-1 kHz, 4- 10 kHz), and cross-correlation factor between optical and electrical fluctuations (d; 20 Hz-20 kHz) dependencies on laser current for FP (on the left) and DFB (on the right) lasers.

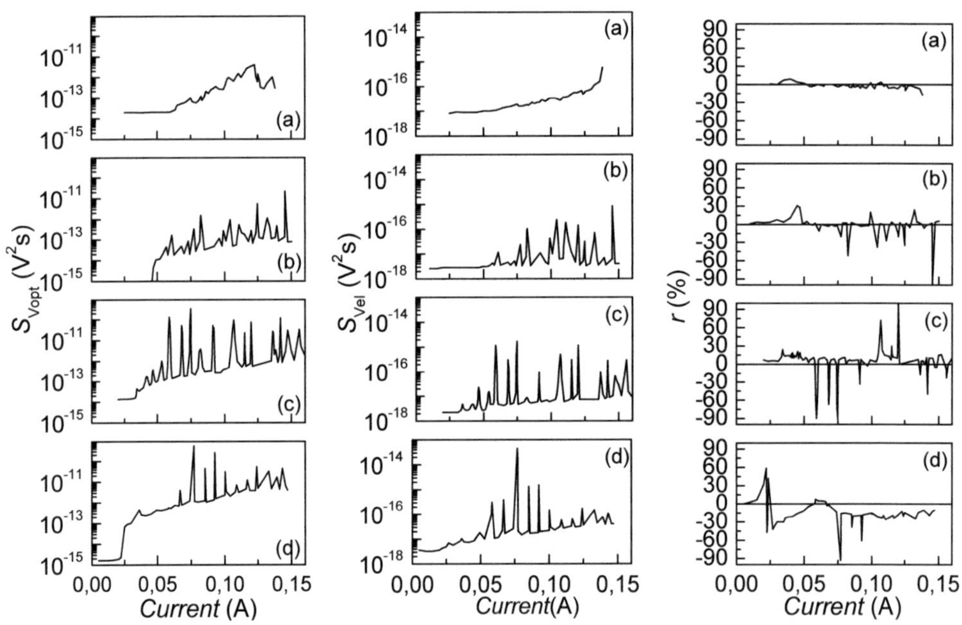

Fig. 4. Optical (on the left) and electrical (in the middle) noise spectral density (10 kHz), and cross-correlation factor between optical and electrical fluctuation (on the right, 1 Hz-1 kHz) dependencies on laser current for different cavity length FP lasers, having barrier layer energy bandgap E_g^b =1.08 eV: a) 1000 μm, b) 750 μm, c) 500 μm, d) 250 μm.

Fabry-Pérot laser noise characteristics at particular temperature and direct current values distinguish by large intensity highly correlated optical and electrical noise peaks (Fig. 3). These noise peaks are caused by the mode-hopping effect that is discussed in the next section. A batch of Fabry-Pérot lasers with some design differences (LDs with different cavity length ((250-1000) μm) and different barrier layer band-gap ((1.03-1.24) eV) has been investigated in order to find out how, these disparities reflect in LDs operation and noise characteristics. There are no clear differences in the noise characteristics of different structure laser diodes (Figs. 4 and 5) (Palenskis et al., 2003). Except long lasers (with cavity longer than 750 μm) that have worse characteristics (Fig. 4): lower efficiency, larger low frequency noise. Laser diodes with larger threshold current usually distinguish by worse other operation characteristics, too (Palenskis et al., 2003; Vurgafman & Meyer, 1997). Larger threshold current is related with more defective LD structure, what leads to the charge carrier leakage out of the active region: some defects randomly redistribute injected current between current component directly related with stimulated recombination and component associated with other processes (leakage current, non-radiative recombination) and lead to the more intensive $1/f^{\alpha}$-type optical and electrical fluctuations. Due to fabrication problems long laser structures contain more defects.

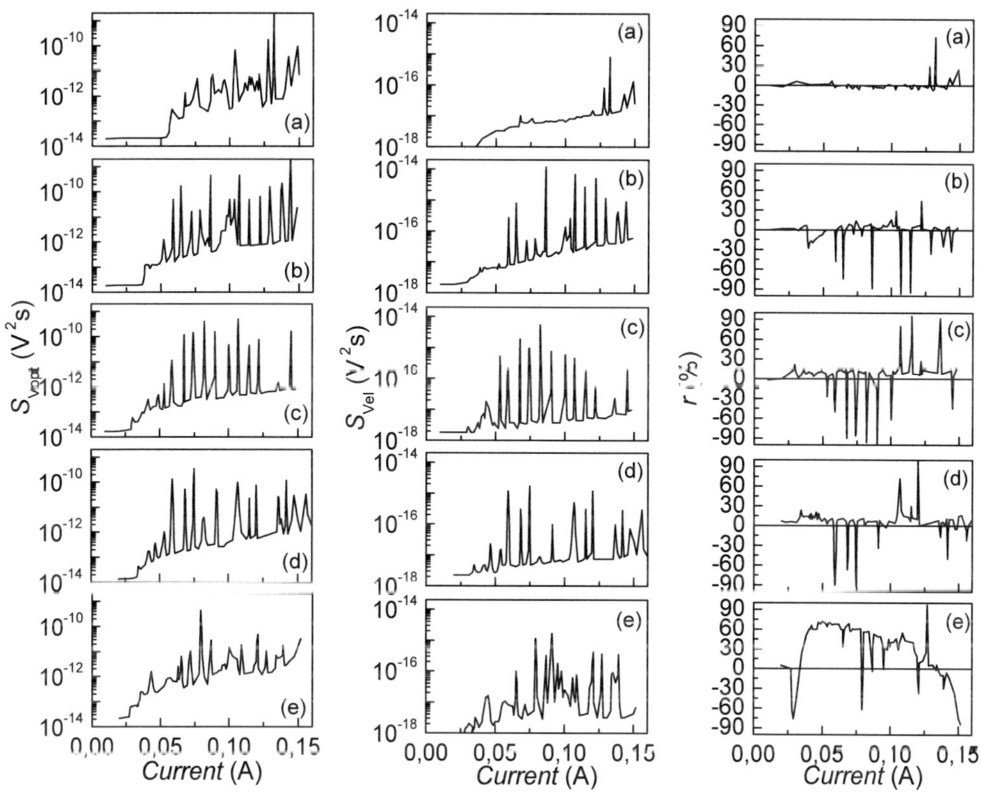

Fig. 5. Optical (on the left) and electrical (in the middle) noise spectral density (10 kHz), and cross-correlation factor between optical and electrical fluctuations (on the right; 1 Hz-1 kHz) dependencies on laser current of FP lasers with cavity length of 500 μm and different barrier layer energy band-gap: a) 1.24 eV, b) 1.18 eV, c) 1.13 eV, d) 1.08 eV, e) 1.03 eV.

In summary, optical and electrical fluctuations of semiconductor lasers at low frequency are $1/f^{\alpha}$-type noise caused by generation-recombination processes through the various defects and imperfections formed generation-recombination centres.

4. Mode-hopping effect in Fabry-Pérot laser diode operation

This Section presents discussion on noise characteristics of Fabry-Pérot lasers during mode-hopping effect, description of physical processes that take place during mode-hopping (Palenskis et al., 2003; Pralgauskaite et al., 2004b, 2011; Saulys et al., 2010).

Increase in the data communication worldwide requires the development of high-speed optical networks. As the number of channel in wavelength division multiplexing system increases and the channel spacing decreases, the requirements for the source wavelength accuracy become increasingly stringent. Multichannel dense wavelength division multiplexing systems require LDs with low-chirp and weak mode competition characteristics (Wilson, 2002). However when temperature varies the mode-hopping occurs in LD operation. Mode-hopping is an abrupt switching from one longitudinal mode to another: the lasing mode hops to another mode in single mode operation (SMO) or the peak mode changes in nearly SMO (Fig. 6), when the current and/or temperature changes, or due to back-reflection influence (Fukuda, 2000; Gity et al. 2006; Paoli, 1975). Mode-hopping noise of LD has been established to be one of the most limiting factors in high-speed lightwave systems. It causes redistribution of light intensity between longitudinal modes (Asaad et al., 2004; Orsal et al., 1994), and leads to an intensive fluctuation of intensity of output light power, to the wider and unstable LD radiation spectrum. During mode-hopping very intensive optical and electrical fluctuations with Lorentzian type spectrum are observed in the Fabry-Pérot laser operation (Figs. 3-5) (Palenskis et al., 2003; Pralgauskaite et al., 2011; Saulys et al., 2010). A detailed investigation of the low frequency noise characteristic of InGaAsP/InP Fabry-Pérot lasers with ten 4 nm thick compressively strained quantum wells in the active region was carried out in order to clear up the reasons of the mode-hopping and describe physical processes that lead to the very intensive optical and electrical noises, and high cross-correlation during mode-hopping.

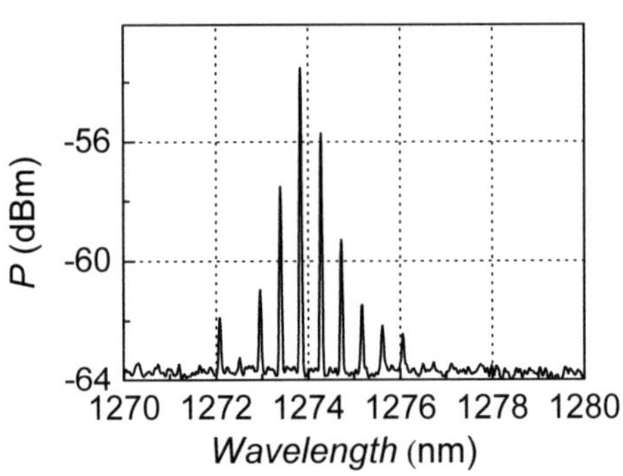

Fig. 6. Light emission spectrum of Fabry-Pérot LD at stable operation (at 72 mA forward current).

Mode-hoping, that took place in FP laser diode operation at particular operation conditions (at particular injection current and temperature), leads to the highly correlated intensive optical and electrical fluctuations (Fig. 7 and 8). Mode-hopping noise peaks are very

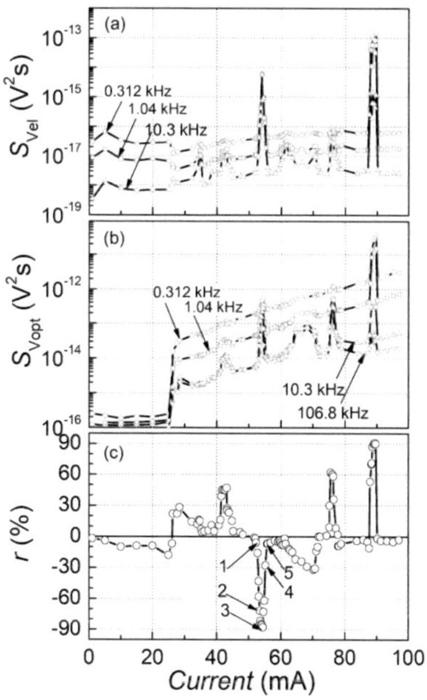

Fig. 7. Spectral density of electrical (a) and optical (b) noises at different frequencies and cross-correlation factor over 10 Hz – 20 kHz frequency range dependencies on injection current (FP laser with 500 μm cavity and 1.03 eV barrier layer band-gap).

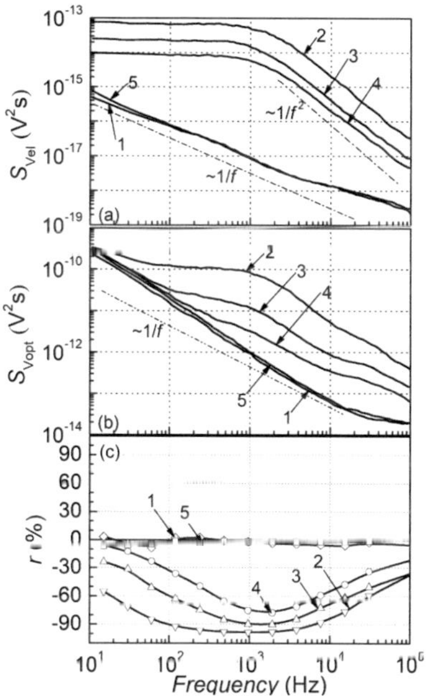

Fig. 8. Electrical (a) and optical (b) noise spectra and cross-correlation factor dependencies on frequency (c) at different injection currents (see numbered points in graph (c) in Fig. 7: 1- 52.1 mA, 2- 53.5 mA, 3- 54.4 mA, 4- 55.2 mA, 5- 55.9 mA) at mode- hopping region (FP laser with 500 μm cavity and 1.03 eV barrier layer band-gap).

sensitive to the laser current and temperature. Noise spectral density at mode-hopping is 1-3 orders of magnitude larger comparing to the stable operation region. Cross-correlation factor between optical and electrical fluctuations at mode-hopping increases up to (60-90) % and could be positive or negative. These correlated noise components distinguish by Lorentzian-type spectrum, and these fluctuations prevail $1/f^{\alpha}$ -type noise at the mode-hopping; also it is seen that high correlation appears only during mode-hopping and is close to zero over all frequency range during stable operation (Fig. 8). As it has been mentioned, optical and electrical noises at some mode-hopping peaks distinguish by positive and at others by negative cross-correlation. It was also observed that not in all peaks the optical and electrical noises are strongly correlated: in some cases correlation factor is small - (10-15) %. Lorentzian type noise at these peaks is weak, too. Different noise characteristics at the mode-hopping show that noise peaks are caused by separate noise sources, i. e. each noise peak may have a different origin. Also noise characteristics (the most evident is cross-correlation factor change) can change, when temperature changes (Fig. 9). Mode-hopping effect is extremely sensitive to the operation conditions: injection current and temperature, and occurs just in particular injection current and temperature ranges - out step of this range mode-hopping (and with it related intensive correlated Lorentzian type optical and electrical fluctuations) vanishes. When temperature increases, the noise peak position shifts to the lower current. The mode hopping position on temperature and current scales changes from sample to sample even for essentially identical devices. As it is shown in Fig. 9, cross-correlation peak related with mode-hopping is negative in particular range of the temperature and injection current, and it becomes positive, when temperature decreases further. This result again shows that there are more than one noise source in a sample related with intensive Lorentzian type noise.

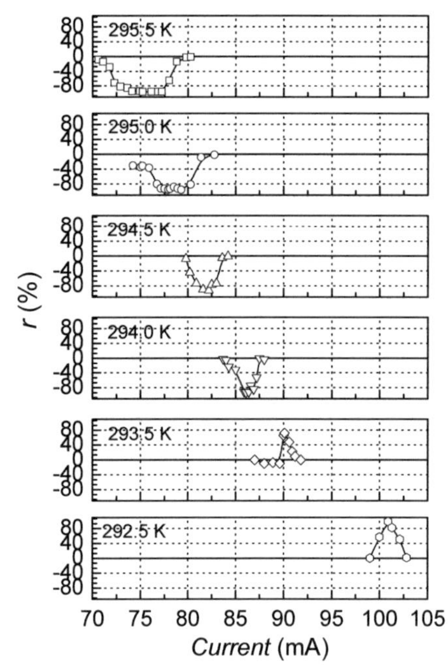

Fig. 9. Cross-correlation factor over 10 Hz – 20 kHz frequency range dependency on injection current at mode-hopping region at different temperatures (500 μm cavity length FP laser with 1.03 eV barrier layer band-gap).

Semiconductor laser radiation spectrum strongly depends on temperature, injection current and optical feedback. For the FP lasers at all laser operation conditions radiation of a few (3-5) longitudinal modes take place (Fig. 6). The main (the most intensive) radiating mode wavelength dependency on laser current is shown in Fig. 10. Laser diode optical gain spectrum moves evenly to the longer wavelength due to Joule heating (lines in Fig. 10) that increases with injection current and temperature increasing. But there are clearly seen uneven changes of the main mode wavelength, when laser current increases: at defined operation conditions these mode intensity redistributions occur in a random manner. Mode-hopping occurs, when laser diode optical gain spectrum (that maximum is determined by the quantum well band-gap energy) moves to the longer wavelength and gain of the adjacent mode exceeds the gain of the radiant peak mode. In this way, the peak longitudinal mode of the laser radiant spectrum changes to the next one (at longer wavelength side). However, there is particular temperature and injection current region, where both peak modes are radiated - laser radiation spectrum is randomly jumping between two radiant modes sets (Fig. 11) - mode-hopping occurs. And observed Lorentzian noise peaks occur during this random modes-hopping.

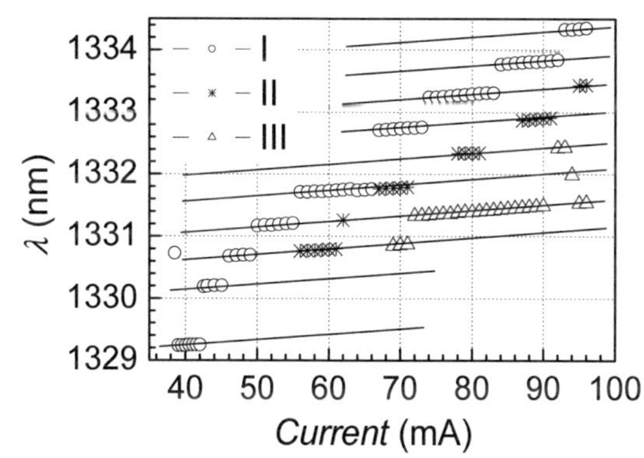

Fig. 10. Three the most intensive modes (I – the most intensive mode, II – the second, and III – the third mode) wavelength dependencies on the laser current at room temperature (FP laser with 500 μm cavity and 1.08 eV barrier layer band-gap).

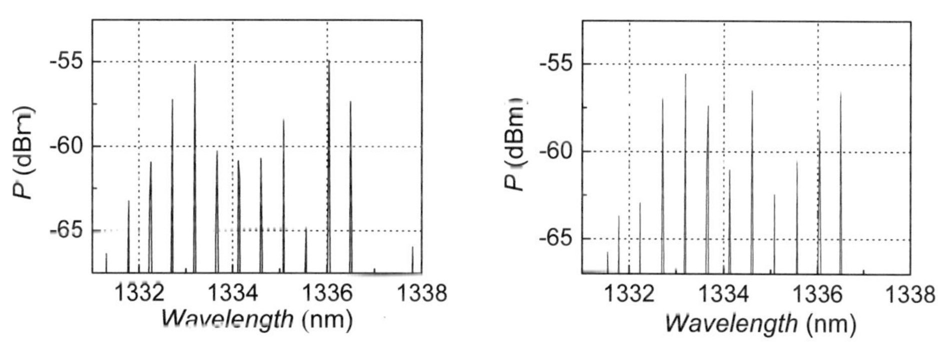

Fig. 11. LD radiation spectra at the mode-hopping operation at differrent time moments (at the same operation conditions: 294.5 K, 81.2 mA) (500 μm cavity length sample with 1.03 eV barrier layer bandgap).

Radiation spectrum of FP lasers is not stable during mode-hopping: optical gain randomly redistributes between several longitudinal mode sets. Consequently, laser diode operation at mode-hopping is unstable. Radiation spectrum of investigated FP LDs consists of two groups of longitudinal modes (in the case presented in Fig. 11 these two groups are around 1333 nm and 1336 nm). At the stable operation one of these groups is clearly more intensive comparing to the second one and total radiation spectrum consists from 5-7 longitudinal modes. When mode-hopping occurs the radiation spectrum visibly outspreads: both modes' groups are intensive, modes' intensity randomly changes in time and spectrum consists of 9-11 modes of almost equal intensity (Fig. 11).

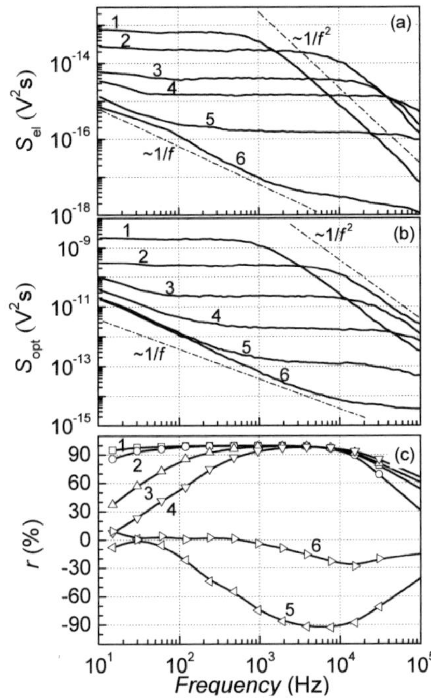

Fig. 12. Electrical (a) and optical (b) noise spectra and cross-correlation factor dependency on frequency at different injection currents and temperatures (1- 95.0 mA, 289.5 K; 2- 82.0 mA; 290.4 K; 3- 72.0 mA, 292.1 K; 4- 64.6 mA, 293.0 K; 5- 52.0 mA, 294.4 K; 6- 49.1 mA, 294.9 K) at the mode-hopping peak maximum (500 μm cavity length sample with 1.03 eV barrier layer band-gap).

Intensive fluctuations observed at mode-hopping peak distinguish by Lorentzian type spectrum and high cross-correlation between optical and electrical noises (Fig. 12). Cross-correlation measurements in one-octave frequency band show that correlation between optical and electrical fluctuations at the mode-hopping depends on frequency region (graph (c) in Fig. 12): the highest cross-correlation factor is found at the frequencies close to the cut-off frequency ($f_c=1/2\pi\tau$, here τ is the relaxation time) observed in noise spectra. When temperature is fixed, the cut-off frequency f_c of the Lorentzian noise does not change at different currents for the same mode-hopping peak. But for different samples, also for the same LD at different operation conditions, different cut-off frequencies are observed: they are in the range from kilohertz to a few megahertz. The characteristic relaxation times for the mode-hopping Lorentzian type noise are distributed from a few milliseconds to a few microseconds (Figs. 12 and 13). The intensive noise during mode-hopping could be related to the recombination processes in the barrier layer that have different capture cross-sections for electrons and, thus, different relaxation times. This leads to different cut-off frequencies of Lorentzian type spectra during mode-hopping for the same sample. As there is large cross-correlation between optical and electrical fluctuations, it shows that these centres are located in the active region. But observed characteristic times indicate that these processes are not related with radiative and non-radiative recombination in the active region (characteristic time of these processes is in the range of nanoseconds). During mode-hopping fulfilment of longitudinal mode radiation threshold condition randomly changes between two mode sets. Injected charge carrier redistribution and lasing energy level position is governed randomly by the state of generation- recombination and carrier capture centres (these centres randomly modulate the barrier height, and therefore the position of lasing levels). Therefore there is strong cross-correlation between optical and electrical fluctuations during mode-hopping. Considering that radiative recombination of charge carriers can occur only in quantum wells, it can be stated that correlated electrical and light intensity fluctuations are related with the random potential height fluctuations of barrier layer due to charge carrier capture and recombination in defects of the barrier layers. These potential fluctuations modulate that part of the carriers that recombines in quantum wells and produce photons.

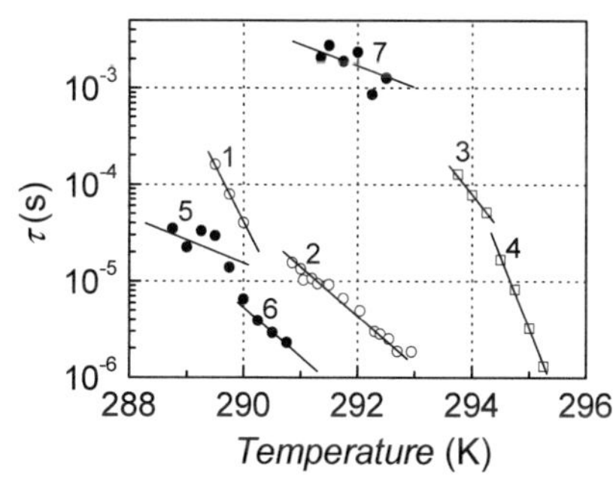

Fig. 13. Characteristic time dependencies on temperature for different Lorentzian noise peaks during mode-hopping for the 1.03 eV barrier layer band-gap sample with 500 μm cavity length (open symbols): 1- current range from 95.0 mA to 89.4 mA, 2- 82.0 mA – 64.6 mA, 3- 88.2 mA – 84.1 mA, 4- 81.9 mA – 75.6 mA; and for the laser with 1.18 eV barrier layer band-gap and 250 μm cavity length (solid symbols): 5- current range from 61.1 mA to 55.3 mA, 6- 55.3 mA – 52.3 mA, 7- 51.5 mA – 43.1 mA.

LD radiation spectrum and mode competition strongly depends on the laser resonator quality and back reflected light (Agrawal & Shen, 1986). Therefore influence of the laser facet reflectivity on the noise characteristics and mode hopping effect has been investigated (Palenskis et al., 2003). Laser facet coating with thin polymer layer changes laser operation (Figs. 14 and 15). After coating (facet reflectivity decreases) LD threshold current slightly increases (from 46.8 mA to 51.0 mA) and efficiency decreases: the optical output power is reduced about 15 % (Fig. 14). In spite of the optical power losses, changes in the noise characteristics given advantages are more significant: after facet coating there are less mode-hopping noise peaks, and noise intensity at the remained peaks is lower about two orders of magnitude (Fig. 14). The mode hopping noise level after coating is substantially reduced, but the background ($1/f^{\alpha}$-type) optical noise level increases about two-three times, while the electrical one decreases by the same degree (Fig. 15). Such small $1/f^{\alpha}$-type noise intensity

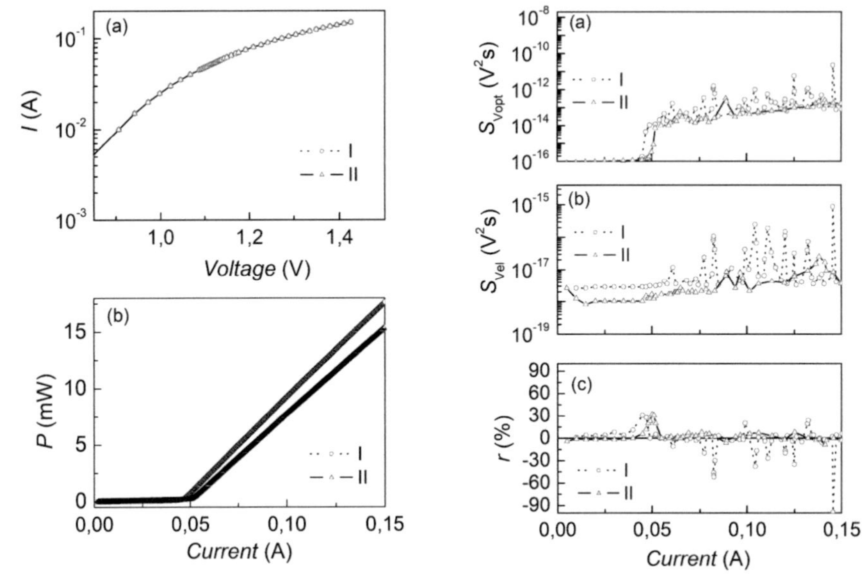

Fig. 14. Current-voltage (a) and LI (b) characteristics before (I) and after (II) antireflection coating (on the left); optical (a) and electrical (b) noise spectral density (10 kHz), and cross-correlation factor between optical and electrical fluctuations (c; 20 Hz-20 kHz) dependencies on laser current before (I) and after (II) antireflection coating (on the right) (FP laser with 750 μm cavity length and 1.08 eV barrier layer band-gap) (Smetona et al., 2001).

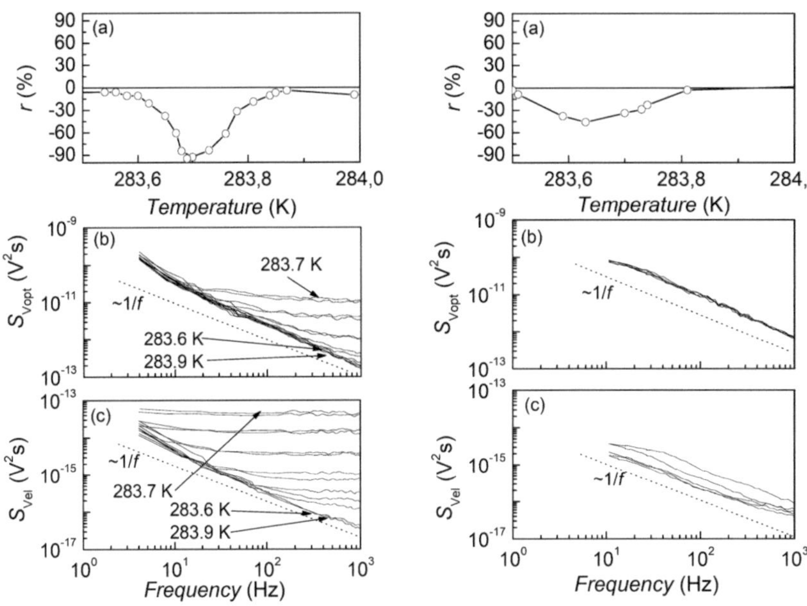

Fig. 15. Cross-correlation factor between optical and electrical fluctuations (a; 1 Hz-1 kHz), optical (b) and electrical (c) noise spectra at the mode-hopping peak before (on the left) and after (on the right) antireflection coating (FP laser with 500 μm cavity length and 1.08 eV barrier layer band-gap).

changes are not important. Lorentzian type noise related with mode-hopping almost disappears after coating: mode-hopping noise becomes lower the background $1/f^{\alpha}$-type noise. But cross-correlation factor still feels existence of the correlated noise related with mode-hopping (Fig. 15). So, cross-correlation is more sensitive to the various operation peculiarities.

Decreasing of the resonator mirror reflection leads to the decrease of the resonator quality, and to the significant reduction of the mode concurrency. Due to the same reasons current and temperature range, where mode hopping event occurs, become wider: weaker mode competition leads to the less strict transition between different radiation spectra. So, after the facet coating FP LD operation becomes more stable.

On the other hand, laser facets coating by antireflection layer results indicate that the level of back-reflected light is very important to the LD operation changing charge carrier and photon distribution and confinement in the active layer and neighbouring regions.

Summary of the Section. Noise characteristics at the mode-hopping point distinguish by highly correlated intensive Lorentzian type optical and electrical fluctuations - mode-hopping noise peaks. Mode-hopping occurs at particular operation conditions: position of these noise peaks is very sensitive to the current, temperature and optical feedback. The charge carrier lifetime in the active region of a few nanoseconds is much shorter than relaxation times of physical processes related with Lorentzian type noise during mode-hopping (that ranges from microseconds to millisecond). This indicates that mode-hopping is not related with charge carrier stimulated or spontaneous recombination in the quantum wells. Mode-hopping effect arises due to charge carrier gathering not in the quantum wells but in the barrier or cladding layers: generation-recombination and carrier capture processes in the centres formed by defects in the barrier and/or cladding layers (random change of state of the centres modulates barrier height and, therefore, lasing conditions for each longitudinal mode). Different cross-correlation factor value and sign, different Lorentzian type spectrum cut-off frequencies at different LD operation conditions show that there are several type recombination centres (with different capture cross-section) that determine FP LD noise characteristics during mode-hopping.

5. Noise characteristics and reliability of laser diodes

Reliability of the laser diodes is a key parameter in various applications, and its prediction in early phase of fabrication could significantly lower the expenses. Low-frequency $1/f^{\alpha}$-type noise is a typical excess noise that is a very sensitive measure of the quality and reliability of optoelectronic devices, as it is related with the defectiveness of the structure of the device (Jones, 2002; Lin Ke et al., 2010; Vandamme, 1994). Measurement of low frequency noise can indicate the presence of intrinsic defects as well as fabrication imperfections, which act as deep traps or recombination centres in semiconductor materials and their structures (Fukuda et al., 1993; Jones, 1994; Palenskis, 1990). This section presents noise characteristic analysis that was performed in order to clear up the origin of laser diode lower quality and degradation, factors that accelerate degradation, to find out noise parameter that could be used for LD quality and reliability prediction. For these purposes special attention was paid to the noise features at the threshold current, aging experiments and measurements at low bias have been carried out (Letal et al., 2002; Matukas et al., 2001; Palenskis et al., 2006; Pralgauskaite et al., 2003, 2004a, 2004b).

On purpose for set tasks specially selected few batches of MQW DFB lasers radiating around 1.55 μm wavelength have been investigated. Sample batches differ by the device reliability: the reliability of the devices was checked on the ground of the threshold current changes with aging time:

- lasers from the batch G are rapidly degrading devices;
- samples from the batch H distinguish by very good quality and reliability.

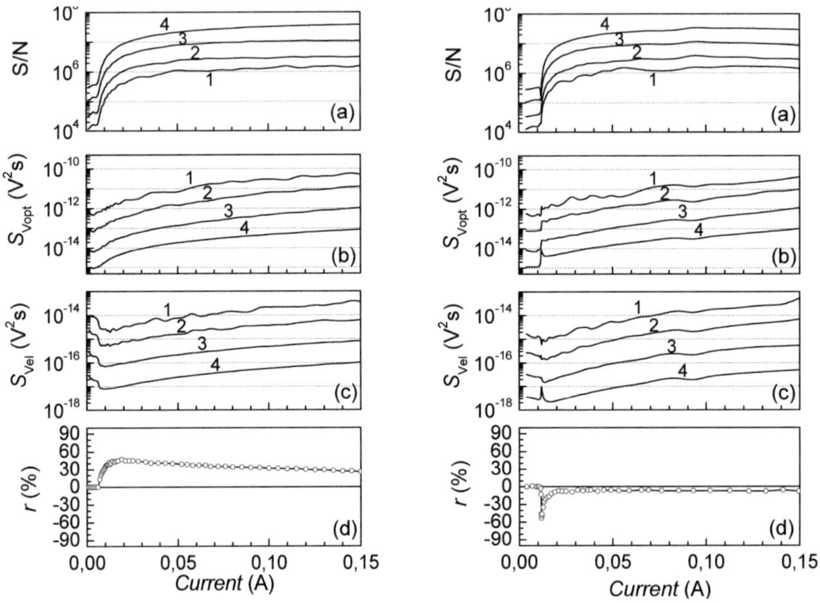

Fig. 16. Signal-to-noise ratio (a), optical (b) and electrical (c) noise spectral density (1- 22 Hz, 2- 108 Hz, 3- 1 kHz, 4- 10 kHz), and cross-correlation factor between optical and electrical fluctuations (d; 20 Hz-20 kHz) dependencies on laser current for reliable laser from batch H (on the left) and quickly degrading one from batch G (on the right) before aging. Reprinted with permission from (Letal et al., 2002). Copyright 2002, American Vacuum Society.

Primarily let us compare initial (of not aged samples) noise characteristics of LDs that would-be of different reliability (Fig. 16). The noise measurement results for stable and degrading not-aged LDs exhibited some differences in noise characteristic. The noise intensity in the lasing operation region of samples from all investigated batches is similar: the optical and electrical noise spectra are of $1/f^{\alpha}$ -type and noise intensity increases approximately proportionally to the laser current. But cross-correlation factor is more sensitive to the laser quality. It is able reflect operation differences that cannot be observed in noise spectral density level, because noise signal is a superposition of a lot of various noise origins and their intensities are averaged. Larger positive correlation at lasing current has been observed for better quality lasers (graphs (d) in Fig. 16): stable devices (batch H) exhibit strong positive cross-correlation factor ((30-60) % for different samples) between the electrical and optical noises above threshold, while rapidly degrading samples (batch G) distinguish by close to zero or even negative cross-correlation factor above threshold. Various semiconductor growth and laser processing defects cause $1/f^{\alpha}$-type noise. Positive correlation in the LD operation indicates that device contains defects in the barrier and other layers close to the active region that substantially modulate current flowing through the active region and lead to the charge carrier and photon number changes in-phase. The defects that cause positively correlated optical and electrical fluctuations distinguish by noise spectra of $1/f^{\alpha}$-type. Injection current fluctuations in the layers remote from the active region are the reason of not 100 % correlation between optical and electrical noises.

The cross-correlation factor behaviour of degrading lasers can be explained as the superposition of two noise sources, the second of which distinguishes by the negative cross-correlation. LD threshold current magnitude is related with laser diode quality: better quality samples usually have lower threshold. Influence on the cross-correlation of the negatively correlated noise components increases with sample threshold current increase. Noise source that causes optical and electrical noises with negative cross-correlation are originated from defects related with leakage current (therefore, larger threshold current is needed to achieve lasing). The existence of such defects is reflected in the cross-correlation between optical and electrical fluctuations.

Accelerated aging experiments have been applied for both groups of lasers: ~500 h aging at 100 ⁰C and 150 mA forward current. Clear noise characteristic degradation of rapidly degrading samples G has been observed after aging (Fig. 17): after aging $1/f^{\alpha}$-type noise intensity of G samples increases about a half of order, while there are no noticeable changes in H LDs operation. Cross-correlation factor between G laser optical and electrical noise changes are especially evident: cross-correlation factor become negative (about 50 % (before aging it was close to zero)) over all lasing current range (graph (d) in Fig. 17). These operation changes during long-time aging indicate that in the G devices structure there are mobile defects that migrate during aging and cause operation characteristic degradation.

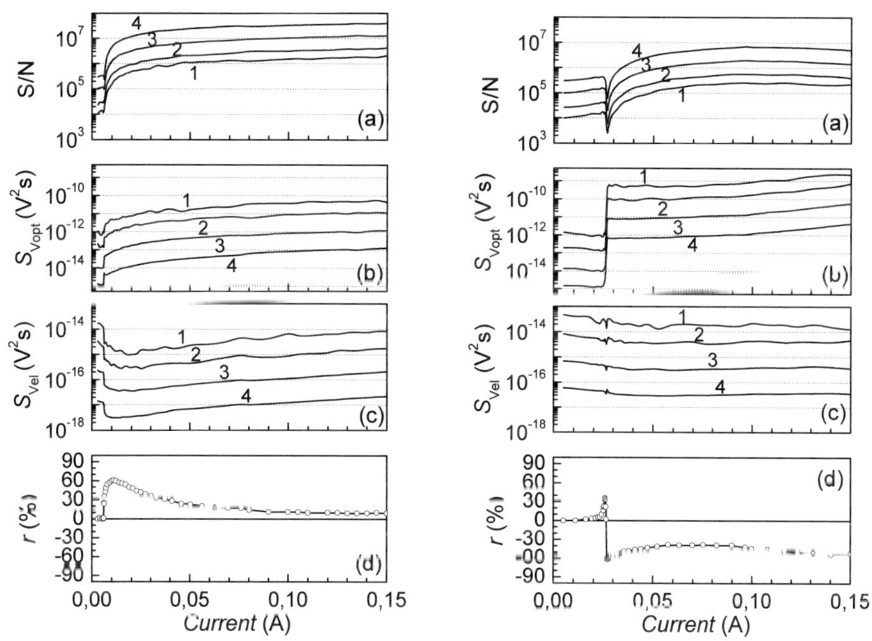

Fig. 17. Signal-to noise ratio (a), optical (b) and electrical (c) noise spectral density (1- 22 Hz, 2- 108 Hz, 3- 1 kHz, 4- 10 kHz), and cross-correlation factor between optical and electrical fluctuations (d; 20 Hz-20 kHz) dependencies on laser current for reliable laser from batch H (on the left) and quickly degrading one from batch G (on the right) after aging (aging conditions: 100 ⁰C, 150 mA, ~500 h). Reprinted with permission from (Letal et al., 2002). Copyright 2002, American Vacuum Society.

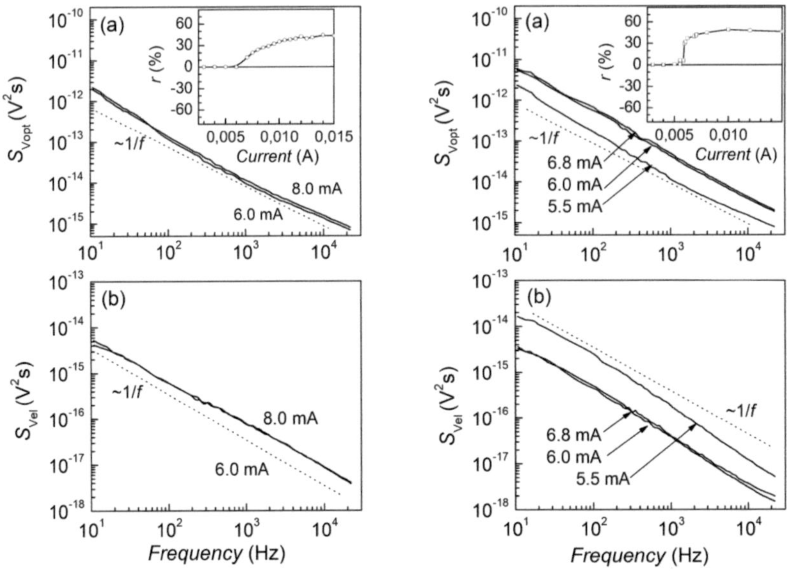

Fig. 18. Optical (a) and electrical (b) noise spectra for reliable laser from batch H around the threshold before (on the left) and after (on the right) aging (aging conditions: 100 ^{0}C, 150 mA, ~500 h; inset: cross-correlation factor dependency on laser current in frequency range 20 Hz-20 kHz). Reprinted with permission from (Letal et al., 2002). Copyright (2002), American Vacuum Society.

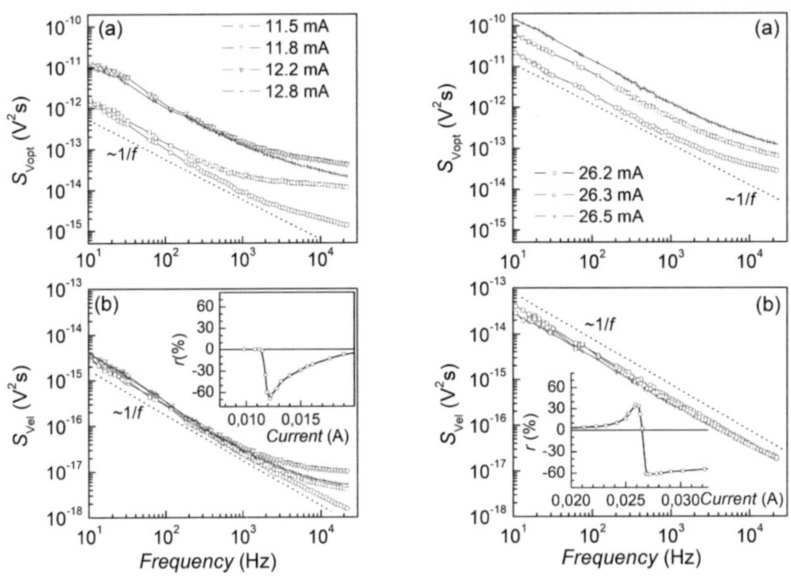

Fig. 19. Optical (a) and electrical (b) noise spectra for quickly degrading laser from batch G around the threshold before (on the left) and after (on the right) aging (aging conditions: 100 ^{0}C, 150 mA, ~500 h; inset: cross-correlation factor dependency on laser current in frequency range 20 Hz-20 kHz). Reprinted with permission from (Letal et al., 2002). Copyright 2002 American Vacuum Society.

Special attention was paid to the noise characteristics in the threshold region (Figs. 18 and 19), as it was found that this transitional operation region is especially sensitive to the device quality. Laser diode threshold characteristics are important due to the threshold parameters that cause the lasing action characteristics.

Lasers from the group G distinguish not only by the larger threshold current, but also by rapid it degradation during aging. Noise characteristics at the threshold after long-time aging have degraded, too. Both optical and electrical noise intensity at the threshold increases after aging by more than an order of magnitude, additional Lorentzian type noise with characteristic negative cross-correlation is observed at the threshold, and pulse noise appearance could be expected below threshold (Fig. 19). Threshold noise characteristics of the stable samples do not change during long-time aging (Fig. 18).

From Fig. 20, it is seen that negative cross-correlation at the threshold region is caused by the higher frequency fluctuation component. At lower frequencies (20 Hz-1 kHz), where $1/f^{\alpha}$-type noise intensity exceeds the Lorentzian type noise level (curve number 1 in Fig. 20) positive cross-correlation factor value is observed. This shows that electrical and optical fluctuations with $1/f^{\alpha}$ -type spectrum distinguish by the positive cross-correlation (from 10 % to 30 % for different samples). These results contrast with that, what is observed at the higher, 1 kHz-20 kHz, frequency range (curve number 4 in Fig. 20), where Lorentzian type noise component dominates in the low quality LD operation. The negative cross-correlation factor is observed at the threshold of the degrading sample before aging. Thus, Lorentzian type noise and negative cross-correlation observed at the threshold are caused by the same origin (i. e. additional "white" electrical and optical fluctuations are negatively correlated). $1/f^{\alpha}$-type and Lorentzian type noises superpose at the threshold current region.

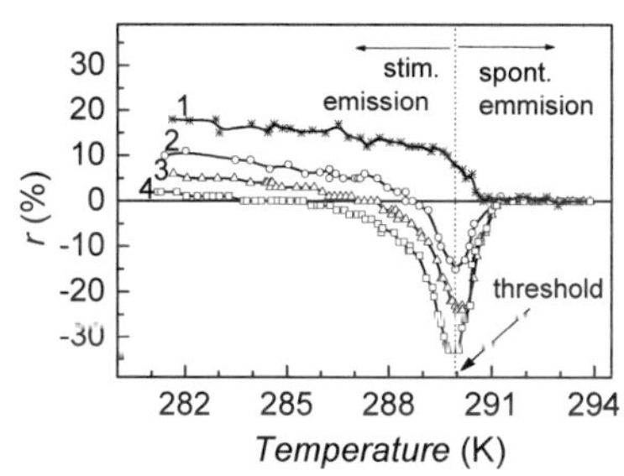

Fig. 20. Typical cross-correlation factor behaviour in the vicinity of the threshold region for the degrading sample: cross-correlation factor dependency on temperature at the threshold current (10.1 mA) (1: 20 Hz–1 kHz, 2: 20 Hz–20 kHz, 3: 200 Hz–20 kHz, 4: 1 kHz–20 kHz; the device was not aged). Reprinted with permission from (Letal et al., 2002). Copyright 2002 American Vacuum Society.

It is observed significant differences in noise characteristic in the vicinity of the threshold between good and poor quality LDs (Figs. 18 and 19). To relate noise characteristic features with LD reliability the detailed low temperature short-time aging experiments have been conducted: 60 ^{0}C temperature and forward current in the range of (100-150) mA have been chosen as aging conditions. Not-burned-in samples from good and poor LD quality groups H and G have been chosen for aging experiments. In Figs. 21 and 22, $I{\times}t$ parameter on x-axes (laser aging current multiplied by aging time) is used in order to avoid confusion due to different current used in different aging steps (at initial aging phase lower current was used because noise characteristic changes have occurred during extremely short time).

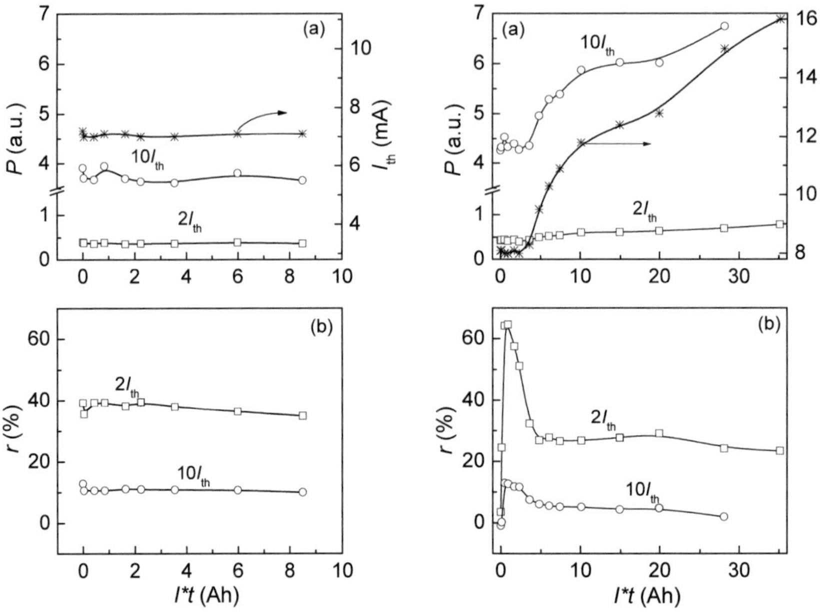

Fig. 21. Optical output power (a, P), threshold current (a, I_{th}) and cross-correlation factor (b; 20 Hz-20 kHz) dependencies on the aging time for different laser currents of reliable (on the left, batch H) and degrading (on the right, batch G) LDs (Pralgauskaite et al., 2004a).

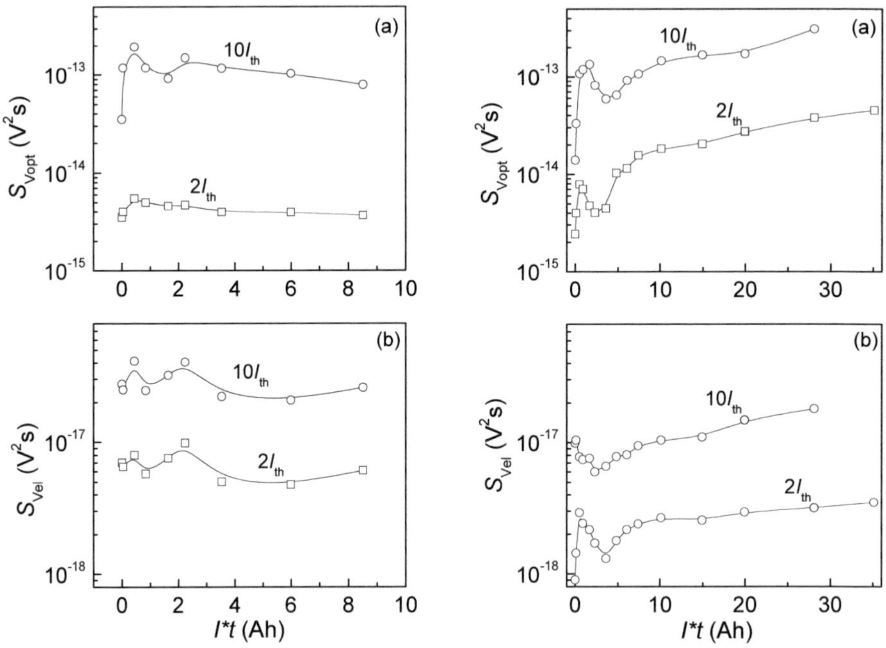

Fig. 22. Optical (a) and electrical (b) noise spectral density (10 kHz) dependencies on the aging time for different laser currents of reliable (on the left, batch H) and degrading (on the right, batch G) lasers (Pralgauskaite et al., 2004a).

Good quality stable devices from group H show neither threshold current and output power, nor noise intensity and cross-correlation factor changes during aging (Figs. 21 and 22). The results indicate that defects that cause $1/f^{\alpha}$-type noise with characteristic positive cross-correlation are stable and do not change their position and parameters during aging. Good quality lasers do not have mobile defects that cause Lorentzian type noise with characteristic negative cross-correlation at the threshold and that deteriorate LD operation characteristics during aging.

Threshold current and output power intensity of degrading devices from group G do not increase in the initial phase of the aging, while noise characteristic changes are the largest in this phase (Figs. 21 and 22). The largest changes have been observed in the cross-correlation between optical and electrical fluctuations factor: e. g., at forward current approximately equal to $2I_{th}$ cross-correlation factor changes from 0 to 68 % (graph (b) in Fig. 21). After further aging stage (next (6-8) h) cross-correlation factor decreases to more moderate positive value, while optical and electrical noise intensities stay at higher level and have tendency slightly increase with aging time. When threshold current of degrading device starts to increase with aging time ($I \times t$>4 Ah, graph (a) in Fig. 21), noise characteristic changes have saturated (Fig. 22). Therefore, noise characteristics are more sensitive to the processes related with short LD lifetime than threshold current or output power, and could let know beforehand about rapid device degradation.

To confirm that worse quality (large threshold current, short lifetime) of laser diodes is related with defects that cause current leakage out of the active region, investigations at low bias (at currents much lower the threshold one) have been carried out. Operation at low bias (<1 mA for investigated laser diodes) is very sensitive to the current flow conditions through the structure as at low current density, if structure is defective, current flows not through the whole device cross-section, but through the defects formed leakage channels (Jones, 2002; Shuang et al., 2007). Therefore this investigation can clear up the diode reliability problems. In Fig. 23, there are compared low bias characteristics of the same design but different quality devices. A non-ideality factor of current-voltage characteristic, *m*, for investigated samples for all groups is larger than 2 (the latter value is a characteristic one for good quality *pn* diodes), what indicates non-uniform current flow and leakage current: the larger value of non-ideality factor the larger leakage current is implied. It is clear seen that larger non-ideality factor is characteristic for samples with larger threshold current and worse reliability; good quality laser diodes from the H batch have much smaller leakage current compare to the degrading samples from the G batch (Fig. 23). Consequently, origin of poor lasing characteristics, worse reliability and leakage currents are related.

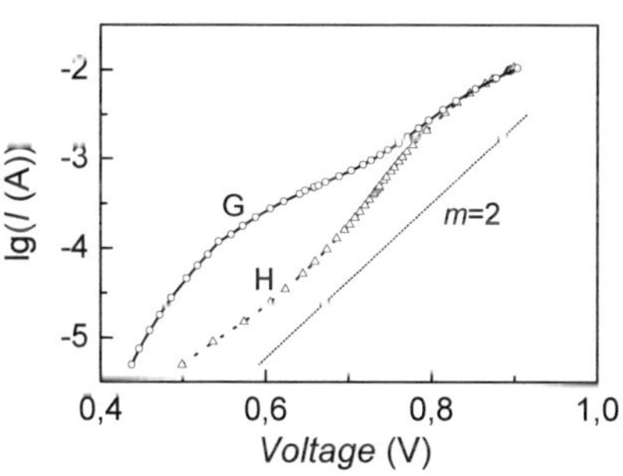

Fig. 23. Current-voltage characteristics for good (H) and worse (G) quality laser diodes.

Summarizing this Section, reasons of worse operation characteristics and poor reliability are common for various design semiconductor lasers: defects at the active region interface lead to the injected charge carrier leakage out of the active region. These defects increase leakage current that leads to the larger threshold current. Defects migrate during aging to form clusters that deteriorate laser characteristics. Low-frequency noise investigation, especially cross-correlation factor behaviour analysis at the threshold, can detect the presence of such defects in the laser diode structure.

6. Noise characteristics and reliability of buried heterostructure laser diodes

Buried-heterostructure (BH) laser diodes distinguish by extremely low threshold current and high efficiency. The advantages are achieved due to strict injection current and optical field confinement in the active region by the current-blocking layers (Lee et al., 2008; Mito et al., 1982; Nunoya et al., 2000; Yamazaki et al., 1999). But additional etching and current-blocking layers regrowth introduce defects at the interface between the active region and current-blocking layers (Fukuda et al., 1993; Krakowski et al., 1989; Oohashi et al., 1999). Such defects, in principle, can lead to the worse LD operation characteristics and quick degradation. It is possible for such lasers to exhibit thyristor turn-on, what does away with laser operation. For the proper LD operation the thyristor turn-on current has to be significantly higher than the operating one and does not reduce with aging. This Section is concentrated on buried-heterostructure laser quality problems (Letal et al., 2002; Matukas et al., 2002; Pralgauskaite et al., 2003).

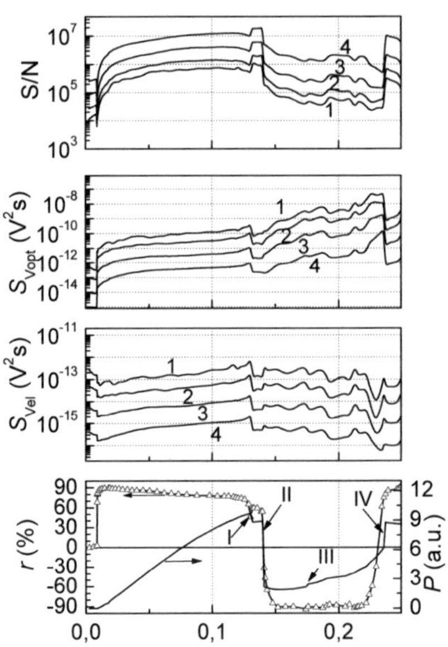

Fig. 24. Signal-to-noise ratio (S/N), optical (S_{vopt}) and electrical (S_{vel}) fluctuation spectral density (1 - 22 Hz, 2 - 108 Hz, 3 - 1.03 kHz, 4 - 10.7 kHz), cross-correlation factor (*r*; 20 Hz–20 kHz) and optical output power (*P*) dependencies on laser current of BH LD with thyristor-like forward breakdown. Reprinted with permission from (Letal et al., 2002). Copyright 2002 American Vacuum Society.

A set of BH DFB devices, where the thyristor turns-on at relatively low currents (~120 mA) due to a non-optimal current-blocking layer growth, has been investigated. The larger injection current, the larger leakage current flows through *pnpn* current-blocking layers. When the particular leakage current is reached the thyristor turns-on. In this regime the resistance of the blocking layers decreases, what causes a sudden decrease of the current density in the active region and the optical output power: the current-voltage and LI characteristics exhibit a kink (Fig. 24). The optical output power and signal-to-noise ratio are very low, when the thyristor has turned-on, and the optical noise is a complex function of the current (Fig. 24). Even so, the noise intensity remains at its characteristic values, and in some current regions is even higher. Cross-correlation factor at current higher than the thyristor turn-on also has a complex behaviour (Fig. 24). Noise characteristics of these samples at lower currents (below the thyristor turn-on current) are typical for the DFB laser diodes and do not have any peculiarities. Noise spectra are $1/f^{\alpha}$-type (graphs (a) in Fig. 25).

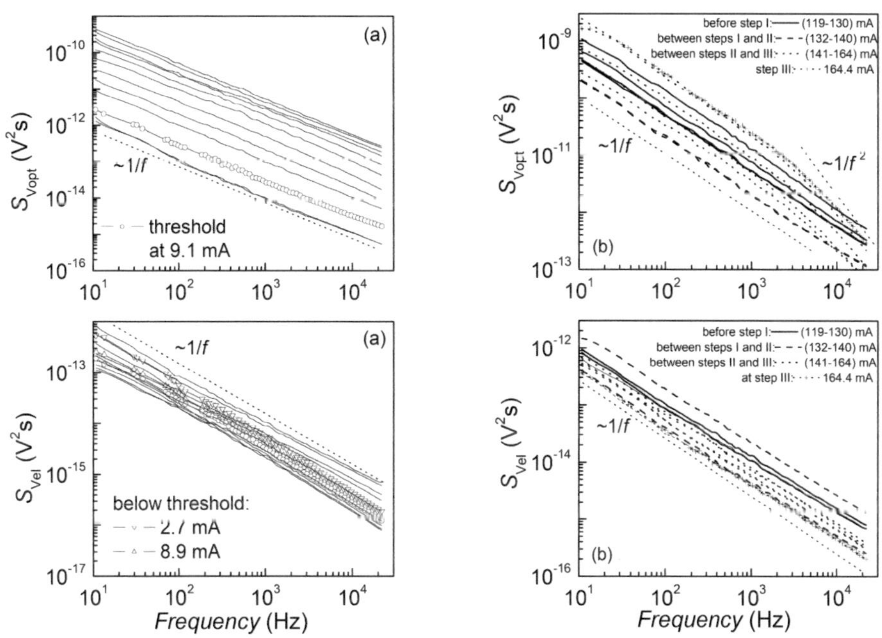

Fig. 25. Optical (S_{Vopt}) and electrical (S_{Vel}) noise spectra of BH laser diode with thyristor-like forward breakdown: (a) below thyristor turn-on current ((3-117) mA), (b) in the current range from 117 mA to 165 mA (for numerated steps look at Fig. 24).

For these samples at lasing operation very large (>90%) positive cross-correlation factor is characteristic (Fig. 24). In addition, optical and electrical noise spectral densities are about 1-2 order of magnitude larger (and the signal-to-noise ratio is about 3 times lower) in the normal lasing region for these devices comparing to the BH lasers without thyristor turn-on effect. These results suggest that in the "thyristor" case, the active region of the BH laser contains additional defects that causes non-uniform current flow and its leakage to the current-blocking layers, and affects both the electrical and optical noise. When coming close to the each kink in the LI characteristic optical and electrical noise intensities increase up to very large values. With sudden optical power decrease noise intensity decreases to the "normal" operation values, and cross-correlation factor changes from positive to negative (Fig. 24). For the low optical power region large negative cross-correlation factor is characteristic. At the "bottom", when optical power is low, optical noise intensity stays large enough, so, signal-to-noise ratio is very low. Cross-correlation factor becomes positive again if optical power increases. After the thyristor turns-on the Lorentzian type component with cut-of frequency of ≈3 kHz has been observed in the optical noise spectra (graphs (b)-(d) in Fig. 25). Therefore, some processes with characteristic time of 0.3 ms decide optical fluctuations. In Fig. 26, there are presented possible charge carrier leakage to the thyristor-type current-blocking layers paths. Current leakage through resistances R_L leads to the lowering of the thyristor forward breakdown voltage. So, accurate controlling of the active region position with respect to the blocking layers *np* junction has to be guaranteed.

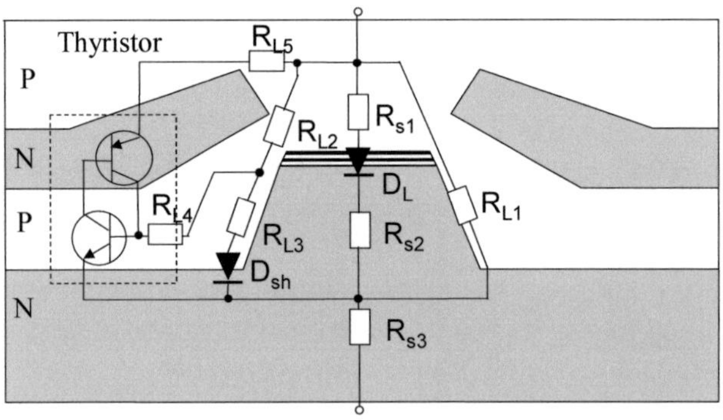

Fig. 26. Schematic diagram of BH LD structure showing leakage paths (R_L - leakage path resistances, R_s - series resistances, D_L is the laser diode, and D_{sh} is the shunt diode in the blocking layers). Reprinted with permission from (Letal et al., 2002). Copyright 2002 American Vacuum Society.

Summary of the Section. Formation of current-blocking layers in buried-heterostructure LDs structure introduces additional defects in the active region interface that lead to the injection current leakage out of the active region and enables thyristor turn-on at low current. Low frequency noise characteristics indicate larger density of defects at the active region interface of BH lasers: devices that exhibit a low-current thyristor turn-on show strongly correlated intensive $1/f^{\alpha}$-type optical and electrical low-frequency noise both before and after the turn-on.

7. Conclusions

Semiconductor lasers at stable lasing operation distinguish by $1/f^{\alpha}$-type partly correlated optical and electrical fluctuations that are caused by the inherent material defects and defects created during the device formation. Lorentzian type noise peaks observed during mode-hopping effect in Fabry-Pérot laser operation are related to the recombination processes in the barrier and cladding layers through the centers formed by the defects (random change of the state of the centers modulates barrier height and, therefore, lasing conditions for each longitudinal mode, that are very sensitive to the light reflection from the facets) involving electrons and holes having different capture cross-sections, what is reflected in different cut-off frequencies of the Lorentzian spectra.

Worse characteristics (larger threshold current, lower efficiency) and poor reliability of laser diodes are caused by the presence of defects at the surface of the active region. These defects increase the leakage currents that lead to the larger LDs threshold. Presence of such defects in LD structure manifests as additional negatively correlated Lorentzian type noise (especially evident at the threshold operation). During ageing, these defects migrate, form clusters, deteriorate the laser characteristics, and, ultimately, cause the failure of LD. Low-frequency noise investigations can detect the presence of these defects.

The presented data suggests that electrical and optical noise tests (especially cross-correlation factor investigation at the threshold), that are quick and undestructive, can be used as the lifetest screen to distinguish reliable and unreliable laser diodes without traditional long-time lifetime tests.

8. References

Agrawal, G. & Shen, T. (1986). Effect of fiber-far-end reflections on the bit error rate in optical communication with single-frequency semiconductor lasers. *Journal of Lightwave Technology*, Vol. 4, No. 1, (January 1986), pp. 58-63, ISSN 0733-8724

Amstrong, J.M. & Smith, A.W. (1965). Intensity fluctuations in GaAs laser emission. *The physical Review C*, Vol. 140, No. 1(A), (January 1965), pp. 155-164, ISSN 0556-2813

Asaad, I. et al. (2004). Characterizations of 980 nm aged pump laser by using electrical and optical noise for telecommunication systems, *Proceedings of 2004 International Conference on Information and Communication technologies: from Theory to Applications*, pp. 173-174, ISBN: 0-7803-8482-2, Damascus, Syria, 19-23 April, 2004

Fronen, R.J. & Vandamme, L.K.J. (1988). Low-frequency intensity noise in semiconductor lasers. *IEEE Journal of Quantum Electronics*, Vol. 24, No. 5, (May 1988), pp. 724-736, ISSN 0018-9197

Fukuda, M. et al. (1993). 1/f noise behavior in semiconductor laser degradation. *IEEE Photonics Technology Letters*, Vol. 5, No. 10, (October 1993), pp. 1156-1157, ISSN 1041-1135

Fukuda, M. (2000). Historical overview and future of optoelectronics reliability for optical fiber communication systems. *Microelectronics Reliability*, Vol. 40, No. 1, (January 2000), pp. 27-35, ISSN 0026-2714

Gity, F. et al. (2006). Numerical analysis of void-induced thermal effects on GaAs/AlGaAs high power quantum well laser diodes, *Proceedings of 2006 IEEE GCC Conference*, pp. 1-6, ISBN: 978-0-7803-9590-9, Manama, 20-22 March, 2006

Jacobsen, G. (2010). Laser phase noise induced error-rate floors in differential n level phase-shift-keying coherent receivers. *Electronics Letters*, Vol. 46, No. 10, (May 2010), pp. 698-700, ISSN 0013-5194

Jones, B.K. (1994). Low-frequency noise spectroscopy. *IEEE Transaction on Electron Devices*, Vol. 41, No. 11, (November 1994), pp. 2188-2197, ISSN 0018-9383

Jones, B.K. (2002). Electrical noise as a reliability indicator in electronic devices and components. *IEE Proceedings on Circuits, Devices and Systems*, Vol. 149, No. 1, (February 2002), pp. 13-22, ISSN 1350-2409

Gray, G.R. & Agrawal, G.P. (1991) Effect of cross saturation on frequency fluctuations in a nearly single-mode semiconductor laser. *IEEE Photonic Technology Letters*, Vol. 3, No. 3, (March 1991), pp. 204-206, ISSN 1041-1135

Krakowski, M. et al. (1989). Ultra-low-threshold, high-bandwidth, very-low noise operation of 1.52 µm GaInAsP/InP DFB buried ridge structure laser diodes entirely grown by MOCVD. *IEEE Journal of Quantum Electronics,* Vol. 25, No. 6, (June 1989), pp. 1346-1352, ISSN 0018-9197

Lau, K.Y. et al. (1993). Signal-induced noise in fiber-optic links using directly modulated Fabry-Pérot and distributed-feedback laser diodes. *Journal of Lightwave Technology,* Vol. 11, No. 7, (July 1993), pp. 1216-1215, ISSN 0733-8724

Lee, F.-M. et al. (2008). High-Reliable and High-Speed 1.3 µm Complex-Coupled Distributed Feedback Buried-Heterostructure Laser Diodes With Fe-Doped InGaAsP/InP Hybrid Grating Layers Grown by MOCVD. *IEEE Transaction on Electron Devices,* Vol. 55, No. 2, (February 2008), pp. 540-546, ISSN 0018-9383

Letal, G. et al. (2002). Reliability and Low-Frequency Noise Measurements of InGaAsP MQW Buried-Heterostructure Lasers. *Journal of Vacuum Science Technology,* Vol. A20, No. 3, (May 2002), pp. 1061-1066, ISSN 0734-2101.

Lin Ke et al. (2010). Investigation of the Device Degradation Mechanism in Pentacene-Based Thin-Film Transistors Using Low-Frequency-Noise Spectroscopy. *IEEE Transaction on Electron Devices,* Vol. 57, No. 2, (February 2010), pp. 385-390, ISSN 0018-9383

Matukas, J. et al. (1998). Low-frequency noise in laser diodes, *Lithuanian Journal of Physics,* Vol. 38, No. 1, (February 1998), pp. 45-48, ISSN 1648-8504

Matukas, J. et al. (2001). Optical and electrical noise of MQW laser diodes, *Lithuanian Journal of Physics,* Vol. 41, No. 4-6, (June 2001), pp. 429-434, ISSN 1648-8504

Matukas, J. et al. (2002). Optical and electrical characteristics of InGaAsP MQW BH DFB laser diodes. *Material science forum: Ultrafast Phenomena in semiconductors 2001,* Vol. 384-385, (2002), pp. 91-94, ISSN 0255-5476

Mito, I. et al. (1982). Double-channel planar buried-heterostructure laser diode with effective current confinement. *Electronics Letters,* Vol. 18, No. 22, (October 1982), pp. 953-954, ISSN 0013-5194

Mohammadi, S. & Pavlidis, D. (2000). A nonfundamental theory of low-frequency noise in semiconductor devices. *IEEE Transaction on Electron Devices,* Vol. 47, No. 11, (November 2000), pp. 2009-2017, ISSN 0018-9383

Nilson, O. et al. (1991). Modulation and noise spectra of complicated laser structures, In: *Coherence, amplification and quantum effects in semiconductor lasers,* Y. Yamamoto (Ed.), pp. 77-96, John Willey & Sons Inc., ISBN 978-047-1512-49-3, New York, USA

Nunoya, N. et al. (2000). Sub-milliampere operation of 1.55 µm wavelength high index-coupled buried heterostructure distributed feedback lasers. *Electronics Letters,* Vol. 36, No. 14, (July 2000), pp. 1213-1214, ISSN 0013-5194

Oohashi, H. et al. (1999). Degradation behavior of narrow-spectral-linewidth DFB lasers for super-wide-band FM conversion. *IEEE Journal of Selected Topics in Quantum Electronics,* Vol. 5, No. 3, (May/June 1999), pp. 457-462, ISSN: 1077-260X

Orsal, B. et al. (1994). Correlation between electrical and optical photocurrent noises in semiconductor laser diodes. *IEEE Transaction on Electron Devices,* Vol. 41, No. 11, (November 1994), pp. 2151-2161, ISSN 0018-9383

Palenskis, V. (1990). Flicker noise problem (review). *Lithuanian Journal of Physics,* Vol. 30, No. 2, (January 1990), pp. 107-152, ISSN 1648-8504

Palenskis, V. et al. (2003). Experimental investigations of the effect of the mode-hopping on the noise properties of InGaAsP Fabry-Pérot multiple-quantum-well laser diodes.

IEEE Transaction on Electron Devices, Vol. 50, No. 2, (February 2003), pp. 366-371, ISSN 0018-9383

Palenskis, V. et al. (2006). InGaAsP laser diode quality investigation and their noise characteristics. *Proceedings of SPIE: Optical Materials and Applications,* Vol. 5946, (June 2006), pp. 290-299, ISBN: 9780819459534

Paoli, L. (1975). Noise characteristics of stripe-geometry double-heterostructure junction lasers operating continuously – I. Intensity noise at room temperature. *IEEE Journal of Quantum Electronics,* Vol. 11, No. 6, (June 1975), pp. 726-783, ISSN 0018-9197

Pralgauskaite, S. et al. (2003). Low-Frequency Noise and Quality prediction of MQW Buried-Heterostructure DFB Lasers. *Proceedings of SPIE: Advanced Optical Devices,* Vol .5123, (August 2003), pp. 85-93, ISBN: 9780819449832

Pralgauskaite, S. et al. (2004a). Fluctuations of optical and electrical parameters of distributed feedback lasers and their reliability. *Fluctuations and Noise Letters,* Vol. 4, No. 2, (June 2004) pp. L365-L374, ISSN 0219-4775

Pralgauskaite, S. et al. (2004b). Fluctuations of optical and electrical parameters and their correlation of multiple-quantum-well InGaAs/InP lasers. *Advanced experimental methods for noise research in nanoscale devices, NATO Science Series: II. Mathematics, Physics and Chemistry,* Vol. 151, (2004), pp. 79-88, ISBN 9781402021695

Pralgauskaitė, S. et al. (2011). Noise characteristics and radiation spectra of multimode MQW laser diodes during mode-hopping effect, *Proceedings of 21th International Conference on Noise in Physical Systems and 1/f fluctuations,* pp. 301-304, ISBN 978-1-4577-0189-4, Toronto, Canada, 12-16 June, 2011

Saulys, B. et al. (2010). Analysis of mode-hopping effect in fabry-Pérot laser diodes, *Proceedings of International Conference on Microwaves, Radar and Wireless Communications,* pp. 1-4, ISBN 978-1-4244-5288-0, Vilnius, Lithuania, 14-16 June, 2010

Shuang, Z. et al. (2007). A Novel Method to Estimate Current Leakage of Laser Diodes, *Proceedings of 8th International Conference on Electronic Measurement and Instruments,* pp. 1-436 - 1-439, ISBN 978-1-4244-1136-8, July, 2007

Simmons, J.G. & Sobiestianskas, R. (2005). On the optical and electrical noise cross-correlation measurements for quality evaluation of laser diodes, *Proceedings of 2005 Conference on Lasers and Electro-optics Europe,* p. 336, ISBN: 0-7803-8974-3, Munich, Germany, 12-17 June, 2005

Smetona, S. et al. (2001). Optical and electrical low-frequency noise of ridge waveguide InGaAsP/InP MQW lasers. *Proceedings of SPIE: Optical organic and inorganic materials,* Vol. 4415, (April 2001), pp.115-120, ISBN: 978-081-9441-20-1

Tsuchida, H. (2011). Characterization of White and Flicker Frequency Modulation Noise in Narrow-Linewidth Laser Diodes. *IEEE Photonic Technology Letters,* Vol. 23, No 11, (June 2011), pp. 727-729, ISSN 1041-1135

Vandamme, L.K.J. (1994). Noise as a diagnostic tool for quality and reliability of electronic devices. *IEEE Transaction on Electron Devices,* Vol. 41, No. 11, (November 1994), pp. 2176 2187, ISSN 0018-9383

Van der Ziel, A. (1970). Noise in solid-state devices and lasers. *Proceedings of the IEEE,* Vol .58, No. 8, (August 1970), pp. 1178-1206, ISSN 0018-9219

Vurgafman, I. & Meyer, J.R. (1997). Effects of bandgap, lifetime, and other nonuniformities on diode laser thresholds and slope efficiencies. *IEEE Journal of Selected Topics in Quantum Electronics*, Vol. 3, No. 2, (April 1997), pp. 475-484, ISSN: 1077-260X

Wilson, G.C. et al. (2002). Long-haul DWDM/SCM transmission of 64- and 256QAM using electroabsorption modulated laser transmitters. *IEEE Photonic Technology Letters*, Vol. 14, No. 8, (August 2002), pp. 1184-1186, ISSN 1041-1135

Yamazaki, H. et al. (1999). Planar-buried-heterostructure laser diodes with oxidized AlAs insulating current blocking. *IEEE Journal of Selected Topics in Quantum Electronics*, Vol .53, No. 3, (May/June 1999), pp. 688-693, ISSN: 1077-260X

9

Investigation of High-Speed Transient Processes and Parameter Extraction of InGaAsP Laser Diodes

Juozas Vyšniauskas, Tomas Vasiliauskas, Emilis Šermukšnis, Vilius Palenskis and Jonas Matukas
Vilnius University, Center for Physical Sciences and Technology, Lithuania

1. Introduction

Fiber optic communication systems rely on the speed of the modulated light sources such as laser diodes (LDs). High speed direct current modulation of the optical signal, narrow spectral width, small dimensions and good light coupling efficiency to the optical fiber are the advantageous features of LD. Especially the long-haul optical links are critical to the chromatic dispersion and thus a single-mode and stable wavelength optical information carrier is needed. Distributed feedback (DFB) LDs are usually used in this situation. These DFB lasers have an almost single-mode optical emission spectrum, a stable optical frequency output, an optical modulation rate in excess of Gb/s and an uninterrupted operation time of up to 10^8 hours. However, the communication system with a high-speed direct modulation is limited by transient processes in the LD.

Optical links continually demand LDs with higher pulse modulation rates. Such techniques as eye diagrams, bit-error rate estimation, power penalty, pseudorandom bits sequences are used as the final and evident proof of LD suitability for specific link and speed applications. LDs with higher pulse modulation rate, higher extinction rate and optical power and also with a low chirp should be employed. Characterisation and modelling of LD dynamic characteristics is then needed. The investigation of small signal and large signal (pulse) characteristics can reveal the LD modulation speed in the fiber links and helps to better understand LD operation in other respects.

Rate equations for electron and photon densities in the active region of LD are widely used as a phenomenological model for the analysis of a small signal, a large signal and noise (Agrawal & Dutta, 1993; Agrawal, 2002; Fukuda, 1999). Despite the simplicity of the model, it predicts the modulation and noise of LD with good accuracy. It is also useful for LD characterisation and parameter extraction.

Small signal modulation equations in the frequency domain are derived directly from the rate equations assuming a small current perturbation at the given DC point. Thus, response spectra of optical power (or electron density) to the applied modulated current are analytically available. Parameters directly or indirectly entering rate equations and small signal modulation equations already comprise material and structural properties of LD. For example, electron lifetime, photon lifetime in the LD cavity, differential gain, nonlinear gain K-factor are only a few parameters which are usually extracted from the small signal modulation measurements (Lu et. al., 1995).

In the case of relative intensity noise (RIN) analysis the situation is similar (Tatham et al., 1992). The similarity of both characterisations becomes obvious if one just notices that the difference is only in the modulation source acting on the same system (LD). First one is the external modulation source probing the response of the system and another one is an internal source of the origin of spontaneous fluctuations of the number of electrons and photons. One can recall the Fluctuation-Dissipation theorem which states that the response of the system at equilibrium to the external small signal perturbation is the same as the response to the spontaneously occurring fluctuations inside the system. However, one applies DC current and thus the system is not at equilibrium. But it means no more that an additional information in the fluctuation spectra could be present in principle. Quantitatively the precision of extracted parameters from RIN measurements should be lower compared with small signal modulation measurements due to the random nature of the detected signal. Qualitatively it may have advantages, since parasitic capacitances are avoided.

Another interesting point is that the equations for small signal modulation and RIN have minor differences in the cases of the LDs of different materials and even structures. One can compare the equations of conventional edge-emitting laser diode (studied here), vertical-cavity surface-emitting laser (VCSEL) (Larsson et. al., 2004) or even quantum-dot (QD) (Bhattacharya et. al., 2000) laser.

Here the case of large signal current modulation is studied. Given dynamic parameters of LD extracted from the small signal measurements (relative intensity noise (RIN) of the optical beam intensity) are compared with the parameters extracted from the large signal measurements (optical power and time-resolved frequency chirp). Contrary to the small signal, the rate equations do not have closed-form expressions for the solutions of large signal (pulse) and thus rate equations should be integrated numerically. If the small signal parameter extraction from the experimental results consists of the standard function fitting procedure, then due to the lack of the closed-form function (in the case of a large signal) one needs other fitting means.

In this work rate equations were numerically solved taking the fixed set of input parameters such as the lifetime of photons, the coefficient of optical amplification, the nonlinear amplification factor, the spontaneous radiation and the optical confinement factor etc. The numerical solution obtained is then compared with the pulse from experiment. The procedure is repeated with a different set of parameters until the best coincidence is obtained. The final results of simulation were compared with experimental ones and the matching accuracy was calculated

2. Large signal: Optical power modulation and chirp

In the time domain, the small signal solutions of the rate equations are in the form of decaying oscillations (oscillation-relaxation solutions) in the densities (or number) of electrons and photons. After a small disturbance, coupled subsystems of electrons and photons oscillate (in number), exchange and dissipate the energy while approaching stationary conditions until the next disturbance. Two terms appear: the oscillation-relaxation frequency (showing the rate of exchange) f_0 and the damping frequency (showing the rate of dissipation) Γ_0. In the frequency domain this process has the form of resonance spectrum.

In the case of a large signal (pulse modulation) oscillation-relaxation rings form after the leading and trailing edges of the rectangular pulse. Thus, the pulsed bias current modulates the number of electrons and photons in the active region of LD (Agrawal & Dutta, 1993). The number of photons is proportional to the lased optical power (Agrawal & Dutta, 1993). which can be measured. The pulsing number of electrons is responsible for the frequency chirp phenomena inherent to LDs. The chirp is a considerable change in the frequency (wavelength) of the optical mode of the LD during the large signal modulation. This frequency chirp can also be measured. The origin of this effect relates to the other phenomena present even at the continuous-wave (cw) operation of LD. It manifests in the enhanced linewidth of the optical mode of the LDs. The principle of that is a change in the real and imaginary parts of the refractive index n' and n'' as electron density is pulsing (chirp) or fluctuates (linewidth enhancement). The phenomenon and the linewidth enhancement factor (or chirp parameter) $a = \Delta n' / \Delta n''$ showing the intensity of the phenomenon itself were correspondingly explained and introduced by C. H. Henry (Henry, 1982, 1983, 1986). Measurements of the frequency chirp during large signal modulation is a well-known technique for the estimation of the optical signal transmission in the fiber communications. The chirp intensity is proportional to the electron density, and thus also features oscillation-relaxation behaviour during the pulse modulation. This can be seen in the time-resolved frequency chirp measurements.

There are several chirp parameter extraction methods and the time-resolved chirp method is one of them (Harder et al., 1983; Arakawa & Yariv, 1985; Kikuchi et al., 1984; Kikuchi & Okoshi, 1985; Linke, 1985; Bergano, 1988). In fact, this method is more frequently used for electro-optical modulators, where the time-resolved chirp is a little simpler than that of LD, because there is only a transient term (Jeong & Park, 1997). Time-resolved frequency chirp measurements reveal more detailed information about LDs optical beam frequency behaviour by comparing with ordinary averaged optical beam spectrum measurements, because the time-resolved representation of frequency deviations shows specific chirp details, such as frequency oscillations and the amplitude of these oscillations (Welford, 1985; Olsen & Lin, 1989; Tammela et al., 1997).

2.1 Devices under investigation

A schematic diagram of the finished laser structure (around the mesa) is shown in Fig. 1.

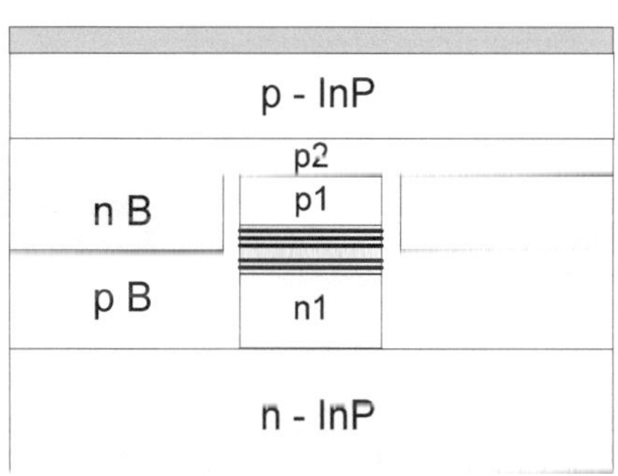

Fig. 1. Schematic diagram of the finished laser structure in the vicinity of the mesa. p-InP p:10^{18} cm^{-3}, p2-InP p:10^{18} cm^{-3}, p1-InP p:$5{\cdot}10^{17}$ cm^{-3}, n1-InP n:10^{18} cm^{-3}, n -InP n:10^{18} cm^{-3} nB-InP n:10^{18} cm^{-3}, pB-InP p:$6{\cdot}10^{17}$ cm^{-3}. Active region of the InGaAsP MQWs is formed between p1 and n1 layers.

The laser structure is epitaxially grown on (100) oriented n-doped InP substrates. The residual reflectivity of the front facet coating is within 3 to 8% as measured at the lasing wavelength. The rear facet is coated with a high reflectivity coating of 60 to 80%.

The active region of the investigated laser diodes is buried between two side layers which helps to make the active region bounded in transversal direction. Six strained quantum wells are formed in the active region. In addition, there are periodically truncated quantum wells along the longitudinal axis of the active region, which helps to create a periodical refractive index (due to different refractive indexes of the material in the truncating region) and a periodical optical gain (due to non-uniform carrier density); that forms gain-coupled (plus index-coupled) distributed feedback.

2.2 Time-resolved frequency chirp measurement setup and methodology

Time-resolved frequency chirp measurements setup under pulse operation is shown in Fig. 2. LD bias current is modulated by using a rectangular pulse generator together with a direct current source connected through the wideband bias tee. When the bias current was modulated, the optical beam power and optical wavelength became modulated as well. The optical beam is coupled to the tapered fiber (Shani et al., 1989; Alder et al., 2000), which is connected to a Mach-Zehnder fiber interferometer (Saunders et al., 1994; Jeong & Park, 1997; Laverdiere et al., 2003). An optical amplifier can be used if the interferometer output signal is too weak. Then an optical signal is passed to an optical input of the sampling oscilloscope connected to the PC. The most important part of the setup is the automated Mach-Zehnder fiber interferometer (Fig. 2). The main advantages of the interferometer are a quick time-domain chirp measurement, which takes only 30 seconds or less, high-resolution (< 20 MHz) and high measurement frequency (50 GHz or more). This allows to measure transmission signals greater than 10 Gbps with the measurement repeatability of ~5%.

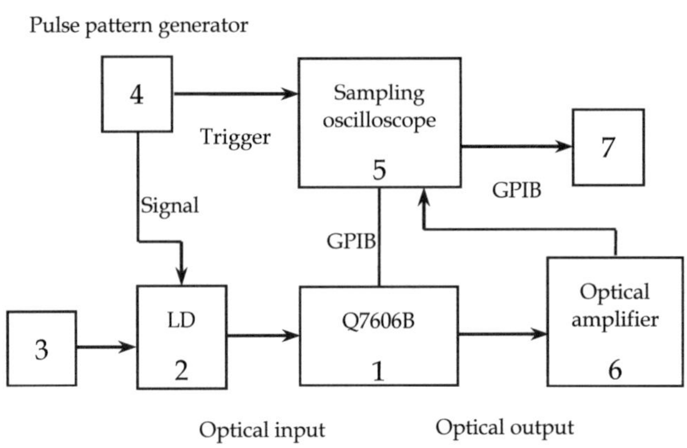

Fig. 2. Chirp measurement setup. (1) Mach-Zehnder fiber interferometer (Advantest Q7606B Optical Chirpform Test Set), (2) semiconductor laser, (3) direct current source (ILX Lightwave LDC-3900 modular laser diode controller), (4) pulse pattern generator (Agilent HP-70843B error performance analyzer), (5) signal analyzer - digital sampling oscilloscope (Tektronix TDS8000 Digital Sampling Oscilloscope), (6) erbium doped fiber amplifier, (7) personal computer with integrated General Purpose Interface Bus (GPIB) interface

An automated polarization controller helps to fix optimal polarization direction. That increases measurement accuracy. By changing the optical path delay time the interferometer can automatically fit the frequency transfer function to a particular carrier frequency of optical signal. The frequency transfer function of the Mach-Zehnder fiber interferometer is shown in Fig. 3. As can be seen, it is a sinusoidal function and its period is called a free spectral range (FSR). The optical signal frequency (chirp) variation should be lower than FSR/4=35 GHz. The frequency transfer function can be arbitrarily shifted with respect to the optical signal carrier frequency. And thus, during the first phase the frequency transfer function is shifted to the carrier frequency at the point "A", during the second phase - at the point "B" Fig. 3.

The frequency transfer function can be written as:

$$K_{total} = \beta K(f) = \beta \frac{A}{2} \cos\left(\frac{2\pi f}{FSR}\right). \tag{1}$$

where β is a optical loss in the system and A is amplitude of transfer function.

It follows from the equation (1) that the transfer coefficient depends on carrier frequency. During the pulse modulation the carrier frequency varies in time as:

$$f(t) = f_{cw} + \Delta f(t) . \tag{2}$$

The last term in equation (2) is a frequency chirp. At first, in order to measure the chirp of the LD or any other device, the optical beam should be passed to the interferometer not modulated, so that the carrier frequency and optical power should be constant. Then the carrier frequency is f_{cw}. As was mentioned earlier, the transfer function is shifted to match f_{cw} to the point "A" and after - to the point "B".

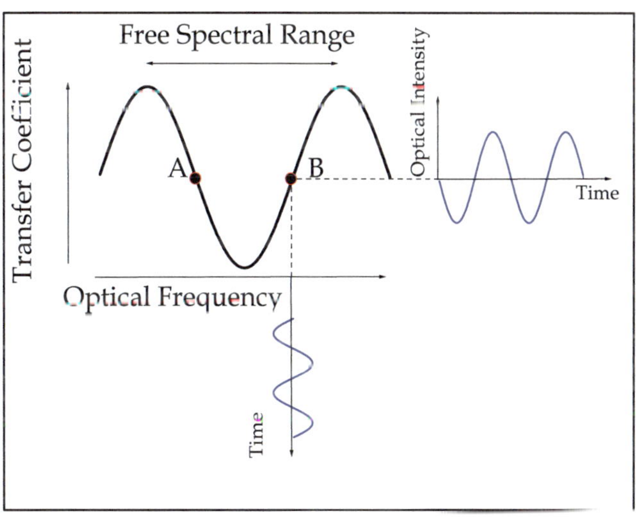

Fig. 3. Mach-Zehnder interferometer frequency transfer function

In both cases the transfer function will be shifted and the transfer coefficient will depend on $\Delta f(t)$.

$$K_A = \beta\left[\frac{A}{2} + \frac{A}{2}cos\left(\frac{2\pi\Delta f}{FSR}\right)\right], \tag{3}$$

$$K_B = \beta\left[\frac{A}{2} - \frac{A}{2}cos\left(\frac{2\pi\Delta f}{FSR}\right)\right]. \tag{4}$$

Thus the output signal power in both cases is:

$$P_{outA}(t) = P_{in}(t)\beta\left[\frac{A}{2} + \frac{A}{2}cos\left(\frac{2\pi\Delta f(t)}{FSR}\right)\right], \tag{5}$$

$$P_{outB}(t) = P_{in}(t)\beta\left[\frac{A}{2} - \frac{A}{2}cos\left(\frac{2\pi\Delta f(t)}{FSR}\right)\right]. \tag{6}$$

In literature these two equations are usually called IM+FM and IM-FM, because in both of them there are two terms which correspond to power modulation and frequency modulation (chirp). So, in both measurement phases (points "A" and "B") the output power depends not only on the chirp but on the input signal power as well. However, after both measurements are done, the power modulation term can be cancelled:

$$\frac{P_{outA}(t) - P_{outB}(t)}{P_{outA}(t) + P_{outB}(t)} = cos\left(\frac{2\pi\Delta f(t)}{FSR}\right) = P_{calc}(t). \tag{7}$$

There $P_{calc}(t)$ only contains information about chirp (carrier frequency deviations) but there is no information about power deviations. Further, the chirp can be easily obtained:

$$\Delta f(t) = \frac{FSR}{2\pi} arccos\left(P_{calc}(t)\right). \tag{8}$$

Or, in case when $\Delta f(t)<<FSR/4$, cosine can be treated as a line and the equation (8) becomes:

$$\Delta f(t) = \frac{FSR}{2\pi} P_{calc}(t). \tag{9}$$

In fact, the LD length is only 300 µm, so in order to make various measurements handily the LD needs to be put on some device carrier (Fig. 4). The device carrier is useful to easily connect a coaxial microwave probe because the LD is located on the microwave microstrip line. There is a serial resistor implanted in the microstrip line in order to match the microstrip line and LD impedances, so that the reflected parasitic signals are cancelled. The microwave cable is connected to a microstrip line by using the microwave probe (RF probe). The cut-off frequency of the microwave probe is ~ 50 GHz. For high frequency measurements any other DC interfaces composed of any LRC circuits should be disconnected.

The detected signal at the optical input of the oscilloscope is quite noisy Fig. 5. The optical pulse is attenuated and cumulates enough noise while it propagates throughout the system. Noise types and sources are different. There is a white amplitude noise, which continuously distributes through the length of the pulse; there is also a phase noise or jitter, which appears because the pulse length or the beginnings of the pulse edges fluctuates. The fiber interferometer and the scope also contribute to the noise level. As was explained, the output signal of the Mach-Zehnder interferometer is displayed on the digital scope screen as well as in the scope's internal memory. It is possible to make inter sample averaging (ensemble averaging); the scope can average up to 1024 samples. That possibility is very important for the periodic signal, because it helps to increase the S/N (signal to noise ratio) (no noise in Fig. 6).

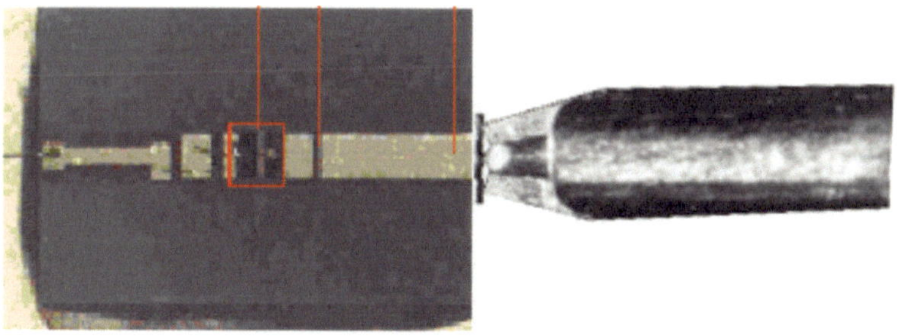

Fig. 4. LD ceramic carrier together with LD, microstrip line and a microwave probe on it

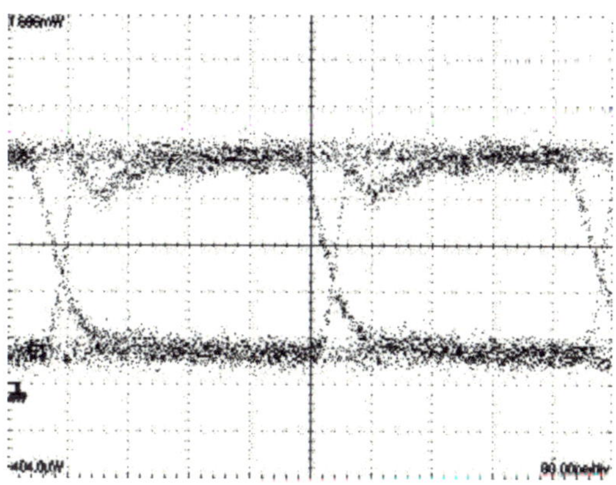

Fig. 5. Eye diagram of a not averaged (noisy) pulse train

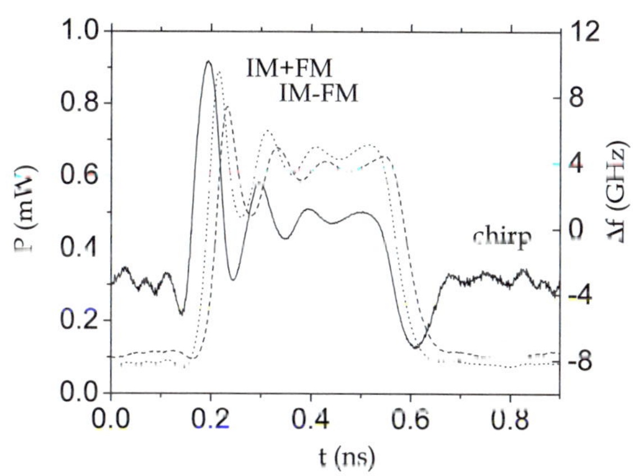

Fig. 6. Chirp pulse plotted together with interferometer output pulses corresponding to both frequency transfer function points

Another effective method is a software high frequency filtering with f_c=50 GHz which should be applied to the already averaged signal in the scope's memory. Such a filter is admissible because the cut-off frequency of the microwave system is also ~50 GHz, and so, all higher frequencies could be considered as parasitic.

After all these noise-related measurement problems were resolved, it became possible to make high-quality time-resolved chirp measurements. After both power pulses IM+FM and IM-FM were measured, by using the equation (9), the chirp pulse was obtained. All three pulses are shown in Fig. 6.

2.3 Time-resolved frequency chirp measurement results

A typical chirp pulse is shown in Fig. 7a, which is referred to as total chirp. The relatively plain region in the chirp waveform is referred to as adiabatic chirp and oscillations (blue or red shifts) corresponding to both power pulse edges are referred to as transient chirp. There the total chirp is the actually observed chirp, which can be expressed as the sum of transient and adiabatic chirp. While the pulse form of the adiabatic chirp is the same as the form of the optical power pulse, the transient chirp is proportional to the optical power derivative (derivative of ln(P) (Kikuchi, 1990; Koyama & Suematsu, 1985):

$$\Delta f(t) = \frac{a}{2\pi}\left[\frac{\dot{P}(t)}{P(t)} + G_P P(t)\right], \tag{10}$$

where P is the photon number, G_P is the nonlinear gain, α is the chirp parameter. The chirp parameter affects both the transient and the adiabatic chirp, while the nonlinear gain affects only the adiabatic chirp (equation (10)). The presence of the adiabatic term distinguishes the LD from the electro-optical modulators (EOM) due to nonlinear gain (Jeong & Park, 1997). While the electro-optical modulator chirp has only a transient part, the LD chirp has a transient and an adiabatic term (Fig. 7).

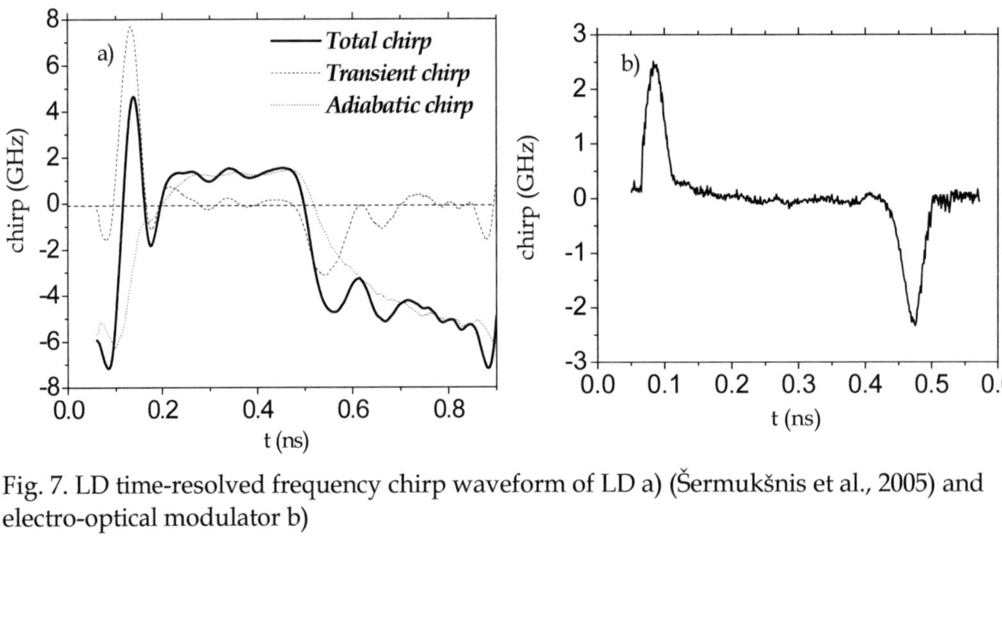

Fig. 7. LD time-resolved frequency chirp waveform of LD a) (Šermukšnis et al., 2005) and electro-optical modulator b)

The chirp pulse can be further used for LD analysis. Chirp and power pulses are shown below in Figs. 8a and 9a. The chirp pulse begins earlier than the power pulse, thus there is some time or phase shift between them. As current pulse is applied, the optical beam carrier frequency (as well as the wavelength) starts to change but the power still stays constant. The chirp parameter is extracted from both measured pulses, as explained in detail below.

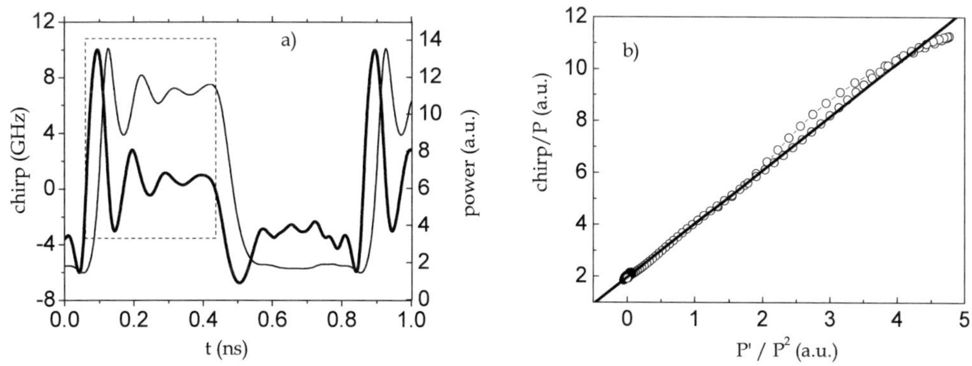

Fig. 8. Chirp parameter extraction in case of a 2.5 Gbit/s modulation rate. Frequency chirp and optical beam power waveform a), linearised time-resolved frequency chirp waveform b) (Šermukšnis et al., 2005)

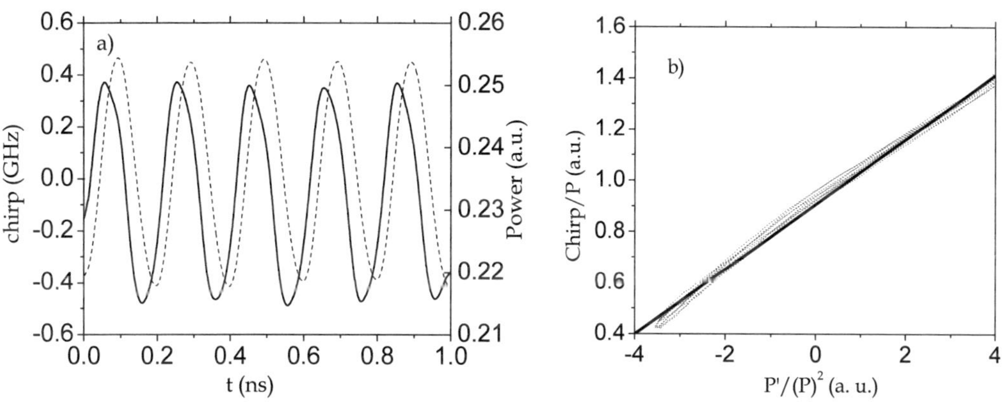

Fig. 9. Chirp parameter extraction in case of a 10 Gbit/s modulation rate. Frequency chirp and optical beam power waveform a), linearised time-resolved frequency chirp waveform b) (Šermukšnis et al., 2005)

In optoelectronic devices such as LDs or electro-optical modulators the chirp parameter should preferably be small. Besides, the chirp parameter extraction with good accuracy is not always possible. Dividing both sides of the equation (10) by the photon number P and after plotting $\Delta f / P$ versus $\dot{P} / P^2$, linear dependence of these two terms is expected. It was observed that in most cases such linear dependence actually occurs, furthermore, the linearity is of good quality (Figs. 8b and 9b). From the slope of that linear plot, the chirp parameter can be extracted. Considering that the photon number is proportional to the optical beam power P_W, the term $\dot{P} / P$ in the equation (10) is equal to $\dot{P}_W / P_W$. So, in real experiment conditions not the photon number P but the optical beam power P_W was used.

For simplicity and accuracy, only part of the single pulse, which is shown in Fig. 8a (area in the dash square), was taken for chirp parameter extraction. The other part of the pulse has a considerable noise and distortion at the „0" level of the chirp pulse. Better accuracy is also reached when a 10 Gbit/s bias current pulse sequence is generated and the corresponding chirp waveform is used (Fig. 9). There, in Fig. 9b, good linearity, as well as good chirp parameter extraction precision, can obviously be seen. Though bias current pulses were almost of rectangular form, as can be seen in Fig. 9a, both power and chirp pulses are of semi-sinusoidal form. Of course, that occurs when the modulation rate becomes comparable to the LD direct current modulation speed limits. By comparing these two cases it was observed that the chirp parameter extraction error does not exceed 10 %. Several tens of lasers were measured in such manner and the chirp parameter for all of them was extracted. It was found that this parameter of these lasers is dispersed between 2 and 2.8.

Besides the chirp parameter, time-resolved chirp pulse amplitude is also important. As it was shown above, there are large damped oscillations at the leading and trailing edges of both power and chirp pulses Figs. 8a and Fig. 10b. At a high modulation rate, these oscillations can increase several times, so that the ring amplitude becomes considerably larger than the distance between „1" and „0" levels. In that case, the rings comprise the large part of the total chirp, as well as have the considerable effect on the pulse propagation in the fiber.

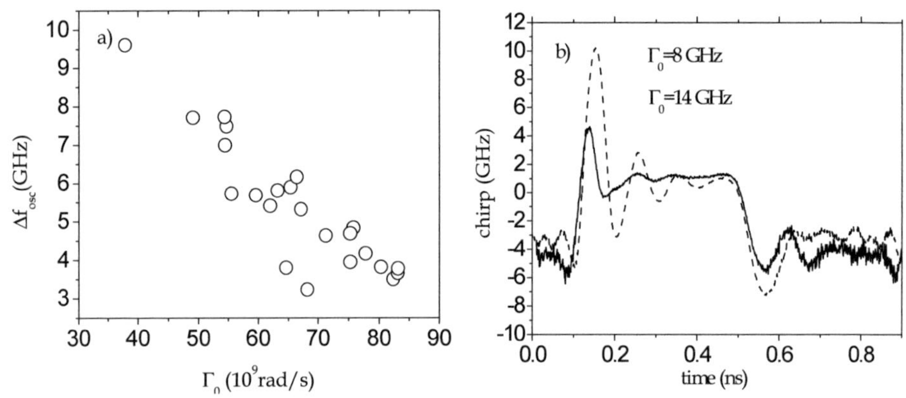

Fig. 10. Chirp pulse ring amplitude dependence on damping rate a) (Šermukšnis et al., 2005) and chirp pulse comparison for two lasers with different damping rate b)

It is always desirable to have chirp rings smaller, particularly in these cases, when rings make the largest contribution to the pulse power. In order to find the chirp ring amplitude analytically, one should account for many factors. Besides, the analytical analysis of a large signal is especially complicated. The oscillation frequency and the oscillation amplitude damping rate nearly correspond to the oscillation-relaxation and damping frequency which are used in small signal analysis. Indeed, it was observed that the amplitude (amplitude of the first ring, which largely contributes to the transient chirp) and the damping rate of chirp rings strongly depends on a small signal damping frequency (extracted from RIN measurements) at constant power for different LDs (Fig. 10). It is smaller in lasers with a large damping rate (Fig. 10b).

To summarise, the adiabatic chirp of different LDs was in the region of 4.2 GHz to 5.6 GHz and the dynamic chirp in the region of 3.2 GHz to 10 GHz. Finally, it was observed that in regular direct current modulation conditions the frequency chirp variation does not exceed 16 GHz in average. For a semiconductor laser with the wavelength of 1550 nm, that corresponds only to about ~0.1 nm variation in wavelength. Such wavelength variation compared to the total wavelength 1550 nm is negligible but even such small variation can noticeably affect the optical pulse propagation in the optical fiber due to chromatic dispersion, particularly when the data transmission rate is high.

3. Theoretical models

In this work four physical models were used. At first, a simplified physical model based on rate equations for carrier and photon densities was used for computer simulation (Šermukšnis et al., 2004a, 2004b). However, though some of the physical factors were not included in this model, the accuracy of modelling results for some DFB lasers was sufficiently good. In the later models, the rate equations were supplemented with additional physical parameters.

For the second physical model the nonlinear amplification factor ε was included in the rate equations. The optical amplification coefficient, α [cm^3/s] was assumed to be equal to the product of the optical differential amplification, g [cm^2], and the group velocity, v_g ($\alpha = v_g g$) in the earlier model of simplified rate equations (Šermukšnis et al., 2004a, 2004b). Including that the optical amplification depends on the photon density:

$$g(N,P) = g_0(1-\varepsilon P) = g(N-N_0)(1-\varepsilon P), \tag{11}$$

where N is the density of electrons, P is the density of photons, N_0 is the carrier density at transparency (corresponding to the onset of population inversion) and ε is the nonlinear amplification factor. Then the term of the number of photons generated by spontaneous emission in the rate equations was supplemented with the factor (1– εP).

In the third model, the term N/τ_s in the rate equation (Šermukšnis et al., 2004a, 2004b) is changed to $N\gamma_e(N)$, because $\gamma_e = 1/\tau_s$, where $\gamma_e(N)$ is the rate of electron recombination (Carroll et al., 1998):

$$\gamma_e(N) = A + BN + CN^2, \tag{12}$$

where the first term A of the equation describes the non-radiative recombination, B is the coefficient of radiant interband recombination and C is the coefficient of Auger recombination. The carrier lifetime τ_s can be described by these parameters in such a way:

$$\tau_s(N) = (A + BN + CN^2)^{-1}, \tag{13}$$

and can be used in the rate equations (14).

In the fourth model we used the following rate equation for electron density (Carroll et al., 1998):

$$\frac{dN}{dt} = \frac{J}{ed} - N\left(A + BN + CN^2\right) - a(N-N_0)(1-\varepsilon P)P, \tag{14}$$

where J is the density of injection current, e is the electron charge, d is the width of the semiconductor laser active region, t is the time. The first term of the rate equation (14) describes the number of injected carriers that passed the unit volume per unit time interval, the second one describes the spontaneous emission and the third term defines the number of photons generated by stimulated emission per unit time interval. An analogical equation was used for photon density:

$$\frac{dP}{dt} = \Gamma a(N - N_0)(1 - \varepsilon P)P - \frac{P}{\tau_p} + \Gamma \beta B N^2, \tag{15}$$

where τ_p is the lifetime of photons, β is the spontaneous emission factor, Γ is the optical confinement factor. The first term of the rate equation (15) is the number of photons generated by stimulated emission per unit time interval, the second term is the number of photons emitted from a resonator (output of the laser). The third term of the equation (15) shows the spontaneous emission contribution to the generated mode.

4. Used experimental results

Experimental results of DFB semiconductor lasers were used for computer simulation. Not all laser parameters needed for simulation were known and some parameters were estimated from experimental results. The product of photon and electron lifetimes, the threshold current and the current at the 5 mW average optical power were known. The extinction ratio (ratio of the optical power level "1" to the level "0") was equal to 8.5 dB:

$$10\lg(P_1/P_0) = 8.5\,. \tag{16}$$

Injection current values at levels "0" and "1" were found from the optical power dependence on injected current by using the condition (6) for the optical power of 5 mW. The active region width was estimated from the TEM image. We also had a possibility to calculate the injection current density from current measurements.

The experimentally obtained parameters of semiconductor DFB lasers, which were used for calculations (see Figs. 11-15):

- product of lifetimes, $\tau_s\tau_p$ (different for each laser);
- threshold current, I_{th} (different for each laser);
- mean current (current, when the average optical power is equal to 5 mW), I_{op} (different for each laser);
- optical powers: P_1 = 8.59 mW; P_0 = 1.41 mW (the same for all lasers);
- injection currents: I_1 (level „1") and I_0 (level „0"), found from the optical power dependence on injected current (see Fig. 11);
- dimensions of the active region of the DFB laser: L = 300 μm (channel length); w = 1.5 μm (width); d = 0.1 μm (thickness); $S = L\cdot w = 4.5\cdot10^{-6}$ cm^2 (area) (the same for all lasers);
- injection current density (calculated by using active region dimensions);
- experimental pulse characteristics: optical power and chirp (different for each laser).

These experimental results for all investigated lasers are shown in Table 1.

No.	DFB laser	$\tau_s \tau_p$ [s^2]	I_{th} [mA]	I_{op} [mA]	J_{th} [A/cm^2]	J_1 [A/cm^2]	J_0 [A/cm^2]
1.	O6j5	$1.91 \cdot 10^{-20}$	12.31	33.09	2736	10711	4066
2.	H3k7	$1.63 \cdot 10^{-20}$	11.56	33.07	2569	10756	3958
3.	G3k4	$1.64 \cdot 10^{-20}$	11.89	32.94	2642	10638	3982
4.	J5k7	$1.65 \cdot 10^{-20}$	11.15	34.26	2478	11300	3927
5.	U2i3	$2.26 \cdot 10^{-20}$	8.19	30.10	1820	10184	3282
6.	R4j5	$1.72 \cdot 10^{-20}$	11.29	32.33	2509	10542	3827
7.	E3k4	$1.68 \cdot 10^{-20}$	11.36	34.38	2524	11313	3967
8.	G1l1	$1.60 \cdot 10^{-20}$	11.74	32.80	2609	10649	3929
9.	P5i3	$2.12 \cdot 10^{-20}$	8.72	30.65	1938	10309	3311
10.	T2s2	$2.06 \cdot 10^{-20}$	8.85	31.46	1967	10767	3449
11.	S2i3	$2.37 \cdot 10^{-20}$	8.09	28.94	1798	9758	3104
12.	R2r3	$1.72 \cdot 10^{-20}$	11.44	35.80	2542	11842	4069
13.	T2i3	$2.25 \cdot 10^{-20}$	8.02	31.63	1782	10796	3262
14.	Q1t1	$1.62 \cdot 10^{-20}$	11.12	36.82	2471	12282	4082
15.	O1k1	$1.63 \cdot 10^{-20}$	10.50	35.32	2333	11809	3871
16.	O2s2	$1.99 \cdot 10^{-20}$	8.74	32.34	1942	10951	3389

Table 1. Experimental results used for simulation (Vasiliauskas et al., 2007)

5. Simulation results

Computer simulation was based on the fourth-order Runge-Kutta method for the system of differential equations. The values of photon and electron densities for different time moments were found from initial values of densities, determining other parameters and choosing the time step and the number of calculation loops.

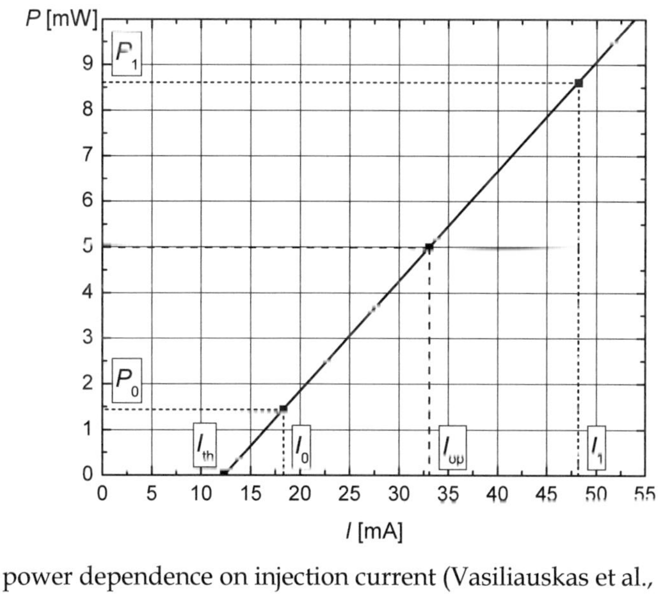

Fig. 11. Optical power dependence on injection current (Vasiliauskas et al., 2007)

In the computer simulation the following unknown parameters were selected: the optical confinement factor Γ, the reflection coefficients from mirrors at the ends, R_1 and R_2 (Fukuda, 1999), the lifetime of photons, τ_p, the coefficient of optical amplification, α, the spontaneous emission factor, β, the carrier density at transparency, N_0, the nonlinear amplification factor, ε, and the coefficients of recombination A, B and C.

The investigation of laser parameter interplay was done. Based on the simulation results obtained, the following technique for the estimation of semiconductor laser parameters was suggested:

- the frequency of the relaxation oscillation is defined by changing the coefficient of optical amplification, a;
- the calculated optical power values are approached to the experimental ones by changing the optical confinement factor Γ;
- by changing the nonlinear amplification factor, ε_1, and the spontaneous emission factor, β, the relaxation oscillation amplitudes can be corrected; in addition, the shape of oscillations can also be corrected by changing the injection current front duration;
- the recombination coefficient, A, and the density of electrons at transparency, N_0, was changed to optimize the static characteristics of the semiconductor laser;
- the recombination coefficients B and C were chosen last. These coefficients have insignificant influence on laser dynamics, they only change the shape of the oscillations a little.

The calculations for model verification were done, one of them being the investigation of watt-ampere characteristics (see Fig. 12). The threshold current was found as well. The results obtained were also compared to those of other authors (Agrawal & Dutta, 1993).

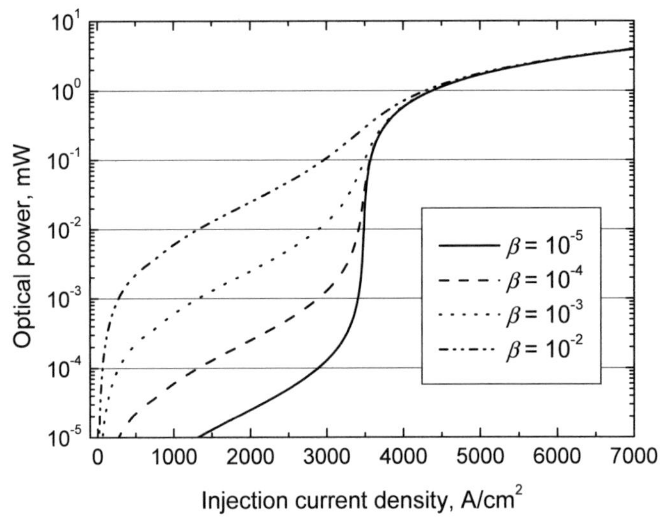

Fig. 12. Watt-ampere characteristics of a laser diode near the threshold of injection current for different spontaneous emission factors β (Vasiliauskas et al., 2007)

The output optical power of the semiconductor laser was calculated by using these equations (Agrawal & Dutta, 1993):

$$P_1 = \frac{hc^2}{\lambda\mu_g}\frac{1}{2L}log\frac{1}{R_1R_2}Ldwn_{ph}\frac{(1-R_1)R_2^{1/2}}{\left(R_1^{1/2}+R_2^{1/2}\right)\left(1-R_1^{1/2}R_2^{1/2}\right)}, \quad (17)$$

$$P_2 = \frac{hc^2}{\lambda\mu_g}\frac{1}{2L}log\frac{1}{R_1R_2}Ldwn_{ph}\frac{(1-R_2)R_1^{1/2}}{\left(R_1^{1/2}+R_2^{1/2}\right)\left(1-R_1^{1/2}R_2^{1/2}\right)}, \quad (18)$$

where R_1 and R_2 are the reflection coefficients of end mirrors, h is the Planck's constant, c is the speed of light, λ is the wavelength of radiated light, μ_g is the group refraction index, L is the length of the laser active region, d is the thickness of the active region, w is the active region width and n_{ph} is the density of photons.

The best results were obtained based on the fourth physical model (equations (14) and (15)). The most precise match of experimental and simulation results for all investigated lasers with different characteristics was obtained. The mismatch of analyzed laser parameters does not exceed the limit of 10%.

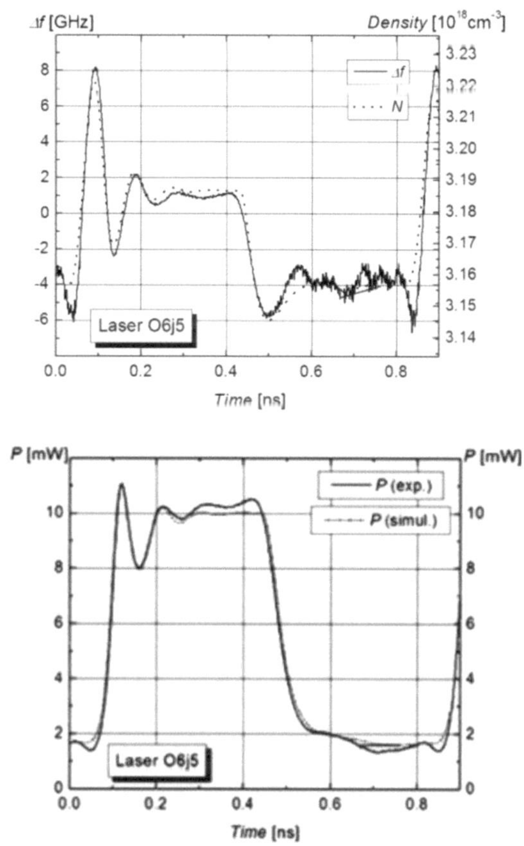

Fig. 13. Simulated electron density (N) and experimental chirp (Δf) pulse characteristics (Vasiliauskas et al., 2007) a), and experimental P(exp.) and simulated pulse characteristics P(simul.) of optical power (Vasiliauskas et al., 2007) b)

No.	DFB laser	Γ	R_1	R_2	τ_p [ps]	α [cm^3/s]	β	N_0 [cm^{-3}]
1.	O6j5	0.55	0.04	0.664	1.5	$16.8 \cdot 10^{-7}$	$1 \cdot 10^{-4}$	$24.3 \cdot 10^{17}$
2.	H3k7	0.49	0.08	0.80	2.0	$24.2 \cdot 10^{-7}$	$1 \cdot 10^{-3}$	$27.0 \cdot 10^{17}$
3.	G3k4	0.49	0.08	0.80	2.0	$23.6 \cdot 10^{-7}$	$1 \cdot 10^{-3}$	$27.0 \cdot 10^{17}$
4.	J5k7	0.33	0.08	0.80	3.0	$29.6 \cdot 10^{-7}$	$1 \cdot 10^{-3}$	$26.0 \cdot 10^{17}$
5.	U2i3	0.37	0.08	0.80	2.0	$21.6 \cdot 10^{-7}$	$1 \cdot 10^{-4}$	$14.0 \cdot 10^{17}$
6.	R4j5	0.67	0.08	0.80	1.8	$21.0 \cdot 10^{-7}$	$1 \cdot 10^{-2}$	$25.0 \cdot 10^{17}$
7.	E3k4	0.32	0.08	0.80	3.4	$29.0 \cdot 10^{-7}$	$1 \cdot 10^{-3}$	$27.0 \cdot 10^{17}$
8.	G1l1	0.42	0.08	0.80	3.0	$26.0 \cdot 10^{-7}$	$1 \cdot 10^{-2}$	$28.0 \cdot 10^{17}$
9.	P5i3	0.61	0.08	0.80	2.0	$20.0 \cdot 10^{-7}$	$1 \cdot 10^{-2}$	$23.0 \cdot 10^{17}$
10.	T2s2	0.57	0.08	0.80	1.5	$17.0 \cdot 10^{-7}$	$1 \cdot 10^{-3}$	$19.0 \cdot 10^{17}$
11.	S2i3	0.38	0.08	0.80	1.9	$22.0 \cdot 10^{-7}$	$1 \cdot 10^{-3}$	$17.0 \cdot 10^{17}$
12.	R2r3	0.66	0.08	0.80	2.1	$19.0 \cdot 10^{-7}$	$1 \cdot 10^{-3}$	$24.0 \cdot 10^{17}$
13.	T2i3	0.41	0.08	0.80	2.0	$29.0 \cdot 10^{-7}$	$1 \cdot 10^{-2}$	$15.0 \cdot 10^{17}$
14.	Q1t1	0.31	0.08	0.80	2.3	$29.0 \cdot 10^{-7}$	$1 \cdot 10^{-2}$	$25.0 \cdot 10^{17}$
15.	O1k1	0.48	0.08	0.80	2.5	$26.0 \cdot 10^{-7}$	$1 \cdot 10^{-2}$	$29.0 \cdot 10^{17}$
16.	O2s2	0.35	0.08	0.80	1.9	$28.0 \cdot 10^{-7}$	$1 \cdot 10^{-3}$	$28.0 \cdot 10^{17}$

Table 2. Fourth model: computer simulation results (1) (Vasiliauskas et al., 2007

No.	DFB laser	ε	τ_{j1} [ps]	τ_{j2} [ps]	A [cm/s]	B [cm^3/s]	C [cm^6/s]
1.	O6j5	$9.5 \cdot 10^{-18}$	54	50	$2.5 \cdot 10^{8}$	$9.1 \cdot 10^{-12}$	$3.80 \cdot 10^{-29}$
2.	H3k7	$11 \cdot 10^{-18}$	56	53	$2.8 \cdot 10^{8}$	$8.4 \cdot 10^{-12}$	$3.25 \cdot 10^{-29}$
3.	G3k4	$11 \cdot 10^{-18}$	56	53	$2.8 \cdot 10^{8}$	$8.4 \cdot 10^{-12}$	$3.25 \cdot 10^{-29}$
4.	J5k7	$15 \cdot 10^{-18}$	50	50	$3.6 \cdot 10^{8}$	$4.7 \cdot 10^{-12}$	$2.44 \cdot 10^{-29}$
5.	U2i3	$17 \cdot 10^{-18}$	66	66	$4.1 \cdot 10^{8}$	$5.5 \cdot 10^{-12}$	$4.00 \cdot 10^{-29}$
6.	R4j5	$16 \cdot 10^{-18}$	55	50	$4.8 \cdot 10^{8}$	$9.6 \cdot 10^{-12}$	$3.00 \cdot 10^{-29}$
7.	E3k4	$19 \cdot 10^{-18}$	75	60	$5.0 \cdot 10^{8}$	$6.0 \cdot 10^{-12}$	$2.00 \cdot 10^{-29}$
8.	G1l1	$14 \cdot 10^{-18}$	60	40	$2.0 \cdot 10^{8}$	$3.0 \cdot 10^{-12}$	$4.00 \cdot 10^{-29}$
9.	P5i3	$15 \cdot 10^{-18}$	60	50	$3.0 \cdot 10^{8}$	$2.0 \cdot 10^{-12}$	$4.00 \cdot 10^{-29}$
10.	T2s2	$14 \cdot 10^{-18}$	50	50	$2.0 \cdot 10^{8}$	$3.0 \cdot 10^{-12}$	$4.00 \cdot 10^{-29}$
11.	S2i3	$18 \cdot 10^{-18}$	60	55	$4.0 \cdot 10^{8}$	$7.0 \cdot 10^{-12}$	$3.00 \cdot 10^{-29}$
12.	R2r3	$11 \cdot 10^{-18}$	50	50	$4.0 \cdot 10^{8}$	$5.0 \cdot 10^{-12}$	$4.00 \cdot 10^{-29}$
13.	T2i3	$17 \cdot 10^{-18}$	65	45	$4.9 \cdot 10^{8}$	$1.0 \cdot 10^{-12}$	$4.50 \cdot 10^{-29}$
14.	Q1t1	$19 \cdot 10^{-18}$	50	40	$2.0 \cdot 10^{8}$	$3.0 \cdot 10^{-12}$	$4.00 \cdot 10^{-29}$
15.	O1k1	$17 \cdot 10^{-18}$	50	45	$4.0 \cdot 10^{8}$	$2.0 \cdot 10^{-12}$	$2.00 \cdot 10^{-29}$
16.	O2s2	$19 \cdot 10^{-18}$	50	45	$2.0 \cdot 10^{8}$	$4.0 \cdot 10^{-12}$	$2.00 \cdot 10^{-29}$

Table 3. Fourth model: computer simulation results (2) (Vasiliauskas et al., 2007)

The most precise match of laser characteristics was obtained for the laser with the characteristics shown in Fig. 13a (chirp and density of electrons; the chirp is proportional to the carrier density (Henry, 1982)) and Fig. 13b (optical power pulse characteristic).

Selected values for the matching of unknown parameters were: $\Gamma = 0.55$; $R_1 = 0.041$; $R_2 = 0.661$; $\tau_p = 1.5\cdot10^{-12}$ s; $\alpha = 16.8\cdot10^{-7}$ cm^3/s; $\beta = 1\cdot10^{-4}$; $N_0 = 24.3\cdot10^{17}$ cm^{-3}; $\varepsilon = 9.5\cdot10^{-18}$; $\tau_{j1} = 54\cdot10^{-12}$ s; $\tau_{j2} = 50\cdot10^{-12}$ s; $A = 2.5\cdot10^8$ cm/s; $B = 9.1\cdot10^{-12}$ cm^3/s; $C = 3.80\cdot10^{-29}$ cm^6/s.

The values of parameters for other lasers, the comparison of simulation and experimental results are represented in Tables 1, 2 and 3. In order to evaluate the accuracy of simulation results the following parameters, such as the frequency (υ) of relaxation oscillations, the time of oscillation decrement (T), "1" and "2" levels proportion (l) and the proportion of amplitude of first peak and level "1" (a) were compared (see Table 4). The averaged value of the mismatch (M) between the simulation and experimental results for these parameters was calculated (see Figs. 14 and 15).

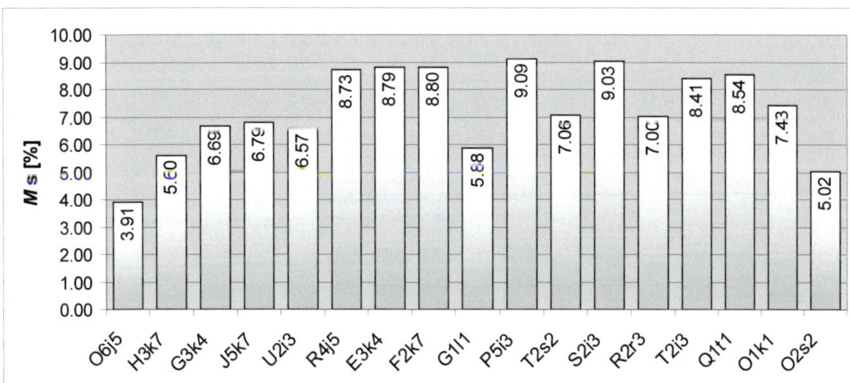

Fig. 14. Averaged value of the mismatch of simulated laser parameters for different lasers: *Ms* is for electron density characteristic (Vasiliauskas et al., 2007)

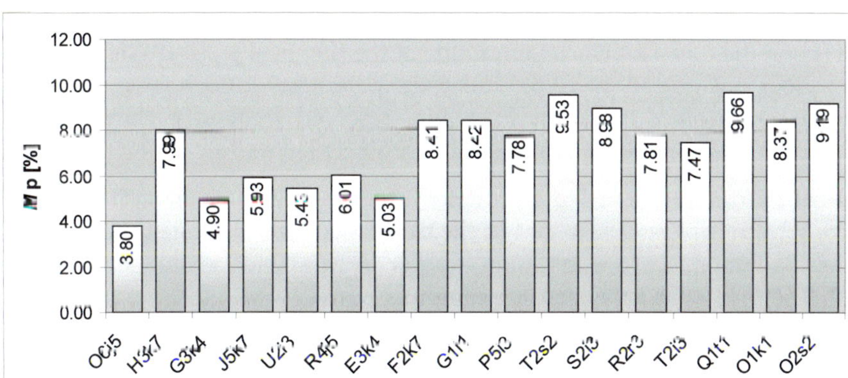

Fig. 15. Averaged value of the mismatch of simulated laser parameters for different lasers: *Mp* is for optical power characteristic (Vasiliauskas et al., 2007)

No.	DFB laser	Mismatch of simulation parameters [%]									
		v_s	v_p	T_s	T_p	l_s	l_p	a_s	a_p	M_s	M_p
1.	O6j5	2.33	2.02	8.60	7.43	1.33	4.85	3.39	0.90	3.91	3.80
2.	H3k7	1.35	7.44	9.72	9.90	9.61	7.33	1.74	7.29	5.60	7.99
3.	G3k4	1.49	5.38	9.84	8.22	9.36	4.08	6.04	1.92	6.69	4.90
4.	J5k7	2.14	9.98	9.97	8.25	8.37	4.50	6.67	0.98	6.79	5.93
5.	U2i3	0.90	4.81	9.75	9.84	9.62	3.94	6.02	3.12	6.57	5.43
6.	R4j5	9.32	6.56	9.84	9.21	6.40	6.20	9.35	2.08	8.73	6.01
7.	E3k4	5.51	4.96	9.87	9.76	9.85	2.19	9.93	3.19	8.79	5.03
8.	G1l1	3.43	8.01	9.78	9.97	1.59	8.96	8.72	6.74	5.88	8.42
9.	P5i3	6.59	9.89	9.97	9.68	9.89	4.57	9.93	6.98	9.09	7.78
10.	T2s2	9.82	9.33	9.43	9.71	2.89	9.83	6.11	9.26	7.06	9.53
11.	S2i3	8.07	7.97	9.51	9.75	9.09	8.92	9.43	9.28	9.03	8.98
12.	R2r3	9.96	9.91	6.64	8.74	1.79	3.59	9.62	8.99	7.00	7.81
13.	T2i3	9.65	9.82	8.60	9.09	9.63	6.67	5.77	4.30	8.41	7.47
14.	Q1t1	9.55	9.93	9.74	9.87	4.95	9.86	9.93	8.99	8.54	9.66
15.	O1k1	8.46	9.83	9.12	9.78	9.80	4.36	2.34	9.52	7.43	8.37
16.	O2s2	3.88	9.81	9.71	8.31	4.40	9.54	2.10	9.09	5.02	9.19

Table 4. Mismatch (%) of simulation data to experimental results (Vasiliauskas et al., 2007)

6. Conclusion

In this work the time-resolved frequency chirp characteristics under pulse operation of DFB LD were investigated. The optical beam carrier frequency (wavelength) deviations were measured by using a fiber Mach-Zehnder interferometer. The pulse current modulation at the 5 GHz and 1.25 GHz rate (10 Gbps and 2.5 Gbps) was achieved and the optical power and time-resolved frequency chirp pulses were investigated. LDs with a higher damping rate are preferable because of lower oscillation amplitudes at pulse edges, both for power pulses and for chirp pulses. It was found that at common operation conditions, i.e. average optical power P_w=5 mW and extinction ratio of 8.5 dB, the adiabatic chirp does not exceed 4.2-5.6 GHz and the transient chirp (amplitude of the first peak)- 3.2-10 GHz. To summarize, total chirp does not exceed 16 GHz. The chirp parameter (also known as the linewidth enhancement factor) was extracted from time-resolved frequency chirp measurements, which was found to be in the range of 2-2.8 for various devices.

The transient characteristics of semiconductor lasers were investigated. The unknown laser parameters which were not measured in the experiment were obtained by using computer simulation. Based on these simulation results a parameter estimation technique was suggested. The rate equations used for computer simulation were improved by including new terms which were not used in earlier simulations (Šermukšnis et al., 2004a, 2004b). Transient processes were simulated for a large number of similar DFB lasers. The best results were obtained by using the fourth physical model. The mismatch between the simulation and experimental results for all analyzed laser parameters in this case did not exceed the limit of 10%.

7. References

Agrawal, G.P. & Dutta, N.K. (1993). *Semiconductor Lasers, 2nd Edition,* Van Nostarnd Reinhold Co., ISBN 0-442-01102-4, New York, USA

Agrawal, G.P. (2002). *Fiber-Optic Communication Systems, 3nd Edition,* John Wiley & Sons, Inc., ISBN 0-471-21571-6, New York, USA.

Alder, T.; Stohr, A.; Heinzelmann, R. & Jager, D. (2000). High-Efficiency Fiber-to-Chip Coupling Using Low-Loss Tapered Single-Mode Fiber. *IEEE Photonics Technology Letters,* Vol.12, No.8, (August 2000), pp. 1016-1018, ISSN 1041-1135

Arakawa, Y. & Yariv, A. (1985). Fermi energy dependence of linewidth enhancement factor of GaAlAs buried heterostructure lasers. *Applied Physics Lett*ers, Vol.47, No.9, (November 1985), pp. 905-907, ISSN 0003-6951

Bergano, N. S. (1988). Wavelenght discriminator method for measuring dynamic chirp in DFB lasers. *Electronics Lett*ers, Vol.24, No.20 (September 1988), pp. 1296-1297, ISSN 0013-5194

Bhattacharya, P; Klotzkin, D.; Qasaimeh, O.; Zhou, W.; Krishna, S. & Zhu, D. (2000). High-Speed Modulation and Switching Characteristics of In(Ga)As-Al(Ga)As Self-Organized Quantum-Dot Lasers. *IEEE Journal of Selected Topics in Quantum Electronics,* Vol. 6, No. 3, (May 2000), pp. 426-438, ISSN 1077-260X.

Carroll, J.E.; Whiteaway, J.E.A. & Plumb, R.G.S. (1998). *Distributed feedback semiconductors lasers (IEE Circuits, Devices and Systems),* IEE, ISBN 0852969171, London, UK

Fukuda, M. (1999). *Optical semiconductor devices,* John Wiley & Sons, ISBN 978-0-471-14959-0, New York, USA

Harder, C.; Vahala, K. & Yariv, A. (1983). Measurement of the linewidth enhancement factor α of semiconductor lasers. *Applied Physics Lett*ers, Vol.42, No.4 (February 1983), pp. 328-330, ISSN 0003-6951

Henry, C.H. (1982). Theory of the line width of semiconductor lasers. *IEEE Journal of Quantum Electroni*cs, Vol.18, No.2, (February 1982), pp. 259-264, ISSN 0018-9197

Henry, C. H. (1983). Theory of the Phase Noise and Power Spectrum of a Single Mode Injection Laser. *IEEE Journal of Quantum Electronics,* Vol.19, No.9, (September 1983), pp. 1391-1397, ISSN 0018-9197

Henry, C. H. (1986). Phase Noise in Semiconductor Lasers. *IEEE Journal of Lightwave Technology,* Vol.4, No.3, (March 1986), pp. 298-311, ISSN 0733-8724

Jeong, J. & Park, Y. K. (1997). Accurate Determination of Transient Chirp Parameter in High Speed Digital Lightwave Transmitters. *Electronics Letters,* Vol.33, No.7, (March 1997) pp. 605-606, ISSN 0013-5194

Kikuchi, K. & Okoshi, T. (1985). Estimation of linewidth enhancement factor of AlGaAs lasers by correlation measurement between FM and AM noises. *IEEE Journal of Quantum Electronics,* Vol.21, No.6 (June 1985), pp. 669 - 673, ISSN 0018-9197

Kikuchi, K.; Okoshi, T. & Kawai, T. (1984). Estimation of linewidth enhancement factor α of CSP-type AlGaAs lasers from measured correlation between AM and FM noises. *Electronics Lett*ers, Vol.20, No.11 (May 1984), pp. 450-451, ISSN 0013-5194

Kikuchi, K. (1990). Static Frequency Chirping in λ/4-Phase-Shifted Distributed-Feedback Semiconductor Lasers: Influence of Carrier-Density Nonuniformity Due to Spatial Hole Burning. *IEEE Journal of Quantum Electronics,* Vol.26, No.1, (January 1990), pp. 45-49, ISSN 0018-9197

Koyama, F. & Suematsu, Y. (1985). Analysis of Dynamic Spectral Width of Dynamic-Single-Mode (DSM) Lasers and Related Transmission Bandwidth of Single-Mode. *IEEE Journal of Quantum Electronics,* Vol.21, No.4, (April 1985), pp. 292-297, ISSN 0018-9197

Larsson, A.; Carlsson, C.; Gustavsson, J.; Haglund, A; Modh, P. & Bengtsson, J. (2004). Direct High-Frequency Modulation of VCSELs and Applications in Fibre Optic RF and Microwave Links. *New Journal of Physics,* Vol. 6, No. 1, (November 2004), pp. 1-17, ISSN 1367-2630

Laverdiere, C.; Fekecs, A. & Tetu, M. A. (2003). New Method for Measuring Time-Resolved Frequency Chirp of High Bit Rate Source. *IEEE Photonics Technology Letters,* Vol.15, No.3, (March 2003), pp. 446-448, ISSN 1041-1135

Linke, R. A. (1985). Modulation Induced Transient Chirping in Single Frequency Lasers. *IEEE Journal of Quantum Electronics,* Vol.21, No. 6, (June 1985), pp. 593-597, ISSN 0018-9197

Lu, H.; Makino T. B & Li, G. P. (1995). Dynamic Properties of Partly Gain-Coupled 1.55-µm DFB Lasers. *IEEE Journal of Quantum Electronics,* Vol. 31, No. 8, (August 1995), pp. 1443-1995, ISSN 0018-9197

Olsen, C. M. & Lin, C. (1989). Time-Resolved Chirp Evaluations of Gbit/s NRZ and Gain-Switched DFB Laser Pulses Using Narrowband Fabry-Perot Spectrometer. *Electronics Letters,* Vol.25, No.16, (August 1989), pp. 1018-1019, ISSN 0013-5194

Saunders, R. A.; King, J. P. & Hardcastle, I. (1994). Wideband Chirp Measurement Technique for High Bit Rate Sources. *Electronics Letters,* Vol.30, No16, (August 1994), pp. 1336-1338, ISSN 0013-5194

Shani, Y.; Henry, C. H.; Kistler, R. C.; Orlowski, K. J. & Ackerman, D. A. (1989). Efficient Coupling of a Semiconductor Laser to an Optical Fiber by Means of a Tapered Waveguide on Silicon. *Applied Physics Letters,* Vol. 55, No.23 (December 1989), pp. 2389-2391, ISSN 0003-6951

Šermukšnis, E.; Vyšniauskas, J.; Vasiliauskas, T. & Palenskis, V. (2004a). Computer Simulation of High Frequency Modulation of Laser Diode Radiation. *Lithuanian Journal of Physics,* Vol. 44, No.6, (December 2004), pp. 415-420, ISSN 1648-8504

Šermukšnis, E.; Vyšniauskas, J.; Palenskis, V.; Matukas, J. & Pralgauskaitė, S. (2004b). Dynamic characteristics of gain-coupled InGaAsP laser diodes and their reliability. *Kwartalnik elektroniki i telekomunikacji,* Vol.50, No.4, (April 2004), pp. 591-603, ISSN 0867-6747

Šermuksnis, E.; Vyšniauskas, J. & Palenskis, V. (2005). Dynamic characteristics and reliability of gain-coupled InGaAsP laser diodes. *Proceedings of SPIE, Optical Materials and Applications,* Vol.5946, No.8, (August 2005), pp. 300-308, ISBN 9780819459534

Tammela, S.; Ludvigsen, H. & Kaivola, M. (1997). Time-Resolved Frequency Chirp Measurement Using a Silicon-Wafer Etalon. *IEEE Photonics Technology Letters,* Vol.9, No.4, (April 1997), pp. 475-477, ISSN 1041-1135

Tatham, M. C.; Lealman, I. F.; Seltzer, C. P.; Westbrook, L. D. & Cooper, D. M. (1992). Resonance frequency, damping, and differential gain in 1.5 µm multiple quantum-well lasers. *IEEE Journal of Quantum Electronics,* Vol. 28, No.2, (February 1992), pp. 408-414, ISSN 0018-9197

Vasiliauskas, T.; Butkus, V.; Šermukšnis, E.; Palenskis, V. & Vyšniauskas, J. (2007). Computer Simulation of Transient Processes in DBF Semiconductor Lasers. *Lithuanian Journal of Physics,* Vol. 47, No.4, (December 2007), pp. 397-402, ISSN 1648-8504

Welford, D. A. (1985). Rate Equation Analysis for the Frequency Chirp to Modulated Power Ratio of a Semiconductor Diode Laser. *IEEE Journal of Quantum Electronics,* Vol.21, No.11, (November 1985), pp. 1749-1751, ISSN 0018-9197

10

Spectral Narrowing and Brightness Increase in High Power Laser Diode Arrays

Niklaus Ursus Wetter
Centro de Lasers e Aplicações,
Instituto de Pesquisas Energéticas e Nucleares de São Paulo,
Brazil

1. Introduction

Diode laser arrays, also called diode bars, are very important light sources that are generally used for pumping of other solid-state lasers and in medical and industrial applications that require high power but low intensity. Although they are very interesting from a commercial point of view, they did not find their way into many important applications because of their very low beam quality and spectral brightness.

This book chapter will explain why diodes have such a low spectral and spatial beam quality and give an overview on the different techniques to improve it.

In recent years, technology has achieved advances that permit to compare the spatial and spectral brightness of diode lasers to that of crystal solid-state lasers. Recent results of the literature are spatial brightness of 10^2 MW/cm²str with a beam quality product of $M_x^2 M_y^2 = 20$ and narrowing of a diode-arrays' emission bandwidth to the GHz or even MHz range.

Diode laser arrays are nowadays the basic building blocks of high power semiconductor lasers. They can continuously emit up to approximately 100 W at specific visible to near infrared wavelengths and can be stacked to give even much higher output powers of the order of several thousand watts. They find their main applications in materials processing, medical applications, printing and optical pumping of solid state lasers. Amongst the many industrial processing applications are welding, hardening, brazing and sintering. Medical and dental applications for diode bars include surgery, hair- and tattoo removal, photodynamic therapy and tooth whitening. Diode bars are also widely used in the military for targeting applications and therefore the distribution of some high power diode arrays is controlled.

Diode arrays present small size, efficiency in excess of 50 %, directly convert electricity into light and are the cheapest lasers in terms of watts per dollar on the market. For these reasons it would be of great benefit for commercial and scientific applications if they could substitute the common and expensive gas and crystal lasers that use resonators which are cumbersome to align and generally require highly skilled assembly and maintenance personnel. The reason that diode arrays so far did not substitute gas and crystal lasers in almost all applications lies in their low spectral and spatial beam quality.

A single diode bar is a linear array of approximately 19–70 individual diode emitters. The overall emitting area of a diode bar has generally a width of 1 cm and is 1 µm high. Due to the small height, comparable to the emission wavelength, the emitted beam diverges strongly in this direction (fast divergence axis) whereas in the other direction, given by the array of emitters, the divergence is smaller (slow divergence axis). As a result, the combined output beam is almost diffraction limited in the fast direction but has divergence that is more than thousand times that of a diffraction limited beam in the other direction.

This very poor beam quality hampers the usefulness of diode bars in many applications. Specifically, when trying to focus the beam tightly, very different waist sizes, waist positions and depth of foci are obtained for the fast and slow axis, making it almost impossible to tightly focus the diode's beam and making them resemble more flash light lamps than collimated laser beams.

Another reason that hinders the usefulness of diode beams is their spatial characteristics. From an application point of view it is in most cases interesting that the emission frequency is stable and as narrow as possible. Applications include pumping of solid-state lasers and laser cooling of gases for atomic clocks. Diodes not only change their emission wavelength strongly as a function of temperature but they also show linewidth broadening as a function of pump current. Typical linewidth are of the order of several nanometers and changes with temperature are around 0.25 nm/°C.

2. Diode bar fabrication and design

The main difference between a semiconductor laser and a conventional solid state laser is that the diode laser converts directly electric power into light without the need of an optical pump. It therefore immediately achieves much higher wall-plug efficiency than any other solid-state laser (around 50% wall-plug efficiency, although more than 70 % have been achieved at room temperature (Knigge et al., 2005)). That combined with its ability to be easily fabricated in mass production makes the semiconductor laser the cheapest laser per Watt of output power and the preferred laser candidate whenever its beam and spectral characteristics are apt for the desired application.

The constructive basics of a simple laser diode are well known. A semiconductor diode laser, or simply laser diode, is formed by doping thin layers on a direct band-gap semiconductor crystal wafer to produce n-type and p-type regions on top of each other, resulting in a p-n junction. The favored crystal growth techniques for semiconductors are liquid phase epitaxy and chemical vapor deposition using metal-organic reagents (MOCVD) but also molecular beam epitaxy (MBE) is frequently used. Forward electric bias across the junction forces holes and electrons to enter the depletion region (a region void of carriers that forms close to the junction because of the difference in electric potential between the n-type and p-type semiconductor). There, the carriers recombine generating preferentially spontaneous emission.

However, in order to achieve better efficiency than with this simple model, the commonly preferred constructive method since the 1990s is the separate confinement heterostructure (SCH) quantum well laser diode. This laser has a quantum well layer at the junction center made of high index material that serves to concentrate electrons in energy states that are prone to the emission of light. Additionally, the quantum well is flanked by two layers of different p-type materials on one side and two layers of n-type materials on the other side. As one moves farther away from the junction, each layer is made of a material with less high refractive index in order to confine the light generated by the recombination process into a very thin layer around the junction called active area. The refractive index graduation can be achieved for example by co-doping the GaAs semiconductor with small amounts of aluminum (wavelength range 630 nm to 1 µm). Typical dimensions for these layers are 0.01 µm for the quantum well layer and 1 µm for the layers of high index p-type and n-type materials closest to the quantum well. Due to the small dimension of the active area perpendicularly to the junction, all semiconductor diodes are essentially single transverse mode in that direction.

As with other lasers, laser action is achieved by giving a preferred direction to the emission of light generated in the active region and by building a resonator. In a laser diode this is generally done by fabricating the metal contact on top of the diode in the form of a long and thin ribbon. Recombination occurs within the semiconductor underneath this contact and preferential emission is achieved in the strip's direction, much as with a solid-state laser rod. Next, the semiconductor crystal is cleaved at its two ends to form a Fabry-Perot resonator. Depending on desired gain and spectral characteristics of the laser diode the end-facets maybe anti-reflection coated or uncoated, in which case they have a reflectance of approximately 30% in the visible region. The width of the active region parallel to the junction is generally of several microns and the length is approximately one millimeter.

Due to the small height of the active region, the stimulated emission perpendicular to the layers is strongly diffracted at the exit facet into an angle of approximately 25 degrees (FWHM) with Gaussian beam quality (fast axis direction). Parallel to the layers, the beam quality depends on the width of the electric contact strip and on the number of strips whereas the divergence angle remains generally around 10 degrees (slow axis direction). The fast axis divergence angle is always larger than in the slow direction because of its smaller dimension and also because of the larger index variation along the fast axis direction which causes larger numerical aperture (NA). Very small strips (width of 2 - 7 µm) lead to single transverse mode with overall diffraction limited beam quality and elliptical beam proportions but have also small power, not suitable for high power beam applications. Furthermore these devices can be made single frequency by limiting the longitudinal modes using suitable wavelength selective devices such as gratings that can be incorporated on chip or externally to the diode. The maximum output power for these devices is around a couple of hundreds of milliwatts for commercial devices although close to one watt has been achieved (Wenzel et al., 2008). Further power enhancement is obtained using an amplifier at the end of the diode in the form of a MOPA (master oscillator and power amplifier), which is a series of more or less independent devices. But also tapered amplifiers are common that employ the whole device including amplifier as resonator. The latter device achieved recently single transverse mode (TEM_{00}) output powers in excess of 10 W at slightly reduced beam quality, but such devices are mostly only experimental (Muller et al., 2011).

Single mode diode lasers have the highest spectral and spatial beam brightness. If higher powers are required the fabrication technology goes different ways and one has to decide what is more important, spatial or spectral brightness or a combination of both.

Brightness B is defined as:

$$B = \frac{P(watt)}{A(cm^2)\cdot\Omega(sterad)} = \frac{P}{w_x w_y \theta_x \theta_y \pi^2} = \frac{P}{\lambda^2 M_x^2 M_y^2} \quad (1)$$

Where P is the total power, A is the area and Ω and θ are the solid angle and half angle divergence, respectively, at $1/e^2$ of the peak power. λ is the wavelength and M^2 is the beam quality factor, which is 1 for Gaussian beams and gets bigger as quality decreases. Spectral brightness is defined as the brightness divided by the diodes bandwidth.

As the contact strip is made wider the output power increases and with it the locally generated heat (about 50 % of the consumed energy) whereas beam quality degrades. 20 W of output power have been achieved at 980 nm emission wavelength in 100 µm wide strip (Crump et al., 2009). These devices that go by the name of "broad area" diodes (BAL) are commercially available and have output powers of up to more than ten watts with beam quality factors in the directions parallel and perpendicular to the junction of typically 20 and 1, respectively. The width of the broad area can be increased and 500 µm is common but without considerable increase in output power and at a considerable loss of beam quality (M^2 parallel to the junction of about 100).

At even larger strip width the transverse direction can develop transverse lasing, amplified spontaneous emission (ASE) or even filamentation. These parasitic phenomena do not only cause losses but also dynamical instabilities and inhomogeneities in the beam profile. It is therefore more convenient to break the active region into a series of parallel strips. The drawback is a strong decrease in beam brightness that is proportional to the amount of "dead space" that is introduced in between the adjacent strips.

Generally there are 19 or 24 emitters in the case of continuous (cw) diode lasers and up to 60 for quasi-continuous (qcw) diode lasers. This high power diode laser (HPDL), which has total dimensions of approximately 1 millimeter length, 10 mm width and about 0.1 mm height is called a "diode array" or "diode bar". Individual emitter width is generally 100 µm to 200 µm and emitter spacing is 500 µm or 400 µm for the 19 and 24 emitter diode bars, respectively, with fill factors of 20% to 30%. The spacing between emitters, which is mainly dictated by cooling requirements, can be made much smaller for qcw diode bars reaching fill factors of up to 90%. Maximum output power for commercial cw diode bars is above 100 W and for qcw diode bars more than 300 W at less than 500 µs pulse duration and less than 10% duty cycle can be achieved. Above values are for conduction cooled diodes. For water cooling the heat removal is more efficient and gets even better with micro-channel water cooling which permits up to 80% fill factor in continuously operated diode bars. As can already be seen and will be shown also further ahead, efficient heat removal is a key factor in obtaining higher brightness for the same amount of output power. Several groups have reported in excess of 700 W cw per bar with production prototypes mainly by increasing the cooling efficiency using different means such as extending the emitter length from 1 mm to more than 3 mm. In all cases the beam quality perpendicular to the junction is close to diffraction limited whereas parallel to the junction the beam quality parameter M^2 is well in excess of 1000. At the junction exit the beam is a linear array of individual beamlets that overlap after a couple of millimeters. In the far field this elliptical beams' divergence angles are 30 degrees along the fast axis and 10-12 degrees along the slow axis.

Output power scales with fill factor in part because the maximum power density inside a diode bar is limited by the onset of catastrophic failure on the diodes exit facet. This failure is due to residual absorption at the beam exit facet caused by surface states at the cleavage plane that do not exist within the bulk. This problem hinders the maximum achievable power particularly in GaAs-based semiconductors and a great deal of effort has been spent in recent years to overcome this residual absorption by "passivation" of the exit facets.

The diode bar is normally soldered onto a thin submount which also serves as electrical contact. This submount is then attached to a large heatsink for effective removal of the heat. During this process the very thin (0.1 mm) but large (1 cm) diode array is easily bend by the pressure applied to bond it to the submount and heatsink but also by the differential expansion during the solder process. The overall effect of this process causes a curvature of the array called "smile" which amounts to about 1 -10 μm difference in height at the center of the bar with respect to the bar ends. Some manufactures also bond the bar directly to the copper heat-sink, which is referred to as "direct bonding" and causes generally less smile.

The problem with array curvature is that it degrades the beam brightness. Once the diode is manufactured, the bar curvature is fixed. Therefore it is of importance to understand the constructive details of diode bar manufacturing to apply effective countermeasures. The deviation from linearity of the smile makes it impossible for a single collimation lens in front of the emitters to be perfectly positioned for all emitters along the bar. This first lens, which serves to collimate the fast-axis (FAC lens), is attached directly in front of the diode bar and is generally a plano-aspherical cylindrical lens for low aberration that resembles approximately a 1 cm long piece of fiber. The focal length of this type of lenses varies from 0.3 to 1.5 mm and the numerical aperture is high, usually between 0.5 and 0.8 to accommodate the strong divergence in the fast axis direction. After the FAC lens, the beam resembles a sheet of light with 1 cm width and of approximately one millimeter height. In the far field the divergence angles are 4-6 degrees along the fast axis and continue 12 degrees along the slow axis.

After attaching the collimation optics to the diode bar the beam errors are set and the rays emitted by the diode occupy a larger volume than the theoretical minimum. But the errors can be fixed if it is possible to design and introduce a suitable optical element or optical system.

2.1 Diode bar beam quality assessment

The beam errors increase the difficulty to design the subsequent optics and reduce the efficiency of the whole optical system in terms of spatial and spectral brightness. It is therefore of great importance to assess the specific nature of these imperfections in order to design effective countermeasures that can in many cases completely eliminate the consequences of these errors. The normal assessment technique for the beam quality of any laser is by means of measuring the quality parameter M^2. This is a simple measurement that can give data on beam divergence, focal spot size and confocal parameter but it is not enough information to characterize which is the appropriate correction technique in the case of beam errors and more elaborate beam assessment methods have to be applied.

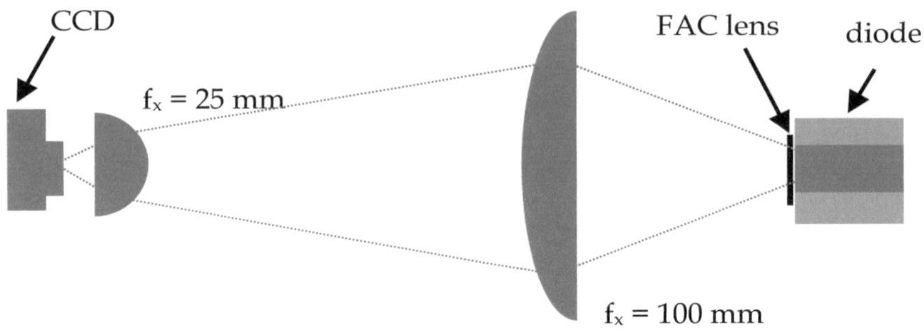

Fig. 1. Simple imaging system to visualize the beam errors introduced by the diode and the FAC lens: the slow x-axis direction is collimated by two cylindrical lenses.

Although there are nowadays many assessment techniques available, one can obtain most required information by using a simple imaging system that projects the individual beamlets onto a CCD camera (Wetter, 2001). If the diode bar has a FAC lens incorporated and therefore has its fast axis already collimated, this system may comprise simply two cylindrical lenses to image the slow axis direction.

Imprecision in the alignment and positioning of the FAC lens at the micron level further reduces the brightness. Failure to align the lens parallel to the junction causes "skew" that projects the individual beamlets from the emitters at different heights. The problem with skew is that it causes astigmatism and coma as shown in figure 2. Another common problem is to keep the precise distance of the fiber to the array along the whole width of 1 cm. This relative focus error manifests itself in different spot sizes of the beamlets at the image or focus plane, further increasing astigmatism and decreasing the brightness. Earlier FAC lenses were simple glass fibers attached in front of the bar. These fiber-FAC lenses could easily be bent and introduce further curvature into the highly astigmatic exit beam. Nowadays they have mostly been replaced by bigger glass lenses that have additional support structure to decrease the possibility of bending.

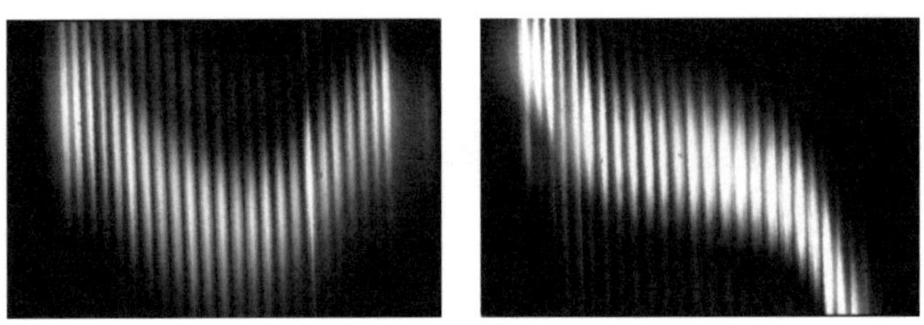

Fig. 2. Two beam errors that are encountered with diode bars; from left to right: smile and skew. The images have been taken with the system in figure 1.

3. Beam-shaping techniques

Beam-shaping techniques applied to diode arrays have two purposes. The first goal is to circularize the highly astigmatic beam of uneven beam quality factors in such a way that it can be used in most practical applications that require generally focusing or collimation and round spot sizes. The second goal is to increase brightness. Brightness increase augments local absorption at the samples surface and inside the sample which leads to higher temperatures and therefore to more efficient material removal and faster processing speeds. As a result, it increases the range of applications the diode bars can be used for. For instance in medical application, low brightness is enough for hair removal whereas high brightness is necessary for surgery. Both applications may use a laser system composed of the same diode bar component. The difference being that the later application requires additional beam shaping with post-bar additional optical components inside the beam path.

Most practical beam-shaping techniques used for beam circularization are based on methods that geometrically transform the diode radiation into a rounder spot size. It is convenient for this purpose that the diodes beam consists of individual beamlets which can be rearranged by suitable optical devices. One of the first and still used methods of beam-shaping consists of individual fibers, only slightly larger than the individual emitter width, which are precisely positioned in front of each of the diode bar's emitters at a very short distance. The fibers can then be bundled into the desired geometrical form which is for most applications of circular shape. Many other beam-shaping systems consist of optical components that chop the large beam coming from the diode bar into a number of sub-beams. These sub-beams are then reflected, diffracted or refracted by clever positioned mirrors, prisms, gratings and micro-lenses in such a way that the sub-beams are stacked on top of each other in order to generate a round spot size.

In figure 3 one can see such a beam-shaper composed of two plane parallel mirrors that are inclined sideways (around a vertical y-axis) and backwards (around the horizontal x-axis). The incoming light sheet from the diode bar passes just above the first mirror where one fifth of the beam is cut-off by the second mirror passing on directly whilst four fifths get reflected slightly downward and to the left (negative x-axis direction) back onto the first mirror. This scheme is repeated until all five parts emerge from the two mirror beam shaper, stacked on top of each other and having ideally a square shape and equal quality factors in the x- and y-directions.

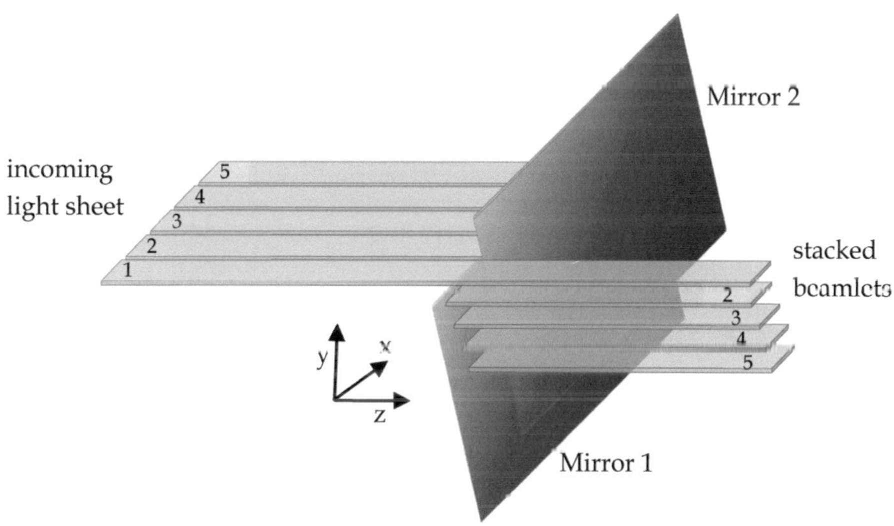

Fig. 3. Two mirror beam shaper used to cut a homogenous light sheet into five sub-beams that are stacked on top of each other. The goal is to achieve a new beam with equal dimensions and quality factors in the horizontal and vertical directions.

3.1 Correction of diode beam errors

Most beam-shaping techniques depend upon a good quality light sheet being emitted by the bar. For instance, if the diode shows smile as in figure 2, the middle part of the beam will get clipped by mirror 1 in figure 3 upon the first passage of the light sheet above the mirror. This is not only an issue for the two-mirror beam shaper but almost all beam-shaping methods rely on good alignment of the individual beamlets in order to reconfigure the beam without losses. In these cases skew, smile and relative focus errors should be compensated for.

One correction technique that can to some degree diminish these errors is the introduction of a plano-convex cylindrical lens in front of the beam shaper (Wetter 2001). The lens should be oriented with its axis perpendicular to the diodes slow divergence direction as shown in figure 4. Rotating the lens by a small negative angle α around the x-axis causes a displacement of parts of the light sheet in the vertical direction. This translation in the positive y-direction depends upon the thickness of the traversed lens material and is therefore largest for that sub-beam that enters the lens at its center. The shift of the center relative to the border of the diodes light sheet is proportional to α and inversely proportional to the focal length of the cylindrical lens. It follows that for a given smile there is always an inclination angle that is capable of lifting the central part of the light sheet to the same height as its borders, provided that the focal length of the lens is short enough.

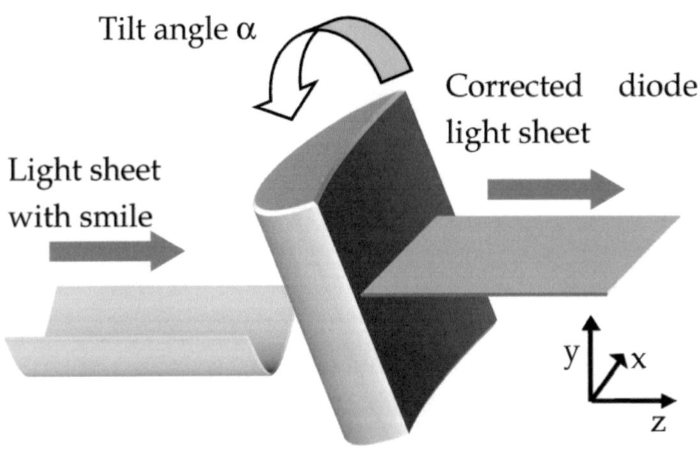

Fig. 4. Schematic diagram of the inclined cylindrical lens.

Given the cylindrical lens shape, this technique works best for a smile of parabolic curvature and with the center of the curvature in the middle of the light sheet. It has been shown that in such cases the height of the smile can be decreased by up to 68% reducing power loss due to clipping on mirror 1 by almost 27%. Skew may be corrected in a similar way by tilting the lens slightly around its z-axis.

More complex systems use arrays of lenslets to control the divergence and beam propagation direction of the emitters in the bar. The number and spacing of the lenslets is such that they match the emitters of the bar. Each micro-lenslet has square dimensions and cylindrical shape in the horizontal and vertical direction in order to collimate the fast and slow axis at the same time achieving a divergence of less than 2 degrees in both directions. The focal length is very small, of the order of 100 µm, reducing greatly the size of the macroscopic optic of the beam shaper that follows. Smile reduces significantly the efficiency of such a system and therefore several lenslet arrays are mounted on top of each other with different row shapes that are described by polynomials which match different smile shapes. A lens array row shape can then be assigned to the specific smile of the diode bar (Hamilton et al., 2004). This technique allows the lenslets to be centered in front of each emitter and to carry out their function of beam collimation and aligning the beam into the right direction.

All of above mentioned beam-shaping techniques are intended to generate square beams of equal beam quality factors in both directions. However a square shape is not ideal for coupling into a fiber. To achieve better beam circularization one must change the individual emitter width. If the emitter in the middle of the bar is the largest and if the emitters become progressively smaller as one moves out to the last emitters on opposite ends, stacking as shown in figure 3 will provide for a round spot composed of stacked emitter beamlets.

3.2 Brightness Increase

So far only beam circularization has been achieved. As stated by equation (1), brightness is inversely proportional to the area occupied by the source. In other words, high brightness is achieved with 100% fill factor and the "dead space" between the emitters of the bar has to be removed in order to increase the brightness. As shown earlier, this space cannot be eliminated at the source for cooling reasons and therefore it is necessary to eliminate it afterwards. This can be achieved for instance with the beam-shaper of figure 3 by adjusting the mirrors' positions and angles. In this figure, the light sheet emitted by the bar is sectioned into five sub-beams as also shown in figure 5.

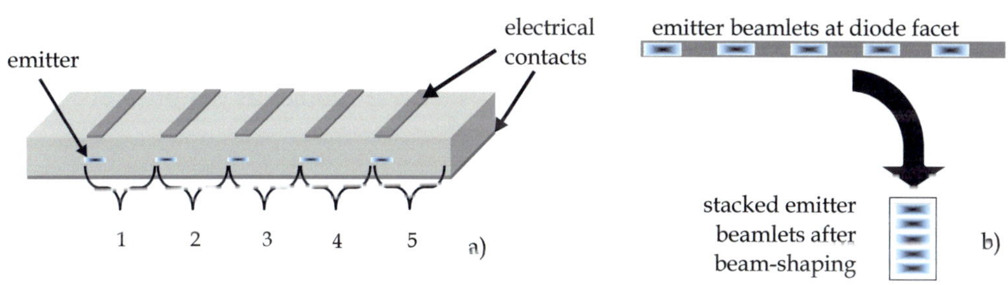

Fig. 5. a) Schematic of a diode bar with 5 emitters. The width of the brackets corresponds to the emitter spacing or emitter pitch. b) Schematic description of the "dead space" being removed by the beam-shaper.

Each sub-beam is composed of one part useful laser radiation and one part dead space. By moving mirror 2 in figure 3 along the negative x-direction the width of the sub-beam 1 will be decreased until only the emitter beamlet component from the sub-beam remains. If then the mirrors' spacing and the inclination angle are readjusted properly, all other four sub-beams will undergo the same procedure and finally, only the 5 emitter beamlets will emerge from the beam-shaper, as shown in figure 5.

Using the simple beam correction and shaping methods outlined in item 3, a power density of 46 kW/cm^2 can be achieved with standard diode bars accompanied by a loss of less than 20 percent in output power. The brightness achieved with this side-by-side beam combining techniques cannot be greater than the brightness of a single emitter because, at best, the power increases at the same rate as the area of the reconfigured beam.

A standard method to couple two diodes that have undergone beam-shaping is the polarization coupling with beam-splitters. In this method, diode 1 and diode 2 have orthogonal beam polarizations that are coupled by a special beam-splitter that is 100% transmissive for the polarization of diode 1 and reflects 100% of diode 2 at 45 degrees incidence, overlaying ideally both beams completely at the output, without loss of spectral brightness. The method is often used to increase the power launched into a fiber by a factor of two.

New developments in the area of facet coatings and passivation techniques have permitted power scaling using longer resonators and narrower emitters (with lower divergence). These bars have the same output power at only 25% of the total emitter facet area of the common 1 cm wide bars. They are called T-bars, super-bars or also mini-bars because their width is of the order of only 3 mm. The reduced width not only decreases greatly the beam divergence and the quality factor in the slow axis direction but also increases its flatness and therefore decreases smile, skew and relative focus errors.

3.3 Spectral beam combining

Spectral beam combining, also called wavelength combining or incoherent beam combining, is a very efficient beam-shaping method that permits to overlap almost completely the beamlets of all emitters thereby increasing greatly the spatial brightness. The result is the same as if in figure 5b) only one beamlet would emerge from the beam-shaper with the power of all 5 beamlets together and hence, the brightness increase is equal to the number of emitters. But this power increase comes at the price of reduced spectral brightness.

The working principle of spectral beam combining is to assign to each emitter a different wavelength with exclusive power-spectra that is then imaged by an optical system onto some wavelength-selective device capable of combining the different beamlets into one single output beam (Daneu et al., 2000). An example of such a device is shown in figure 6. In this particular beam-combining beam-shaper the front-facet of the diode bar is anti-reflection coated and hence, the resonator is formed between the bar's back-facet and the output mirror. The transform lens, the grating and the flat output coupler serve to assign to each emitter on the bar a different incidence angle on the grating and therefore a different center wavelength. The overall gain bandwidth of the diode and the dispersion of the grating must be adjusted such that the spectral emission bandwidth of each diode is exclusive.

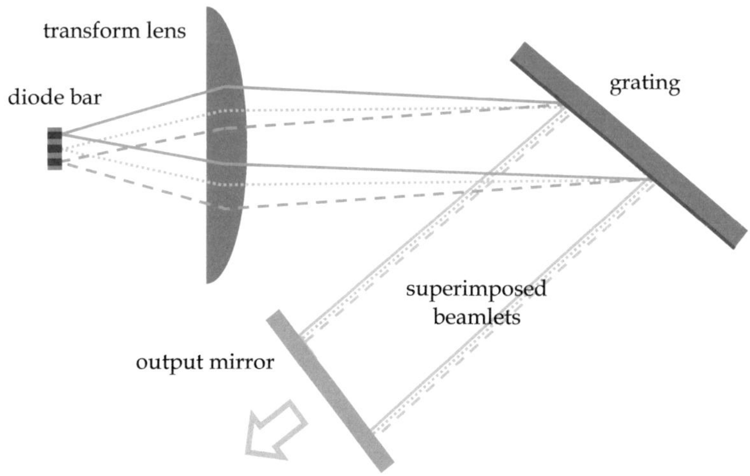

Fig. 6. Schematic of the working principle of beam-shaper of class "incoherent beam combiner".

The beam quality factors of the bar after beam-shaping achieved with above device are the same as of a single emitter, which is 10-20 in the slow axis direction and approximately 2 in the fast-axis direction. The measured brightness increase is about 20x, achieving close to 100 MW/cm^{2}str.

It should be noted that this design is different from the well known wavelength-division-multiplexing (WDM) used to combine different wavelength in a waveguide in as much as this is a purely passive design whereas in WDM one generally tunes first each emitter independently and then combines them with the wavelength-selective device. The wavelength combination may occur also by means of prisms and volume-Bragg gratings (VBG) whilst for only a few beamlets one may also use dichroic mirrors.

One of the interesting features of this technique is that it is further power scalable because diodes can be stacked into a 2-dimensional array with much higher output power than a single bar and chirped VBGs can be used to assign individual wavelengths to the emitters which then are superposed by the grating (Chann et al., 2006).

3.4 Coherent beam combining

Coherent beam combining (CBC), also called coherent addition or phase-locked arrays, is a technique that permits joining of several beams increasing the spatial brightness. It differs from spectral beam combining in as much as there is no need to assign different wavelengths to each beamlet, however, the phase relation between the beamlets has to be well determined such that they all combine coherently in the output beam. As a result, this technique permits increase in spatial brightness and spectral brightness at the same time because the total bandwidth can be as small as the bandwidth of a single emitter. Because of this prospect, this class of beam-shaping has received a lot of attention. The method is very well known in microwave and radio-frequency applications as phased-array transmitters.

However, this technique has so far not been able to give the expected results in laser applications, mainly because of the shorter wavelength, and it has been proven to be very challenging to maintain the phase relationships within a fraction of 2π outside the laboratory.

The first step in coherent beam combining is to achieve beamlets that have the same polarization and the same phase and therefore may interfere coherently. Most methods used to obtain phase-locking of the beamlets require further that all beamlets operate at the same frequency, or single-frequency, although there are some important exceptions to this rule.

Two beams of equal power that travel side-by-side and interfere constructively already have a brightness increase of a factor two because the power doubles and the beam quality remains the same. This side-by-side combining (also called tiled-aperture combining or phased array) has been used since the 1970's mostly with semiconductor arrays whose emitters are closely spaced in a way that they optically couple through evanescent waves and excite some higher order transverse mode that incorporates all emitters (Scifres et al., 1978). Aside from this passive method of phase-locking (that may be applied also to fiber lasers) exist a series of active methods that generally result in a much more robust phase-lock amongst emitters (or channels). For example, one may use a periodic 2-D array of semiconductor emitters inside a Talbot cavity (a Talbot cavity images the near filed image of the array onto itself) and use a phase-plate (for example a spatial light modulator - SLM) between the array and the output coupling mirror that adjusts the optical path difference of each individual emitter. To calculate the phase introduced for each emitters' optical path sophisticated feedback-loops are required that may act upon a SLM or the emitter current or temperature to adjust the emitter's phase. If the emitters' resonator lengths are different and if there are several channels, as it is for example with combining of fiber lasers, it might become difficult to find longitudinal modes that are common to all resonators. In such cases one can use a single-frequency laser at the origin that becomes amplified in several separate channels and then apply active phase correction at the output. Up to 64 fibers have been combined in this manner (Bourderionnet et al., 2011).

In a next step one may further increase the spectral and spatial brightness by combining the separate channels into one single channel using for example diffraction gratings or beam splitters (this is called filled-aperture combining). A 50/50 beam-splitter normally has two outputs unless both beams are adjusted in phase such that one output interferes destructively. The correct path difference may be obtained actively or passively by careful positioning and calculation of the beam splitters' dimensions as shown in figure 7. An additional benefit of this scheme is that because all beamlets share the same resonator at the output coupler, there is generally no need for active phase control at the output end. A total of 16 beamlets originating from the same laser rod have been added in this manner generating an overall spectral brightness increase of thirty times (Eckhouse et al., 2006).

There is also a class of beam CBC methods that, instead of comprising arrays or beam splitters, use non-linear effects such as non-linear mirrors based on second harmonic generation (SHG) or Brillouin scattering.

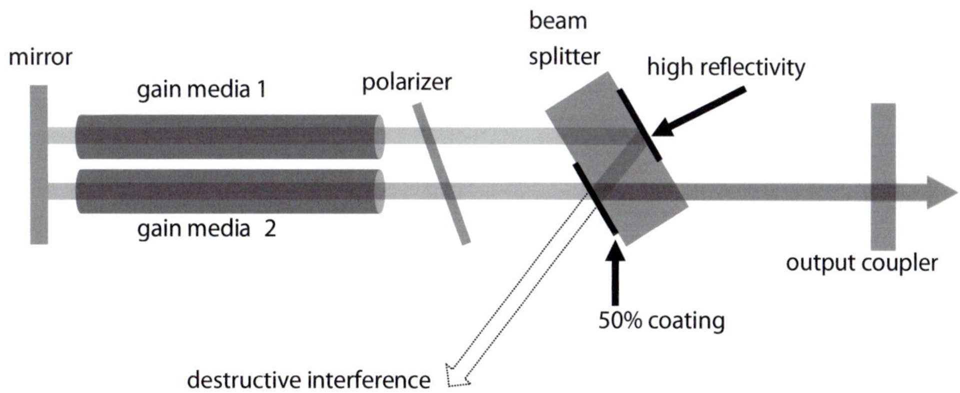

Fig. 7. Schematic of two channels being coherently added by means of a beam splitter.

4. Spectral narrowing

Many applications would benefit from the advantages of diode lasers mentioned in this chapter, if the linewidth were smaller than the 3 nm (approximately half a THz) of standard high–power diode arrays. LIDAR detection techniques used for trace gas detection and wind speed measurements, optical pumping of rare-earth solid state-lasers and mm-wave lasers, laser cooling for atomic clocks and WDM in optical communication systems are all applications that could use much smaller linewidth. These diverse applications have been traditionally served by tunable broad emission bandwidth lasers such as Ti:Sapphire and Dye lasers.

The high power diode lasers (HPDLs) can be easily tuned with external cavities (as already seen in the former chapter) and depending on cavity parameters and reflectivity of the emitter facets' AR coating, the tuning range is naturally around 30 nm even at very high output powers. It therefore represents in principle an ideal substitute for the Ti:Sapphire or dye laser if the bandwidth can be made small enough as required by the application.

Several injection schemes have been tested nowadays and they can be divided into self-injection locking and master-injection locking. The former usually requires an external cavity that supplies frequency-selective feedback and the latter relies usually on a single-frequency master oscillator. In self-injection schemes the frequency-selective device for the feedback from the external cavity comes normally from a diffraction grating but also a VBG, an etalon or a prism can serve to this purpose.

The most common external cavity schemes employing gratings are the well known Littrow configuration, which directly injects the first diffraction order back to the diode, and the Littman-Metcalf configuration, which uses an additional mirror to reflect the first order back and therefore passing the beam twice through the grating. The advantage of the former method is higher output power (throughput of the system is typically 80%) whereas the latter method shows generally more spectral narrowing because of the double pass through the grating and also a fixed output beam direction.

Narrowing can only be achieved if the beamlets from each emitter overlap at the grating (Wetter, 2007). It is therefore essential that smile is minimized because in the end it will increase the linewidth by approximately 0.9 nm per μm of smile. But overlap can be increased if a telescope is used in the beam path of magnification M. The better overlap at the grating of the magnified images will reduce angular spread and decrease the influence of smile by a factor of 1/M and hence further narrow the diodes' spectra. Additionally the light re-injected from the external cavity will match better each emitter's position and therefore less re-injected power is necessary to force the emitters to oscillate at the desired wavelength. The amount of light that gets re-injected can be adjusted for example with a half-wave plate.

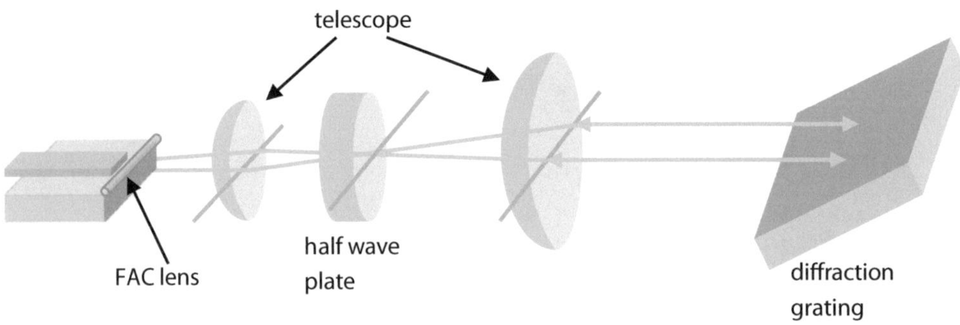

Fig. 8. Schematic of a Littrow configuration using an intracavity telescope and a half-wave plate to adjust the retro-injected power.

Wavelength narrowing down to less than 40 pm, corresponding to a linewidth of about 10 GHz at NIR wavelength, is achievable using VGBs and Littrow set-ups, respectively (Zhdanov et al., 2007) (Vijayakumar et al., 2011) , and spectral brightness increase of up to 86 times has been obtained using HPDL with up to 21 emitters. This is enough narrow for optically pumping alkali vapor lasers such as potassium vapor and cesium.

The self-injection narrowing shown in figure 8 is proportional to the number of emitters and better results of spectral brightness increase are achieved with fewer emitters because the overall width is smaller and therefore smile can be reduced. Even with new production methods, smile continues at about 1 μm height. Increasing the magnification augments the size and decreases the mechanical stability of the set-up. It seems therefore, that the most promising approach so far is to use BALs of short cavity width. As was shown earlier, these broad area diode lasers can have quite small width and present very high output powers. Using a 1000 mm wide BAL with maximum output power of up to 25 W in a Littman-Metcalf configuration, linewidth narrowing from 2 nm down to 1.8 MHz has been achieved at output powers up to 10 W (Sell et al., 2009). This very small linewidth is not frequency stable and a measurement over 0.5 seconds results in a linewidth of 130 MHz, which shows the influence of temperature and mechanical instabilities.

5. Conclusions

Standard HPDL bars are nowadays factory equipped with beam-shapers for beam circularization with only a minor increase in cost. This kind of beam-shaper generates a circular beam of equal quality factors in the horizontal and vertical direction which can be used to achieve, for example, very efficient fiber-coupling. Up to 60 W of output power from fibers of 100 μm fiber core diameter and NA of 0.22 are standard using a single bar. Polarization coupling of two bars into the same fiber diameter with total output power of 120 W is also commercially standard. Systems using spectral and coherent beam combining are still relatively expensive but are bound to become important players on the market of high-power, high-brightness lasers. Better facet passivation techniques have also boosted lately the competition between BALs and diode bars. Broad area diodes have a series of advantages such as better spatial and spectral brightness as outlined in this chapter and if powers should become comparable to diode HPDL arrays then they are certainly advantageous for many applications.

6. References

Bourderionnet, J., Bellanger, C., Primot, J. & Brignon, A. 2011. Collective coherent phase combining of 64 fibers. Optics Express, 19, 17053-17058.

Chann, B., Goyal, A. K., Fan, T. Y., Sanchez-Rubio, A., Volodin, B. L. & Ban, V. S. 2006. Efficient, high-brightness wavelength-beam-combined commercial off-the-shelf diode stacks achieved by use of a wavelength-chirped volume Bragg grating. Optics Letters, 31, 1253-1255.

Crump, P., Blume, G., Paschke, K., Staske, R., Pietrzak, A., Zeimer, U., Einfeldt, S., Ginolas, A., Bugge, F., Hausler, K., Ressel, P., Wenzel, H. & Erbert, G. 20W continuous wave reliable operation of 980nm broad-area single emitter diode lasers with an aperture of 96 mu m. Conference on High-Power Diode Laser Technology and Applications VII, Jan 26-27 2009 San Jose, CA. Spie-Int Soc Optical Engineering.

Daneu, V., Sanchez, A., Fan, T. Y., Choi, H. K., Turner, G. W. & Cook, C. C. 2000. Spectral beam combining of a broad-stripe diode laser array in an external cavity. Optics Letters, 25, 405-407.

Eckhouse, V., Ishaaya, A. A., Shimshi, L., Davidson, N. & Friesem, A. A. 2006. Intracavity coherent addition of 16 laser distributions. Optics Letters, 31, 350-352.

Hamilton, C., Tidwell, S., Meekhof, D., Seamans, J., Gitkind, N. & Lowenthal, D. High power laser source with spectrally beam combined diode laser bars. Conference on High-Power Diode Laser Technology and Applications II, Jan 26-27 2004 San Jose, CA. 1-10.

Knigge, A., Erbert, G., Jonsson, J., Pittroff, W., Staske, R., Sumpf, B., Weyers, M. & Trankle, G. 2005. Passively cooled 940 nm laser bars with 73% wall-plug efficiency at 70 W and 25 degrees C. Electronics Letters, 41, 250-251.

Muller, A., Vijayakumar, D., Jensen, O. B., Hasler, K.-H., Sumpf, B., Erbert, G., Andersen, P. E. & Petersen, P. M. 2011. 16 W output power by high-efficient spectral beam combining of DBR-tapered diode lasers. Optics Express, 19, 1228-1235.

Scifres, D. R., Burnham, R. D. & Streifer, W. 1978. Phase-Locked Semiconductor-Laser Array. Applied Physics Letters, 33, 1015-1017.

Sell, J. F., Miller, W., Wright, D., Zhdanov, B. V. & Knize, R. J. 2009. Frequency narrowing of a 25 W broad area diode laser. Applied Physics Letters, 94, 51115 - 51115-3.

Vijayakumar, D., Jensen, O. B., Barrientos-Barria, J., Paboeuf, D., Lucas-Leclin, G., Thestrup, B. & Petersen, P. M. 2011. Narrow line width operation of a 980 nm gain guided tapered diode laser bar. Optics Express, 19, 1131-1137.

Wenzel, H., E. Bugge, M. Dallmer, F. Dittmar, J. Fricke, K. H. Hasler, and G. Erbert. 2008. Fundamental-lateral mode stabilized high-power ridge-waveguide lasers with a low beam divergence. Ieee Photonics Technology Letters, 20, 214-216.

Wetter, N. U. 2001. Three-fold effective brightness increase of laser diode bar emission by assessment and correction of diode array curvature. Optics and Laser Technology, 33, 181-187.

Wetter, N. U. 2007. Tunable dual wavelength emission and bandwidth narrowing of a laser diode array with a simple external cavity. Applied Physics B-Lasers and Optics, 86, 515-518.

Zhdanov, B. V., Ehrenreich, T. & Knize, R. J. 2007. Narrowband external cavity laser diode array. Electronics Letters, 43, 221-222.

11

Tunable Dual-Wavelength Laser Scheme by Optical-Injection Fabry-Perot Laser Diode

Chien-Hung Yeh
Information and Communications Research Laboratories, Industrial Technology Research Institute (ITRI), Chutung, Hsinchu, Taiwan

1. Introduction

Tunable and stable multiwavelength fiber lasers are important and attractive in recent years because of their potential applications in wavelength-division-multiplexed (WDM) technique (Das et al., 2002; Slavik et al., 2002), optical code-division multiple-access (OCDMA) technique (Barmenkov et al., 2008; Alvarez-Chavez et al., 2007), fiber sensor system (Chou et al., 2008; Chen et al., 2007), and optical instrument measuring (Talaverano et al., 2001; Nilsson et al., 1996) and testing (Liu et al., 2004; Li et al., 1998). Different techniques for the reduction of wavelength competition have been used to achieve stable multiwavelength oscillations. However, the homogeneous gain broadening of erbium-doped fibers (EDFs) would lead to the wavelength competition [2]. Many previously reports have been focused on the fiber laser technique by inserting the optical filter, such as the tunable bandpass filter (Barmenkov et al., 2008), Fabry-Perot tunable filter (Li et al., 1998) and fiber Bragg grating (Alvarez-Chavez et al., 2007), into the EDF laser cavity for single or multiwavelength oscillations (Chen et al., 2007). In such configurations, the cavity losses corresponding to the different wavelengths have to be balanced with the cavity gains simultaneously. Therefore, it is difficult to control the lasing wavelength output.

Recently, the self-injection Fabry-Perot laser diodes (FP-LDs) and distributed feedback laser diode (DFB-LD) with mode-locked operation using Bragg grating or optical filter to generate tunable single-wavelength (Schell et al., 1995), dual-wavelength (Kim et al., 2005; Yang et al., 2002) or multiwavelength short pulses (Peng et al., 2003; Fok et al., 2007) have been proposed and experimentally analyzed. Mode-spacing and wavelength tuning laser using bismuth-oxide fiber or photonics crystal fiber with nonlinear effect have been studied and reported (Liu et al., 2008; Chen et al., 2007; Al-Mansoori et al., 2008). Comparing with our proposed dual-wavelength laser schemes as following (Yeh et al., 2008; Yeh et al., 2009), we only need commercially available and standard components to achieve the two-mode lasing and also has a benefit of cost-effective.

2. Dual-wavelength laser diode with fixed mode-spacing lasing

Here, first of all, we will introduce the dual-wavelength tuning semiconductor laser scheme via optical injection technology with fixed mode-spacing ($\Delta\lambda$) output. Thus, in this experiment, Fig. 1 shows the proposed structure of tunable dual-wavelength EDF ring laser (Yeh et al., 2008). The proposed fiber laser is consisted of an erbium-doped fiber amplifier (EDFA), a 2×2 and 50:50 optical coupler (OCP), a FP-LD and a polarization controller (PC). In Fig. 1, the EDFA constructed by an 10 m long EDF (Fibercore DC1500F), a 980/1550 nm WDM coupler, an optical isolator (OIS) and a 980 nm pumping laser. In this proposed laser scheme, when the pumping power exceeds 100 mW, it would saturate the lasing output. Thus, the 980 nm pump LD is set at 100 mW. The 3 dB bandwidth and average insertion loss of TBF used are nearly 0.4 nm and 3.5 dB respectively. The TBF also has a 40 nm tuning range from 1520 to 1560 nm.

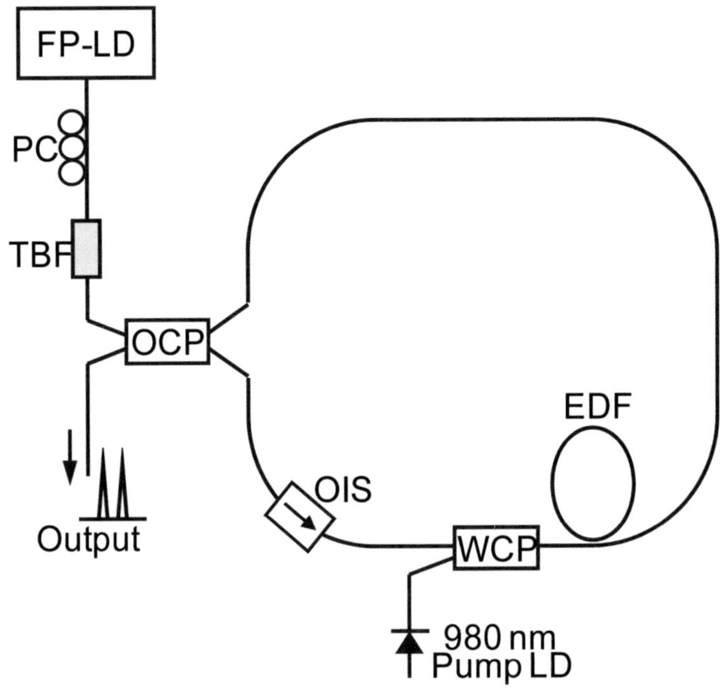

Fig. 1. Experimental setup of the proposed tunable and stable dual-wavelength fiber laser.

The PC between FP-LD and OCP is used to control the polarization state of the feedback injection light into the FP-LD. According to the past self-injected report (Liu et al., 2004), only one polarized direction being parallel to the TE-mode of FP-LD of feedback wavelength leads to the maximum self-injected efficiency. In the experiment, the mode spacing ($\Delta\lambda$) and threshold current of FP-LD are 1.3 nm and 10 mA, respectively. In the measurement, the threshold pumping power in this experiment is around 20 mW while the FP-LD is operated at 18 mA and 25 ^{o}C. Therefore, we set the bias current of the FP-LD at 18 mA at the temperature of 25 ^{o}C. To measure and analyze the output power and wavelength of the proposed dual-wavelength laser, an optical spectrum analyzer (OSA) with a 0.05 nm resolution is used for the measurement. Fig. 2 presents the output wavelength spectrum of the FP-LD without self-seeding operation in the wavelength range of 1520 to 1570 nm. Fig. 2 also shows that the output power level of FP-LD is above −20 dBm around the wavelengths between 1545.0 and 1550.0 nm.

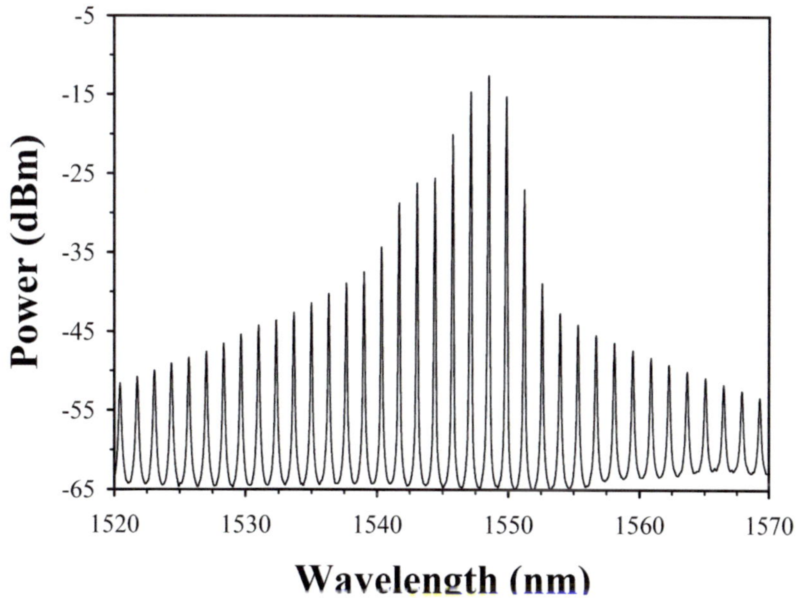

Fig. 2. Output wavelength spectrum of the FP-LD without self-seeding operation in the wavelengths of 1520 to 1570 nm when the LD operates at 18 mA and 25 °C.

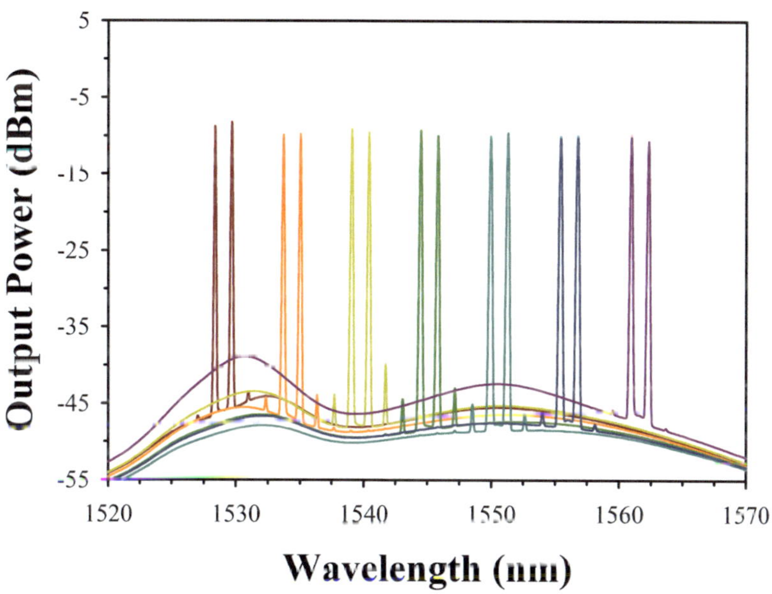

Fig. 3. Output wavelength spectra of the proposed dual-wavelength fiber laser with 100 mW pumping power and 18 mA bias current, in the operating wavelengths of 1523.08 to 1562.26 nm.

Then, by using the proposed self-injected structure, Fig. 3 displays the output wavelength spectra of the proposed dual-wavelength fiber laser when the 980 nm pump LD power and bias current of the FP-LD are 60 mW and 18 mA, respectively, in the operating wavelengths of 1523.08 to 1562.26 nm with 1.3 nm tuning step. In Fig. 3, the mode spacing of the dual-wavelength laser is measured at nearly 1.3 nm. The minimum side-mode suppression ratio (SMSR) is larger than 36.5 dB over the operating range. The maximum and minimum output powers of −9 and −14.5 dBm are also observed in the tuning range. Besides, the dual-wavelength can be slightly tuned by adjusting the temperature of the FP-LD. While the temperature difference (ΔT) of the FP-LD is ±5 °C, the central wavelength variation also shifts at ±0.2 nm. Therefore, the dual-wavelength can be tuned continuously by controlling the temperature. Based on the proposed laser architecture, the fiber laser not only can lase dual-wavelength but also enhance the tuning wavelength range to 39.18 nm. As a result, Fig. 2 is a free-run FP-LD, with very small output powers at shorter wavelength (< 1545.0 nm) and longer wavelength (>1555nm). However, in our proposed scheme, we simultaneously achieve two-mode laser with even and high output power across the wavelength range of 39.18 nm. The significantly improvement can be seen in Fig. 3. We have tried several coupling ratios (such as 90/10, 80/20, 70/30, 60/40 and 50/50) and we found that the 50/50 is the optimum case.

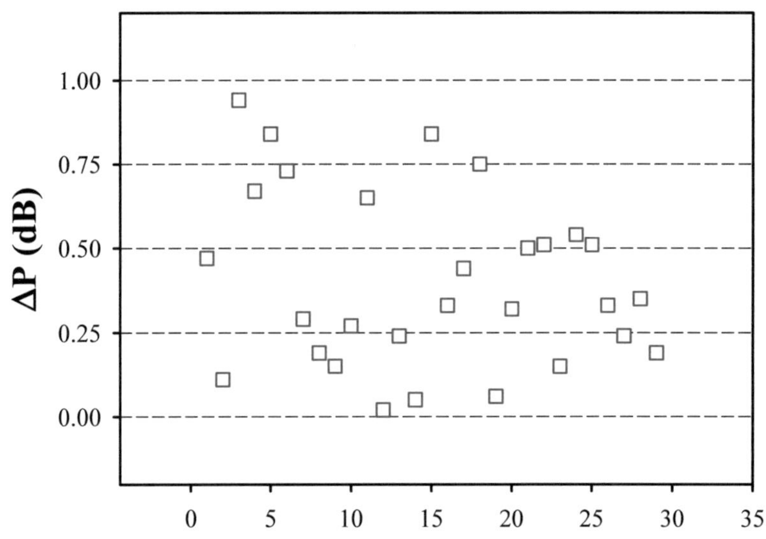

Fig. 4. Power difference (ΔP) of the lasing dual-wavelength in the operating range with ~1.3 nm tuning step.

Figure 4 shows the output power difference (ΔP) of the lasing dual-wavelength in the operating range with ~1.3 nm tuning step. The output power difference is defined to $\Delta P = |P_1 - P_2|$. The maximum and minimum ΔP of 0.9 and 0.1 dB are measured in Fig. 4. As a result, the dual-wavelength also presents a good equalizing output power over the tuning range.

Fig. 5 shows the average output power and SMSR of the lasing dual-wavelength with 1.3 nm tuning step in the wavelengths of 1523.08 to 1562.26 nm. In the tuning step 7 to 28 (from

1530.88 to 1558.18 nm), the average output power and SMSR could be larger than –10 dBm and 40.1 dB, respectively. For the whole tuning range, in the wavelengths of 1523.08 to 1562.26 nm (step 1 to 29), the minimum average output power and SMSR of –14.6 dBm and 36.4 dB can still be achieved. Fig. 5 also shows a flat average output power in the tuning range in the tuning step 4 to 27 (ΔP_{max} = 0.9 dB). Moreover, the average SMSR spectrum in Fig. 5 presents two peaks at the step 15 and 26, respectively. And the maximum difference of average SMSR is ~5.4 dB in the tuning range.

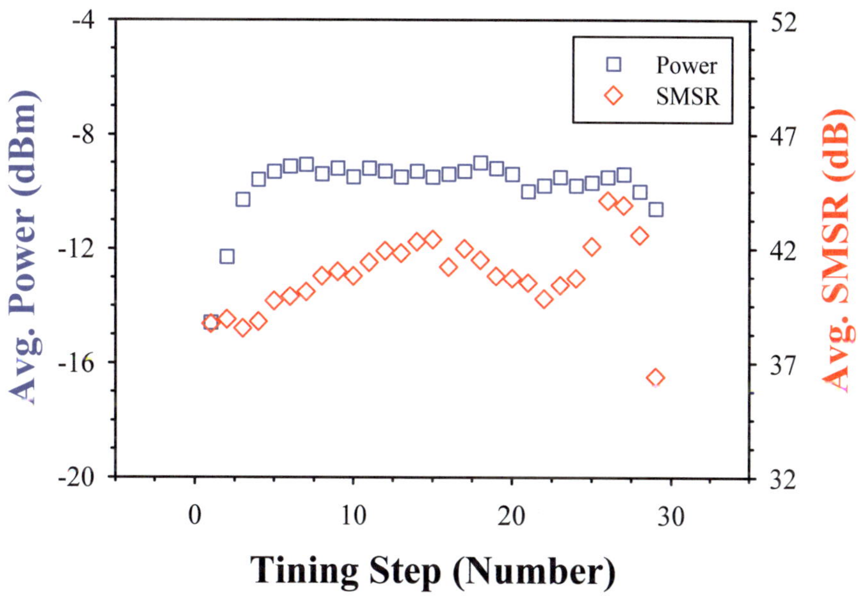

Fig. 5. The average output power and SMSR of the lasing dual-wavelength with 1.3 nm tuning step in the wavelengths of 1523.08 to 1562.26 nm.

In order to investigate the output stabilities of the proposed dual-wavelength laser, a short-term stability of output power and wavelength is measured and observed. In the measurement, the lasing two wavelengths are 1528.36 and 1529.67 nm with output power of –9.0 and –8.4 dBm initially over 20 minutes observation time. The output wavelength variations of the two wavelengths are zero and the maximum power fluctuation of –1 and –2 are 0.5 and 0.4 dB, respectively, as shown in Fig. 6. Then, in one hour observing time, the output stability is also maintained as mentioned before.

In summary, we have proposed and investigated a stable and tunable dual-wavelength erbium-doped fiber ring laser employing a self-injected FP-LD. By adding an FP-LD incorporated with a tunable bandpass filter within a gain cavity, the fiber laser can lase two wavelengths simultaneously due to the self-injected operation. The proposed dual-wavelength laser shows a good performance of output power and optical side-mode suppression ratio. The laser also presents a 39.18 nm wide tuning range from 1523.08 to 1562.26 nm. As a result, our proposed dual-wavelength fiber laser not only has the better optical output efficiency, but also has a wide tuning range of 39.18 nm. Besides, it has the advantage of simply architecture, low cost and better output efficiency.

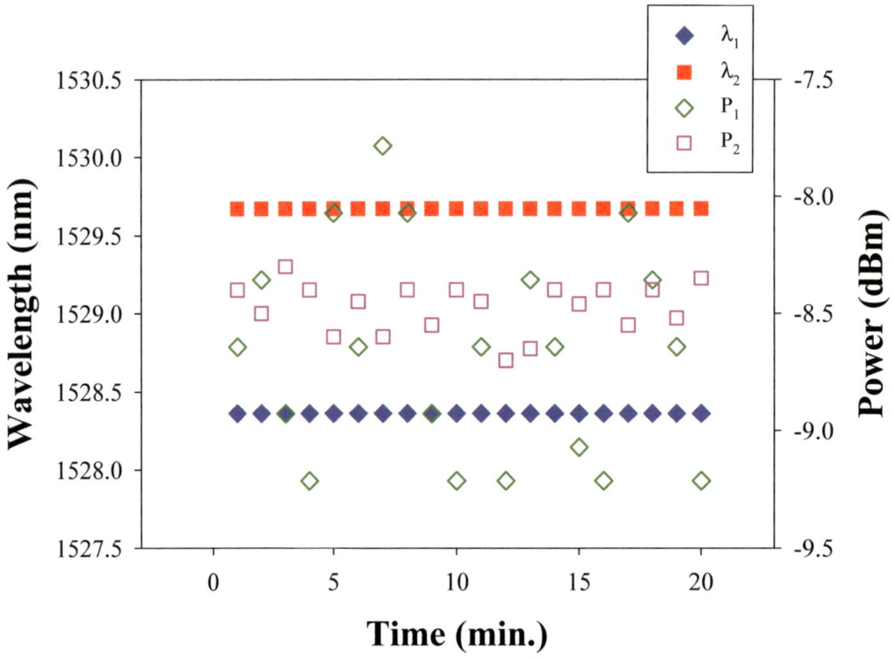

Fig. 6. The average output power and SMSR of the lasing two wavelengths with 1.3 nm tuning step from 1523.08 to 1562.26 nm.

3. Dual-wavelength laser diode with adaptive mode-spacing tuning

Then, to achieve a dual-wavelength lasing together with adaptive mode-spacing tuning, we can design a new optical-injection erbium-doped fiber (EDF) ring laser structure and use two tunable bandpass filters (TBFs) inside gain cavity loop for dynamically dual-wavelength selection. Here, the proposed laser has the following advantages: (i) two wavelengths can be tuned separately using the two tunable bandpass filters (TBFs); (ii) the mode spacing can also be tuned by the TBFs; (iii) the laser has a broadly tuning range; and (iv) the laser structure is relatively simple.

Fig. 7 shows the experimental setup for the tunable dual-wavelength fiber ring laser (Yeh et al., 2009). The proposed fiber laser consisted of one 2×2 3-dB coupler (CP), two 1×2 3-dB CPs, two TBFs, two polarization controllers (PCs), a FP-LD, and an erbium-doped fiber amplifier (EDFA). The EDFA was constructed by a 980/1550nm WDM coupler (WCP), 980 nm pump laser diode (LD), an optical isolator (OIS) and a 10 m EDF.

When the pumping power exceeds 100 mW in the proposed laser, the output power will be saturated. Thus, the 980 nm pump LD was set at 100 mW in the experiment. The 3 dB bandwidth and insertion loss of the TBF used was 0.4 nm and 4 dB respectively. The tuning range of the TBF was 40 nm (1520 to 1560 nm). The PCs were used to adjust the polarization states of the feedback lightwave into the FP-LD. According to the past study (Schell et al., 1995), injected optical signal at TE-mode of FP-LD can result in maximum injection locking efficiency.

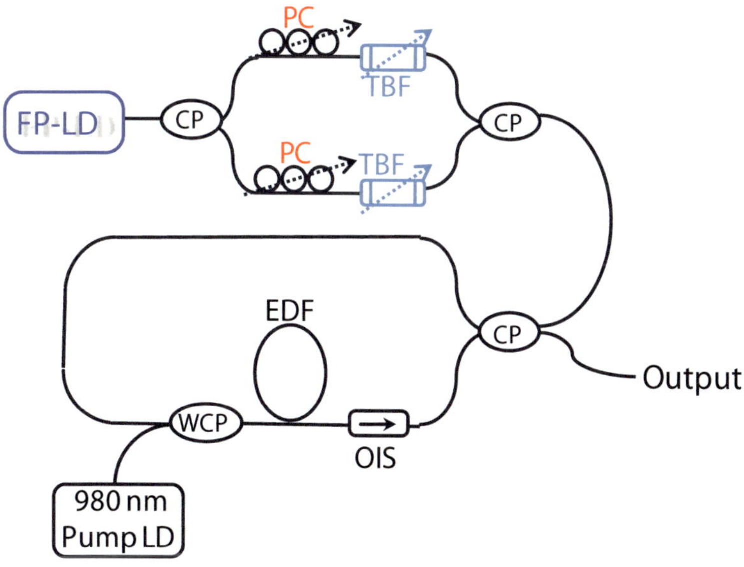

Fig. 7. Experimental setup for the tunable dual-wavelength fiber ring laser scheme.

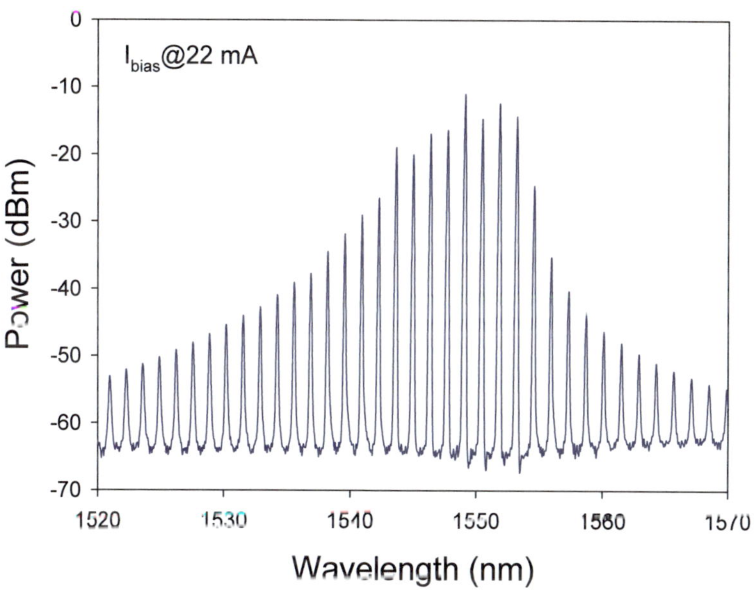

Fig. 8. Original output spectrum of FP-LD operates at 22 mA on 25 °C without self-seeding.

The threshold pumping power in this experiment was around 14 mW while the FP-LD was operated at 22 mA and 25 °C. To measure the output power and wavelength of the proposed dual-wavelength laser, an optical spectrum analyzer (OSA) with a 0.05 nm resolution was used in the measurement. Fig. 8 shows the original output spectrum of the free-run FP-LD without self-injection. The mode spacing ($\Delta\lambda$) and threshold current of FP-LD are 1.38 nm and 9.5 mA, respectively.

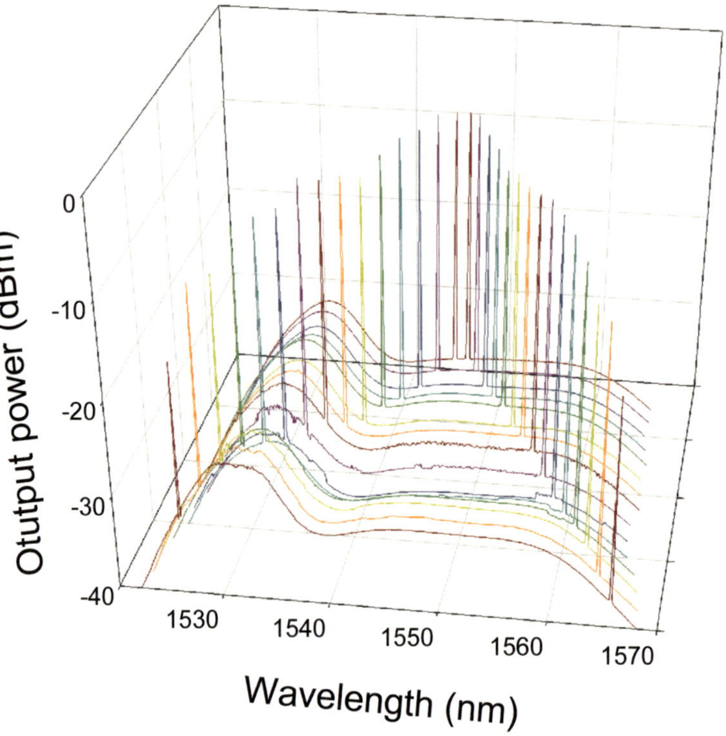

Fig. 9. Output wavelength spectra of the proposed tunable dual-wavelength fiber laser while the pumping power of 980 nm LD and bias current of the FP-LD are 110 mW and 22 mA, respectively, in the wavelength range of 1526.3 to 1565.8 nm with different mode spacing.

The two TBFs inside the ring cavity were used to align and filter the corresponding modes for two wavelengths lasing simultaneously. Fig. 9 shows the output spectra (wavelength range from 1526.27 to 1565.76 nm with different mode spacing) of the proposed tunable dual-wavelength fiber laser at the pumping power = 100 mW and bias current of the FP-LD = 22 mA. The mode spacing can also be determined by adjusting the two TBFs to properly align and match the filtering mode of the FL-LD. In Fig. 9, the maximum and minimum mode spacing of the dual-wavelength laser was 39.49 nm and 1.32 nm, respectively. Over the tuning range, the maximum and minimum power variation (ΔP_{max} and ΔP_{min}) was equal to 0.98 and 0.01 dB when the mode spacing is 36.69 and 17.65 nm, respectively. The proposed laser is limited to step tuning (in this case, it is 1.32 nm). We can also retrieve the continuous wavelength tuning by adjusting the temperature of the FP-LD. While the temperature difference (ΔT) of the FP-LD was ± 5 °C, the central wavelength variation was ± 0.2 nm. Thus, the dual-wavelength can be tuned continuously by controlling the temperature.

Fig. 10 shows the output power differences and the average output powers under various mode spacing of the fiber laser. The average output power is between –13.56 and –4.33 dBm, as shown in Fig. 10. By increasing of the mode spacing gradually, the average output power is decreasing, as illustrated in Fig. 10. Thus, when the mode spacing is 39.49 nm, the output power will drop to –13.56 dBm due to the smaller gain in both shorter and longer wavelength sides of the FP-LD (in Fig. 8). Besides, the output power differences of the dual-wavelength can be less than 1 dB due to the proper adjusting of the PCs. The nonlinear filter function of the self-injected FP-LD can also improve the noise performance of the proposed laser (Zhao et al., 2002). We also studied the stability of the proposed laser, and the power variation during a 30 minutes observation time was negligible. We believe that the inhomogenously broadened FP-LD gain saturation plays a part in suppressing the erbium gain competition. The stability of the proposed laser can be further enhanced by using all polarization maintaining (PM) components inside the laser cavity.

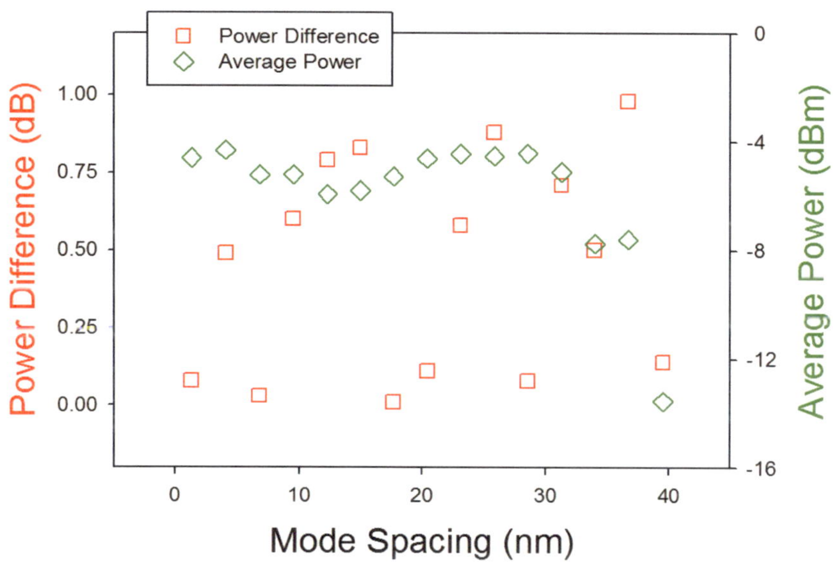

Fig. 10. Output power difference and average output power under various mode spacing for the tunable dual-wavelength fiber laser.

Fig. 11 shows the optical spectra of the dual-wavelength fiber laser lasing at wavelengths of 1545.08 and 1546.40 nm, when the pumping power was adjusted from 14 to 120 mW. The lasing output power will start to saturate at 100 mW, as shown in Fig. 11. The threshold pumping power of the proposed dual-wavelength laser was 28 mW, as also illustrated in Fig. 11.

In summary, we proposed and experimentally demonstrated a tunable dual-wavelength fiber laser based on a self-injected FP-LD and an EDF. The dual-wavelength output of the proposed fiber laser is widely tunable, and the mode spacing of the two lasing wavelengths can also be adjusted within the tuning range. Two TBFs were used inside the laser cavity to generate the dual-wavelength output. The dual-wavelength tuning range can be achieved up to 39.49 nm from 1526.27 to 1565.76 nm. The mode spacing of the dual-wavelength can be tuned by using the two TBFs to align the corresponding longitudinal-mode of the FP-LD. The maximum and minimum mode spacing of the laser is 39.49 and 1.32 nm, respectively. The proposed laser is limited to step tuning (in this case, it is 1.32 nm). Moreover, the threshold pumping power and saturated pumping power of the laser are 28 and 100 mW, respectively. The output power difference of dual-wavelength can be controlled to smaller than 1 dB.

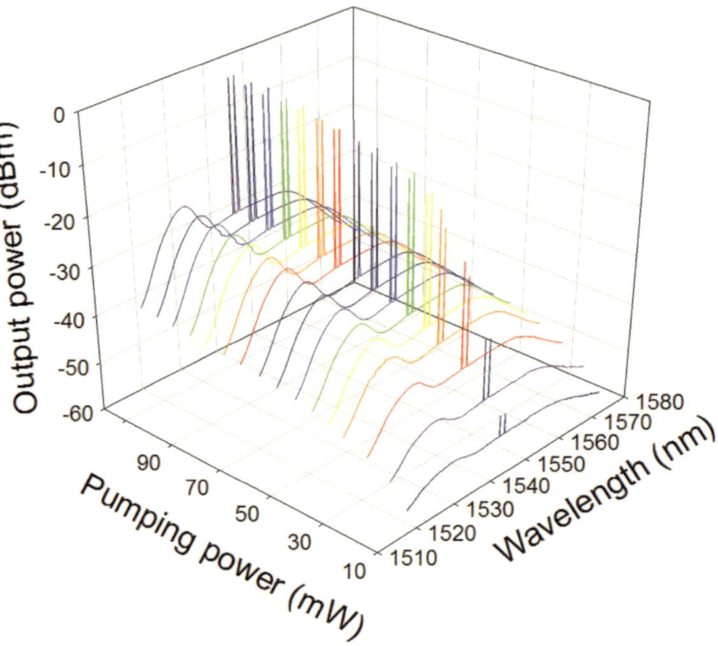

Fig. 11. Output wavelength spectra versus the different pumping power under 14 to 120 mW at the two wavelengths of 1545.08 and 46.40 nm initially.

4. Conclusion

For the first proposed laser scheme, we have proposed and investigated a stable and tunable dual-wavelength erbium-doped fiber ring laser employing a self-injected FP-LD. By adding an FP-LD incorporated with a tunable bandpass filter within a gain cavity, the fiber laser can lase two wavelengths simultaneously due to the self-injected operation. The proposed dual-wavelength laser shows a good performance of output power and optical side-mode suppression ratio. The laser also presents a 39.18 nm wide tuning range from 1523.08 to 1562.26 nm. When compared with the previously proposed schemes, our proposed dual-wavelength fiber laser not only has the better optical output efficiency, but also has a wide tuning range of 39.18 nm. Besides, it has the advantage of simply architecture, low cost and better output efficiency.

For the second proposed laser structure, we have proposed and experimentally demonstrated a tunable dual-wavelength fiber laser based on a self-injected FP-LD and an EDF. The dual-wavelength output of the proposed fiber laser is widely tunable, and the mode spacing of the two lasing wavelengths can also be adjusted within the tuning range. Two TBFs were used inside the laser cavity to generate the dual-wavelength output. The dual-wavelength tuning range can be achieved up to 39.49 nm from 1526.27 to 1565.76 nm. The mode spacing of the dual-wavelength can be tuned by using the two TBFs to align the corresponding longitudinal-mode of the FP-LD. The maximum and minimum mode spacing of the laser is 39.49 and 1.32 nm, respectively. The proposed laser is limited to step tuning (in this case, it is 1.32 nm). Moreover, the threshold pumping power and saturated pumping power of the laser are 28 and 100 mW, respectively. The output power difference of dual-wavelength can be controlled to smaller than 1 dB. As a result, the proposed two dual-wavelength fiber laser configurations not only are simple, but also are easy to setup for the future applications of WDM communications and optical sensor.

5. References

Al-Mansoori, M. H., Mahdi, M. A., and Zamzuri, A. K. (2008): Tunable multiwavelength Brillouin-Erbium fiber laser with intra-cavity pre-amplified Brillouin pump. *Laser Phys. Lett.* 5, 139-143.

Alvarez-Chavez, J. A., Martinez-Rios, A., Torres-Gomez, I., and Offerhaus, H. L. (2007) : Wide wavelength-tuning of a double-clad Yb3+-doped fiber laser based on a fiber Bragg grating array. *Laser Phys. Lett.* 4, 880-883.

Barmenkov, Y. O., Kir'yanov, A. V., Perez-Millan, P., Cruz, J. L., and Andres, M. V. (2008): Experimental study of a symmetrically-pumped distributed feed-back Erbium-doped fiber laser with a tunable phase shift. *Laser Phys. Lett.* 5, 357-360.

Chen D., and Shen, L. (2007): Switchable and tunable Erbium-doped fiber ring laser incorporating a birefringent and highly nonlinear photonic crystal fiber. *Laser Phys. Lett.* 4, 368-370.

Chen, D., Ou, H., Fu, H., Qin, S., and Gao, S. (2007): Wavelength-spacing tunable multi-wavelength erbium doped fiber laser incorporating a semiconductor optical amplifier. *Laser Phys. Lett.* 4, 287-290.

Chou, S.-Y., Yeh, C.-H., and Chi, S. (2007): Unitizations of double-ring structure and Erbium-doped waveguide amplifier for stable and tunable fiber laser. *Laser Phys. Lett.* 4, 382-384.

Das, G., and Lit, J. W. Y. (2002): L-band multiwavelength fiber laser using an elliptical fiber. *IEEE Photon. Technol. Lett.* 14, 606–608.

Fok, M. P., and Shu, C. (2007): Tunable dual wavelength erbium-doped fiber laser stabilized by four-wave mixing in a 35-cm highly nonlinear bismuth-oxide fiber. *Opt. Express* 15, 5925-5930.

Kim, Y. J., and Kim, D. Y. (2005): Electrically Tunable dual-wavelength switching in a mutually injection-locked erbium-doped fiber ring laser and distributed-feedback laser diode. *IEEE Photon. Technol. Lett.* 17, 762–764.

Li, S., Chan, K. T., Liu, Y., Zhang, L., and Bennion, I. (1998): Multiwavelength picosecond pulses generated form a self-seeded Fabry-Perot laser diode with a fiber external cavity using fiber Bragg gratings. *IEEE Photon. Technol. Lett.* 10, 1712–1714.

Liu, Y. G., Feng, X., and Yuan, S. (2004): Simultaneous four-wavelength lasing oscillations in an erbium-doped fiber laser with two high birefringence fiber Bragg gratings. *Opt. Express* 12, 2056–2061.

Liu, Z. Y., Liu, Y. C., Du, J. B., Kai, G. Y., and Dong, X. Y. (2008): Tunable multiwavelength erbium-doped fiber laser with a polarization-maintaining photonic crystal fiber Sagnac loop filter. *Laser Phys. Lett.* 5, 446-448.

Nilsson, J., Lee, Y. W., and Kim, S. J. (1996): Robust dual-wavelength ring-laser based on two spectrally different erbium-doped Fiber amplifiers. *IEEE Photon. Technol. Lett.* 8, 1630–1632.

Peng, P.-C., Tseng, H.-Y., and Chi, S. (2003): A tunable dual-wavelength erbium-doped fiber ring laser using a self-seed Fabry-Perot laser diode. *IEEE Photon. Technol. Lett.* 15, 661–663.

Schell, M., Huhse, D., Utz, W., Kaessner, J., Bimberg, D., and Taraov, I. S. (1995): Jitter and dynamics of self-seeded Fabry-Perot laser diodes. *IEEE J. Select. Topics Quantum Electron.* 1, 528–534.

Slavik, R., and LaRochelle, S. (2002) : Multiwavelength single-mode erbium doped fiber laser for FFH-OCDMA testing. *Proc. of OFC*, Paper WJ3.

Talaverano, L., Abad, S., Jarabo, S., and Lopez-Amo, M. (2001) : Multiwavelength fiber laser sources with Bragg-grating sensor multiplexing capability. *J. Lightwave Technol.* 19, 553–558.

Yang, S., Li, Z., Yuan, S., Dong, X., Kai, G., and Zhao, Q. (2002): Tunable dual-wavelength actively mode-locked fiber laser with an F-P semiconductor modulator. *IEEE Photon. Technol. Lett.* 14, 1494–1496.

Yeh, C. H., Chow, C. W., Shih, F. Y., Wang, C. H., Wu, Y. F., and Chi, S. (2009): Tunable dual-wavelength fiber laser using optical-injection Fabry-Perot laser. *IEEE Photon. Technol. Lett.* 21, 125-127.

Yeh, C. H., Shih, F. Y., Wang, C. H., Chow, C. W., and Chi, S. (2008): Tunable and stable single-longitudinal-mode dual-wavelength erbium fiber laser with 1.3 nm mode spacing output. *Laser Phys. Lett.* 5, 821-824.

12

The Coherent Coupled Output of a Laser Diode Array Using a Volume Bragg Grating

Bo Liu[1,2], Qiang Li[2], Xinying Huang[1] and Weirong Guo[2]
[1]*Institute of Laser Engineering, Beijing University of Technology, Beijing,*
[2]*Beijing Kantian Tech.com., LTD, Beijing,*
China

1. Introduction

High-power laser diode array (LDA) is characterized by higher overall efficiency with a longer operating lifetime than any other laser types. The compact construction of LDA's is extremely attractive for applications such as material processing, solid-state laser pumping, free space communications, and numerous medical procedures.

However, LDA's face a number of challenges in terms of beam quality improvement and bandwidth reduction. Among the numerous methods used to improve the performance of high-power LDA's, the external cavity has been accepted as an effective method [1-3]. Apollonov et al. achieved two lobes of the out-of-phase mode in the far field using the Talbot cavity with phase compensation [4, 5]. In-phase mode selection could be achieved by tilting the cavity mirror at a low injection current

(I<20 A)[6] .With amplitude compensation, an in-phase mode that produced a single lobe in the far-field was selected[7], but the phase locking was local and not all of the emitters were completely locked, and this resulted in a high pedestal. However, there was still a problem with phase-locked emitters of LDA with in-phase mode high-power output. In this work, a volume Bragg grating (VBG) and a transforming lens were employed to diffract coupling between emitters of broad area multi-stripe lasers in the external cavity. The in-phase mode output produced a single lobe in the far-field.

The experimental setup is shown in Fig. 1. A C1-60 laser diode array from nLIGHT Corporation was used, consisting of 49 emitters with a diode width of a=100μm and a spacing period of d=200 μm, a free running wavelength at 808nm within a bandwidth of 4 nm. It was packed into a patented sandwich structure in order to reduce the "smile" effect. The back face was coated with a high-reflection coating and the front face of the array was coated with an antireflection coating, so as to eliminate oscillation caused by the internal cavity. A cylindrical micro lens from LIMO, f = 91μm, NA = 0.8, was used to collimate the beam of the laser diode array in the fast-axis direction. A 38mm focal length cylindrical lens was used to transform the beam of the laser diode array in the slow-axis direction. A 0.62mm thickness VBG from PD-LD Corporation, with 15% diffraction efficiency at 807.8nm, was placed in the focal plane of the transforming lens. A convex lens and CCD were used to detect the far-field output beam. The experiment was carried out using a current of 40 A, and a heat sink temperature of 12 °C.

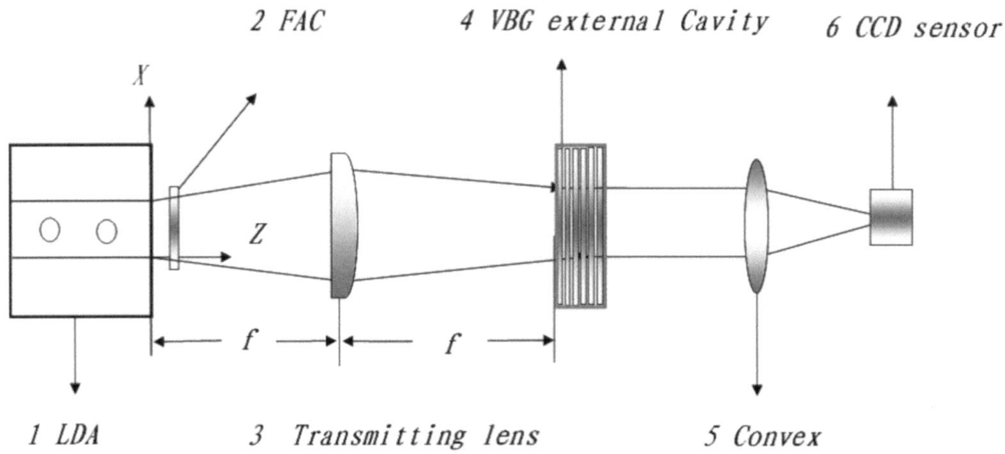

Fig. 1. Experiment setup for VBG external cavity phase-locking.

In the preliminary experiment, an FAC was used to align the output beam in the fast-axis direction. The transform lens was placed one focal distance away from the diode array, putting the front focal plane just on the output surface. A VBG was used as the external cavity and it was set at the rear focal plane of the transform lens so that the front surface of the VBG and the diode array output surface formed a conjugate plane of the transform lens. The emitting surface of the diode array was on waist of the cavity. The transform lens functioned in such a way that the VBG surface was positioned at the waist of the radiated beam. Thus the emitting light diffracted from the VBG directly returned to the diode cavity as self-emitter feedback. Consequently, the output laser distribution in the slow-axis direction was a series of peak values of power, with each peak corresponding to a broad-area laser emitting element, as shown in Fig.2 (a). Due to the spectral selectivity of the VBG, the wavelength of all the emitters diffracted from the VBG was locked at 807.8nm, which was just the center wavelength of the VBG spectral selectivity with normal incidence. The FWHM of the spectral profile was reduced from 1.7nm to 0.2nm, as shown in Fig.2 (b). Because the narrow angular acceptance of the VBG strongly reduced the total amount of the uncollimated light diffracted from the VBG, as calculated in Fig.3, the divergence of the beam in the resonator was suppressed, and the output beam quality was improved. However each emitter was incoherently lasing. This was similar to the wavelength stabilization and spectrum narrowing of high power multimode laser diode arrays by VBG [8].

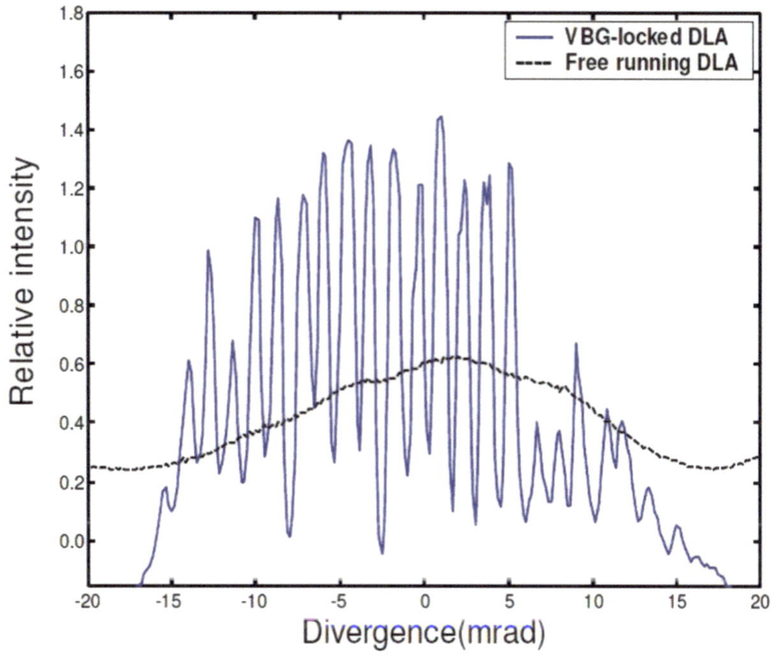

(a)

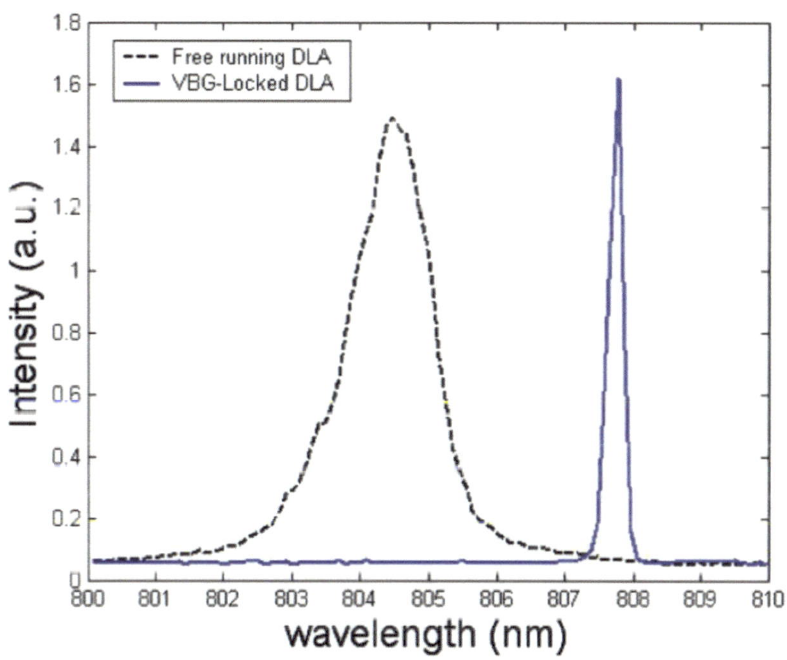

(b)

Fig. 2. The output far field intensity distribution (a), and spectrum (b) of VBG-locked LDA.

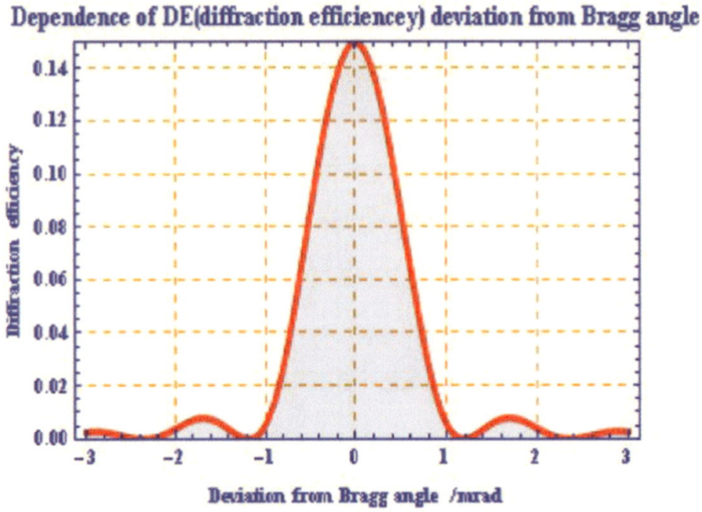

Fig. 3. Dependence of diffraction efficiency on deviation from Bragg angle.

Next, second stage experiments were performed for coherent coupling output. By carefully adjusting the VBG in the slow-axis direction by a small angle, a single lobe coherent light was obtained, as shown in Fig. 4(a). The oblique incidence angle of the VBG was about 1.5 degrees in the slow-axis direction. Measuring the output laser beam, the center wavelength of single lobe coherent light was 801.9nm, the FWHM of the spectral profile was 0.17nm (Fig. 4(b)), the far-field divergence was 1.47mrad (Fig. 4(c)), and power concentrated on the central lobe was 3.67W.

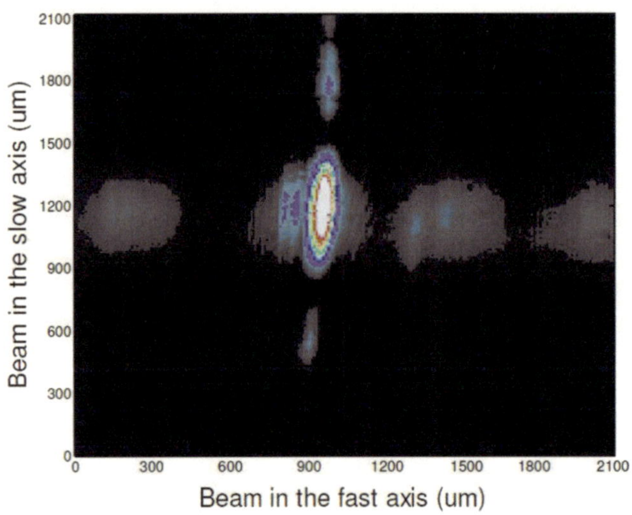

(a)

Fig. 4. The far field intensity distribution of a slope VBG coherent coupled LDA (a), output spectrum (b), and far-field divergence (c).

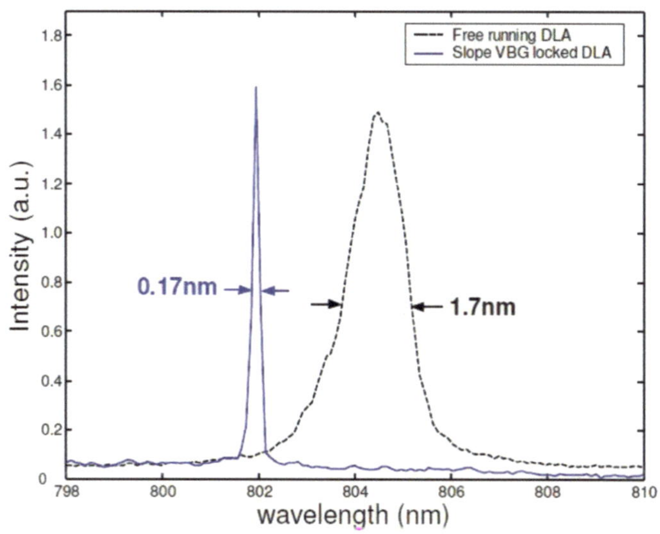

(b)

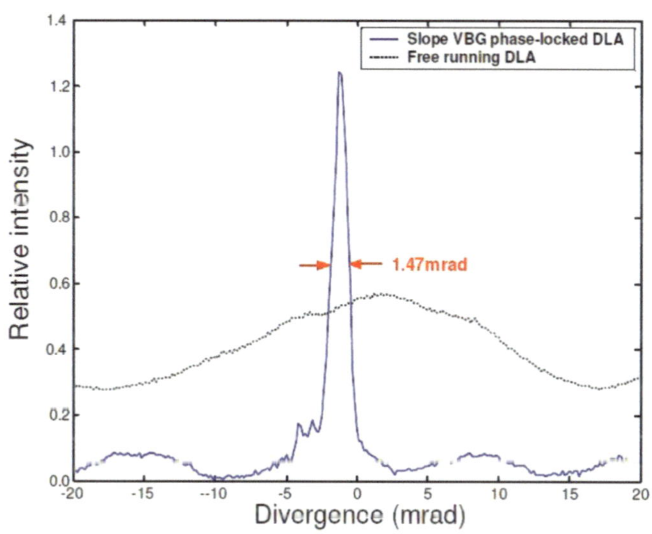

(c)

Fig. 4. The far field intensity distribution of a slope VBG coherent coupled LDA (a), output spectrum (b), and far-field divergence (c). (Continuation)

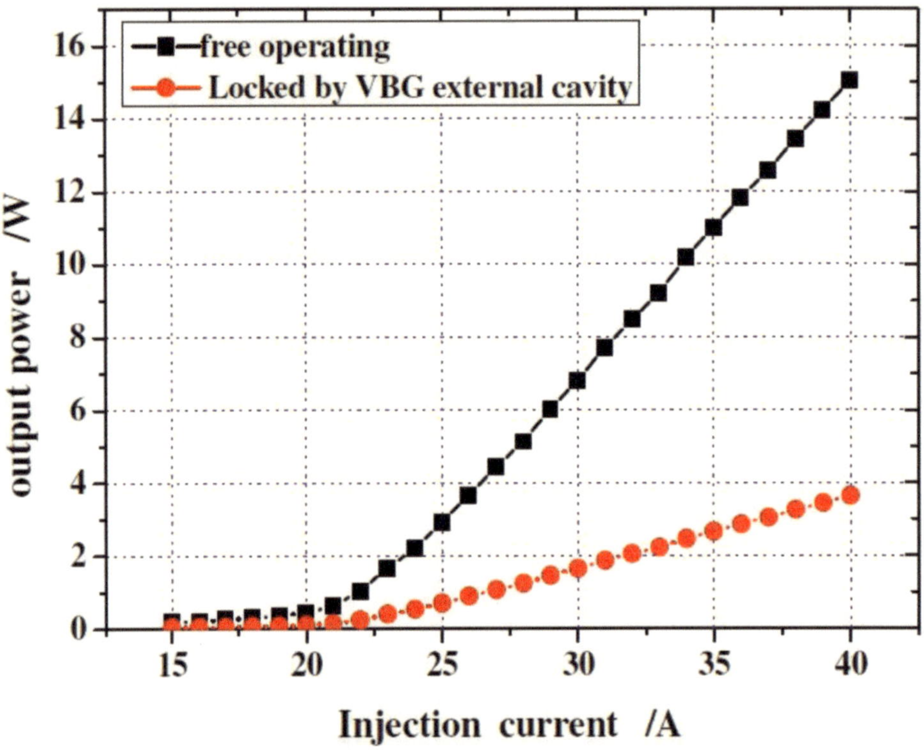

Fig. 5. Output power of DLA with slope VBG locked and free running.

The experimental results can be explained as follows. In the two cases considered, for incident rays that were normal and oblique into the VBG, diffraction followed a different trace pattern. For normal incidence into the VBG, the three rays of incidence (Ii), diffraction (Id) and transmission (It) were coaxial. The wave vectors of the incidence beam (Kim), diffraction beam (Kdm) and grating wave vector (KG) were also coaxial inside the grating medium, as shown in Figure 6(a). So, the exiting light diffracted by the VBG returned directly into its own path and the output wavelength was 807.8nm, which was the grating central wavelength determined by the grating spectral selectivity. For oblique incidence into the VBG, the rays of Ii, Id, It, and the wave vectors of Kim, Kdm, KG were not coaxial, as shown in Figure 6(b). It was easily found that the equivalent grating period was reduced, so the center wavelength of the VBG decreased from 807.8nm to 801.9nm. Because of the non-coaxiality of the incidence ray (Ii) and diffraction ray (Id), there was a space at the grating surface. The space could be estimated by applying energy and momentum conservation of the wave vectors of Kim, Kdm, KG inside the grating medium, and using grating parameters [9]. For a thickness of 0.62mm and a 15% diffraction efficiency of the VBG, the space was about 10.5μm. Also, the diffracted ray (Id) diverged over an extended area. Considering a 1 degree angular acceptance of the VBG with the light path back to the diode array surface, the extended area of the diffraction ray(Id) was about 330μm. So the emitting light diffracted by the VBG could return to the adjacent emitter of the diode array, which had an emitter width of 100 μm and a spacing of 200 μm, and they would couple with each other in the same mode. In this way coherent output was obtained.

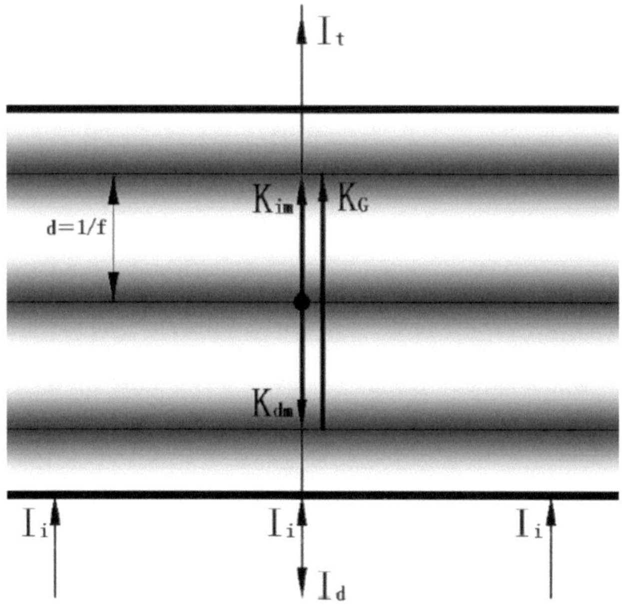

(a)

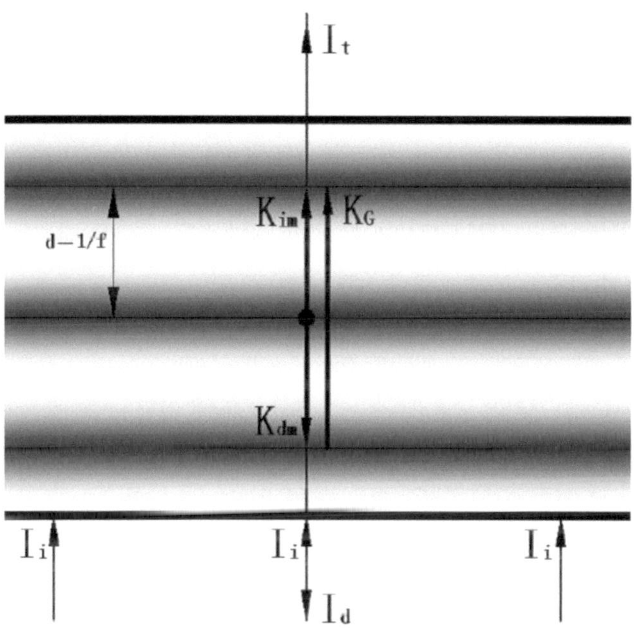

(b)

Fig. 6. The rays and wave vectors of normal incidence (a), and oblique incidence (b) of the VBG.

The measured far-field divergence was larger than the diffraction limit for the array (theoretical value is about 1.0 mrad), indicating that not all of the emitters were completely locked to the in-phase mode. Single lobe coherent coupled output power was only 3.67W out of 15W of free running output power. We suggest that the degree of coherent output was indeed limited by some emitters, which were not locked due to the smile effect or manufacturing variations and thermal gradients of LDA. These resulted in degradation of coupling strength between adjacent emitters, limiting the scalability of the array power.

In conclusion, we demonstrated a method to coherently couple the output and suppress the bandwidth of a LDA, based on external cavity feedback among adjacent emitters with a VBG. One single lobe mode was achieved in the far-field pattern with an output power of 3.67 W, a FWHM spectral profile of 0.17nm and a far-field divergence of 1.47 mrad. We propose that coherent coupling with an external VBG can allow for scaling to a level of 10 W in single lobe mode, improving coupling strength by compressing the coupling zone, and improving manufacturing technology as it decreases the "smile" effect and the thermal gradient. This method provides a potential laser source for applications in material processing and pumping solid-state lasers.

This work was supported by the National Natural Science Foundation of China under Grant No. 10276003, 60678014, and the Natural Science Foundation of Beijing under Grant No. 4051001.

2. References

[1] A. F. Glove, Phase locking of optically coupled lasers, Quantum Electron. 33, 283 (2003).

[2] X. Gao, Y. Zheng, H. Kan, and K. Shinoda, Effective suppression of beam divergence for a high-power laser diode bar by an external-cavity technique, Opt. Lett. 29, 361 (2004).

[3] M. V. Romalis, Narrowing of high power diode laser arrays using reflection feedback from an etalon, Appl. Phys. Lett. 77, 1080 (2000).

[4] V. V. Apollonov, S. I. Derzhavin, V. I. Kislov, A. A. Kazakov, Y. P. Koval, V. V. Kuz'minov, D. A. Mashkovskii, A. M. Prokhorov, Spatial phase locking of linear arrays of 4 and 12 wide-aperture semiconductor laser diodes in an external cavity, Quantum electronics. 28, 257 (1998)

[5] V. V. Apollonov, S. I. Derzhavin and V. A. Filonenko et al., High power laser diode array phase-locking, Proc. SPIE. 3889, 134 (2000)

[6] K. J. Wang, Q. Li, W. R. Guo, P. F. Zhao, T. C. Zuo, In-phase mode selection of diode laser array in external Talbot cavity, High power laser and particle beams. 18, 177 (2006) (in Chinese)

[7] Qiang Li, Pengfei Zhao, Weirong Guo, Amplitude Compensation of a Diode Laser Array Phase locked with a Talbot Cavity, Appl. Phys. Lett. 89(23), 1120(2006)

[8] S. Riyopoulos, G. Venus, and L. Glebov, Mode selection and phase locking of sidelobe-emitting semiconductor laser arrays via reflection coupling from an external narrow-bandwidth grating, J. of Appl. Phys. 103, 113107(2008)

[9] L. B. Glebov, V. I. Smirnov, C. M. Stickley, and I. V. Ciapurin, New approach to robust optics for HEL systems, In Laser Weapons Technology III. W.E. Tompson and P.H. Merritt, Editors. Proceedings of SPIE, 4724, 101-109 (2002)

Section 3

Applications of Laser Diode

13

Laser Diode Pump Technology for Space Applications

Elisavet Troupaki, Mark A. Stephen,
Aleksey A. Vasilyev and Anthony W. Yu
NASA Goddard Space Flight Center,
USA

1. Introduction

In this chapter we will discuss the use of diode pump technology in space LIDAR (Light Detection and Ranging) missions. Space applications have a unique set of engineering requirements. In general, devices need to be reliable, compact, lightweight, efficient and rugged. Satellites need to survive launch and then operate in a vacuum, high-radiation, micro-gravity environment. Because of the high costs associated with launching payloads into space, lift capacity needs to be minimized and instruments need to be lightweight and compact. Because solar arrays and batteries that are necessary to operate powered equipment need to be launched as well, efficiency should be optimized to minimize the on-orbit power draw. Finally once an instrument has been launched into space, it is critical that it operates reliably - usually for several years. For the missions we will discuss in this paper, long-term reliable operation is very important to make comprehensive global measurements needed by the science community. In addition, these missions require customized science measurements, necessitating the design of one-of-a kind instruments

A major challenge is designing and building an instrument that is both unique and reliable.

To date, the space-based LIDAR instruments NASA has flown have been designed to make global (Earth, planetary and lunar) altimetry measurements. By making individual altimetry measurements as the satellite orbits a planet (or moon), the topography can be mapped. This measurement technique uses the time-of-flight of a laser pulse to travel from the spacecraft to the planetary surface and back to determine the distance between the source and target. This means there must be enough energy in the laser pulse to reach the planet and get sufficient return signal to measure the laser pulse. The systems engineering for these measurements is discussed elsewhere [1,2,3,4,5] but for this type of measurement we used repetition rates in the 10's of Hz and pulse energies from a few to hundreds of milli-Joules (mJ). To build this type of laser we used quasi-continuous wave (QCW) laser diode arrays (LDAs) to pump an Nd:YAG laser crystal. Further details of the laser design are given in Section 2. To match the strong absorption in the Nd:YAG crystal, we chose 808 nm arrays. QCW LDAs are well suited for this application because they are efficient, compact and robust and are capable of delivering a high-power optical signal onto the laser crystal. They can also have both high spectral and spatial brightness. For these reasons, a QCW array has excellent performance characteristics and is very well suited to the space-flight application with one caveat: it does not have a well-defined lifetime or reliability. It is this aspect of the diode arrays that we will discuss in this chapter and although space-flight applications are a fairly narrow field, highly reliable operation is required in a broad range of applications and we believe this information is of general utility.

In the telecommunications industry, there is a large market that requires high volume production and supports device development and fabrication that adheres to telcordia requirements, ensuring repeatable, reliable performance. Unfortunately the diode-pumped solid-state laser (DPSSL) market does not have the same amount of investment and the market is much more disparate so there is less synergy around any one product or set of requirements. As a result, the available investments are spread among different devices and packaging schemes. In addition, the DPSSL market does not (in general) require multi-year reliability - bad parts can usually be serviced or replaced. The market is also very dynamic and requires continuous improvement so there is no set of frozen standards as exist in the telecommunications market. This means there is a continuous drive for increased performance so the devices are in a constant state of redesign. Gathering long-term data is not only not supported or required by the industry, in general, it does not make sense because the product whose reliability will be measured is obsolete by the time the data is gathered. So there is little effort to quantify the reliability or measure the lot-to-lot variability of these products.

None of this means the devices are of poor quality or that they do not perform well; on the contrary the devices perform very well, but the reliability of a given device is not well quantified.

So our challenge is to build a reliable laser based on parts that are not well understood. At NASA's Goddard Space Flight Center (GSFC) we have been evolving a process to get high performance, commercial-off-the-shelf (COTS) parts to meet our requirements for long-term space-flight operation.

A typical space flight mission consists of different phases

- Pre-Phase A, Conceptual Study
- Phase A, Preliminary Analysis
- Phase B, Definition
- Phase C/D, Design and Development
- Phase E, Operations Phase

Formal reviews are typically used as control gates at critical points in the full system life cycle to determine whether the system development process should continue from one phase to the next, or what modifications may be required. Only key reviews at various phases of the mission development are shown in Table 1. In the last column, we listed the development phases of the instrument coinciding with various mission reviews. We will briefly describe the space qualification process of a particular component.

Mission Phase	Mission Reviews and Operations	Instrument Development
Pre-Phase A Conceptual Study	• MISSION CONCEPT REVIEW (MCR)	• Conceive Concept and Conduct Trade Studies
Phase A Preliminary Analysis	• SYSTEM REQUIREMENTS REVIEW (SRR) • SYSTEM DESIGN REVIEW (SDR) • NON-ADVOCATE REVIEW	• Preliminary Design • Proof of concept Breadboard
Phase B Definition	• PRELIMINARY DESIGN REVIEW (PDR)	• Define baseline technical solution • Define requirements • Procure parts to begin qualification process • Build Engineering Model (EM)
Phase C/D Design & Development	• CRITICAL DESIGN REVIEW (CDR) • TEST READINESS REVIEW • FLIGHT READINESS REVIEW	• Freeze design • Procure flight and spare parts • Build Flight and Flight Spare System • Environmental Testing • Deliver Flight and Flight Spare • Integrate with Spacecraft
Operations Phase Mission Ops & Data Analysis	• PRIMARY MISSION • EXTENDED MISSION	

Table 1. Full mission cycle with critical reviews within the cycle and corresponding development phase of the instrument.

Well-defined and proven manufacturing processes producing reliable components from space-qualified vendors are key to a successful mission. Previously flown missions provide references to baseline components, which include EEE, optical, photonics and opto-electronics parts for upcoming missions with lessons learned. In practice, these previously flown parts are often obsolete or have enough modification in either the manufacturing process or ingredients that these references can only be used as guidelines. During the Phase A, the proof-of-concept breadboard is being built, components will be chosen with future qualification in mind. Components that were flown previously will have first priority due to heritage along with other qualifiable alternate choices. During this breadboard phase, samples from potential vendors will be purchased and their performances characterized. This information will then be used to select vendors and parts for the engineering model (EM) development in Phase B. Another approach to mitigate risk on key components is to choose multiple vendors for the flight to guard against the risk of manufacturing flaw in the flight lot from a particular vendor. Sometimes, but not often due to schedule constraints, multiple rounds of purchase and test would be conducted before the components are integrated into the instrument. This is typically done within Phase B of the program.

Procurement of flight parts typically occurs at the end of Critical Design Review (CDR). Exception may apply for long lead items such as pump diodes where procurement usually starts after completion of the Preliminary Design Review (PDR). When procuring flight parts, typical practice is to procure a large enough quantity for building the flight, flight spares and additional parts for quantifying the performance through various tests described below.

Upon receiving the flight parts, the performances of the components are characterized to understand how they operate and the part-to-part variability in their performances. Our strategy is to purchase enough quantities to characterize the lot-to-lot variability and obtain insights into the overall instrument performance range with these variances. Extensive optical characterization and other measurements will be described in Section 3.

In addition to measuring the optical characteristics of the device, we also have a suite of high-fidelity tests that are designed to detect characteristics of the device like mechanical stress, manufacturing errors or device damage. These methods can be used to screen the flight devices so that only the best devices are selected for integration onto the satellite.

Once the as-delivered characteristics of the devices are well characterized, we then subject them to a battery of environmental and operational tests to quantify how they will operate in our system. These include life tests, vibration, simulated thermal stresses and vacuum. Again we look for the mean performance as well as the variability so that the on-orbit performance can be adequately predicted. These test procedures and sample results will also be presented in Section 3.

Where possible, we perform accelerated life-tests and derive the acceleration parameters from the collected data. Understanding accelerated lifetest results is extremely difficult and relies heavily on a good design of experiment. Actual lifetime data in the as-used conditions is important even if it means taking data for a long period of time. Where there is uncertainty in results, it is important to be conservative in the conclusions that can be drawn. Once the results of the tests are available, correlating lifetest and environmental data to the screening criteria should be done where possible to validate the methodology.

2. Current NASA space LIDAR missions employing QCW LDAs

There are currently three operational LIDAR systems developed by NASA orbiting the Earth, the Moon and the planet Mercury gathering scientific data to form a better understanding of our Earth and solar system. All these LIDAR systems were built using high power QCW LDA pumped solid-state laser (SSL) architectures. These QCW LDA pumped SSL laser systems are the enabling technology that led to a series of successful spaceflight LIDAR systems for Earth observing and planetary exploration. One of the first DPSSL was the Mars Orbiter Laser Altimeter (MOLA) laser transmitter launched in 1996, which produced a high resolution topographic map of the planet Mars. [6] This was followed by the Geoscience Laser Altimeter System (GLAS) in 2003. Three LIDAR systems currently in operation also use this architecture: the Cloud-Aerosol Lidar with Orthogonal Polarization (CALIOP) launched in 2006, the Mercury Laser Altimeter (MLA) launched in 2006; and the Lunar Orbiter Laser Altimeter (LOLA) launched in 2008. [7,8,9,10] These US-led LIDAR instruments as well as missions from other countries have produced vast amounts of valuable scientific data for the study of our solar system.

QCW LDA pumped Nd:YAG lasers have been used since the late 1990's in all of the spaceborne LIDAR missions for NASA. The MOLA instrument was launched on the Mars Global Surveyor (MGS) in 1996. The MOLA laser was a diode-pumped, Nd:YAG zigzag slab, cross-Porro, electro-optically q-switched laser transmitter. It operated at 1.064 μm wavelength with a pulse energy of 40 mJ, pulse width of 10 ns and repetition rate of 10 Hz. The prime science objective of MOLA was to determine the global topography of Mars at a level suitable for addressing problems in geology and geophysics. Since the launch of MGS in November 1996 until the instrument was commanded off on June 2001, a total of 670 million shots were fired with 640 million measurements of the Mars surface and atmosphere. This represents more than ten times the number of laser measurements than all previous space LIDAR missions combined. The MOLA instrument performance and lifetime surpassed all goals of the MOLA investigation. [11] The Shuttle Laser Altimeter (SLA) I & II flew a copy of the MOLA laser in 1996 and 1997 to produce topographic profiles of the Earth surface and accumulated a total of about 6 million shots in space. [12]

The GLAS laser transmitter is the first passively q-switched, master oscillator power amplifier (MOPA) design that has operated in space. The instrument was launched in 2003 and completed its mission in early 2011 with a total of just under 2 billion shots in space from three transmitters. [13] The GLAS lasers represent a substantial improvement in performance over previous space-based remote sensing laser transmitters. The GLAS lasers simultaneously increased performance in power (110 mJ total with 75 mJ at 1.06 μm and 35 mJ in 0.532 μm), full beam divergence (110 μrad), pulse width (<6 ns), single spatial mode with $M^2 \sim 2$ and repetition rate (40 Hz). The GLAS MOPA's modular laser design also provided the baseline architecture for the next two missions; to the planet Mercury and the Moon.

The Mercury Surface, Space Environment, Geochemistry, and Ranging (MESSENGER) mission to the planet Mercury requires a laser altimeter capable of performing range measurements to the surface of the planet over highly variable distances and with a constantly changing thermal environment. The satellite reached final orbit after more than 5 years of transit through space and began gathering data in March 2011. It is in a highly elliptical orbit with a periapsis of 200 km, an apoapsis of approximately 15,200 km and an orbital period of 12 h. For the altimeter instrument, science observations are taken during the 0.5 h of closest approach to the planet. The laser must support the instrument requirement of a range resolution of less than 40 cm to the surface of the planet. The anticipated radiation exposure over the mission life is 30 krad (Si), total dose, using an effective shielding of 0.1 cm of aluminum. During the close approach to the day side of the planet, the satellite is heating and it is not possible to fully isolate the laser subsystem from the rest of the satellite. Mission requirements on the laser include more than 18 mJ of output energy in a near-diffraction-limited beam with 6 ns pulses at an 8 Hz repetition rate, while the laser bench temperature is executing a thermal ramp from 15 to 25°C at a rate of approximately 0.4°C/min. The MLA laser transmitter was based on the GLAS MOPA laser design but with only a single amplifier stage. The oscillator was changed from a Porro/Mirror resonator to a cross-Porro resonator to ensure that the laser will maintain its alignment over the harsh environment of Mercury.[4]

The first spaceflight LIDAR, coming just ten years after the demonstration of the first laser in 1960, was flown on Apollo 15, then subsequently on Apollo 16 and 17, provided the first topographic mapping of the Moon from orbit. It was a flash lamp pumped, mechanically q-switched ruby laser built by RCA. [14] The Lunar Reconnaissance Orbiter (LRO) marked our return to the moon after more than four decades since the Apollo era with the objectives to conduct scientific investigations and prepare for future human exploration of the Moon, Mars and beyond. The overall mission objective for LRO is to find and locate landing sites, identify potential resources, characterize the radiation environment and demonstrate new technology. LRO was launched in 2009 and has completed the exploration phase of the mission, which lasted for 1-year after first turn-on. The mission is now in its science phase. The LOLA instrument is the first multi-beam altimeter system in space; a diffractive optical element (DOE) at the exit aperture of the transmit telescope produces five beams that illuminated the lunar surface. For each beam LOLA measures time of flight (range), pulse spreading (surface roughness), and transmit/return energy (surface reflectance). With its two-dimensional spot pattern, LOLA unambiguously determines slopes along and across the orbit track. Analysis of the data at cross-over points from multiple orbital tracks will also provide insight on the gravitational field of the moon and its center of mass. The LOLA laser transmitter, using only a modified master oscillator from the GLAS modular design, is a diode pumped, Cr:Nd:YAG slab with passive q-switch and a cross-Porro resonator configuration. The laser transmitter consists of a beryllium (Be) flight laser bench housing two oscillators (a primary oscillator and a cold spare), an 18X transmit beam expander and a DOE. [15]

3. Space qualification process at NASA's goddard space flight center

As already mentioned, there are currently no universal military or NASA standards to follow in order to space-qualify the laser diodes that are used to pump spaceborne lasers.

Over the last decade, our group at GSFC has been working to quantify the reliability of COTS parts and address any issues with their operation in space. Our work has focused primarily on the 808nm QCW diode stacks.

Characterizations of samples and flight parts, long-term tests and familiarity with the market and diode technology have helped us identify and use quality products. Our practice has been to design a diode qualification and screening process based on the operating conditions and requirements of each mission, the acceptable risks as well as the available time and budget. We present an illustrative example of such a process in Section 4. Table 2 summarizes the types of pump diodes used in recent missions, their requirements and operating conditions such as pump pulse width, pulse repetition rate (PRF) and operating current:

MISSION-DURATION	Instrument	Number of diode bars / wavelength / type	ENVIRONMENT/ OPERATING CONDITIONS (Pump Pulsewidth, PRF, Nominal Operating)
MGS 1996-2001	MOLA	44/808nm/ diode stack	Vacuum 150μs, 10Hz, 60A
ICESat 2003-2010	GLAS	54/808nm/ diode stack	Vacuum 200μs, 40Hz, 100A
CALIPSO 2006 - present	CALIOP	192/808nm/ diode stack	Pressurized 150μs, 20Hz, 60A
MESSENGER 2006 - present	MLA	10/808nm/ diode stack	Vacuum 160μs, 8Hz, 100A
LRO 2008 - present	LOLA	2/808nm/ diode stack	Vacuum 200μs, 28Hz, 90A
ICESat2 2016 launch	ATLAS	NA/880nm/fiber-coupled	Pressurized CW, - TBD

Table 2. Summary of flown types of diode pumps, their requirements and operating conditions

Since 2003, the laser diode group at GSFC has developed the necessary infrastructure and has an on-going program for LDA characterization and long-term testing in air and in vacuum environments. The data collected is used to compare and rank the LDAs. Only LDAs satisfying certain criteria, usually established by the mission requirements, are selected for use in space.

After receiving an LDA from a vendor, we mount it on a specially designed test plate, which can be mounted on all necessary test stations and thus minimize handling of the device and the risk of electro-static discharge (ESD) events (Fig. 1). Then the LDA is characterized. Fig. 2 shows a schematic of the characterization set-up.

Fig. 1. LDA on a test plate mounted on the performance test station.

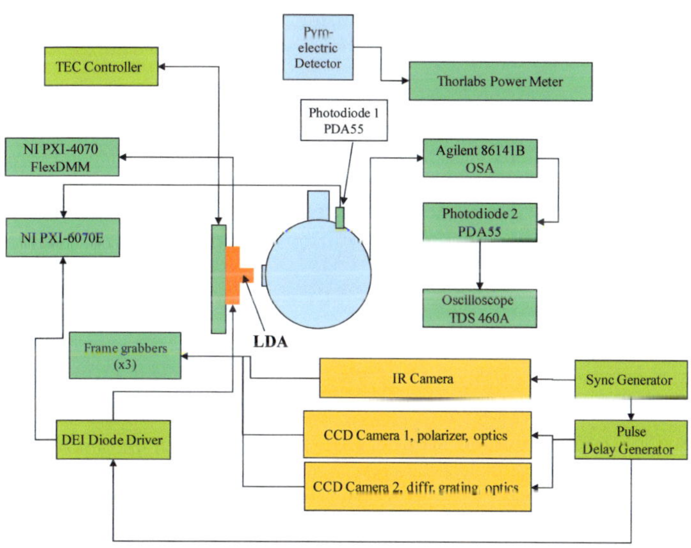

Fig. 2. Schematic of the LDA characterization set-up.

Each standard LDA characterization consists of the following steps:

1. **Microscope inspection of the diode bars' facet and side** - To reveal defects introduced during manufacturing such as minor cracks, solder material on the emitting area, impurities on the diode facet, discoloration and coating defects (see examples in Fig. 3).

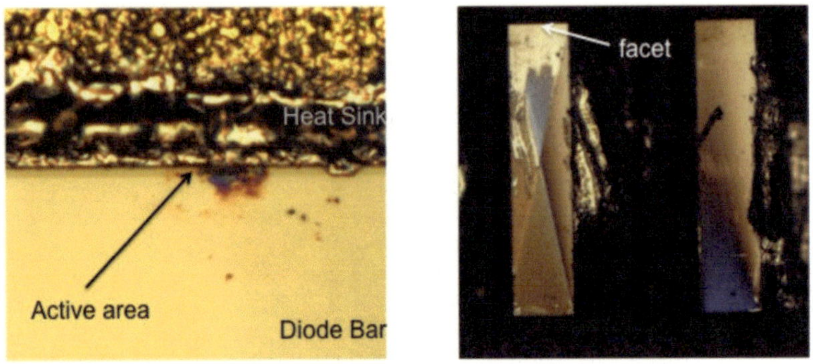

Fig. 3. Samples of high-magnification imaging of the bar facet (left) and side (right).

2. **IR bar facet thermography** - To reveal any hot spots. Temperature stresses may affect device lifetime by generating or accelerating the formation of defects. [16,17] Fig. 4 shows a sample of an IR thermographic image of a two-bar LDA facet operating at 100A, 0.6% duty cycle. The image shows the temperature difference between the LDA when it is powered on and off. The temperature profile is attained by subtracting the IR image of the LDA which is off (background image) from the image when the LDA is on.

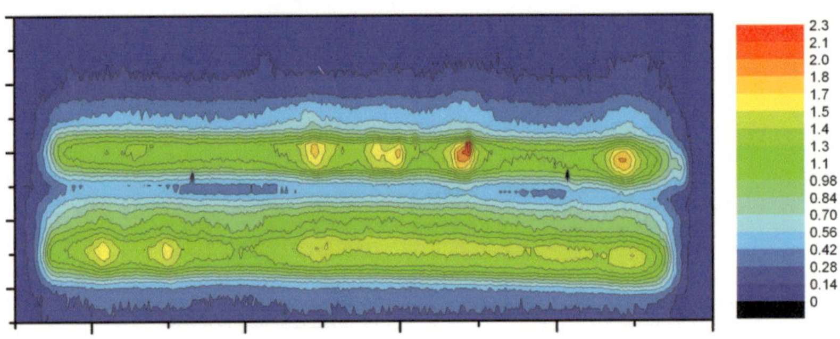

Fig. 4. IR facet thermographic image of a two-bar LDA operated at 100A. In this example, the hot spots have temperatures approximately twice the average bar temperature.

3. **Near Field (NF) imaging** - To determine if there are any dead emitters (Fig. 5).

Fig. 5. Near Field imaging of a 4-bar LDA operating at 100A, 0.6% duty cycle can reveal dead emitters.

4. **Polarization ratio measurements** - Internal defects and stress in the active region of the diode laser can be revealed by a small de-polarization of the output light. [18,19] For that reason, we filter the optical output with a polarizer set to either 0° or 90° with respect to the diode junction plane to capture polarization-resolved near field images. We place the polarizer behind the Neutral Density (ND) filters to prevent polarizer damage or over heating by radiation. To further reduce the heat load on the polarizer, the LDA is triggered at 1Hz. A sample of the profile of the LDA output for TE and TM modes at 100A is shown in Fig. 6. The pink line shows the relative power polarized in the TE mode (parallel to the slow axis of the emitters) across the diode bar and the blue line shows the relative power polarized in the TM mode. Because the light is preferentially polarized in the TM mode, the TE mode line is scaled up approximately 200 times. The normal de-polarization ripples represented by the weak TE mode output and shown along the bar are correlated with the V-grooves that separate the emitters. A more pronounced de-polarization peak is visible in the center of the bar. The measurements of the degree of polarization well below the LDA threshold current of 12A - 14A can also be used to detect laser bar stress. [20,21]

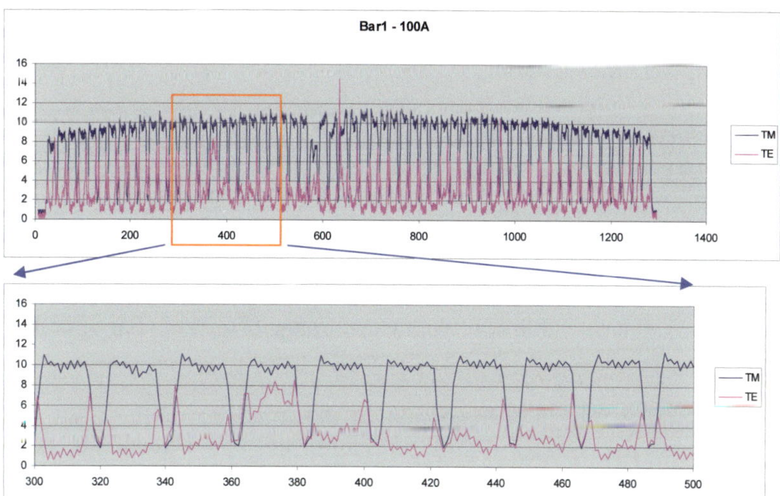

Fig. 6. Intensity profile of an LDA optical output at 100A for TE and TM polarization modes. We filter the optical output with a polarizer set to either 0° or 90° with respect to the diode junction plane to capture polarization resolved near field images. We are using a polarizer with an extinction ratio greater than 10,000 in our wavelength range. The optical signal is attenuated with ND filters to avoid saturating the CCD sensor. We place a polarizer behind ND filters to prevent polarizer damage or over heating by LDA output radiation. To further reduce the heat load on the polarizer, the LDA was triggered at 1Hz.

5. **NF spectrographic imaging**

To identify peaks in the wavelength spatial profile of the diode bar. Spatially resolved bar spectra are acquired by an imaging spectrometer (Fig. 7). The spectrometer has a diffraction grating of 1200 lines/mm and the resolution is better than 0.1 nm. The spectrum is captured by a monochrome CCD camera. A spatially resolved spectrum of the laser diode array bar is shown in Fig. 8.

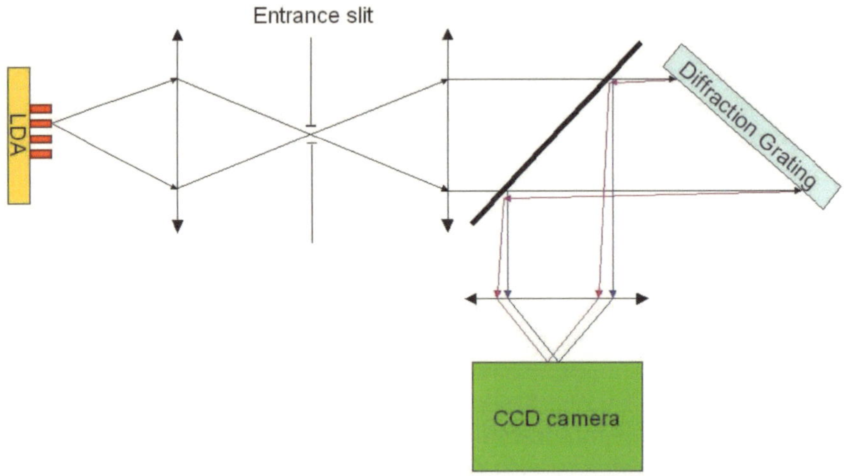

Fig. 7. The diagram of imaging spectrometer. The spectrometer has a diffraction grating of 1200 lines/mm working in the auto collimation regime. The spectrometer dispersion is equal to 4.3 x 10^{-6} and resolution is better than 0.1 nm. The spectrum is captured by a monochrome 1.4Mpixel CCD camera.

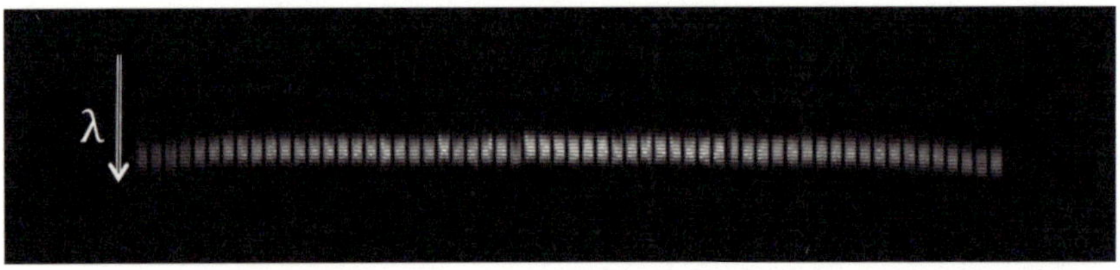

Fig. 8. An example of an acceptable spectrographic image of a diode bar operated at 20A, 0.6% duty cycle.

Each light feature in Fig. 8 corresponds to the intensity of the light emitted by an element; however the vertical distribution indicates the wavelength dispersion. The wavelength increases in the vertical direction of the arrow. The spectrum is taken at 20A current.

In a recent case, the emission wavelength appeared to have a wavy pattern along the bar (Fig. 9). The periodic change of the emission wavelength indicated residual mechanical stress. We worked closely with the vendor who quickly identified and resolved the issue by improving the bar mounting process. The spectrographic images of the new LDAs then looked similar to the one in Fig. 8.

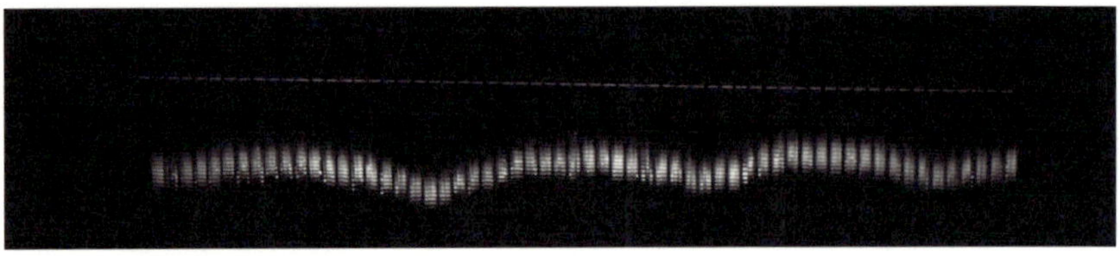

Fig. 9. A "wavy" spectrographic image of a diode bar, measured at 20A, 0.6% duty cycle, indicates residual mechanical stress.

6. **Current and voltage measurements**
We measure current supplied by diode driver and voltage across the laser diode array. These measurements are necessary to calculate the threshold current, slope efficiency, diode array serial resistance, and wall plug efficiency. The LDA threshold current and slope efficiency are derived from optical power vs. current measurement (P-I curve, as in Fig. 10 (left)), LDA serial resistance from voltage vs. current measurement, and LDA efficiency from the previous two. We assess the efficiency of the LDA and observe whether there is any roll-over on the P-I curves, which may suggest thermal issues on the package.
7. **Average spectral measurements**
We typically measure LDA average spectrum at three different currents (Fig. 10 (right)) to get a baseline measurement of the center wavelength and wavelength distribution. We have seen evidence of spectra on a few devices that change shape as the current is increased which is generally a signal that there is uneven heating in the array.

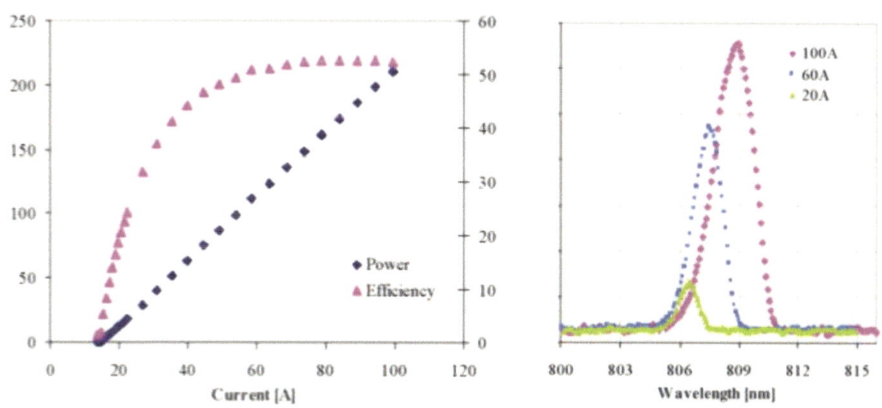

Fig. 10. Samples of P-I and efficiency curves (left) and spectral measurements at three different currents (right). [22]

8. **Temporally and Spectrally Resolved (TSR) measurement at maximum current**
A measurement of the temperature of the active region is an important parameter as the lifetime of LDAs decreases with increasing temperature. [17] We measure the temperature from time-resolving the spectral emission in an analogous method to Voss et al. [23]. Time-resolved spatially averaged emission measurements are made with an optical spectrum analyzer set to "filter-mode", a photodiode fiber-coupled with an integrating sphere and a digital oscilloscope as shown in Fig. 2. In filter-mode the OSA acts as a narrow optical band-pass filter. The center of the filter is stepped through the emission spectrum of the LDA. For each wavelength, the intensity vs. time profile is recorded via the digitizing oscilloscope. The resulting data matrix is then analyzed to determine the peak wavelength as a function of time. The temporally resolved spectra show a temperature induced chirp to longer wavelengths. This chirp is attributed to the current pulse thermally tuning the device. We calculate the thermal rise of the LDA using the typical wavelength shift value of ~0.3 nm/°C. The peak wavelength, full-width at half-maximum (FWHM) and temperature derived from the temporally resolved data matrix are plotted in Fig. 11.

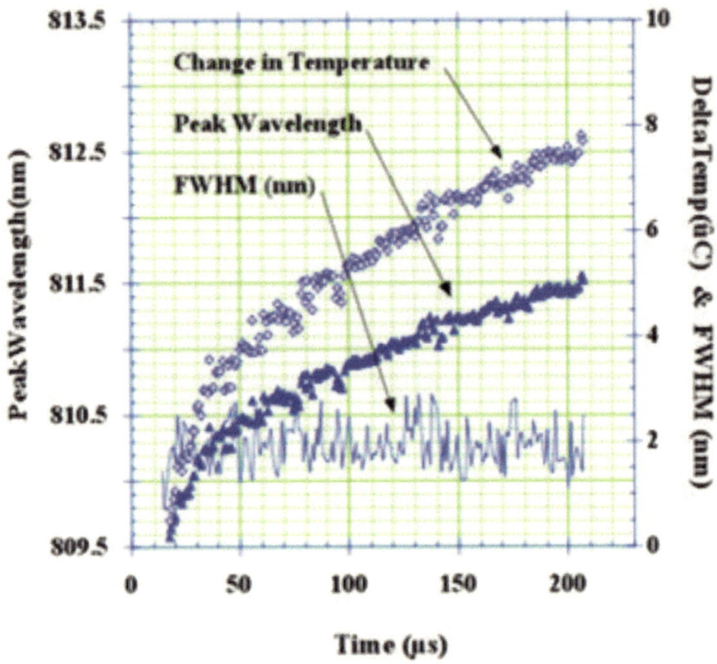

Fig. 11. Change in peak wavelength, full-width at half-maximum and temperature as a function of time at 100A, during a current pulse of 200 μs, derived from the temporally resolved data matrix. [22]

Packaging-induced stress and defects have been found to play a significant role in LDA reliability and therefore their early detection is critical for device screening and space qualification. There are a number of non-destructive techniques available for that purpose such as micro-photoluminescence and photocurrent spectroscopy. [24] Although the testing equipment is available, we have not yet included such measurements into our standard LDA characterization process.

Next, the LDAs undergo aging or accelerated testing in air or in vacuum, depending on the mission requirements, to measure output power degradation rates and verify that they meet the mission lifetime.

The laser diodes, as part of the space laser system, need also to survive mechanical shock and vibration events before and during launch as well as ionizing radiation and temperature extremes during flight. Vibration, shock and temperature extremes can cause mechanical damage to the diode package. Ionizing radiation (gamma, proton) may cause internal damage to the semiconductor resulting in decreased optical efficiency. Durand et al. [25] have reported no apparent changes in LDAs' electro-optical properties under proton irradiation fluences that exceeded the ones experienced at low earth orbit. Our test results have also indicated that the performance of the LDA have not been affected by vibration levels up to 20 grms and total irradiation doses up to 200 krad. [26]

All laser diodes we have tested have survived environmental testing (radiation, vibration, vacuum). The only failures we have experienced were due to Indium solder creeping as seen in the example illustrated in Fig. 12. LDA vendors have recently been using "harder" types of solders with higher melting temperatures such as AuSn and have improved the reliability of their devices.

Fig. 12. Image of a diode bar failure due to solder creeping.

4. Qualification of laser diode arrays for LOLA

LOLA is one of seven instruments aboard the LRO spacecraft, launched on June 18, 2009.

The mission requirement was for more than one billion shots to be made using a maximum of two flight lasers to complete both the exploration and science phases of the mission. LOLA is the first multi-beam space-based, altimeter system. The transmit beam is split into five using a DOE at the exit aperture of the transmit telescope. Since each laser shot will provide five altimetry time-of-flight measurements, over the course of the mission, LOLA will provide more than five billion measurements. To meet the LOLA goal, we built a total of four lasers - two flight lasers for satellite integration and two engineering model lasers. The lasers are side-pumped Nd:YAG oscillators. Each laser requires two, 2-bar arrays. The total required for the flight build was eight 2-bar arrays (including 2 flight spare lasers.)

The laser design has two 2-bar arrays side-pumping a single Nd:YAG crystal. On orbit, degradation in the diode pumps may be compensated by adjustability in both current amplitude and pulse width. In addition, two lasers are being flown where one might be expected to meet the mission requirements. In this way we have built in both derating and redundancy.

For the EM laser, we bought LDA from two different vendors. The operating conditions and specifications for the LDA are listed in Table 3.

Number of bars in array	2
Power per bar	70 Watts
QCW Peak optical power	140 Watts
Pulse width	170 μs
Duty cycle	0.54%
Center wavelength	808.0 nm
Center wavelength tolerance	±2.0 nm
Spectral width	< 3.0 nm FWHM
Fill factor	90%
Slow axis divergence	< 12°
Fast axis divergence	< 40°
Bar pitch	400 um ±25 um
Array size	10 mm x 0.8 mm
Threshold current	< 25 A
Operating current	< 85 A
Operating voltage	< 5.0 V
Operating temperature	25°C
Ambient condition	Vacuum
Lifetime	1.0 billion pulses

Table 3. Operating conditions and performance specifications of the LOLA laser diode arrays. [27]

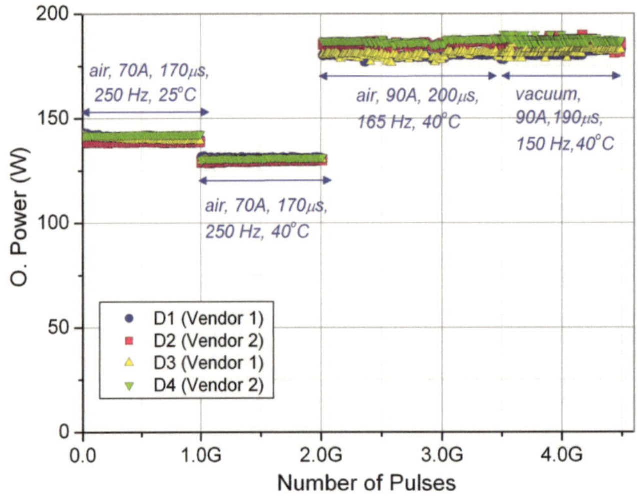

Fig. 13. Test data from LOLA engineering model LDAs operated for 4.5 billion pulses. The power drop at 1 billion pulses was due to the increased operating temperature. [28]

Four EM devices (two per vendor) were tested under various operating conditions, in air and in vacuum, similar to the flight configuration. They were first operated in air, at the beginning-of-life (BOL) operating conditions (70A, 170 μs) for one billion pulses (Fig. 13). The repetition rate was 250 Hz instead of 28Hz due to test schedule constraints. Then, the operating temperature was increased to 40°C for another billion pulses. The LDAs operated for another 1.5 billion pulses at 90A, the highest value expected on- orbit, reaching a total of more than 3.5 billion pulses. They were then operated in vacuum for another billion pulses. The LDAs were fully characterized at intervals during this test and all showed less than 5% total power degradation.

The flight LDA were from a different production run which made this engineering test of no statistical significance to the flight devices. Nevertheless, we found it to be informative, since it gave some indications of the differences between the two different types of LDAs and encouraging since there was not significant power drop even after accumulating 4.5 billion shots.

For the purchase of the flight arrays, a decision was made to buy from and qualify the same two vendors. We purchased 30 LDAs from each vendor. We used 15 arrays for test and qualification and the remaining fifteen as flight devices and flight spares. Because of the extra arrays, it would also be possible to reject units due to poor characteristics. All the arrays were subject to initial characterization and analysis. Two were used for destructive physical analysis to look for latent failure mechanisms. Ten were used for operational testing. The remaining test arrays were reserved as spares to replace failed arrays and maintain statistics.

The LDAs from Vendor 1 (V1) were manufactured using "soft" solder (Indium) while LDAs from Vendor 2 (V2) were made using "hard" solder (AuSn). Both sets of stacks were two-bar assemblies, G-type, with 1 cm long, 400 μm pitch bars. Table 4 summarizes the specifications of the flight LDAs from each vendor.

Property	Vendor 1	Vendor 2
Type of package	G-type package	G-type package
Number of bars	2	2
Bar width	1 cm	1 cm
Number of emitters per bar	60	69
Cavity length	600 μm	600 μm
Solder	In	AuSn
Type of heatsink	CuW on both sides of each bar	CuW on p-side, Au ribbon on n-side
Peak Wavelength	808 ±3 nm at 25°C	808 ±3 nm at 25°C
Typical beam divergence (FWHM)	10 x 40 deg.	10 x 40 deg.
Peak optical power at 100A, 25°C	≥ 200W	≥ 200W

Table 4. Specifications of the LOLA flight LDAs

All LDAs had been burned-in according to each vendor's factory procedures. As soon as they were received, they were fully characterized and then operated at 100A, 2% duty cycle until they accumulated another 100 million pulses. They were then re-characterized, ranked and delivered for flight or testing.

Fig. 14 shows an example of a LDA that was rejected for flight based on IR (bottom) and spectrographic NF imaging results (top). The LDA lost 2% of its power during burn-in and then was delivered to be tested in air. It lost another 10% of its optical power very soon after the beginning of the test.

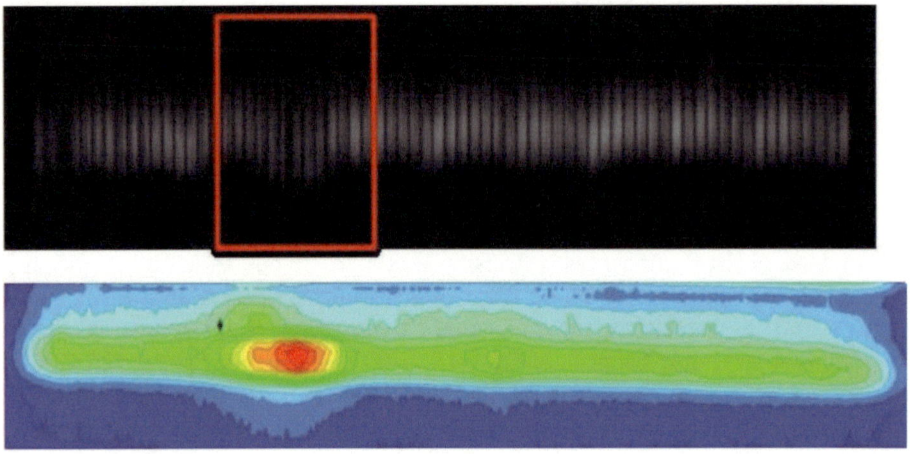

Fig. 14. NF spectrographic image (top) and IR image (bottom) of a LDA from vendor 2 at 100A showed an area of thermal stress on one of its bars.

Some LDAs were operated in vacuum and some in air. Table 5 below summarizes all tests performed on the LDAs.

				Number of LDA	
Environment	Operating conditions	Peak Power Rating	Duty Cycle	Vendor #1	Vendor #2
Vacuum	28 Hz, 70A	70%	0.48 %	3	3
	150 Hz, 70A	70%	2.55 %	3	3
Air	210 Hz, 70A	70%	3.57 %	3	3
	155 Hz, 100A	100%	2.64 %	3	3

Table 5. LOLA performance test matrix. [28]

The purpose of the increased repetition rate was to accelerate the accumulation of pulses. We assumed that as long as the repetition rate is low enough to maintain thermal control, the lifetime of the LDA depends less on operating time and more on the number of accumulated pulses. We tested this assumption by operating a group of LDAs in vacuum at nominal flight conditions and comparing the results to the accelerated vacuum test at 150Hz. In addition, we operated LDAs in air at their maximum current capacity of 100 A. For all tests the current pulsewidth was set at 170 μs and the temperature at 25°C. The vacuum was kept at approximately 1.2E-6 Torr during the course of the tests.

Air and vacuum tests lasted for more than two years and were stopped because of diode driver and other test hardware failures. Except for the group tested in vacuum at 28Hz, the other three accumulated more than 4.5 billion pulses, far exceeding the mission requirement. All tested LDA were re-characterized after the tests ended. None of the LDAs lost more than 20% of its power, which would be considered a diode failure.

In Fig. 15, we have plotted the change in optical power as a function of the pulse count up to 850 x 10^6 shots, which was the total pulse accumulation of the "slow" group tested in vacuum (28Hz). The power drop does not appear to be affected by the repetition rate. The LDAs from Vendor 2 show more power loss than the ones from Vendor 1 as was also the case for the tests in air.

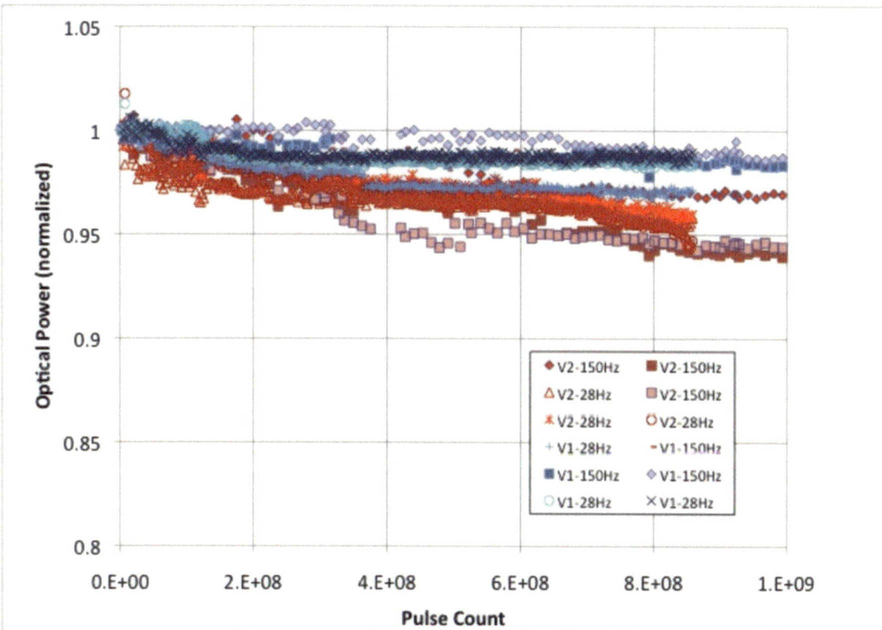

Fig. 15. Optical power vs. pulse count for both groups of LDAs operating in vacuum.

On a parallel effort, two LDAs per vendor were tested during extended testing in vacuum of the LOLA EM laser system. [29] The LOLA EM laser operated in vacuum for over one billion shots. Even though some system degradation was measured during this test it was attributed to other components, not the diode arrays. LDA data helped interpret laser power trends both during the laser EM test and on orbit.

LRO officially reached lunar orbit on June 23, 2009. As of Monday 8/22/2011, LOLA laser 1 (pumped by V1 diode stacks) had accumulated 1,142,583,120 shots and laser 2 (pumped by V2 diode stacks) 656,575,360 shots totaling 1.799158480 x 10^9 shots with no significant degradation, well exceeding the mission requirement. Laser degradation curves are in agreement with the diode aging test results and could be included in a future publication by the LOLA team.

5. ICESat-2 – NASA is transitioning to a new generation of laser remote sensing instruments

Although successfully used for these spaceflight missions, the current DPSSL technology cannot address all the demanding requirements of future lidar missions. New laser transmitter technologies along with highly efficient measurement approaches are necessary to achieve future lidar remote sensing objectives.

Due to some of the fundamental issues with designing an efficient, highly reliable, high pulse energy, low repetition rate laser, NASA is moving in the direction of lasers with lower pulse energy and higher repetition rate with a commensurate change in measurement technique and systems engineering. [29]

Thus far all space flight instruments, except for LOLA, are single beam systems. Future space-based instruments will very likely be multiple beam systems that will improve on pointing and reduce the uncertainty in altimetry measurements introduced by the cross-track surface slope. In addition, for land topography and vegetation, improved pointing will provide observations along exact repeat ground tracks, and sampling along uniformly spaced ground tracks will provide well-sampled grids of topography and biomass. Also to improve on measurement efficiency, a micropulse laser altimetry approach received a lot of interest in the past year. Indeed, ICESat-2, the follow-on mission to ICESat, slated to be launched in 2016, is baselining the micropulse altimetry approach. ICESat-2 will launch a single instrument - the Advanced Technology Laser Altimeter System (ATLAS). In this case, a high repetition rate (10's kHz), lower energy (100's μJ), shorter pulse width (~1 ns), multi-beam laser altimeter system will be used. A single laser transmitter having sufficient laser energy will be split into multiple beams using DOE similar to the one used on LOLA. [16]

The micropulse laser altimeter system for ATLAS represents a new space-based altimeter architecture for NASA GSFC. Similar altimetry systems utilizing high repetition rate, low energy pulses, multiple wavelengths, multiple beams and single-photon ranging were successfully flown on recent airborne platforms. [30,31,32] All previous laser altimeters for space have been QCW LDA pumped solid-state oscillator/amplifier systems with low repetition rate and high energy with a single-beam footprint (except for LOLA). For ATLAS, continuous wave (CW) diode pumping is anticipated with a short pulse (<1.5 ns) high repetition rate (10 kHz) laser. The short laser pulse, high beam quality ($M^2 < 1.6$) and narrow, stable linewidth (<30 pm) drive the laser architecture for ATLAS to a MOPA design. The short pulse width and high beam quality can be met with a lower energy, short cavity master oscillator (MO) and a power amplifier (PA) following the MO can then bring the energy to the required level.

6. Conclusions

Diode pumped solid-state laser transmitter technology has come a long way since the first demonstration in 1968. [33] The performance and reliability of semiconductor laser pump diodes have improved many fold since then. Since the successful launch and operation of the MOLA laser, NASA has flown many more successful missions based upon the use of QCW LDA pumped solid-state lasers. Reliability of the QCW LDAs became an issue on ICESat when several of the LDAs failed due to excessive indium solder during the manufacturing process of these arrays resulting in a metallurgical reaction, which progressively eroded the gold conductors through the formation of a non-conducting intermetallic, gold indide, at a rate dependent on temperature. [34] This focused attention on the use of soft solders that can creep and should be used sparingly for high reliability, multi-year applications. It also pointed out the importance of verifying the workmanship of the specific devices being used for flight - because the fact that the indium solder was in contact with the gold conductors was the result of a breakdown in process control.

The manufacturing processes of these LDAs are critical to the success of our space program. Our goal is to provide feedback to the vendors based on our assessment of their delivered products. We have developed a set of inspection and characterization processes to inspect, quantify and assess the QCW LDAs for space, which began minimally with the MOLA program and has substantially evolved and improved for several missions in the past ten years. This practice was successfully implemented for two flight missions and we will continue to exercise this procedure with necessary modifications and updates for upcoming programs. The difficulties arose when LDAs from previously qualified vendors underwent changes such as different soldering processes, soldering materials, packaging materials or gain medium, which negate the use of existing data to qualify parts for future programs. These new LDAs will then need to be re-qualified before use. This is a time and cost consuming but necessary and unavoidable step to ensure quality and reliable components on an instrument.

808 nm QCW laser pump diodes continue to be a major concern for laser system reliability at NASA. Fortunately, high-reliability, 9xx pump diodes can be used for many NASA systems with appropriate trades. Other NASA laser/lidar systems, however, will require commercial diode arrays at other wavelengths depending on the laser/lidar application (e.g. 792 nm pumps for 2 μm wind lidar). Characterization and lifetime testing of all QCW and CW diode arrays is critical for future missions.

Future missions such as ICESat-2 will be utilizing CW LDAs for pumping solid state lasers to generate higher repetition rate (10's kHz), lower energy (~100's μJ) and shorter pulse width (< 1ns) pulses. This micropulse lidar or altimetry architecture may represent the next generation of remote sensing instrumentation for NASA. The change provides a higher measurement efficiency and also potentially improves the overall wall-plug efficiency of the laser system used in such instruments. To support this, we are currently developing and modifying our procedures to assess CW packaged free-space laser arrays and fiber coupled single emitters for future space missions.

We have summarized our approach to quantify and qualify LDAs for previous missions. Future missions such as spectroscopy from space by measuring CO2 from Earth's orbit, [35] high resolution mapping of Earth, [36] free-space laser communications [37] and gravitational wave detection [38] will need much improved efficiency and lifetime on pump lasers to ensure successes on missions.

7. References

[1] D.E. Smith, M.T. Zuber, and H.V. Frey et al., "Mars orbiter laser altimeter: Experiment summary after the first year of global mapping of Mars," J. Geophys. Res. Planets, vol. 106, no. E10, pp. 23689–23722, Oct. 2001.

[2] H.J. Zwally, B. Schutz, W. Abdalati et al., "ICESat's laser measurements of polar ice, atmosphere, ocean, and land," J. Geodyn., vol. 34, pp. 405- 445, 2002.

[3] D.M. Winker, J. Pelon, and M.P. McCormick, "The CALIPSO mission: Spaceborne lidar for observation of aerosols and clouds," Proc. SPIE, 4893, 1-11, 2003.
[4] J.F. Cavanaugh, et al., "The Mercury Laser Altimeter Instrument for the MESSENGER Mission," Space Sci Rev (2007) 131: 451–479, DOI 10.1007/s11214-007-9273-4.
[5] D.E. Smith, "The Lunar Orbiter Laser Altimeter Investigation on the Lunar Reconnaissance Orbiter Mission," Space Sci Rev (2010) 150: 209–241, DOI 10.1007/s11214-009-9512-y.
[6] R.S. Afzal, "Mars Observer Laser Altimeter: laser transmitter," Appl. Opt. 33, 3184-3188 (1994).
[7] R.S. Afzal, et al., "The Geoscience Laser Altimeter System (GLAS) Laser Transmitter," Selected Topics in Quantum Electronics, IEEE Journal of, 13, 511-536, (2007).
[8] F. Hovis, "Qualification of the Laser Transmitter for the CALIPSO Aerosol Lidar Mission", in Proc. of SPIE, 6100, 61001X (2006).
[9] D.J. Krebs, et al., "Compact, passively Q-switched Nd:YAG laser for the MESSENGER mission to Mercury," Appl. Opt. 44, 1715-1718 (2005).
[10] A.W. Yu, et al., "Laser Transmitter for the Lunar Orbit Laser Altimeter (LOLA) Instrument," in Conference on Lasers and Electro-Optics/Quantum Electronics and Laser Science Conference and Photonic Applications Systems Technologies, paper CMQ2, (2008).
[11] D.E. Smith, et al., "Mars Orbiter Laser Altimeter: Experiment summary after the first year of global mapping of Mars," J. Geophys. Res., 106, (E10), 23, 689–23,722 (2001).
[12] J.L. Bufton, "Shuttle Laser Altimeter," Lasers and Electro-Optics, 1997. CLEO/Pacific Rim '97., 143-144, (1997); J. Garvin, et al., "Observations of the Earth's topography from the Shuttle Laser Altimeter (SLA): laser-pulse echo-recovery measurements of terrestrial surfaces", Phys. Chem. of the Earth, 23, 1053-1068 (1998).
[13] A compilation of the recent scientific results can be found in a special issue on ICESat-I in Geophysical Research Letters, 32, Numbers 21, 22, and 23, (2005).
[14] W.L. Sjogrem, Apollo Laser Altimeter Analysis, Final Report, NASA JPL, (1975).
[15] J.G. Smith, et al., "Diffractive Optics for Moon Topography Mapping," Proc. SPIE, 6223, (2006).
[16] M. Fukuda, Reliability and Degradation of Semiconductor Lasers and LEDs, Artech House, Boston, 1991.
[17] A. Kozlowska, "Infrared imaging of semiconductor lasers", Semicond. Sci. Technol. 22 R27 2007.
[18] P.G. Eliseev, N.B. Sverdlov and N. Shokhudzhaev, "Reduction of the threshold current of InGaAsP InP heterolasers by unidirectional compression," Sov. J. Quantum Electron. 14, 1120–1121, 1984.
[19] N.B. Patel, J. E. Ripper, and P. Brosson, "Behavior of threshold current and polarization of stimulated emission of GaAs injection lasers under uniaxial stress," IEEE J. Quantum Electron. QE-9, 338–341, 1973.

[20] D. Lisak, T.D. Cassidy and A.H. Moore, “Bonding stress and reliability of high power GaAs based lasers,” IEEE Trans. Components Packag. Manuf. Technol. Part A 24, 92–98, 2001.

[21] F.M. Ryan and R.C. Miller, “The effect of uniaxial strain on the threshold current and output of GaAs lasers,” Appl. Phys. Lett. 3, 162–163, 1963.

[22] M.A. Stephen, A. Vasilyev, E. Troupaki, G.R. Allan, and N.B. Kashem, “Characterization of high-power quasi-cw laser diode arrays”, Proc. SPIE 5887, 58870A1 (2005).

[23] M. Voss, C. Lier, U. Menzel, A. Barwolff and T. Elsaesser, "Time-resolved emission studies of GaAs/AlGaAs laser diode arrays on different heat sinks", J.Appl. Phys. 79(2), pp 1170-1172, 1996.

[24] J. Tomm, J. Jimenez, “Quantum-Well Laser Array Packaging”, McGraw-Hill Professional, New York, 2006.

[25] Y. Durand, A. Culoma, R. Meynart, J. Pinsard J., and G. Volluet, "Performance of high-power laser diode arrays for spaceborne lasers," Appl. Opt. 45, 5752-5757 (2006).

[26] E. Troupaki, A. Vasilyev, N.B. Kashem, G.R. Allan, and M.A. Stephen, “Space qualification and environmental testing of quasi-continuous wave laser diode arrays”, J.Appl. Phys., Vol. 100, pp. 063109 (2006).

[27] E. Troupaki, A. Vasilyev, M.A. Stephen, A.A. Seas and N.B. Kashem, "Qualification of laser diode arrays for space applications", Proc. SPIE 7193, 719307 (2009).

[28] G.B. Shaw, M.A. Stephen, E. Troupaki, A.A. Vasilyev and A.W. Yu, "Longevity validation of the LOLA laser design by extended vacuum testing of the LOLA engineering model laser", Proc. SPIE 7193, 719306 (2009).

[29] A.W. Yu, M.A. Stephen, S. X. Li, G.B. Shaw, A. Seas, E. Dowdye, E. Troupaki, P. Liiva, D. Poulios, and K. Mascetti, “Space laser transmitter development for ICESat-2 mission”, Proc. SPIE 7578, 757809 (2010).

[30] D.J. Harding, “The Swath Imaging Multi-polarization Photon-counting Lidar (SIMPL): A Spaceflight Prototype,” Proceedings of the 2008 IEEE International Geoscience & Remote Sensing Symposium, 06-11 March, Boston, MA, (2008).

[31] P. Dabney, et al., “The Slope Imaging Multi-Polarization Photon Counting Lidar: Development and Performance Results,” Paper 4644, Proc. IEEE Int. Geosci. Rem. Sens. Symp., Honolulu, HI, 25-30 July 2010.

[32] J. Degnan, et al., “Photon-counting, 3D imaging lidars operating at megapixels per second,” CLEO/QELS 2009. 2-4 June 2009, Baltimore, MD.

[33] M. Ross, “YAG laser operation by semiconductor laser pumping,” Proc. IEEE vol. 56, pp. 196-197. 1968

[34] R.A. Kichak, IGARB Executive Summary, http://icesat.gsfc.nasa.gov/publications.htm.

[35] J.B. Abshire, H. Riris, G.R. Allan, C.J. Weaver, J. Mao, X. Sun, W.E. Hasselbrack, A.W. Yu, A. Amediek, Y. Choi and E.V. Browell, "A lidar approach to measure CO_2 concentrations from space for the ASCENDS Mission", Proc. SPIE 7832, 78320D (2010); doi:10.1117/12.868567.

[36] A.W. Yu, M.A. Krainak, D.J. Harding, J.B. Abshire, X. Sun, S. Valett, J. Cavanaugh, and L. Ramos-Izquierdo, "Spaceborne laser instruments for high-resolution mapping," Proc. SPIE 7578, 757802 (2010), DOI:10.1117/12.843191.

[37] LCRD - Laser Communications Relay Demonstration Mission, http://www.nasa.gov/offices/oct/crosscutting_capability/tech_demo_missions.html.

[38] LISA – Laser Interferometer Space Antenna, http://lisa.nasa.gov/

14

Monitoring of Welding Using Laser Diodes

Badr M. Abdullah
King Fahd University of Petroleum and Minerals Dhahran, Saudi Arabia

1. Introduction

For traditional manual welding, the positioning of the welding torch and the selection of the welding parameters are controlled by the operator. Hence, the quality of the resultant weld was determined by the practical skill of the welder and their knowledge of the welding process. For automation and control, a sensor or a vision system is required to detect the various deviations and implement the necessary changes. The overall objective of this system is to provide reliable real-time measurements of the molten weld pool for use with a process controller. The weld pool contains important information about the welding process, which can be used by the process controller to adjust the welding parameters and regulate the weld pool width in order to allow for consistent welds. As welding becomes more complex and operates at higher speeds in more difficult environments, automation of welding become an increasingly attractive option to industry. Defects in welding also need to be detected to ensure efficient, high-quality production. All these needs have led to a growing interest in the use of sensors to provide accurate, robust, real-time monitoring where this cannot be achieved by more traditional testing and inspection techniques. Yet, most of these techniques have concentrated on specific applications and have led to the development of semi-automated welding systems rather than fully automated welding systems. The use of sensors monitoring systems has limited and specific applications. However, the use of vision systems has paved the way for a new direction in weld monitoring. This kind of weld monitoring has not been thoroughly investigated, although it has huge number of applications such as the oil industry.

Techniques based on vision systems have so far been affected by the high expense of the components and are not useful for production systems, in many cases because these sensing systems are either bulky, very expensive to produce and only suitable for specific applications. During the previous study, it was demonstrated that the intense arc light can be successfully eliminated and that a substantial amount of information can be obtained in real time, e.g. metal transfer in TIG welding and weld pool geometry, position and development. However, in this project, the focus is on the use of laser diodes as a source of illumination. In this study, a promising alternative low-cost and compact illumination source is used to illuminate the weld pool area with sufficient power is investigated. This illumination source is based on laser diodes, which are generally more affordable and have characteristics that make them more attractive than the bulky and expensive laser systems.

2. Weld monitoring overview

One of the major advantages of using a visual sensing technology to monitor welding operations is the fact that the visual sensor is not touched or interfered with during the welding process and visual images of the weld pool contain more abundant and accurate information about the welding dynamics. The main difficulty encountered in vision-based sensing of the weld pool geometry is the strong interference from the arc light across a wide spectrum. A major difficulty with automation of arc welding is the lack of a suitable weld quality sensor. Typical automated welding systems may be capable of controlling only torch position, travel speed, arc voltage and arc current (Boughton et al., 1978)]. However, these variables are generally controlled only according to some pre-determined conditions rather than according the actual weld condition. Many attempts have been made in the past to find a method of sensing weld penetration so that it can be adequately controlled. Weld pool geometric appearance has bee used to measure weld joint penetration (Kovacevic et al., 1996). Chin used infrared thermography from the backside of the weld sheet to make isothermal maps (Chin et al., 1983). Wickle used infrared sensing of arc welding (Wikle & Kottingam, 2001). Richardson used brightness pyrometry coupled with an axial view torch to measure and control the width of the weld bead (Richardson et al., 1982). Infrared measurements of base metal temperature for online feedback controls were used. However, both arc and electrode radiation interfered very significantly with infrared measurements made at a location on the base metal (Richardson & Edwards, 1995; Farson et al., 1998). Li developed a theoretical model to relate the arc light radiation to welding parameters (Li & Zhang, 2000). Other attempts have also been made on ultrasonic technology and chromatic filtering of thermal radiation (Miller et al., 2002; Baik et al., 2000). Other techniques were developed using artificial intelligence based approach for process parameter prediction (Balfour et al., 2000; Luo et al., 2002). Weld pool monitoring for specific applications has also been attempted, e.g. welding of steel pipes, CO_2 short circuiting arc welding and laser welding of thin sheet metals (Ancona et al., 2004; Sun et al., 2006; Frazer et al., 2002; Du et al., 2000).

The continuous reduction in the cost of cameras and illumination systems over the last five years has allowed vision systems to be increasingly used as sensors to extract information about the weld pool. Intensive research has been done to develop a vision system, which can be used for direct weld pool viewing (Chen et al., 2003; Wu et al., 2000). Some researchers used both camera and an illumination source in their systems, while others just used a camera without any illumination. Y. M. Zhang and R. Kovacevic used a polar coordinate model to characterise the weld pool geometrically. A neural network algorithm was developed to identify the welding parameters in real-time. Although they used pulsed laser illumination to illuminate the weld pool, interference from the arc light is clearly visible in the images obtained (Zhang et al., 1996). L Hong used a high speed camera and a high power pulsed laser for illumination. A LaserStrobe vision system and a neurofuzzy control system were used for arc welding process control (Hong et al., 2000). Other research used only cameras to observe the weld pool without the use of any illumination sources. These vision systems lacked proper illumination, and hence, the images appeared either too dark when the shutter time of the camera is decreased or too bright when the shutter time is increased. For example, K. Y. Bae used a vision system for both seam tracking and weld pool control. A CCD camera was used in the vision system to capture images of the weld pool area, but again the interference from the arc light proved a major problem and the images obtained suffered greatly from the presence of arc light (Bae et al., 2000). C. S. Wu used a CCD camera combined with a light filter to form a vision sensing system. Since no illumination was used, the weld pool images appeared too dark with very bright areas in the centre due to the intense arc light (Wu et al., 2003).

During the past few years more studies have emerged using more complex techniques. A study by Huang and Kovacevic used acoustic signals for online monitoring of weld depth.

Acoustic signals generated during the laser welding process of high-strength steel DP980 were recorded and analysed. A microphone was used to acquire the acoustic signals. A spectral subtraction method was used to reduce the noise in the acoustic signals, and a Welch–Bartlett power spectrum density estimation method was used to analyse the frequency characteristics of the acoustic signals. The study showed that good welds can be distinguished from bad weld and that the acquired signal can be used to control weld depth (Huang & Kovacevic, 2009). Another study used a multiple of sensors to monitor weld penetration. Infrared, ultraviolet and sound sensors were used simultaneously. Infrared sensors were used to detect heat radiation, ultraviolet sensors were used to measure optical radiation and sound sensors were used to measure welding penetration state (Zhang et al., 2008; Allende et al., 2008). Another spectroscopic technique based on the acquisition of the optical spectra emitted from the laser generated plasma plume and their use to implement an on-line algorithm for both the calculation of the plasma electron temperature and the analysis of the correlations between selected spectral lines (Sibillano et al., 2009). Song and Zhang utilized the reflection property of the weld pool surface. In their system, a dot-matrix pattern of structured laser light was projected onto the specular weld pool surface and its reflection was imaged on a self-designed imaging plane. Then the distorted reflected image (pattern) was captured and processed. Based on the obtained information, two reconstruction schemes named interpolation reconstruction scheme (IRS) and extrapolation reconstruction scheme (ERS) were used in order to rebuild the three dimensional weld pool surface off-line (Song & Zhang, 2008). Arc sound has been found to be strongly related to process parameters and weld quality (Arata et al., 1980; Futamat, 1983; Pal et al., 2010). Welding arc light can be seen as a signal carrying essential information about the welding process. The information contained in the signal can be exploited in the monitoring of the welding process. A recent study by the Institute of Welding in Poland employed this idea for monitoring MIG welding (Weglowski, 2009). A review was proposed by TWI to review the available literature on sensor systems for the top-face control of weld pool penetration in arc and laser welding, with a view to developing a commercial system for the top-face control of weld penetration (Anderson, 1997). The review concluded that there is currently no system commercially available which satisfies all the requirements of the ideal system, although some sensors have been successfully applied in specific applications. The range of application and accuracy of control of these systems can be summarized as follows:

1. Ultrasonic techniques are only suitable for plate thickness of less than 2mm or greater than 10mm. The accuracy of control needs to be improved before they can be used for practical applications.
2. Acoustic emission monitoring will function independently of welding position, process and joint type, but has still to be effectively demonstrated.
3. Weld pool sag sensing is limited to the flat position, butt welds and requires a highly accurate power source.
4. Weld pool oscillation frequency monitoring by optical systems can only be used for full penetration butt welds, and requires the back surface of the weld to be free of oxide and contamination.
5. Direct weld pool viewing functions independently of position, process and joint type, but requires accurate control of component thickness and joint fit-up.
6. Thermal sensing functions independently of position, process and joint type, but its accuracy is reduced by variations in the surface emissivity of the component material.

3. Basis of vision system

3.1 Welding spectra

Arc light emissions have the potential as a welding information source. The emission changes with weld parameters, these parameters include current, voltage, electrode diameter and type, and shielding gas. Some researchers have applied welding spectra for weld monitoring (Zackenhouse & Hrdt, 1983; Sorensen & Eagar, 1990; Sforza & de Blasiis, 2002). In these studies, it was shown that the emission spectra can be used to detect changes in the welding arc. Other studies including this one incorporated welding spectra to enhance the image quality of the weld pool. Further studies investigated the possibility of sensing welding behaviours based on arc light (Li & Zhang, 2000). In another research work, an arc light sensor was used to monitor the droplet transfer mode in GMAW (Madigan et al., 1989). To determine the optimum spectral window to operate at, welding trials were carried out on stainless steel with argon and helium as shielding gasses. An optical spectrum analyser with a wavelength range from 350nm to 1750nm was used to obtain the spectra of TIG welding shown in Fig. 1.

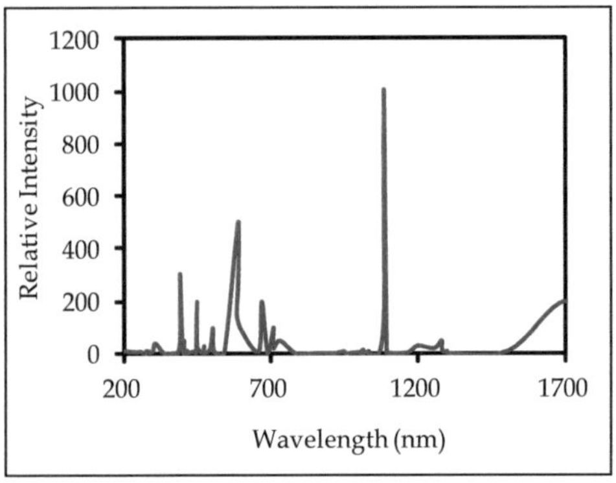

a. Argon spectrum

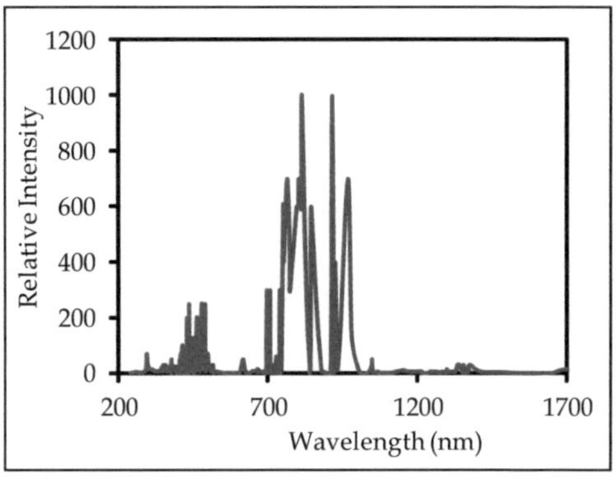

b. Helium spectrum

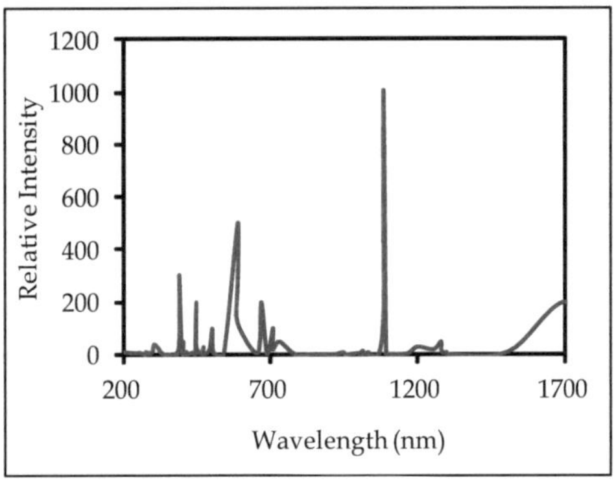

c. TIG welding on stainless steel, 100% argon, 150A

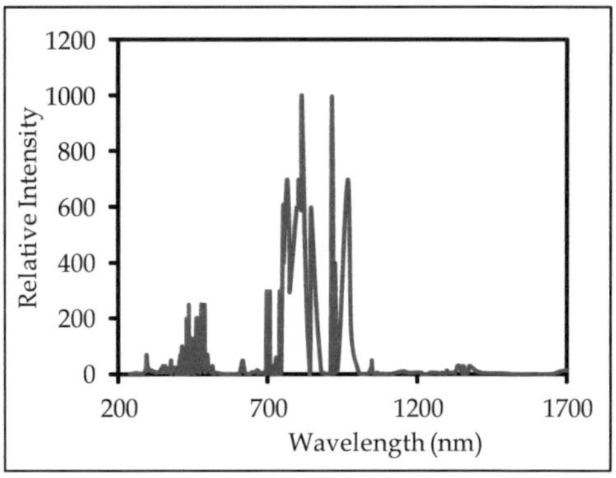

d. TIG welding on stainless steel, 100% argon, 150A

Fig. 1. Welding spectra

Strong emission lines from the shielding gas can be seen at different wavelengths in each of the two spectral distributions of shielding gases shown. Strong emission lines appear at around 800nm in the spectral distribution of argon, whilst with helium, strong lines appear at around 400nm to 600nm with the exception of a very strong line at around 1100nm. It has also been found from this study that most of the strong emission lines appear at all current settings on the same base material and shielding gas but with varying relative intensities. Higher current settings result in higher relative intensities, and lower current settings result in lower relative intensities for all shielding gases and base materials. When helium is used, the relative intensity across the spectrum is greatly reduced especially at around 800nm, where it was at its peak with argon and the argon mixture. Some strong emission lines are still abundant at around 400nm but with greatly reduced intensities. It has also been found that the base metal also affect the emission intensities in both shielding gases. The strong lines that appeared at 800nm disappeared when mild steel is used. However, intensity peaks have shifted towards the UV region. Since arc light is abundant in the visible and UV regions and weld pool emissions are abundant in the IR region, the optimum spectral window to operate at is in the IR region. Therefore, taking into consideration the limitations of the camera above 1000nm, the best wavelength to operate at is around 800nm - 950nm.

3.2 Spectral filtering

During arc welding arc light is emitted over a range of wavelengths. Only parts of the wavelengths will be detected by a given camera. A typical spectral response range for a standard visible light camera is 350nm - 850nm, falling sharply outside this range. However, laser diode illumination is at a very specific wavelength, with a typical wavelength of a few nanometres. This is illustrated in Fig. 2, which shows how the light from a laser source might compare with the arc light. If we illuminate the bright welding area with a laser diode source and place a narrow band pass filter with the same wavelength as the laser diode in front of the camera, then only the laser diode light will pass with a fraction of the arc light. The result is a much-attenuated arc light with little effect on the laser diode light as illustrated.

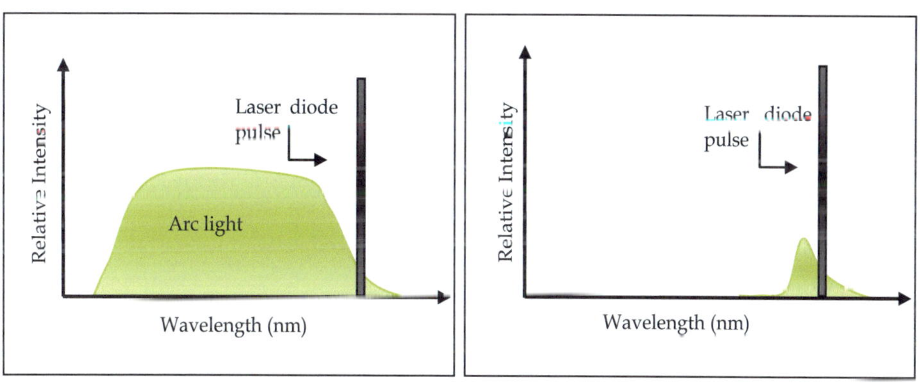

a. Pre-spectral filtering b. Post-spectral filtering

Fig. 2. Effect of spectral filtering on light absorbed by the camera

3.3 Temporal filtering

Spectral filtering attenuates the arc light but it is not enough alone to totally eliminate the arc light. Temporal filtering provides further arc light reduction without attenuating the laser diode light as long as the pulse width remains shorter than the camera's shutter exposure time. More unwanted arc light will be captured if a longer exposure time is used. This has no effect on how much of the desired laser diode light is captured as long as the pulse width remains shorter than the exposure time. This is always the case for most laser sources which have pulse width durations in the nanosecond region, whereas the camera exposure time is normally in the microsecond region. Laser light will only be attenuated if the pulse width is greater than the exposure time of the camera or if the laser is a continuous wave laser rather than a pulsed laser. The reduction of the arc light when a shorter exposure time is used is illustrated in Fig. 3.

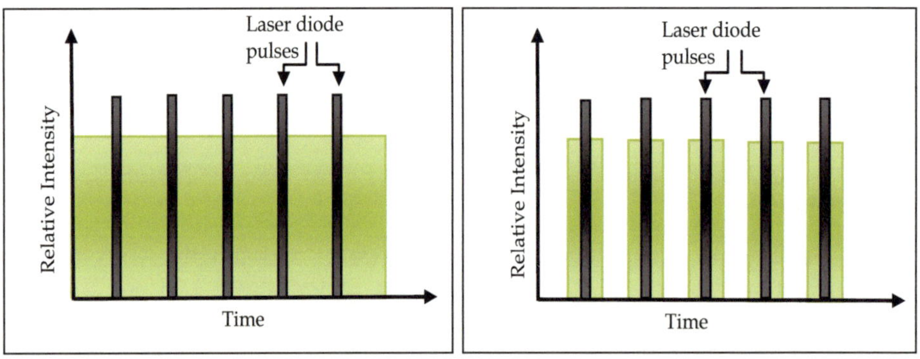

a. Pre-temporal filtering b. Post-temporal filtering

Fig. 3. Effect of temporal filtering on light absorbed by the camera

3.4 Camera-laser synchronisation

3.4.1 Trigger methods

The use of triggering is a powerful feature to the capabilities of image acquisition systems. It gives us the ability to efficiently capture a short duration and high speed events by eliminating the need to continuously acquire images while waiting for the event to occur. An external trigger is an event that starts an exposure. The trigger signal is either generated on the frame grabber (soft-trigger) or comes from an external trigger between the laser diode and the camera is essential for obtaining high quality images without the need for a high power illumination system.

Figure 4 shows four synchronisation methods that can be used to capture the laser pulse. In Fig. 4a, the camera is driving the laser. A pulsed output from the back of the camera is connected to the laser external input. This guarantees a good synchronisation. However a pulse delay is introduced which can be overcome by delaying the camera strobe signal. In Fig. 4b, the laser is driving the camera. The external output of the laser device is connected to the external trigger input of the camera. This will also introduce a delay to the camera shutter that can be overcome by introducing a pulse delay. Both the camera and the laser can be driven by an external master-triggering unit which consists of either one trigger source or two triggering sources with one of the trigger sources acting as the master trigger and the other as a slave so that a delay can be introduced as shown in Fig. 4c. The frame rate, laser diode frequency and timing of the system can be controlled when connected to a pulse generator. The fourth method is used mainly when using a continuous wave laser as an illumination source as shown in Fig. 4d. Both the camera and the laser are free running and are independently controlled. The idea is to capture as much as possible of the CW laser power, which can be achieved by adjusting the exposure time of the camera.

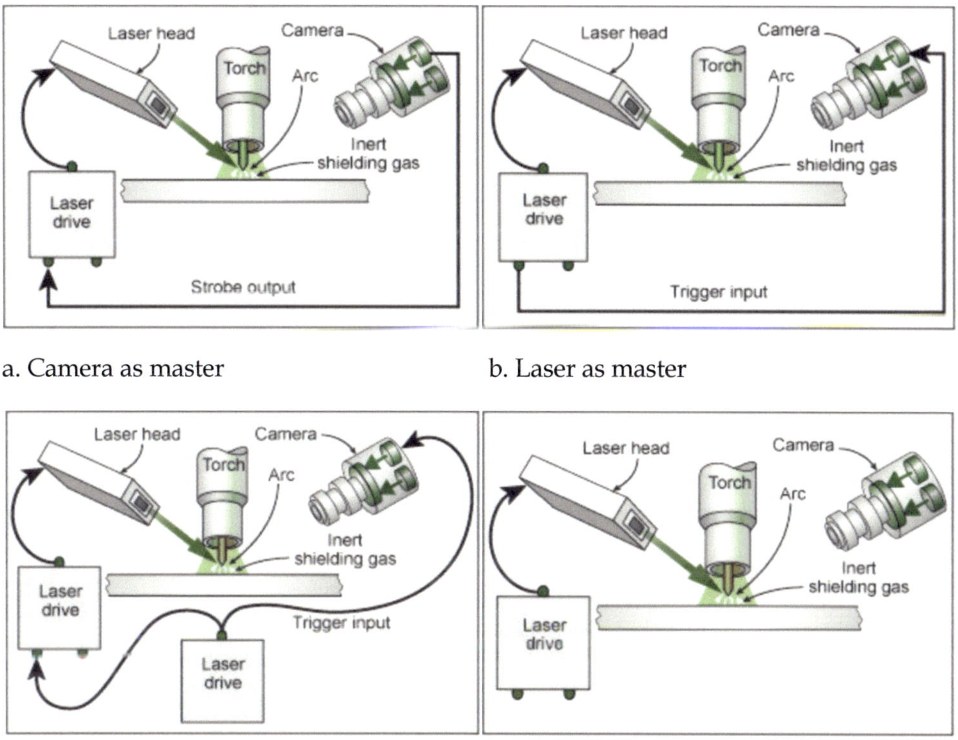

a. Camera as master

b. Laser as master

c. Laser and camera driven by an external driver

d. Laser and camera independently driven

Fig. 4. Laser camera synchronisation methods

4. Light capturing

The light source can either be mounted on the same side as the camera i.e. backward reflection or opposite the camera i.e. forward reflection as shown in Fig. 5. More light is reflected to the camera when the laser is on the opposite side. Hence, less laser energy is required. However, uneven reflection from the weld pool occurs. Varying the angle between the laser beam and the work-piece can improve this. On the other hand, mounting the light source on the same side as the camera gives natural lighting with little glare but requires more laser energy as most of the laser light reflects away from the camera.

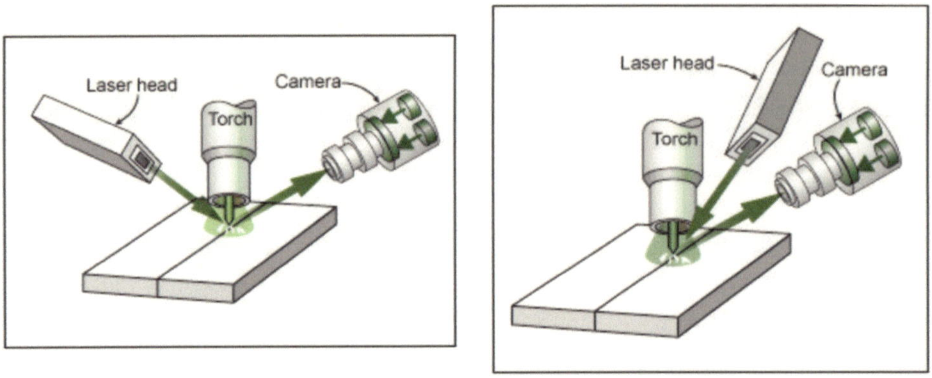

Forward reflection Backward reflection

Fig. 5. Light capturing methods

5. Laser diode vs. LED

Illumination technology plays a major role in the design of vision systems, especially for industrial applications where real-time processing is a crucial requirement. The region of interest can be made significantly clear by the proper choice of lighting techniques.

There are five types of illumination that are commonly used for machine vision;

- Halogen (using fibre optic light guides)
- Incandescent
- Fluorescent tube technology
- LEDs
- Lasers

Halogen, incandescent and fluorescent lighting is poor for this type of application. They have very limited use in machine vision applications which require pulsing at short widths and suffer from inherent flicker and can not be easily controlled to maximise their intensity.

Practically, LED technology now provides the most appropriate solution for most machine vision applications. This is because LEDs produce a relatively high light intensity at a relatively low cost compared to halogen, incandescent and fluorescent technologies. However, applications such as real-time weld pool monitoring require much more power than a commercial LED can afford and require narrow bandwidth and very short pulsing widths. In this section we compare the two main rivals, LEDs and laser diodes.

5.1 Light emitting diodes

The structure of an LED consists of a chip, made of semiconductor material, impregnated with impurities to create a junction. The semiconductor diode chip is mounted in the reflector cup of a lead frame that is connected to electrical (wire bond) wires, and then encased in a solid epoxy lens. Electrical current flows easily from the positive anode to the negative cathode of the junction but not in the reverse direction. Electrical charge-carriers, electrons, and electron holes flow into the junction from electrodes with different voltages.

When an electron meets a hole, it falls into a lower energy level and releases light energy in the form of a photon. In other words, when electrical current passes through them, LED chips illuminate, creating fast direct light that lasts much longer, about 50 times as long, and burns much brighter than conventional incandescent light bulbs. LEDs emit light when energy levels change in the semiconductor diode. This shift in energy generates photons, some of which are emitted as light. The specific wavelength of the light depends on the difference in energy levels as well as the type of semiconductor material used to form the LED chip. LEDs are made from compound semiconductor materials such as gallium arsenide (GaAs), gallium phosphide (GaP), gallium arsenide-phosphide (GaAsP), silicon carbide (SiC) and gallium indium nitride (GaInN). The material used in the construction of an LED's PN junction determines the energy of the emitted photons and hence the colour.

5.2 Laser diodes

Laser diodes are complex semiconductors that convert an electrical current flowing through a P-N junction into light. Electrons and holes combine, releasing the energy as photons. This process can be spontaneous, but can also be stimulated by incident photons leading to amplification. The conversion process is fairly efficient in that it generates little heat in the process compared to incandescent lights. Laser diodes are normally built as edge emitting lasers, where the laser cavity is formed by coated or uncoated end facets (cleaved edges) of the semiconductor wafer. They are often based on a double hetero-structure, which restricts the generated carriers to a narrow region and at the same time serves as a wave guide for the optical field. This arrangement leads to a lower threshold pump power and better efficiency. The active region also often contains quantum wells or quantum dots. Some modern kinds of laser diodes are of the surface emitting type, where the emission direction is perpendicular to the wafer surface. Laser diodes are typically constructed of GaAlAs (gallium aluminium arsenide) for short-wavelength devices. Long-wavelength devices generally incorporate InGaAsP (indium gallium arsenide phosphide).

5.3 Main characteristics

Table 1 offers a quick comparison of some of the main characteristics for laser diodes and LEDs.

Characteristic	LED Diodes	Laser Diodes
Current	Up to 2.5A	Up to 40A
Power	Less than 2.5W	Up to 250W
Emission pattern	Larger than 40°	10° to 25°
Wavelengths available	0.66µm to 1.65µm	0.78µ to 1.65µm
Spectral width	Wide (20nm -190nm FWHM)	Narrow (3nm to 10nm FWHM)
Cost	Less than £2	£15 to £400
Pulse width	100us (minimum)	40ns (minimum)

Table 1. Main characteristics of laser diodes and LEDs

5.4 Spectral width

Ideally, all the light emitted from a laser diode would be at the peak wavelength, but in practice the light is emitted in a range of wavelengths centred at the peak wavelength. This range is called the spectral width of the source. Laser diodes have very narrow spectral width ranging from 3nm to 10nm but not as narrow as other commercial laser systems like the Nd:YAG laser, which normally have very narrow spectral width ranging from 0.00001nm to 1nm. Meanwhile, LEDs have a wider spectral width, typically over 20nm. Spectral width is a crucial factor for the vision system. Wide spectral width means a wider band pass filter will be used and hence more the unwanted arc light will be captured by the camera.

Most laser diode wavelengths emit in the near infrared spectral region, but some can emit in the visible or infrared regions. However, laser diode wavelength is being pushed further and further into the visible spectrum. The latest generation of Visible Laser Diodes (VLD's) operate at or near 635nm which is highly visible to the human eye. VLD's in the range from 635nm to 685nm are replacing the traditional HeNe laser in many commercial products for good reasons: lower cost, compact size, and superior long-term reliability. Another intrinsic benefit, laser diodes are generally better suited for battery operated devices and other low voltage applications. LEDs are available in both visible and infrared wavelengths. Infrared LEDs reach wavelengths of 830nm to 940nm. Visible colours include red, yellow, orange, amber, green, blue/green, blue, and white. These fall into the spectral wavelength region of 400nm to 700nm. The coloured light of a LED is determined exclusively by the semiconductor compound used to make the LED chip and independent of the epoxy lens colour. The FWHM bandwidth of a typical LED ranges from 20nm to 190nm as shown in Fig. 6.

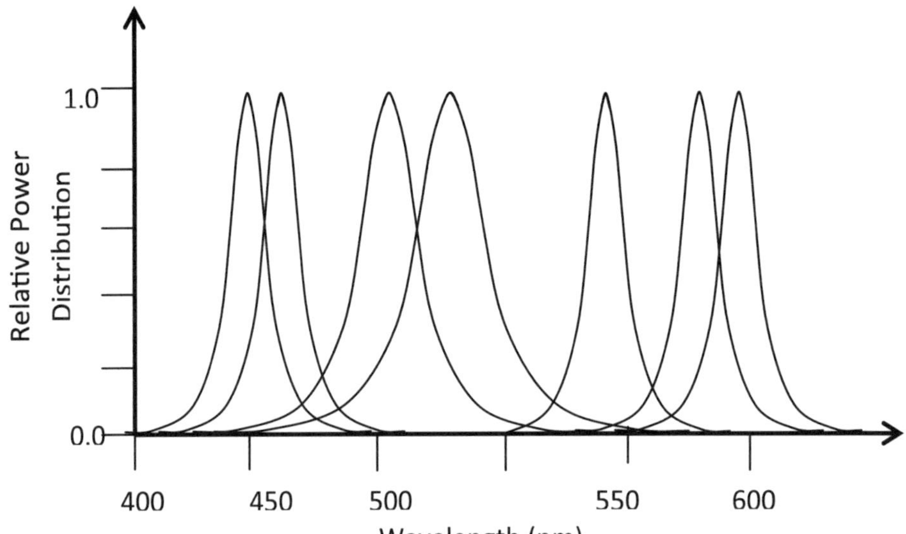

Fig. 6. Bandwidth of typical commercial LEDs

5.5 Pulse width

Laser diodes can be pulsed at very short pulse widths typically around 40ns - 100ns (minimum). The pulse width can be increased to around 10us as long as the duty cycle is kept low and lower currents are used. Meanwhile, LEDs have a much longer pulse width, typically around 100us. Pulsing laser diodes at very short pulse widths also means that they can be overdriven to obtain more power as long as the duty cycle is not exceeded.

There are two main reasons for the use of pulsed operation with LED illuminators intended for use in machine vision applications. The first is to freeze action to acquire an image with the camera shutter. The second is to increase the effective brightness of the illuminator during the pulse by using a higher pulse current than the CW rating, since the luminance is proportional to current. As the pulse width (i.e. when the LED is on) decreases, more current can be used to drive the LED, and therefore more power is produced. The duty cycle also decreases with a decreasing pulse width. If low duty cycles are combined with short pulse widths, so that the junction temperature of the LED is kept close to ambient, then the effective operating lifetimes can be extended.

5.6 Emission pattern

The pattern of emitted light affects the amount of light that can be either directed onto the welding area either by coupling it into an optical fibre or by placing it close enough to the weld scene and focussing the beam without fibre coupling.

Figure 7 illustrates the emission pattern of a laser diode and LEDs. The angle at which light comes out from the laser diode is much smaller than that of the LEDs as shown below. Since the weld pool area to be illuminated is small, an illumination source with a narrow emission pattern is needed.

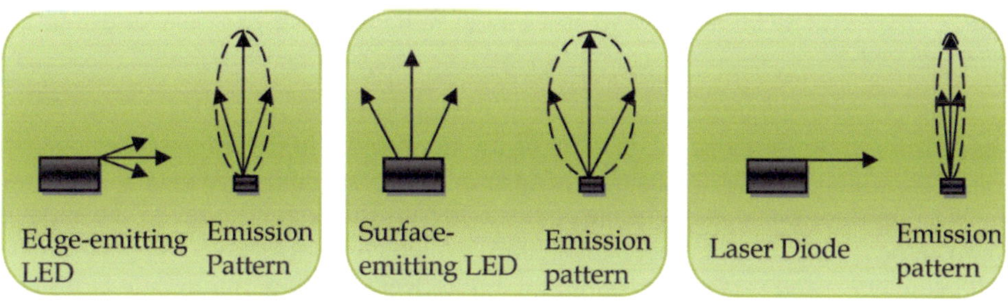

Fig. 7. Emission patterns for LEDs and laser diodes

There are two types of LEDs, edge emitters and surface emitters. Edge emitters are more complex and expensive devices, but offer high output power levels and high speed performance. The output power is high because the emitting spot is very small, typically 30-50μm, allowing good coupling efficiency to similarly sized optical fibres. Edge emitters also have relatively narrow emission spectra. The full-width half-maximum (FWHM) is typically about 7% of the central wavelength. Another variant of the edge emitter is the super radiant LED. These devices are a cross between a conventional LED and a laser diode. They usually have a very high power density and possess some internal optical gain like a laser diode, but the optical output is still incoherent, unlike a laser diode. Super-radiant LEDs have very narrow emission spectra, typically 1-2% of the central wavelength and offer power levels rivalling a laser diode.

Surface emitters on the other hand have a comparatively simple structure and are relatively inexpensive. They offer low to moderate output power levels, and are capable of low to moderate operating speeds. The total LED chip optical output power is as high or higher than the edge emitting LED, but the emitting area is large, causing poor coupling efficiency to the optical fibre. Adding to the coupling efficiency deficit is the fact that surface emitting

LEDs are almost perfect lambertian emitters. This means that they emit light in all directions. Thus, very little of the total light goes in the required direction.

5.7 Power

In general, laser diodes are a lot more powerful than LEDs. Laser diodes can be driven with more current and hence an increased output power. For example the Osram SPL PL90_3 can be pulsed with a maximum pulse current of 40A, which produces a peak power of up to 90W. Whereas, LEDs can only be pulsed at currents much less than 40A and can only produce a few watts.

5.8 Luminous efficacy

Luminous efficacy is the ratio between the total luminous flux emitted by a device and the total electrical power consumed by it. It is a measure of the efficiency of the device with the output adjusted to account for the spectral response curve (luminosity function). When expressed in dimensionless form this value may be called luminous efficiency. Luminous efficiency is a measure of the optical power produced by the source to the actual rated electrical power. The radiant power can be found by dividing the total lumens produced by the luminosity function. For example, a 100W tungsten incandescent light source with a luminous efficacy of 17.5lumens/watt has a total radiant power of 2.56W and a luminous efficiency of 2.56%.

The radiant power is the total radiated power in watts which is also called radiant flux. This power must be factored by the sensitivity of the human eye to determine luminous flux in lumens. Today's high power LEDs operating at 1W or more deliver 50 - 60lm, enabling just 50 LEDs to produce the same light output as a 3000lm fluorescent tube. At the same time, luminous efficacy has risen to more than 60lm/W, which far surpasses the performance of incandescent bulbs and is fast approaching the energy efficiency of fluorescent lighting. As a result, high-power LED manufacturers are now working to develop products addressing the general illumination market, which is currently valued at around £7 billion. However, real success in the mainstream lighting market will require manufacturers to reduce the price of LEDs, while continuing to deliver steady improvements in the device's luminous efficacy and total lumen output.

6. System design

The systems described below are designed and fabricated based on two types of laser diodes, the first system is based on the Osram SPL PL90_3 high power laser diode and the second system is based on the Osram SPL LL90_3 high power laser diode.

6.1 SPL PL90_3

The illumination system is made of a cluster of low-cost nano-stack InGaAs/GaAs pulsed laser diodes. Each laser diode is stacked in a plastic package with a maximum peak output power of 90W and a typical output power of 75W per diode. Each individual laser diode is stacked with three emitters with a laser aperture of 200μm × 10μm. The maximum power is achieved when the diode is driven by a pulse width of 100ns (FWHM) with a maximum forward current of 40A. The pulse width can be increased to 1μs or even 10μs with a reduced forward current. The diode was tested with a reduced current of 20A and a pulse width of 1μs, an optical power output of 43W was achieved. The wavelength of the laser diode is 905nm and its spectral width is 7nm (FWHM).

6.2 Beam profile

The photo detector circuit shown in Fig. 8 is used to obtain the beam profile of the laser diode. The light output beam from the laser diode is detected by the photo diode and converted into voltage by the photo detector circuit. Fig. 9a shows the relative intensity output against the range when the horizontal and vertical displacements are kept at zero. Meanwhile, Fig. 9b shows the relative intensity against horizontal displacement at different range distances. From the beam profile results, we can conclude that the emission pattern of the laser diode is narrow, and hence, it is possible to use a cluster of laser diodes for illumination without the need to deliver the optical output to the weld pool via optical coupling.

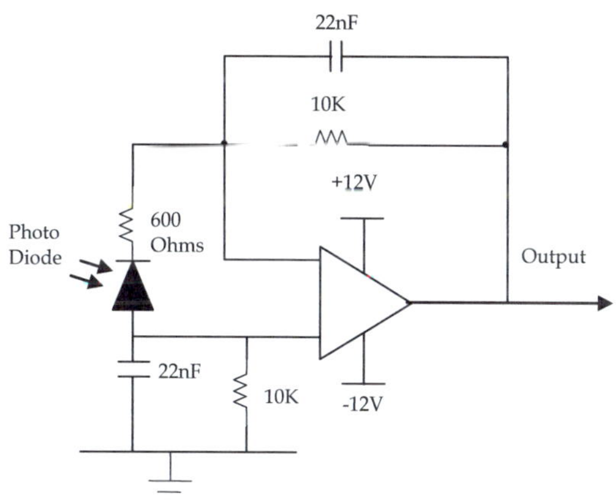

Fig. 8. Photo detector circuit

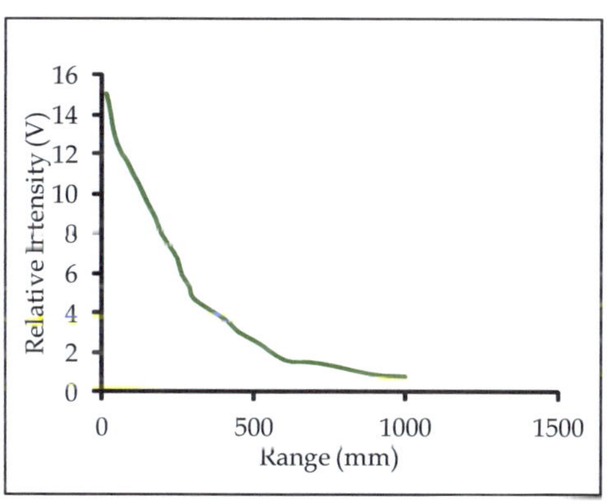

a. Relative intensity vs. range, vertical and horizontal distances are kept at zero

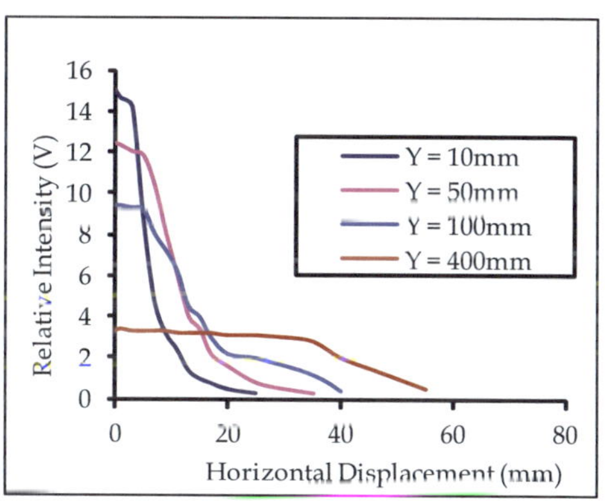

b. Relative intensity vs. horizontal displacement

Fig. 9. Laser diode beam profile

6.3 Driver circuits

The circuit shown in Fig. 10 is constructed using N-Channel enhancement mode MOSFET. This MOSFET is designed to minimize on-state resistance, which is 0.01Ohms for this MOSFET, and provide superior switching performance and withstand high current pulse in the avalanche mode. A MOSFET driver is also used to rectify the trigger signal.

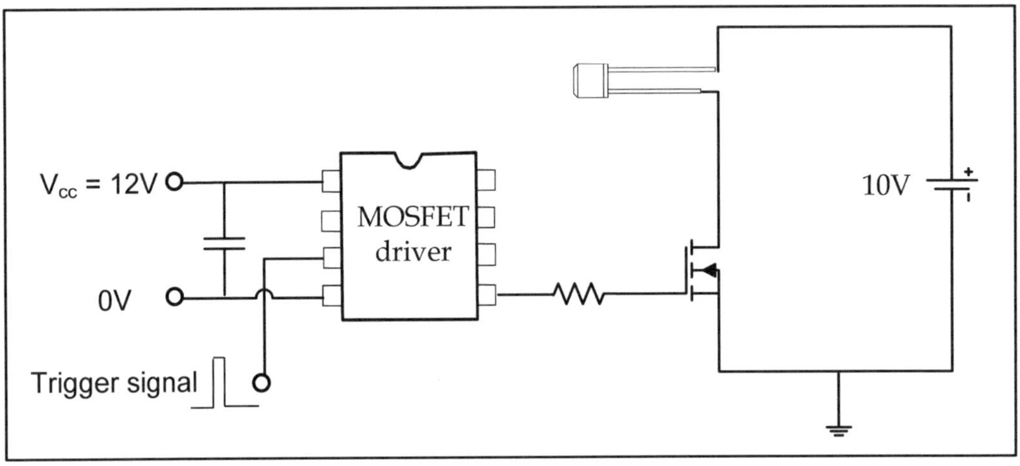

Fig. 10. Schematic diagram of a laser diode driver using a MOSFET as switch

6.4 SPL LL90_3

This system is constructed from a low-cost high power laser diode, Osram SPL LL90_3, with integrated driver stage for pulse control and a maximum peak output power of 80W. The SPL LL90_3 is a hybrid laser module with a lasing wavelength of 905nm. It also contains two capacitors and a MOSFET, which act as a driver stage. The two capacitors are connected in parallel to sum their individual capacitance of 47nF. The capacitors are charged using a constant DC voltage. Each time the gate of the MOSFET is triggered, the capacitors are uncharged via the laser chip leading to a short and high amplitude current pulse. Each individual laser diode is stacked with three emitters with a laser aperture of 200μm × 10μm. The maximum power is achieved when the diode is driven by a pulse width of 40ns. The pulse width can be increased to 1μs or even 10μs with a reduced forward current.

6.5 Driver circuit

In principle the width of the laser pulse is determined by the value of the capacitors inside the plastic package and the pulse width of the MOSFET trigger (gate) signal. The maximum pulse width is 30ns. By increasing the trigger pulse widths beyond 30ns, the FWHM width of the optical pulse and the peak power remain constant but the pulse energy increases. To obtain short optical pulses, the MOSFET gate has to be charged very fast. The MOSFET has a gate capacitance of 300pF. To obtain the required gate-source threshold voltage of 5V the gate must be charged with about 7nA within several nanoseconds. Therefore a pulsed trigger current of about 1A is required. Such a signal can be generated by a high speed power MOSFET driver IC which itself is triggered by a TTL level voltage signal.

Figure 11 shows the block diagram of the SPL LL90_3 together with the MOSFET driver. The MOSFET inside the hybrid package is the Infineon BSP318S. To operate the SPL LL90_3, two DC voltages are needed namely the supply voltage V_{CC} for the MOSFET driver IC and the charge voltage V_C for charging the capacitors. The charging resistor determines the charging current and therefore the time necessary to charge the capacitors i.e. the maximum lasing repetition rate. Problems that can occur are CMOS latch-up, over-voltage spikes, insufficient overdrive and thermal overload. These phenomena can be prevented by using bypassing capacitors and by using clamping schottky diodes and external resistors.

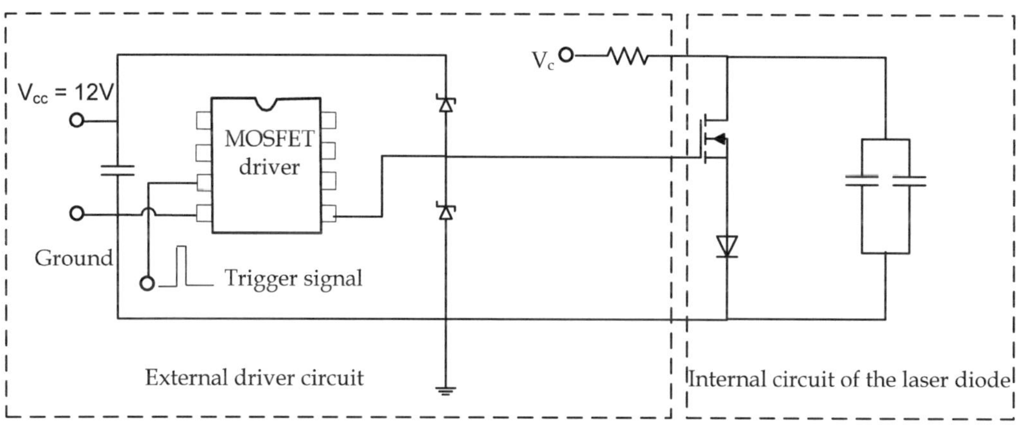

Fig. 11. Schematic of driver circuit of the SPL LL90_3 laser diode

7. Results

Figure 12a and Figure 12b show the images obtained using the low-cost Osram SPL PL90_3 laser diode system (type 1) as an illumination source with an emission wavelength of 905nm. Two diodes were used at a pulse width of 5us and a duty cycle of 0.01%. Diodes are placed 20cm away from the weld pool.

Figure 13a and Figure 13b show the images obtained using the low-cost Osram SPL LL90_3 laser diode system (type 2) as an illumination source with an emission wavelength of 905nm. The arc light is almost totally eliminated by the use of spectral and temporal filtering and the weld pool area is illuminated by only two laser diodes with an approximate peak power of 150W. Using more laser diodes will evenly illuminate the weld pool area and produce better images.

Figure 14a shows an image with arc the arc light totally eliminated. However, the image suffers from uneven distribution of illumination on the weld scene. This can be solved by several methods, e.g. using more diodes to illuminate the dark area, spreading the light across the weld pool, varying the angle at which light is focussed onto the weld pool or by relocating the illumination source. Figure 14b shows an arc without any weld pool illumination. This emphasises the importance of the illumination source. Almost no information can be extracted about the welding process and the image is totally black as the arc light is eliminated by the filtering techniques employed. Figure 14c shows an arc with laser diode illumination. Two SPL PL90_3 diodes are used to obtain this image at a pulse width of 5us.

Figure 15a and Figure 15b show more TIG welding when two SPL PL90_3 diodes are used but with slightly different angle. Figure 16a and Figure 16b show images of TIG welding on stainless steel and mild steel when three laser diodes are used. As can be seen from all the images shown below, weld arc is totally eliminated and the weld pool area is clear and the information that can be extracted for welding process control are in abundance due to the absence of the arc light. The results obtained demonstrate the potential and the effectiveness of the laser diode as an illumination source for monitoring arc welding. The attributes and low cost of the laser diode makes an attractive option compared to the bulky and expensive illumination systems.

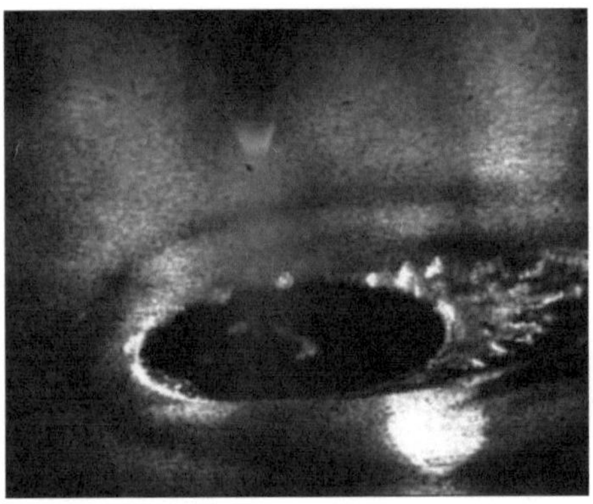

a. 80A, 5 us exposure time, forward reflection

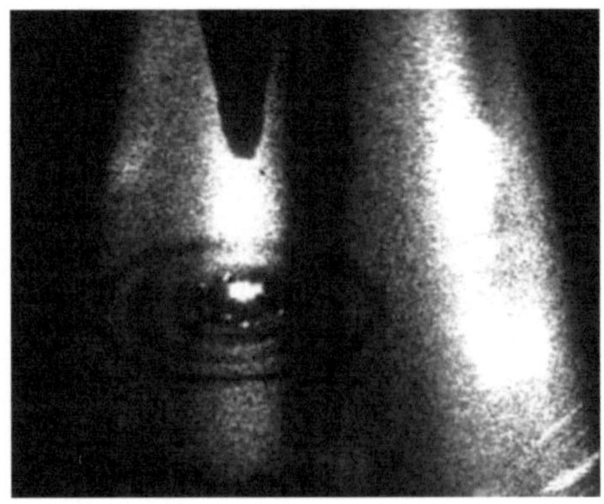

b. 50A, 5 us exposure time, forward reflection

Fig. 12. TIG welding on stainless steel, type 1

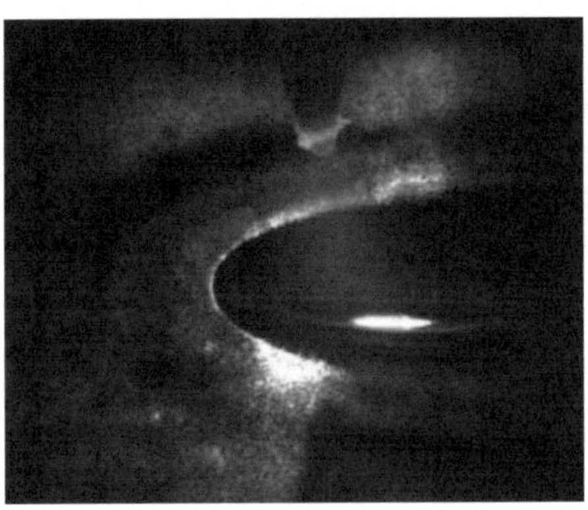

a. 100A, 5 us exposure time, forward reflection

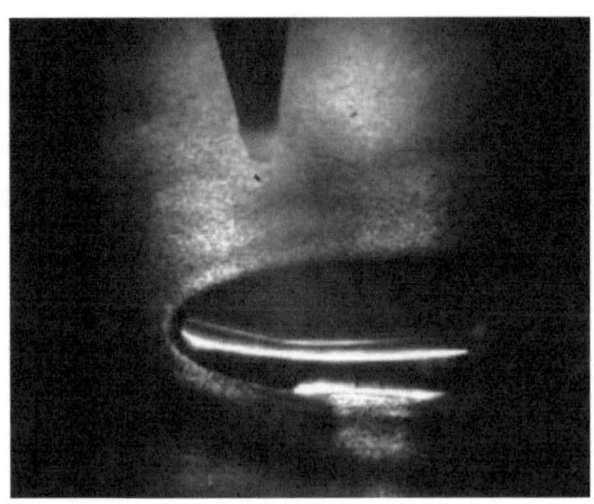

b. 80A, 5 us exposure time, forward reflection

Fig. 13. TIG welding on stainless steel, type 2

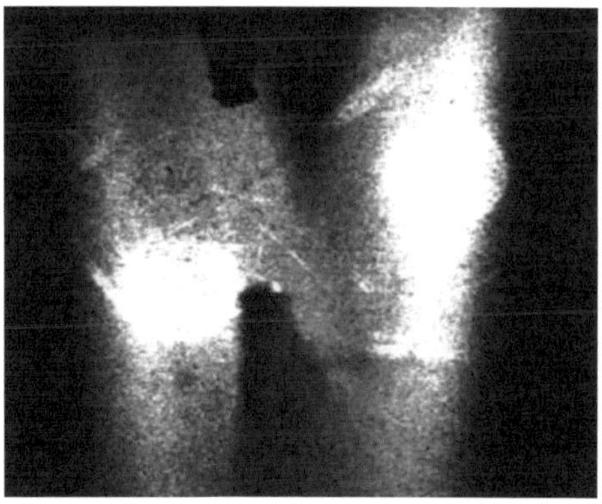

a. Illumination from two laser diodes, 50A

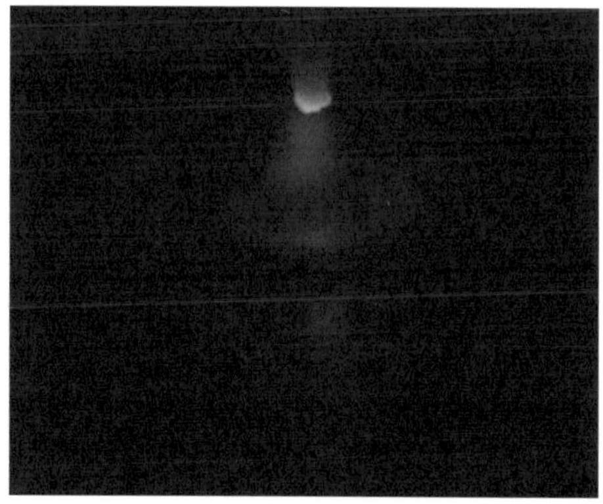

b. Arc light is totally cut off by spectral and temporal filtering, no illumination used, 50A

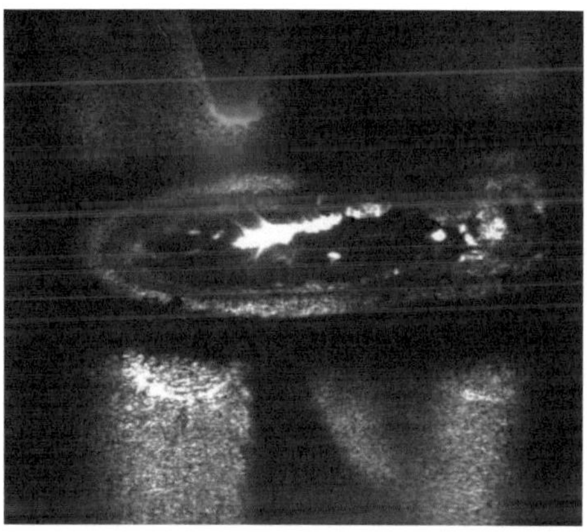

c. Illumination from two laser diodes, 80A

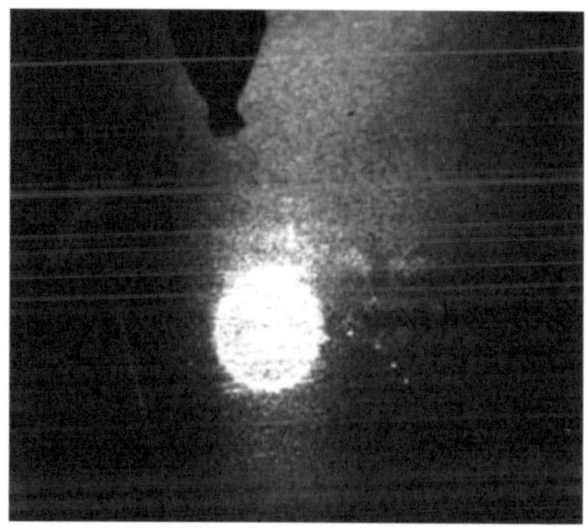

d. Illumination from one laser diode, 50A

Fig. 14. TIG welding, uneven light distribution

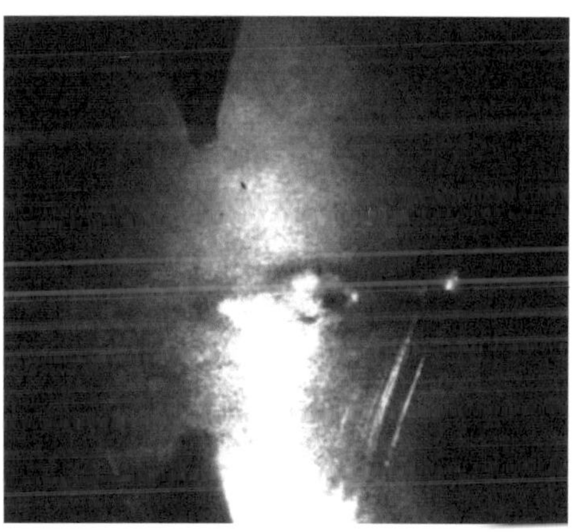

a. 50A

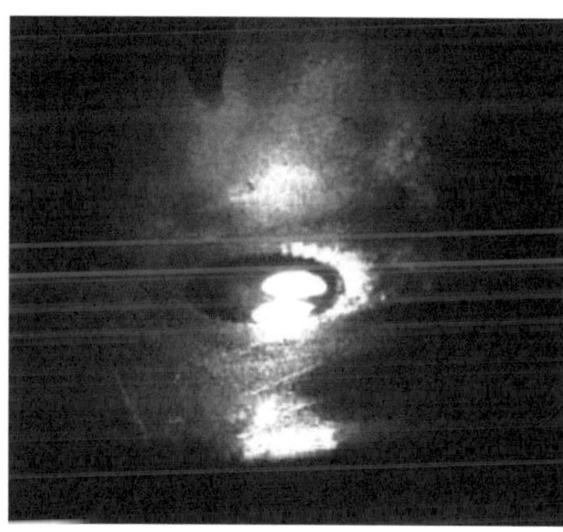

b. 100A

Fig. 15. TIG welding, different angle, type 1

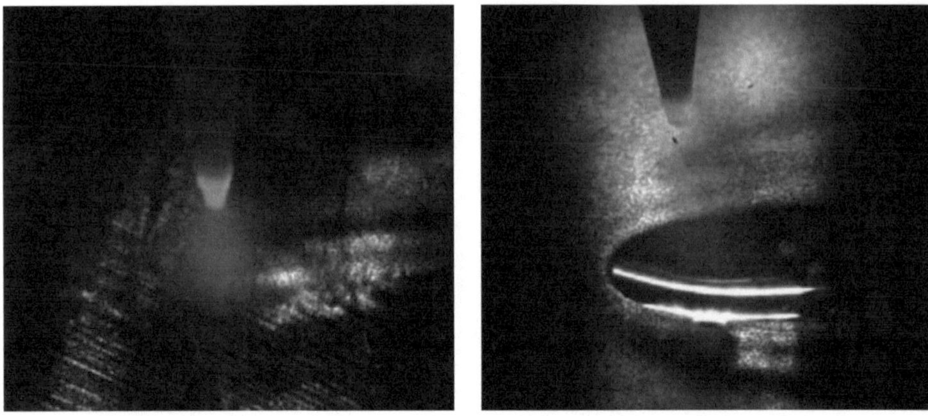

a. TIG welding on mildsteel, 124A b. TIG weldign on stainless steel, 80A

Fig. 16. TIG welding, better angle, better light distribution, 3 laser diodes, type 1

8. Conclusion and future work

In this study a novel low cost vision system has been developed to image the topside of the weld pool. The vision system described has been shown to effectively remove arc light and produce reliable and high quality real-time welding images. The system incorporates a CMOS camera, a laser diode based illumination source, a lens and a narrow band pass filter. The system was tested by conducting some TIG welding trials on stainless steel and mild steel and employing only one, two and three laser diodes at a time. Some of the images obtained are included in this chapter.

Arc light is extremely intense that it is impossible to see through the arc to the weld pool, so for effective viewing the arc light must be reduced or even eliminated. The selection of the specific wavelength at which the illumination source and the band pass filter will be based on is a crucial aspect for blocking the arc light and providing a high quality weld pool image. It was found that arc light radiations are dependent on the shielding gas and work piece material as well as the current levels. TIG welding experimental results obtained show that the spectral distribution of arc light is at a minimum level in the infrared region. Hence, operating at the infrared region seems to be the ideal case so that a low power illumination source can be used. However, operating at this region would require an infrared camera and an infrared illumination source and this option would be commercially unattractive. Therefore, an optimum operating wavelength has been chosen based on the welding spectra obtained and taking into consideration the limitations of both the camera and the illumination source. This wavelength or spectral window where the camera will have the least disturbance from welding arc emissions was found to be between around 800nm - 950nm. Improved images with relatively low power illumination source were obtained at this spectral window.

The illumination achieved by using laser diodes has proved very successful. This is the first time that laser diodes are used as an illumination source for weld pool monitoring. Experiments were carried out with just a few diodes to produce sufficient power to illuminate the weld pool. More diodes and better beam delivery to the weld scene would increase the quality of the images dramatically. For example, the use of 10 laser diodes would produce a peak power output in the range of 750W - 900W at a low cost of under £150. This is a major achievement and would surely revolutionise the illumination industry since its output power is comparable to that of laser illumination but it is much more affordable and economical as a commercial system. The cost of a laser system is in the range of £10K and above, but the cost of this laser diode system is very small (less than £100). More work is required on the delivery of the power onto the weld area. Beam delivery could either be fibre coupling or focussing optics. This would be required when a large number of laser diodes are used. However, a system such as the one developed in this research project which uses only a few laser diodes does not suffer from power loss due to the short distance from the illumination source to the work piece. The developed vision system is compact and light and can therefore be situated near the work piece without the need for a beam delivery method.

In general, laser diodes are a lot more powerful than LEDs. Laser diodes can be driven with more current and hence an increased output power. For example the Osram SPL PL90_3 can be pulsed with a maximum pulse current of 40A, which produces a peak power of up to 90W. Whereas, LEDs can only be pulsed at currents much less than 40A and can only produce a few watts. Laser diodes can be pulsed at very short pulse widths typically around 40ns - 100ns. The pulse width can be increased to around 10us as long as the duty cycle is kept low and lower currents are used. Meanwhile, LEDs have a much longer pulse width, typically around 100us. Since laser diodes can be pulsed at shorter pulse widths, they can be overdriven to obtain more power as long as the duty cycle is not exceeded. Although high power, super-radiant and high flux LEDs are widely available, the power output is still low for weld pool monitoring. It can be concluded that due to several factors such as, emission pattern, spectral width and pulse width limitations, it is unlikely that LEDs will be the illumination source for such applications in the near future. LEDs have wide spectral width, wide emission pattern and inadequate pulsing capabilities as well as low power outputs compared to laser diodes. Therefore, the low cost laser diode illumination system developed is an extremely efficient and economic alternative illumination system to the bulky and expensive laser systems usually adopted.

9. Acknowledgment

This project is supported by the Deanship of Scientific Research of King Fahd University of Petroleum and Minerals under project number JF090016. Part of this work was conducted at the Electrical Engineering Department, University of Liverpool, UK.

10. References

Allende, G.; Mirapeix, J.; Cobdo, A.; Conde, O. M. & Lopez-Higuera, J. M. (2008). Arc welding quality monitoring by means of near infrared imaging spectroscopy. *Society of Photo-Optical Instrumentation Engineers*, ISBN 978-0-8194-7130-7, 18-20, Orlando, Florida, USA, March 2008

Ancona, A.; Lugara, P. M.; Ottonelli, F. & Catalano, I. M. (2004). A sensing torch for on-line monitoring of gas tungsten arc welding processes of steel pipes. *IOP Publishing, Mea. Sci. Technol.*, Vol. 15, pp 2412-2418

Anderson, P. C. J. (1997). A review of sensor systems for the top face control of weld penetration. *The TWI Journal*, Vol. 6, No 4, pp 654-697

Arata, Y.; Inoue, K.; Fuatama M. & Oh, T. (1980). Investigation of Welding Arc Sound. *Transactions of JWRI*, Vol. 9, pp 25-30

Bae, K. Y.; Lee, T. H. & Ahn, K. C. (2002). An optical sensing system for seam tracking and weld pool control in gas metal arc welding of steel pipe. *Journal of Materials Processing Technology*, Vol. 120, pp 458-465

Baik, S. H.; Kim, M. S.; Park, S. K.; Chung, C. M.; Kim, C. J. & Kim, K. J. (2000). Process monitoring of laser welding using chromatic filtering of thermal radiation. *IOP Publishing, Meas. Sci. Technology*, Vol. 11, pp 1772-1777

Balfour, C.; Lucas, J. Maqbool, S.; Smith, J. S. & Mcilroy, L. (2000). A Neural Network Model for MIG welding Parameter Prediction. *Proc. 10th Int. Conf. on Computer Technology in Welding*, Copenhagen, Denmark

Boughton, P.; Rider, G. & Smith, C. J. (1978). Feedback Control of Penetration in 1978. *Advances in Welding Processes*, The Welding Institute, Cambridge, England, pp 203-309

Chen, S. B.; Zhang, Y.; Qiu, T. & Lin, T. (2000). Robotic welding systems with vision sensing and self learning neuron control of arc welding processes. *Journal of Intelligent and Robotic Systems*, Vol. 36, pp 191-208

Chin, B A.; Madsen, N. H. & Goodling, J. S. (1983). Infrared Thermography for Sensing the Arc Welding Process. *Welding Journal*, Vol. 62, pp 227-234

Du, J.; Longobardi, J.; Latham, W. P. & Kar, A. (2000). Welding geometry and tensile strength in laser thin sheet metals. *Science and Technology of Welding and Joining*, Vol. 5, No. 5, pp. 304-309, October 2000

Farson, D.; Richardson R. W. & Li, X. (1998). Infrared measurement of base metal temperature in gas tungsten arc welding. *WJ supplement, AWS*

Frazer, I.; Fyffe, L.; Gibson, O. J. & Lucas, W. (2002). Remotely operated underwater thermal cutting processes for decommissioning of large North Sea platforms. *Proc. of OMAE, 21st International Conf. on offshore Mechanics and Artic Engineering*, Norway, June 2002

Futamata, M. (1983). Applications of arc sound for detection of welding process. *Journal of the Japan welding society*, Vol. 1, No. 1, pp 11-15

Hong, L.; kee, L. F. M.; Yu, J. W. J.; Mohanamurthy, P. H.; Devanathan, R.; Xiaoqi, C. & Piu, C. S. (2000). Vision based GTA weld pool sensing and control using neurofuzzy logic. *SIMTech Technical Report (AT/00/011/AMP). Singapore Institute of Manufacturing Technology*

Huang, W. & Kovacevic, R. W. (2009). Feasibility study of using acoustic signals for online monitoring of the depth of weld in the laser welding of high-strength steels. *Proc. IMechE Vol. 223 Part B: J. Engineering Manufacture*

Kovacevic, P.; Zhang, Y. M. & Li, L.(1996). Monitoring of weld joint penetration based on weld pool geometrical appearance. *Welding Journal*, Vol. 75, No. 10, pp 317-329

Li, P. J. & Zhang, Y. M. (2000) Analysis of an Arc Light Mechanism and its Applications in Sensing of the GTAW Process", *Welding Research Supplement*, Research and Development, pp 252-260, September 2000

Luo, H.; Devanathan, R.; Wang, J.; Chen, X. & Sun, Z. (2002). Vision based Neurofuzzy logic control of weld pool geometry. *Sci. and Tech. of Welding and Joining*, Vol. 7, No. 5, pp321-325

Madigan, R. B.; Quinn, T. B. & Siewert, T. A. (1989) Sensing Droplet Detachment and Electrode Extension for Control of Gas Metal Arc Welding. *Proceedings of the 2nd International Conference on Trends in Welding Research*, pp 999-1002, Gatlinburg, Tenn, 1989

Miller, M.; Mi, B; Kita, A. & Ume, I. C. (2002). Development of automated real-time data acquisition system for robotic weld quality monitoring. *Mechatronics*, Vol. 12, pp 1259-1269

Pal, K.; Bhattacharya, S. & Pal, S. (2010). Investigation on arc sound and metal transfer modes for on-line monitoring in pulsed gas metal arc welding. *Journal of Materials Processing Technology*, Volume 210, Issue 10, Pages 1397-1410.

Richardson, R. W. & Edwards, F. S. (1995) Controlling GT arc length from arc light emissions. *Proceedings of the 4th Trends in Welding Research*, Gatlinburg, Tenn, Vol. 6, pp 715-720, 1995.

Richardson, R. W.; Gutow, D. A. & Rao, S. H. (1982). A Vision Based System for Arc Weld Pool Size Control. *Measurement and Control for Batch Manufacturing, The American Society of Mechanical Engineers*, New York, pp 65-75

Sforza, P. & de Blasiis, D. (2002). On-line optical monitoring system for arc welding", *NDT&E International*, Vol. 35, pp 37-43, 2002

Sibillano, T.; Ancona, A.; Berardi, V. & Lugarà, P. M. (2009). A Real-Time Spectroscopic Sensor for Monitoring Laser Welding Processes. *Sensors Review*, 9, 3376-3385; doi:10.3390/s90503376, 2009

Song, H. S. & Zhang, Y. M. (2008). Measurement and Analysis of Three-Dimensional Specular Gas Tungsten Arc Weld Pool Surface. *Supplement to The Welding Journal*, pages 85-95, April 2008

Sorensen, C. D. & Eagar, T. W. (1990) Measurement of oscillations in partially penetrated weld pools through spectral analysis. *Journal of Dynamic Systems*, Measurement and Control, Transactions ASME, Vol. 112, No. 9, pp 463, September 1990

Sun, Z.; Chen, Q.; Zhang, W.; Cao, Y. & Liu, P. (2006). A novel visual image sensor for CO2 short circuiting arc welding and its applications in weld reinforcement detection. *IOP Publishing, Mea. Sci. Technol.*, Vol. 17, pp 3212-3220

Węglowski, M. S. (2009). Measurement of Arc Light Spectrum in the MAG Welding Method. *Institute of Welding , Błogosławionego Czesława* 16/18, 44-100 Gliwice , Poland

Wikle, H. C. & Kottingam, S. (2001). Infrared sensing techniques for penetration depth control of the submerged arc welding processes, *Journal of materials processing technology*, Vol. 113, pp 228-233

Wu, C. S.; Gao, J. Q.; Liu, X. F. & Zhao, Y. H. (2003). Vision-based measurement of weld pool geometry in constant-current gas tungsten arc welding. *Proc. Instn. Mech. Engrs, Vol. 217, Part B, J. Engineering Manufacture*

Wu, C. S.; Polte, T. & Rehfeldt, D. (2000). Gas metal arc welding monitoring and quality evaluation using neural network. Science *and Technology of Welding and Joining*, Vol. 5, No. 5, 2000, Vol. 38, pp 131-141.

Zackenhouse, M. & Hrdt, D. E. (1983). Weld Pool Impedance Identification for Size Measurement and Control. *ASME Journal of Dynamic Systems*, Measurement and Control, Vol. 105, pp179-184, 1983

Zhang, P.; Kong, L; Wenzhong, L, Jingjing, C & Zhou, K. (2008). Real-time Monitoring of Laser Welding Based on Multiple Sensors. *Control and Decision Conference*, CCDC, page 1746-1748, China, July 2008

Zhang, Y. M.; Kovacevic, R. W. & Li, L. (1996). Characterisation and real-time measurement of geometrical appearance of the weld pool. *Int. J. Mach. Tools Manufact.*, Vol. 36, No. 7, pp. 799-816

15

The Development of Laser Diode Arrays for Printing Applications

O. P. Kowalski
Intense, Inc.,
USA

1. Introduction

Printing is one of the developed world's oldest industries, currently commanding worldwide annual revenues in excess of $700bn (Global Print, 2009). In many ways it can be regarded as a gauge of the health and development of local economies, as it serves a diverse spread of business sectors as well as local and national governments. Within the printing industry lasers are ubiquitous, from small desktop laser printers to large scale industrial printing presses. This chapter will provide a brief summary of the history of printing technology, concentrating on the development and implementation of lasers to accommodate specific needs within the printing sector.

There is a widespread and continually expanding use of laser diodes, and, in particular, laser diode arrays in printing applications. Laser diode arrays are enabling revolutionary developments in a range of printing applications which have generally been driven by the requirements to improve printing speed, reduce cost and increase quality. As a result of the growing demand from commercial printing suppliers, arrays of laser diodes have been developed for all areas of the printing market.

The printing market is expected to undergo continued growth in coming years, despite the global economic slowdown and growing competition from emerging multi-media technologies. This growth will largely be met by increasing demands from the emerging markets of developing nations in the Far-East and South America, and will provide a significant opportunity for printing press manufacturers. In particular, there will be a growing range of applications that can be successfully addressed by laser diode arrays, which provide a low-cost yet highly efficient means to improve print speeds and print resolution.

This chapter describes the development of novel laser diode arrays designed to meet specific requirements for state-of-the-art printing presses. The bulk of this work has been carried out by Intense, Inc., the world's leading supplier of laser diode array print products. Intense utilises a proprietary quantum well intermixing (QWI) process, which provides a means for post-growth band gap engineering. This can be used to develop a range of novel optoelectronic devices, including photonic integrated circuits. However, the company's main focus is the development of high power single and multimode lasers for a range of markets, including printing, defence, medical and industrial applications. Through the use of QWI and the development of high yielding wafer fabrication and assembly processes, the company has pioneered the development of a range of laser diode array products that exhibit market-leading performance and reliability. These arrays have been designed and manufactured to meet the demands of a range of diverse applications within the printing market. This chapter will address each of these applications, discussing the market needs and the solutions that have been engineered to fulfil these requirements.

The future generation of laser diode arrays will be enabled by the development of novel device designs and processing technologies which yield significant improvements over existing technologies. The use of one such technology, Quantum well intermixing (QWI) is discussed in Section 2. Section 3 will discuss the development of laser diode arrays in high quality commercial print applications with a print resolution of 3600 dots per inch (dpi). The emergence of new digital printing presses in variable printing applications with resolutions of 1000-2500 dpi will be discussed in section 4, in particular describing the development of ultra-fine pitch laser arrays for next generation printing presses. Section 5 will discuss the application of laser arrays in forming large size print-heads for a range of coding and marking applications, where lasers will provide improvements in cost, quality and versatility compared to incumbent technologies, such as inkjet, thermal and gas laser printers.

2. Quantum well intermixing in printing applications

Many of the modern printing applications in which lasers are deployed require a laser device with high output power, good beam quality, high reliability and a long lifetime. Achieving the output powers required in certain applications is not a trivial task, due to the occurrence of catastrophic optical damage (COD), the key failure mechanism in short wavelength single mode lasers.

COD occurs due to the predominance of surface states at the laser facet. These defects act as non-radiative recombination centres during operation and lead to an increase in temperature in the vicinity of the facet, and a commensurate reduction in band gap at the facet. This then results in additional optical absorption at the facet, and ultimately induces a thermal run-away cycle which culminates in the sudden failure of the facet.

Quantum well intermixing (QWI) provides a simple and inexpensive, yet powerful technique for the prevention of premature failure through COD. It has thus become a key enabling technology for the realization of high power laser devices that are required in many printing applications. The technology enables localized post-growth modulation in the band gap of III-V materials which can be applied to the facet regions of laser devices in order to suppress COD.

QWI generally involves the localized introduction of defects into the semiconductor crystal, followed by a high temperature anneal. The defects enable an increase in the atomic diffusion rate, resulting in enhanced inter-atomic diffusion, and a change in the composition of the quantum well, which is generally manifested as an increase in the quantum well band gap. By controlling the defect density it is thereby possible to control the rate of intermixing.

Spatially selective QWI can be achieved by patterning the surface of the semiconductor wafer with different dielectric caps and subjecting the wafer to a high temperature anneal (600-1000 °C) (Kowalski et al., 1998). During the high temperature anneal, defects are generated within the semiconductor wafer, with the level of defect generation determined by the specific properties of the dielectric capping layer. By employing two different cap layers, one which enhances defect generation rate and one which suppresses defect generation, it is possible to locally increase the band gap in certain regions of the wafer whilst retaining the original band gap in other regions. This band-gap engineering process can be used advantageously to provide a non-absorbing mirror (NAM) function. This is achieved using QWI to locally increase the band gap in the facet region, thereby rendering the facets non-absorbing. In this way it is possible to prevent the thermal runaway process that stems from increased absorption at the facet and ultimately results in premature failure of the diode. This enables high power operation to be sustained with greater robustness.

This is demonstrated in Figure 1, which shows typical LI curves for 8xx nm lasers developed for printing applications. All laser configurations employ standard ridge waveguide geometry, using a double trench design for planarisation purposes. The ridge width is 2 µm, defined using an inductively coupled plasma (ICP) dry etch process, with a very simple device geometry as illustrated in Figure 1(a). Data is shown in Figure 1(b) for lasers with and without NAMs, where lasers with NAMs possess varying degrees of QWI band gap shift. All the lasers underwent an identical premature aging cycle prior to this LI measurement.

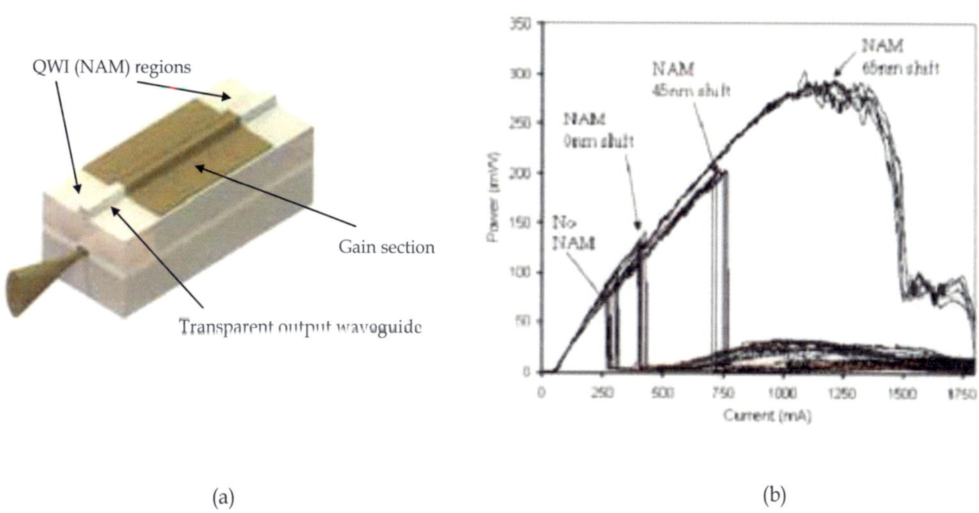

Fig. 1. (a) Schematic 3-D illustration of the chip layout used to produce high power single mode laser diodes using intermixed NAMs to suppress COD, and, (b) L-I curves for 8xxnm ridge waveguide lasers with and without NAMs, where devices with NAMs have undergone a range of QWI band gap shifts.

The laser without a NAM undergoes catastrophic failure at under 100 mW output power. For a laser with a long unpumped section, but without QWI shift in band gap, there is an approximate 50% increase in output power, simply due to the reduction in carrier diffusion from the active to passive sections. Using QWI to induce a 45 nm band gap shift in the NAM regions, a significant further increase in output power is observed, to 200mW, more than double the output power of the standard laser. By further increasing the QWI shift in the

NAM section to 65 nm, the laser does not fail. It instead undergoes thermal rollover, illustrating the significant improvement in high power operation that can be achieved by suppressing band-to-band absorption and carrier diffusion at the facet.

The manufacturing benefits enabled by NAMs are key to enabling the emergence of laser diode arrays for printing, by providing the technical means to achieve the requisite performance levels together with considerable improvement in manufacturability, as further described in section 3.

3. Computer to plate applications

3.1 Introduction

A key development within the printing industry has been provided by computer automation, in particular the development of desktop publishing software. This has led to substantial improvements in the efficiency and throughput of printing systems which has been complemented by the development of increasingly efficient printing presses, in particular the use of modern computer to plate (CtP) printing technology.

CtP printing is the unequivocal standard for high quality, high volume commercial printing applications. It superseded earlier computer to film and phototypesetting techniques, by using laser sources to directly image a photosensitive plate. This enabled the elimination of film preparation steps used in earlier technologies, providing substantial improvements in printing productivity and quality together with reductions in costs.

The printing process itself involves a procedure known as offset printing, in which a printing plate is wrapped around a rotating drum. The plate is first processed to create image areas which are ink-receptive and non-image areas that are ink-repellent. The plate is then coated with ink which adheres only to the pre-defined image areas. The overall image is then transferred via an offset drum with a rubber coating to the print medium, as illustrated in Figure 2. The printing plate is repeatedly coated with ink and used to typically

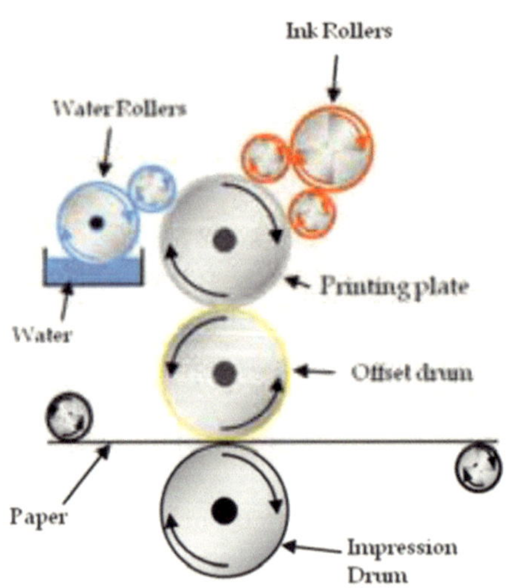

Fig. 2. Illustration of basic working principles employed in offset printing system.

reproduce many thousands of identical impressions. Originally the printing plate would be defined by hand, but in the late 20th century photographic techniques were developed to create film negatives which could be used to expose printing plates, inducing a chemical change in the plate coating, and thereby generate image and non-image areas. More recently, with the advent of computer automation, it became possible to carry out much of the editing and graphic design digitally, and thereby create computer-generated film negatives, in a process known as computer to film (CtF).

3.2 Modern developments

CtP technology has emerged since then, superseding CtF as the current state-of-the-art technique for high quality commercial printing. CtP has eliminated the film preparation stage and the associated plate exposure steps used in CtF. This has significantly improved the efficiency and quality of the printing process, as defects generated within the film are not transferred to the final print product and misalignment errors are eradicated. Instead the printing plate is directly exposed by illuminating its surface with a laser source which is scanned across the surface. In turn, the laser output is modulated by digital data output from the desktop publishing software to create image areas in-line with the finished product.

The first CtP systems employed a low power UV laser beam (300-450 nm) to expose the printing plate, compatible with the chemistry of existing plate coatings. Subsequently, a range of printing plate coatings has been developed to improve printing quality and efficiency, including the development of thermal printing plates.

Thermal CtP plates possess a thermally sensitive coating layer, the properties of which are altered by exposure to a near infra-red laser beam, enabling a laser with appropriate wavelength (800-1050 nm) to directly image the printing plate. This provides some significant advantages in reducing printing times and increasing print quality, however, a higher power laser source (~ 200 mW) is required to adequately expose the plate.

A key advantage of the thermal imaging system is that costs associated with the development of the printing plate can be significantly reduced. For UV systems the development process often requires specific lighting conditions, and a greater dependence on chemical development. With thermal printing, the demands on developing solutions and environment are significantly reduced, leading to more efficient, lower cost printing with reduced ecological impact. In addition, print quality can be significantly improved through the use of thermal systems, as the plate cannot be overexposed - providing it is subjected to a threshold energy exposure, increased exposure does not alter the plate properties. This leads to higher print quality with improved image clarity and sharpness.

A disadvantage of using a thermal imaging laser is the requirement for increased laser power in order to image the printing plate. For this reason it is necessary to use a multiple laser source to image a plate efficiently. The print-head comprising the multiple laser source is scanned across the surface of the printing plate as the plate is rotated in an orthogonal direction, as shown in Figure 3. The output of each laser element within the print-head is independently modulated to selectively expose the printing plate in designated image areas, as dictated by the output of the editing software.

Several approaches have been utilised in order to attain a printing system with multiple imaging sources. These include the use of:

- Discrete laser arrays comprising multiple laser sources, the output of which are combined through optical fibres to a single optical header comprising a bundled fibre-array
- Laser lightvalve technology to generate multiple beams from a single high power multi-mode laser.
- Monolithic single mode laser diode arrays comprising a single chip containing multiple individually addressable laser diode emitters, the output of which can be independently modulated to completely expose the printing plate.

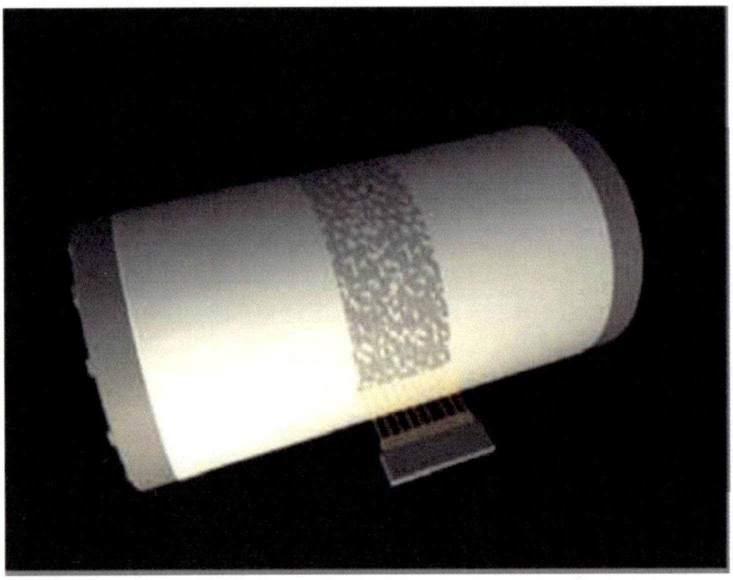

Fig. 3. Illustration of a laser array used to image a CtP printing plate.

Discrete laser arrays were first employed in laser printing systems in the mid 1990s. Generally such systems use a large number (up to 100) of fibre pig-tailed individual laser sources. The output of these are coupled together in a single optical head, in which each output fibre is accurately positioned to form a linear fibre array with a well defined pitch, of the order of 150-200 μm. This requires very accurate alignment of the individual fibres to form the optical head, often involving a v-groove submount. This can prove time-consuming and costly, and produces a relatively large and bulky laser array source, with added system complexity. In addition, the minimum pitch is determined by the fibre diameter, which precludes the use of fine-pitch arrays and limits system specifications.

A more recent development has involved the use of novel technology to modulate the output of a single high power laser source using laser light-valve technology (Tamaki et al. 2004). This involves the use of a high power, multi-mode laser diode source, either a single emitter or bar, the output of which is collimated to create a uniform line of light. This is then spatially modulated through the use of an array of modulators, to produce a multiple-beam output which is then used to image a printing plate. The modulator array tends to use either ferroelectric modulators or grating light-valve technology. This approach provides some advantages over discrete arrays, in particular, the fact that only one laser source is required, reducing costs and inventory control. In addition, the system can be designed to accommodate the failure of individual elements within the laser bar, and the pitch of output beamlets can be precisely controlled in the modulator array manufacturing process. However, a disadvantage is that significant light loss tends to occur due to inefficient coupling to the modulator array. In addition, the use of a multi-mode laser source introduces further complications and costs in terms of the high resolution imaging of the printing plate. The reduced depth of field available when imaging with a multi-mode source, combined with imperfections in the dynamic operation of a printing system means that an auto-focussing system is required to maintain high resolution imaging across the printing plate. Such a system, capable of responding at high speed, leads to considerable increase in overall costs. Utilising single mode laser sources, which provide a much higher depth of field, allows the self-focussing system to be dispensed with. This reduces system complexity, as well as cost.

By employing monolithic arrays of single mode lasers in CtP imaging over discrete laser arrays or systems using modulator arrays, numerous benefits can be realised. These include a reduction in form-factor, reduced inventory, improved print quality and reduced costs. The development of commercially deployed monolithic arrays is discussed in the following section.

3.3 Monolithic laser array development

This section describes the development of commercially available high power monolithic laser arrays for CtP, comprising x64 individually addressable elements, operating over the 800-840 nm (8xx nm) wavelength range. As described above, such arrays have distinct advantages over previously developed laser array solutions.

As these arrays do not provide any scope for failure during operation, their commercial development requires a very high yielding manufacturing process, with unheard of single emitter yields. The array yield itself is determined by the single emitter yield raised to the power N, where N is the number of array elements. Thus for an array comprising 64 elements, a single emitter yield greater than 99 % would be required to achieve an array yield in excess of 50%.

Such yields have been made possible through the use of QWI technology which provides an inexpensive and high-yielding process for the manufacture of laser diodes with NAM facet passivation. NAMs have proven critical for the manufacture of high power laser diode arrays for print applications. This is not only due to the improved power that can be attained through COD prevention but also due to the improved manufacturing tolerance that can be realized through the use of NAMs in laser diode array fabrication. The incorporation of a relatively long passive NAM section reduces the tolerance to processing pattern misalignment errors. During bar cleaving, such misalignment can result in a variation in the distance between the edge of the active region and the facet, which can lead to variable parametric performance across the bar. The NAM also provides a reduced tolerance to placement errors in die-bonding of the laser array. As the NAM region is not pumped and therefore remains cool during operation, it does not require heatsinking. This enables the packaging process to be simplified, as it is possible to deliberately overhang the NAM of a diode array with respect to the heatsink on which it is bonded. This allows for adequate heat-sinking, whilst simultaneously facilitating optimum optical access to the laser facet. This is required for optical alignment, for example to collimating optics, and also eliminates bonding issues such as solder overspill at the facets (Yanson et al., 2007).

In addition to providing high output power in a single mode, it is vitally important that the laser array provides a highly uniform spot size for each array element, in order to uniformly image the printing plate with high reproducibility. To achieve this, custom designed optical components are required, including micro-optic arrays. However it is also necessary that the laser output is consistent in terms of both vertical and horizontal far-field distributions. In order to control far-field, a novel epitaxial layer design has been developed which incorporates a 'V'-profile farfield reduction waveguide layer (Najda et al., 2005), and has been shown to reduce the vertical far-field divergence and suppress higher order mode oscillation with no reduction in the optical overlap factor. In addition, this epitaxial design has been shown to provide significantly improved tolerance to variations in epitaxial layer thickness and composition, plus variations in laser processing, e.g. etch depth variation.

Despite the advantages that are provided by QWI technology and customised epitaxial design, the achievement of the extraordinary yields required to realise commercially viable x64 array chips remains a substantial technical challenge. This necessitates a high level of attention to detail in manufacturing processes and concerted yield improvement activities. In particular, sources of defectivity in wafer growth, processing and assembly must be extremely low. By implementing such continuous process improvement activity along with a rigorous defect reduction philosophy, the levels of uniformity and performance required for a CtP laser array have been demonstrated, along with exceptionally high yield.

Figure 4 shows output power as a function of emitter number for a x64 element array, driven at 250mA CW, illustrating that there is little variation in output power across the array.

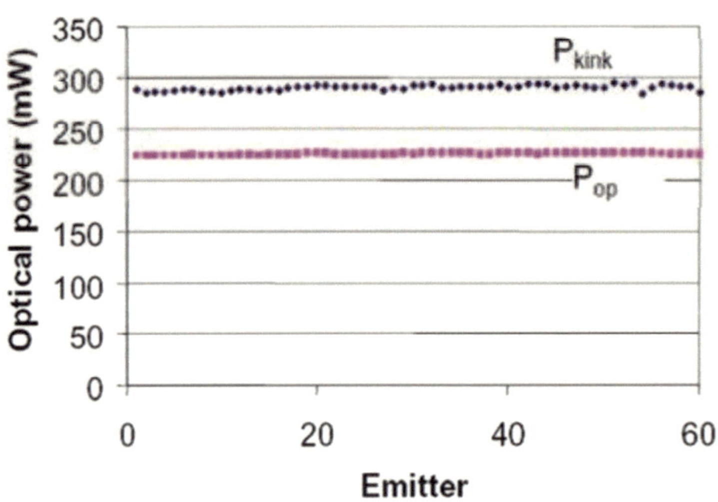

Fig. 4. Variation in output power at fixed drive current, and variation in kink power across a x64 array operating at 8xx nm.

Figure 5 illustrates the variation in far-field for both fast and slow divergence axes as a function of emitter position within the array.

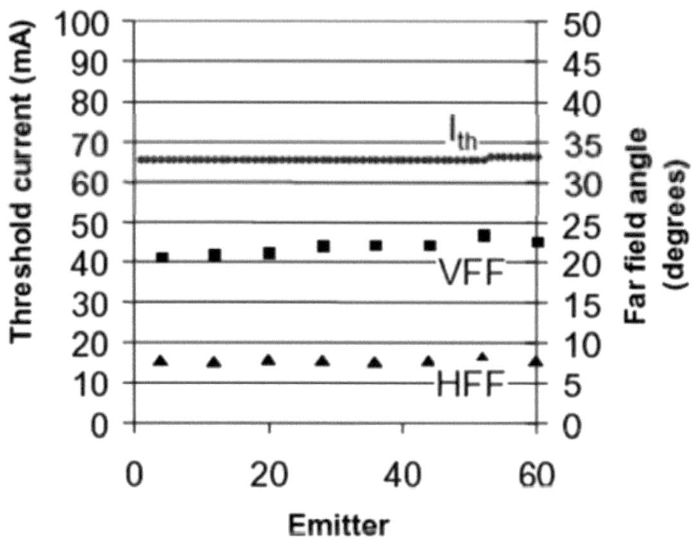

Fig. 5. Typical array uniformity in threshold current and far-field divergence (FWHM) for x64 monolithic 8xx nm laser diode array.

Although the laser array is a key part of the overall CtP system that determines the system's quality, speed and cost, it is only one component among many. The array chip itself must be integrated with the overall control system in order to effectively image print plates. The array chip in this case is integrated into a compact module that includes drive electronics and imaging optics. This is delivered in a compact and low cost platform using an ASIC to independently control the drive current to each array emitter and perform any necessary signal processing tasks. The output of the array is conditioned using custom optical components that control the image spot size on the printing plate. A single cylindrical lens optic is used to collimate the array output in the fast axis direction, while a micro-optic array having the same pitch as the laser array is used to control the output of each individual emitter in the slow divergence axis.

The packaging demands for such an array product are also extremely challenging, given the large number of arrays and high output powers required. As the laser performance is highly sensitive to any changes in the thermal environment, the packaging process must minimise local variations in heat generation and dissipation, e.g. hotspots, which would alter the array's uniformity and ultimately affect its lifetime. This requires careful packaging design in terms of the heatsink and solder materials (finishing, composition) and thorough control over the assembly process in order to ensure adequate heatsinking of the chip along with low electrical, optical and thermal crosstalk between individual array elements.

In addition to obtaining high output power in a single mode with well controlled beam divergence parameters, and high uniformity across the array, it is critical that any array deployed in CtP systems is robust and can withstand many thousands of hours of continuous operation. There is no scope for failure of any of the array channels during the lifetime of the array module, and extension of the lifetime is vital in order that monolithic arrays can compete with alternative technology. Each emitter in the array must perform within the tight system specifications throughout the long operational lifetime of the product.

Through the use of QWI, in addition to well controlled facet coating, the use of tailored epitaxial designs and tightly controlled manufacturing processes, excellent lifetime performance has been obtained for array devices operating at 8xx nm. Figure 6 shows monitored lifetest of fully-assembled array modules with each emitter operating at 200mW in a single mode over a 22,000 hour period. The data shows no catastrophic failure for any of the array channels and minimal degradation over the monitored lifetest period. Together with similar data accrued for many array modules, totalling more than 30 million device hours, the failure in time (FIT) rate is 118 (10^9 device hours), with a resultant mean-time to failure (MTTF) for an array module of > 17,000 hours. This represents an unheard of reliability for such a monolithic array, and illustrates the growing attraction for integrating monolithic arrays into next generation CtP systems.

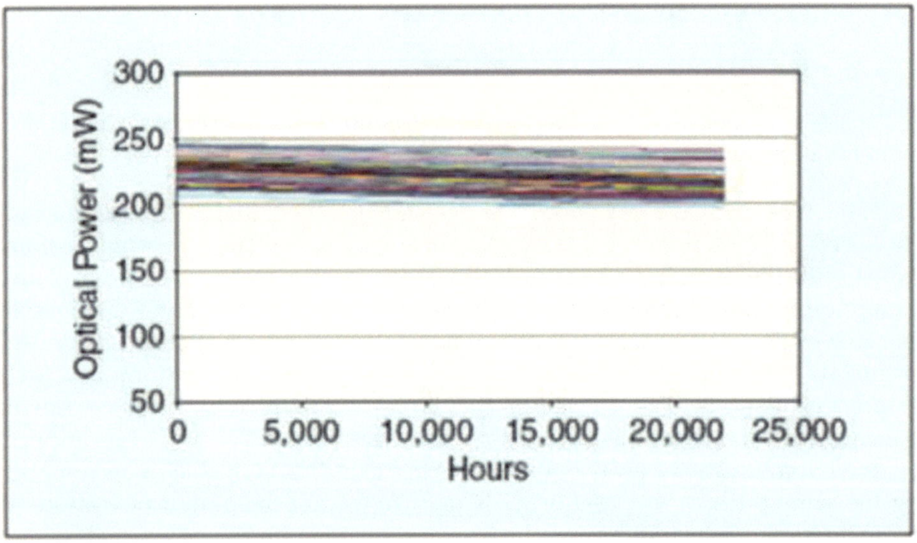

Fig. 6. Monitored lifetest data for a x64 element array operating at 8xx nm. The array module exhibits highly robust performance over 20, 00 hour CW operation.

While these devices are deployed in current CtP printing systems, the need for higher power to enhance printing speed will drive the development of next generation devices in which the output power is likely to be increased to 300 mW and beyond.

4. Digital printing applications

4.1 Introduction

Digital printing is a rapidly growing sector of the commercial print market, which until recently commanded only a small portion (~ 5-10%) of the total output of the global printing market. However recent rapid advances in the quality of digital printing presses have led to significant improvements in the speed, quality and operating costs. This means that modern digital printing presses are competitive with more traditional offset printing methods, at least for small to medium sized print jobs.

Due to the ongoing technical improvements, trends towards greater print-on-demand and variable printing, and increased ecological concerns (and associated legislation), it is anticipated that the flexibility offered by digital printing presses will lead to considerable growth within this area. Market analysts have predicted that the digital printing market will continue to grow over the next 5-10 years, by a 24% compound annual growth rate. The market accorded to digital printing might therefore be expected to reach ~30-50% market share by 2020 (Fleming, 2003).

The printing process utilised in digital printers is that of electrophotography, which is used in virtually all laser desktop printers. Figure 7 illustrates the key components used in electrophotographic printing systems.

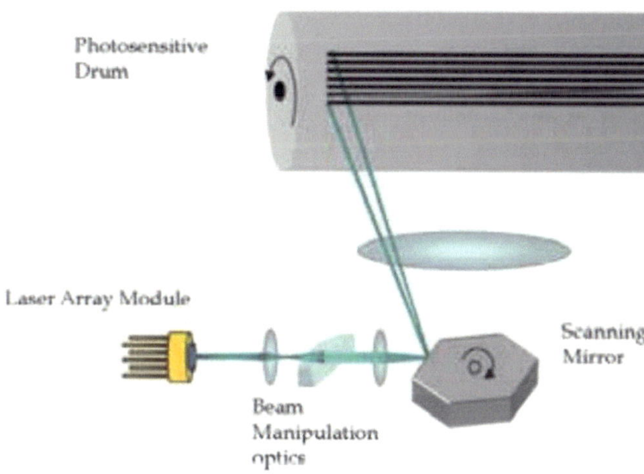

Fig. 7. Schematic illustration of the key elements of an electrophotographic printing press.

The key component in such systems is the photoreceptor, a revolving drum which is uniformly coated with a layer of charge as it rotates, via a corona wire. The charged drum is then irradiated with a laser (or LED) beam which is scanned across the drum as it rotates using a rotating polygonal mirror. The laser output is modulated so that certain areas are selectively exposed to the beam, as dictated by the data supplied to the laser drive circuit via the raster image processor. These areas are discharged, whereas areas on the drum that are not exposed remain charged, creating an electrostatic latent image corresponding to the data supplied from the editing software.

The rotating drum is then coated with positively charged toner, which adheres only to the discharged surfaces of the drum. As the drum rotates, a sheet of paper is brought into contact with it and a large negative charge is applied to the paper to transfer the charged toner particles to the paper. For colour prints, generally a composite image is formed from a combination of images produced in cyan, magenta, yellow and black. The photoreceptor must be imaged once for each of the four component colours.

Photoreceptor coatings are generally sensitive in the wavelength range from the visible to the short IR wavelengths, generally covering the spectral band 500-900 nm and therefore laser sources with this output range are used. There is an advantage in using shorter wavelength sources as the focussed spot size is smaller, allowing higher print resolutions to be achieved. However currently, cost-effective sources are only available from the red to the infra-red part of this wavelength band. The output powers required to successfully discharge a photoreceptor vary dependent on the type of photosensitive coating used, but is typically under 10 mW.

Key drivers in the development of modern digital printing presses are the desire for increased printing speed and increased printing quality.

The printing speed can be enhanced by scanning several independent laser sources across the drum simultaneously. In this way several lines of data can be written simultaneously, thereby reducing the time it takes to fully expose the drum and complete a given print job. Rather than utilising multiple discrete lasers with the complexity and cost that this would incur, the preferred approach is to employ a monolithic laser array to image the photoreceptor. In order to achieve a high resolution, and thereby improve print quality, the array pitch should be minimised. In conventional Fabry-Perot laser arrays the minimum pitch is controlled by the limitations of wire bonding. For standard wire bonding technology, a ball bond is formed that is roughly three times the wire diameter, setting a minimum wire-bond pad dimension of 45 µm. For conventional double-trench ridge waveguide devices, the minimum trench width is ~ 25 µm. Taking these dimensions into consideration, together with the expected accuracy of bond positioning, the minimum pitch that can generally be achieved using a p-side up individually addressable laser diode array is ~ 60-80 µm. Such laser diode arrays are commercially available and used in modern digital print systems.

The application requires very tight control over a number of device characteristics, including output power, resistance, crosstalk, farfield, polarisation and packaging smile. In order to produce a high quality image, it is necessary that there is minimal variation in any of these parameters across an array. This requires a very strict adherence to design and manufacturing tolerances, not only in the epitaxial growth and device fabrication stages, but also in any packaging processes, to ensure that effects of strain induced during chip bonding and any optics attach steps are minimised.

Despite the commercial deployment of fine pitch (~ 60-80 µm) laser diode arrays in digital printing applications, future generation presses will require improved printing speed and increased resolution. This has spurred the development of ultra-fine pitch laser arrays, (~ 20 µm pitch) requiring the development of novel components to ensure each array element can be independently driven. Such an array pitch is difficult to realize in an individually addressable array and requires the development of novel fabrication and/or laser bonding techniques.

A number of approaches have been conceived to produce an individually addressable ultra-fine pitch laser array, including the use of p-side up approaches, such as air bridge technology to provide individual electrical contacts, and specially designed contact pads. However these have certain disadvantages surrounding their uniformity, overall performance and manufacturability, and are unlikely to provide a highly scalable solution for developing a large element array. Until recently, commercially available array sources have been limited to x4 and x8 arrays emitting at 780 nm.

Recently, an ultra-fine pitch (20 μm) visible laser diode array has been demonstrated that utilises a p-side down flip-chip bonding approach to enable the individual array elements to be driven independently. This array incorporates 22 independent emitters housed within a custom designed butterfly package. Using a p-side down array provides increased scope for scaling the size of the array to a higher number of elements, but also provides additional benefits, in reducing thermal cross-talk between adjacent array channels.

4.2 Submount design and manufacture

The design and manufacture of suitable submounts to enable p-down bonding together with individual addressability has been key to achieving such a fine pitch laser array with many elements. In order for the array to be individually addressable, it is required that the ceramic submounts possess an electrical circuit that enables each individual array element to be bonded to a separate electrical track. This involves patterning the submount to provide independent current injection paths for each laser element, on the same 20 μm pitch as the laser array. A narrow track-and-gap arrangement (of the order of 10 μm) is required for such interconnects. Also, to enable the bonding process, it is required that the electrical tracks on the ceramic submounts are coated with a eutectic solder. The manufacture of such a complex submount with narrow solder/metal tracks (~10 μm) is beyond the current capability of most commercial suppliers. Therefore the realisation of a p-side down ultra-fine pitch array requires the custom design and development of a suitable submount.

This has been achieved using standard photolithography and etch processes on metal layers deposited on a BeO substrate. Design of the custom array submount is illustrated in Figure 8. This shows the conductive tracks with a width of 10 μm, laid out on a 20 μm pitch in the chip bonding region. In this region the electrical interconnects are coated with a AuSn layer to facilitate the flip-chip bonding of the laser to the submount. Away from the die-bond area at the rear of the submount, the conductive tracks fan-out, increasing their effective pitch, then terminate in a staggered array of wire-bond pads with a 70 μm pitch that is compatible with the limitations of the wire bonding process.

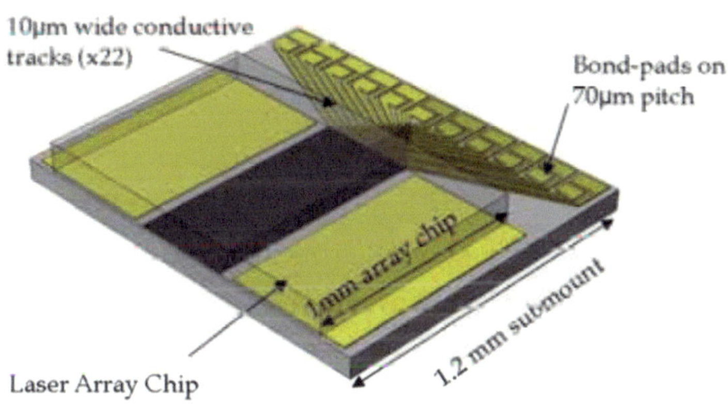

Fig. 8. Illustration of the submount design used to demonstrate p-side down ultra-fine pitch array performance.

The initial step in the submount manufacture involves the blanket deposition of a Ti/Pt/Au layer used to provide the electrical contact. Then, a novel jet vapour deposition™ (Gorski & Halpern, 2003) process is employed to deposit a layer of 4-5 μm thick AuSn solder on the metallised regions. Photolithography combined with ion milling is then used to define the 10 μm wide tracks with 10 μm separation.

4.3 Chip design

Figure 9 is a SEM image illustrating the array chip design. Each individual element is a single-mode ridge waveguide laser formed by dry etching, with a 2 μm ridge width. The array is fabricated in a 3" wafer, employing a standard separate confinement heterostructure (SCH) GaInP-AlGaInP red laser design, operating in the wavelength range 630-680 nm. Between the array elements, deep isolation trenches are etched into the substrate to increase inter-element resistance and thereby reduce electrical crosstalk.

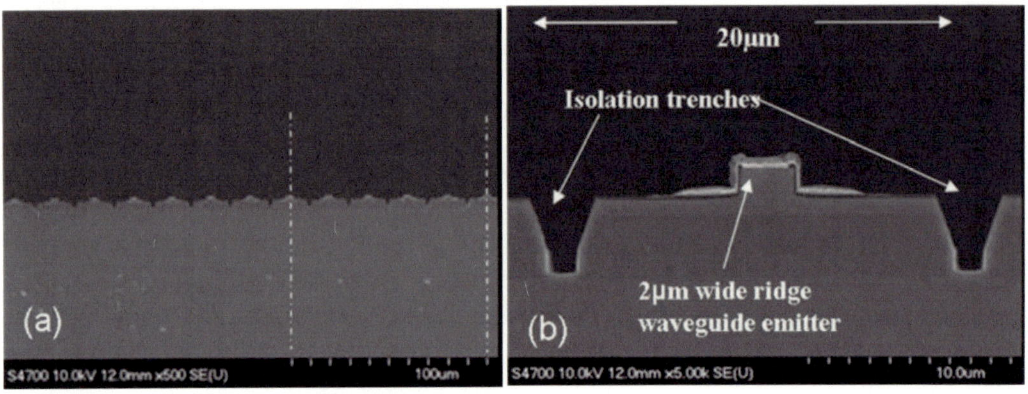

Fig. 9. SEM images of a 20μm pitch red laser array, showing (a) x500 magnification of half the array chip, (b) x5000 magnification of a single emitter within the array.

4.4 Array die bonding

The array chips were die bonded onto the customised submounts, by first carefully aligning each of the ridge waveguides with its associated submount track. The chip was then placed in contact with the submount and the assembly taken through a heating cycle up to ~ 320 °C to reflow the solder and form a eutectic bond. Flip-chip bonding was performed using a Palomar 3500-II die-bond tool which enables precision alignment of the array chip to the submount. An illustration of the chip-carrier alignment prior to bonding is shown in Figure 10.

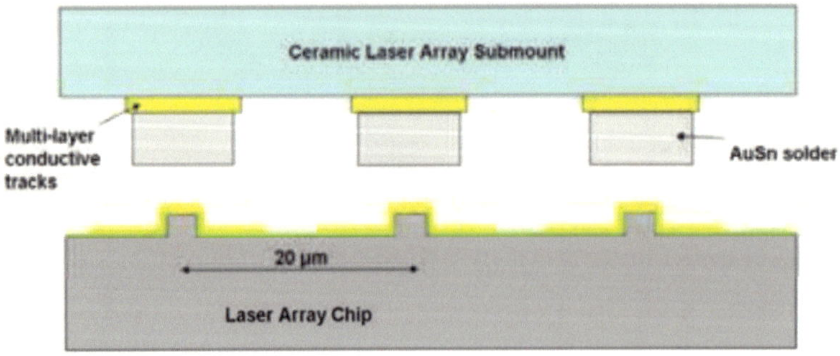

Fig. 10. Schematic illustration of the flip-chip die bonding process used to manufacture a 20 μm monolithic laser array.

The array on submount sub-assemblies were then bonded onto a second larger ceramic (commercially available) specifically designed for integration of p-side up array chips in a butterfly package. These were then integrated into the 26-pin butterfly package with a hermetically sealed sapphire window (Kowalski et al, 2008).

4.5 Array test

Bonded devices were tested CW to determine key characteristics for each emitter, including threshold current, slope efficiency, far-field, resistance, and uniformity across the array. Figure 11 shows LI curves for all 22 elements of a typical array module, illustrating the high level of uniformity in LI performance. The lasers exhibit far-field divergence angles of 8° in the slow (horizontal) axis and 40° in the fast (vertical) axis.

Although it is vital that a single laser emitter conforms to the system specifications in terms of its individual performance, the overall performance of the array is determined by how the different array elements interact with one another - the array crosstalk, which is a primary parameter determining the ultimate image quality. When a number of laser elements are operated simultaneously, their close proximity leads to significant alteration in output power characteristics, e.g. reduction in slope efficiency, due predominantly to the thermal crosstalk between devices. By utilising a p-side down approach described above, the junction temperature for a given emitter is reduced due to the improved heat dissipation. This leads to a significant drop in thermal crosstalk.

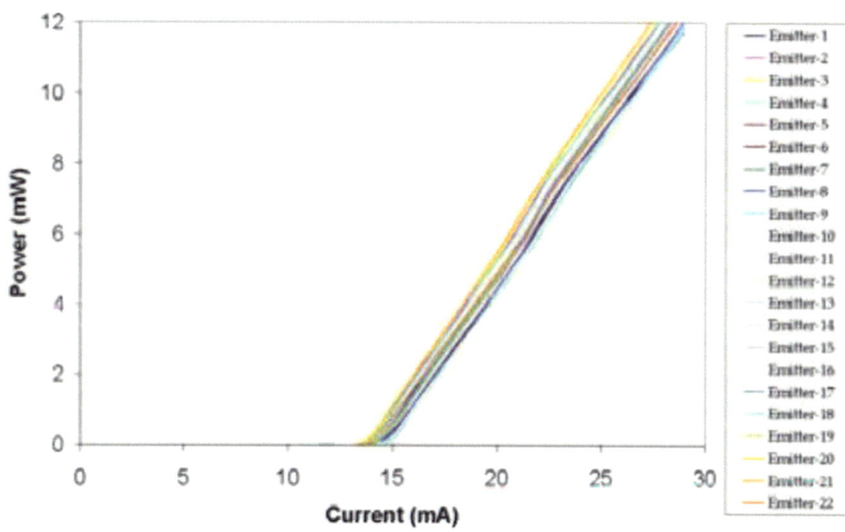

Fig. 11. LIV curves obtained for fully packaged 20 μm pitch x22 element laser array, illustrating high uniformity achieved in output power.

Crosstalk is determined by measuring the time-varying output from the array when all elements are initially powered up to their operational drive current.

Figure 12 shows a typical cross-talk measurement for a fine-pitch red laser array in which the integrated output power over all channels is measured as a function of time following initial power-up. The measured power droop after ~ 1s is 15%, in-line with application requirements.

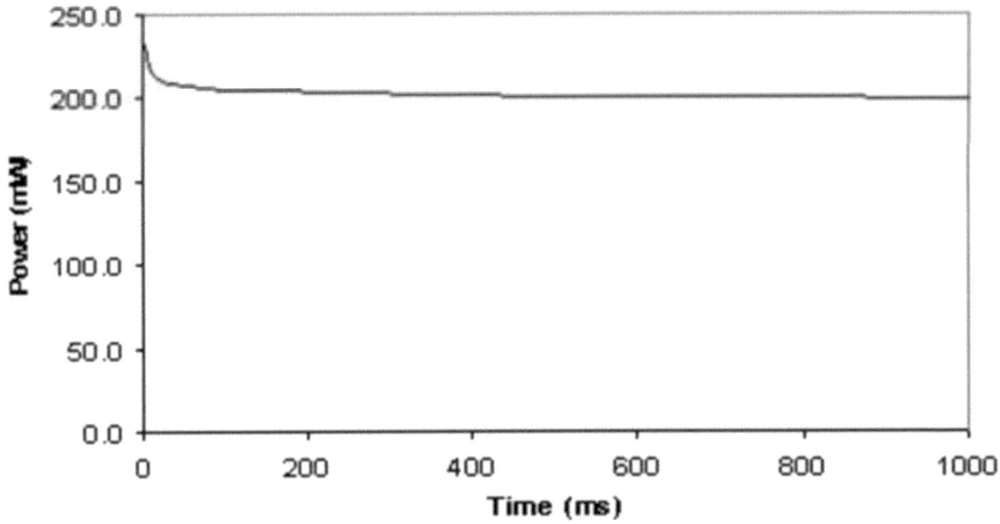

Fig. 12. Measured power-droop across fully packaged array module.

Reliability measurements for these ultra-fine pitch modules are currently limited to burn-in data accrued over 500 hrs operation in application mode. Over this time period no device failures were observed, whilst power degradation was negligible. Wider pitch red laser arrays produced in similar material and operated at higher drive currents (equivalent to greater thermal and optical stress than the finer pitch module) have been placed on monitored lifetest for several 1000 hrs with no sudden failures and a typical degradation of 2%. This illustrates that the product is robust and reliable at the operation levels likely to be required for future generation digital presses.

4.6 Alternative array designs

While the above approach has demonstrated excellent performance, a number of alternative means to produce ultra-fine pitch laser arrays have been proposed and partially demonstrated. An alternative approach to arraying the active laser elements close together can be achieved using ring laser technology, in which a series of ring lasers are arrayed such that their output waveguides are arranged in parallel with close proximity to each other.

The device architecture is illustrated in Fig. 13 (a). This has a number of potential advantages in that the effective pitch size can be further reduced to the order of 5 μm or less. As the output waveguides are passive, the inter-element crosstalk should be significantly reduced. The only notable disadvantage may be the need to significantly increase the size of the chip in order to accommodate the array of active ring elements, and this may ultimately limit the number of array elements that can be integrated into a single fine-pitch array.

Work on such a device has been carried out through the IOLOS program, a sixth framework program funded by the EU to explore various physical phenomena occurring in ring lasers and their applications. Work on ring laser arrays has been limited, but initial attempts to realise a x4 element array have been successful. Fig. 13 (b) shows a microscope image of a x4 ring laser array. Further work is required to improve the design and processing in order to reduce bending losses in the ring elements, which will then be used to determine the performance compared to a similar Fabry-Perot array chip and assess potential for further pitch reduction and scalability.

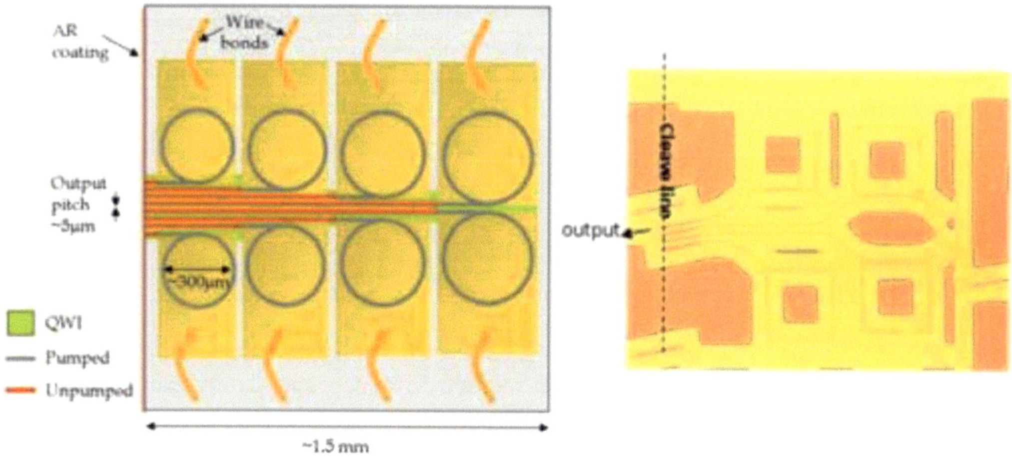

(a) (b)

Fig. 13. (a) Illustration of chip layout for ultra-fine pitch array of ring lasers and (b) microscope image of actual chip, using square rings with etched turning mirrors.

Another approach towards providing laser array solutions for future generation digital print systems is to use VCSEL arrays. Using VCSEL elements it is possible to construct 2-D arrays and thereby enable higher printing speeds to be attained. Such lasers have been demonstrated operating at ~ 780 nm, but although relatively large arrays of 32 elements have been fabricated, their output power is curently limited to 3 mW with array pitch of 30 μm (Mukoyama et al. 2008). Nevertheless, VCSEL arrays do provide another route towards fulfilling the requirements of future generation digital print presses.

5. Coding and marking applications

5.1 Introduction

Coding and marking is a relatively low-resolution application, which generally implements low-technology and low-cost solutions. However, within this sector, laser sources are becoming more desirable compared to existing solutions, e.g. thermal and inkjet printers. The emergence of laser sources has been driven in response to the requirements of customers within this market who require more flexible printing systems, including a print-on-demand capability, together with lower running costs. This is led by a number of factors, including increasingly stringent health and safety regulations, and the need for improved traceability of product, optical identification of goods at the point of sale and counterfeit security measures. Marking systems, by definition, require a high printing speed together with high reliability; whilst, in addition, there is a strong desire to reduce costs through a reduction in downtime and maintenance, and a decrease in consumable expenditure.

Within certain markets there is a recognised need to introduce print-on-demand technology - the ability to instantaneously alter the content of any labelling to reflect changes in product look-up-codes, bar codes, and packing date. In addition, within the food production market, following a number of highly publicised food scares, there is increasing demand for implementing greater detail in product traceability and tracking. A print-on-demand capability provides significant advantages to the producer and retailer in reducing label inventory and consumables cost, whilst also providing improvements in operating efficiency.

At present there are a number of competing technologies within the coding and marking sector, including:

- Inkjet printing
- Thermal transfer
- Scanning laser marking

Each printing solution possesses their own relative advantages and disadvantages relating to the need for rapid printing with high quality and low running costs.

Inkjet solutions are widely available and relatively low cost. However, although there is no direct contact between the print-head and the print medium, there is a tendency for the inkjets to become clogged by dirt and debris, particularly in poorly controlled environments such as packing houses. Such blockages can lead to incomplete printing, which, e.g. in the case of bar-code scanning, is unacceptable. Also, although inkjets are capable of sustaining a high printing rate there are associated restrictions relating to drying time, in order that the ink is not smudged and made illegible. Ink usage is also unsuitable in many production environments, e.g. food packing, due to potential cross-contamination with product. In addition, many inks are solvent based, and can therefore be considered as atmospheric pollutants, whilst spillages can lead to ground water contamination. As a result, there is increasing restriction placed on the ink usage in certain environments.

Thermal transfer systems use a consumable ribbon, which is heated to transfer wax or resin to the substrate. The ribbon itself is relatively expensive and the process requires a finite cooling time, which can significantly reduce print quality at higher printing speeds. However the key disadvantage of thermal printing systems is the direct contact required between the thermal printhead and the marking substrate, which tends to reduce the reliability of the printing components, requiring frequent maintenance together with associated costs and management.

Both solutions rely on heavy use of consumables, which are undesirable on the production line and which require careful inventory and cost control. Neither solution is well-suited to applications requiring non-planar surfaces.

In contrast, laser marking overcomes many of these disadvantages, allowing consumables to be removed from the production environment, simplifying consumable management and providing a high speed solution with low contamination risks. One of the main advantages is the non-contact nature of laser marking, which brings increased printing speed together with improved reliability, as mechanical wear-and-tear is eliminated. Another advantage is that laser marking presses are effectively ink-free; reducing consumable costs and relaxing many environmental and contamination concerns.

Single beam solutions are widely used in laser marking applications, and until recently have utilised a single high power (~10-30 W) laser source, e.g. CO_2 or solid-state laser, which is scanned across the substrate using galvanometer mirrors. The combined system can be cumbersome and expensive and the requirement for scanning a single source tends to limit the ultimate printing speed that can be attained. Also, CO_2 lasers are electrically inefficient, which raises running costs, and in many applications the marking is performed using an ablation approach, which can lead to particulate generation, a particularly undesirable feature in production environments.

A more recent development, which provides a more elegant solution, has again involved the use of laser diode arrays. In this case the laser diode arrays act as a source for direct, single-pass marking, which does not require scanning of the laser source. Instead, by using a laser array, multiple lines of data can be written simultaneously using a marking medium which incorporates a light sensitive coating. This generally utilises pre-existing thermochroic coatings in which a colour change is induced in the print media by the heat generated through absorption of laser energy, usually at infra-red wavelengths (~700-1500 nm). Figure 14 illustrates the basic mechanisms involved in this method of laser marking.

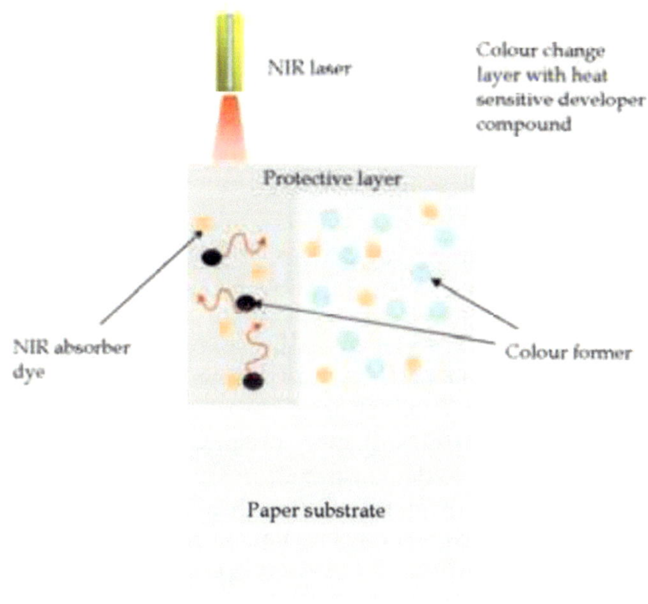

Fig. 14. Thermally driven colour change induced by laser absorption in thermochroic print media.

This provides a high speed, low cost approach, with high print quality, no consumable costs and high reliability, as there is no contact between the print media and the marking laser. This solution is particularly suited to clean environment production lines, as required in food and pharmaceutical industries, where potential contaminants, such as ink, are highly undesirable.

Laser diode arrays can also address emerging trends to develop secure coding and marking capabilities for brand protection and counterfeit prevention measures. In these applications, they can be used to encode data which is concealed in sub-surface layers, for example as a layer embedded within a laminated package. In this way it is possible to add further tamperproof coding, providing additional protection to forgery.

The following section will describe the commercial development of a direct laser marking system which provides a print head with a width of up to 75 mm with 203 dpi resolution. This consists of ~ 300 individually addressable laser elements, on a constant pitch of 125 μm, comprising a number of separate array chips, each containing ~ 100 laser elements.

5.2 Laser diode array solutions in coding and marking

Achieving such a laser diode array cost effectively represents a significant technical challenge. The print width needed for a single pass marking system requires widths of several cm. Print speeds required are of the order of 1.5 m/s, determined by the sensitivity of the print medium and the output fluence of the laser.

In order to provide a laser diode array solution for this market place, it is essential to understand the physical interaction of the laser source with the print medium. This requires accurate knowledge of the physical properties of the print medium so that an appropriate imaging laser can be designed. A good understanding of the absorption coefficient for the imaging laser, the thermal properties of the coating and its thickness is also necessary.

However the key technical challenge is the construction of an array with a pitch of the order of 100 μm which comprises up to several hundred individually addressable elements. This is a considerable challenge to meet with current manufacturing capabilities. Forming a single monolithic laser diode array of several cm width is not practically feasible with currently available wafer processing capabilities. It would require the growth of large diameter III-V substrates together with compatible processing equipment and unrealistic device yields.

Nevertheless, it is possible to manufacture a print-head in which the laser diode array comprises a number of separate monolithic array chips. In this case, each array chip can be fully tested and screened prior to subsequent packaging steps. Using this approach, large (75 mm) print heads have been demonstrated, comprising up to 300 emitters, each producing up to 500 mW output at 980 nm, with an inter-element pitch of 125 um, equivalent to 203 dpi. The product, known as a DLAM (Direct Laser Array Marking®) has been developed by Intense Inc. to meet specific needs within the food labelling market. Similar technology has also been used to realise super-arrays comprising up to 800 elements for a range of print and other applications.

For this product, the benefits of QWI technology are again essential in enabling the requisite output power to be sustained with the necessary levels of reliability. The additional benefits provided by incorporating long NAM sections into the laser cavity, in relaxing processing and packaging tolerances, are also an important contributor to the success of this approach. Again, successful development of a cost-competitive multiple-array product requires an extremely high single chip yield to be sustained throughout the wafer fabrication process. This requires high uniformity and conformity to specifications through all process steps and a rigorous effort to minimise wafer defectivity levels throughout the manufacturing line.

The primary engineering challenge, however, is in constructing a 'super-array' comprising several discrete monolithic array chips, aligned in such a way that the inter-element array pitch is maintained from one array chip to another. This requires extremely accurate alignment between chips before and during the die-bonding step, to ensure that the edge emitters of all adjacent array chips maintain the same pitch as that within each monolithic array, as illustrated in Figure 15.

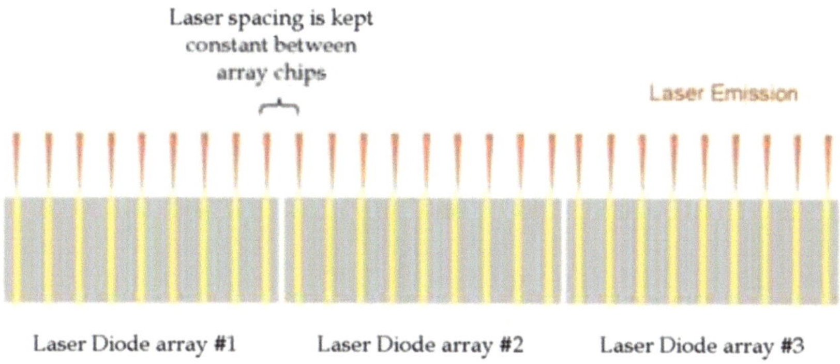

Fig. 15. Schematic layout of super array comprising 3 monolithic array chips.

In order to achieve this, the edge emitters on each array chip are arranged close to the cleaved edge, providing improved tolerance to cleave variations. New high-precision die-bonding techniques have been developed to enable multiple chips to be bonded in close proximity (+/- 3 um) to one another as illustrated in Figure 16. The bonds are formed using hard eutectic solder, without compromising the bond characteristics, enabling optimum thermal performance and reliability to be achieved. A novel chip layout is also required to ensure separate wire bond pads are provided for each individual emitter within the array, including the edge emitters.

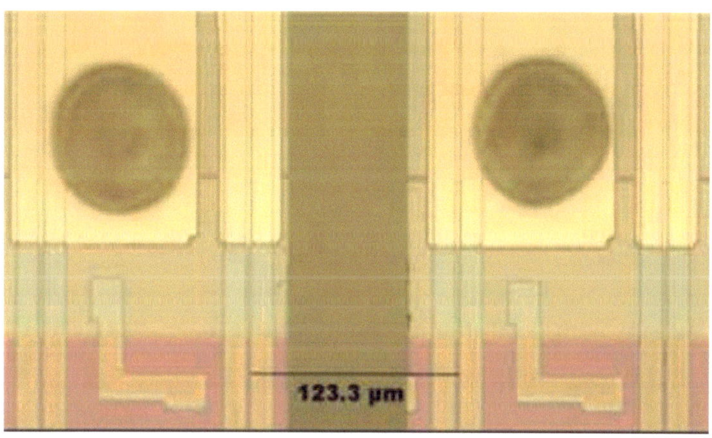

Fig. 16. Microscope image showing the precision alignment between two neighbouring array chips within a DLAM array.

Figure 17 is a photograph image of a laser print-head comprising a number of arrays 'stitched together' to form a 'super-array'.

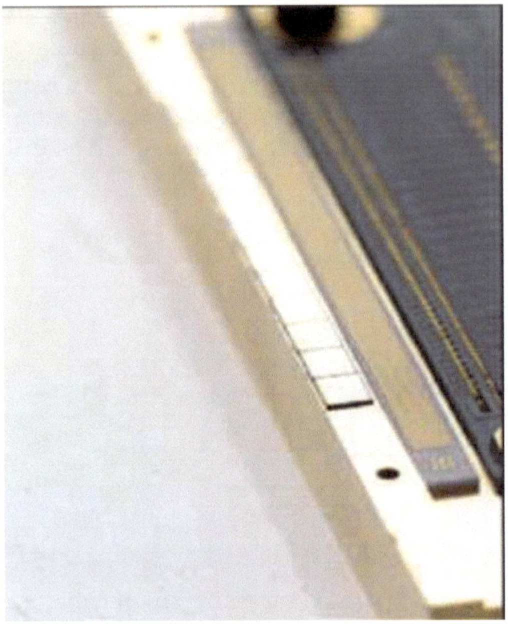

Fig. 17. Photograph of a DLAM 'super-array' comprising a number of monolithic array chips bonded on a common carrier, with the pitch between arrays maintained across the entire print head.

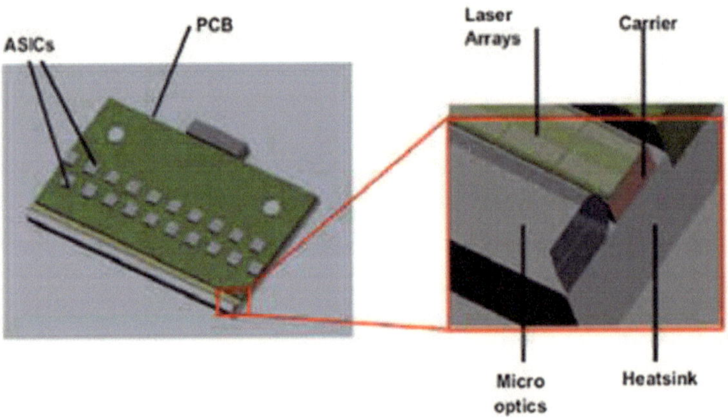

Fig. 18. Illustration of a DLAM module with 'super-array' chip bonded to a carrier and mounted on a heatsink with ASIC and micro-optic array.

In order to minimise packaging induced strain, the laser chips are mounted onto a thermally conductive substrate fabricated from material with a closely matching thermal expansion coefficient. This is integrated into a custom designed submount as shown in Figure 18. This provides a heatsink and integrated ASIC which controls the current supply to each individual array element. In addition, a custom designed micro-lens is incorporated to provide adequate beam manipulation in line with system specifications. The alignment, in particular of the micro-optics to each laser element is critical, placing stringent demands upon the flatness of the substrate and the overall assembled module. This has required extensive process development and refinement to enable highly precise alignment between the system components. Additional imaging optics are incorporated into the package to provide the required spot size characteristics at the imaging plane. The optics themselves are designed to provide a high depth of focus, producing good image quality over a range of relative displacements to the print media. They also impart added flexibility to the overall print-head characteristics, allowing different application requirements to be addressed by tuning the print-width, resolution and power density.

Figure 19 shows the measured L-I curves for each of the 270 elements of an entire DLAM array, illustrating the high uniformity achieved in output power across the array.

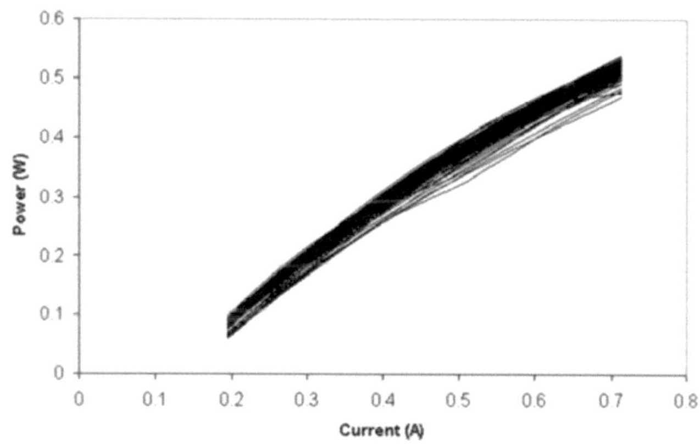

Fig. 19. L-I curves for 270 elements comprising a super-array operating at 980 nm for coding and marking applications.

DLAM array products have undergone successful application trials and are currently deployed in food labelling applications. Given the advantages provided, there is likely to be a growing market for laser array products within coding and marking, however one where the demands on the product are likely to grow in order to increase printing speed and quality whilst reducing overall system cost.

6. Conclusions

Laser diode arrays provide an increasingly attractive solution for a range of technical issues that have to be overcome in existing and future generation printing presses. As a result of the advantages they provide over existing non-laser and single laser solutions, their use within commercial printing is continuing to grow and expand into different areas of the commercial printing market. Laser arrays are now widely used in high resolution CtP applications, and their deployment in highly versatile variable printing applications, e.g. digital printing and coding and marking is becoming more widespread.

QWI has proven an essential tool in the development of high power laser devices for printing applications. It has provided the technological breakthrough that has made large monolithic laser diode arrays a commercially viable product by enabling the required yields and reliability to be realised.

As the demand for laser diode arrays expands, the demands placed on the technical performance and pricing will continue to grow, requiring the development of increasingly inventive and novel diode array solutions. The expected growth will nevertheless provide a substantial business opportunity for laser diode manufacturers and spur competition between suppliers to provide technically superior and cost effective products.

7. Acknowledgement

I would like to thank all those at Intense, past and present, for their enormous hard work, innovation and expertise that has provided the wealth of data, knowledge and experience reflected in this text. Particular mention and gratitude should go to Berthold Schmidt, Steve Gorton and Prof John Marsh for their assistance through the provision of data, insight and inspiration.

8. References

Fleming, M. (2003), "Armed with competitive technology and a more mature sense of the market, suppliers and print providers are turning digital printing into a high-volume business", Digital Output, May 2010, Available from: <http://www.digitaloutput.net/content/contentct.asp?p=101.>

Global Print (2009), Available from: http://www.global-print.org/gpmarket/

Gorski, M. (2003). Halpern, B., "Jet vapor deposition for Au/Sn solder applications" Advanced Packaging. Vol. 12, no. 2, pp. 31-32,34.

Kowalski, O. P. (1998). Hamilton, C. J., McDougall, S. D., Marsh, J. H., Bryce, A. C., De La Rue, R. M., Vögele, B., Stanley, C. R., Button, C. C. & Roberts, J. S. "A universal damage induced technique for quantum well intermixing" Applied Physics Letters vol.72 no.5 1998 pp.581-3

Kowalski, O. P. (2008). McDougall, S. D., Qui, B., Masterton, G. H., Armstrong, M. L., Robertson, S., Caldecott, S. & Marsh, J. H. "Ultra-fine pitch individually addressable visible laser arrays for high speed digital printing applications", Proceedings of the SPIE, vol.7230, pp. 72301J-08, 2009

Mukoyama N. (2008). Otoma, H., Sakurai, J., Ueki, N. & Nakayama, H. "VCSEL array-based light exposure system for laser printing", Proc of SPIE Vol.6908, 6908H-11 (2008)

Najda, S. P. (2005). Bacchin, G., Qiu, B., Smith, C. J. M., Vassalli, O., Toury, M., McDougall, S. D., Hamilton, C. J. & Marsh, J. H. "Very large arrays of individually addressable, high power, single mode laser arrays in the 800-1000nm wavelength range obtained by quantum well intermixing techniques", Proceedings of SPIE, Vol. 5738, pp 33-39

Tamaki E. (2004). Hashimoto, Y. & Leung, O. "Computer-to-plate printing using the Grating Light ValveTM device" Proceedings of SPIE, vol. 5348, 5348-08, p. 89

Yanson, D. (2007). Marsh, J. H., Najda, S. P., McDougall, S. D., Fadli, H., Masterton, G., Qiu, B., Kowalski, O. P., Bacchin, G. & McKinnon, G., (2007) "High-power, high-brightness, high-reliability laser diodes emitting at 800-1000 nm", Proceedings of SPIE Vol.6456, 64560L

16

High-Power Pulsed 2-μm Tm^{3+}-Doped Fiber Laser

Yulong Tang and Jianqiu Xu
Key Laboratory for Laser Plasmas (Ministry of Education) and Department of Physics, Shanghai Jiao Tong University, Shanghai, China

1. Introduction

1.1 Research background

Laser beam in the 2~3μm spectral range has wide applications. It is a good candidate in laser microsurgery due to high absorption of water in this spectral region. It can also be used in environment monitoring, LIDAR, as optical-parametric-oscillation (OPO) pump sources, and so on [1-4].

Tm^{3+}-doped fiber is very suitable for producing ~2-μm laser emission due to its several unique advantages. First, Tm^{3+}-doped fiber has a strong absorption spectrum that has good overlap with the emission spectrum of commercially available AlGaAs laser diodes. Second, the specific energy-level structure of Tm^{3+} ions provides the Tm^{3+}-doped fiber laser (TDFL) quantum efficiency close to two through the cross relaxation (CR) process. Thirdly, the Tm^{3+}-doped fiber has a very broad emission band (~400 nm), offering great wavelength tunability for the fiber laser.

For achieving high-power output, Tm^{3+} ions are often doped into silica glass fiber to construct 2-μm fiber lasers. The silicate glass fiber has maximum phonon energy as large as 1100 cm^{-1} (compared to 550 cm^{-1} for fluorides) [5], which limits its infrared transparency wavelength less than 2.2 μm.

In the past, the fiber laser was usually core pumped. Small fiber core area (<100 μm^2) requires high brightness of the pump beam, greatly limiting the pump power than can be launched into the fiber core. With the advent of double-clad fiber configuration, the pumping area was changed to the cladding area (>10000 μm^2), significantly facilitating pump-power launching. From then on, the output power of fiber lasers has been greatly enhanced.

For fiber laser oscillation, the simplest (commonly adopted) laser resonator is defined as Fabry-Perot resonator, which is shown in Fig. 1. The pump light is launched into the fiber through a dichroic mirror, high transmissive for pump beam but high reflective for laser light. Laser oscillation forms between this dichroic mirror and the distal-end fiber facet (~4% Fresnel reflection).

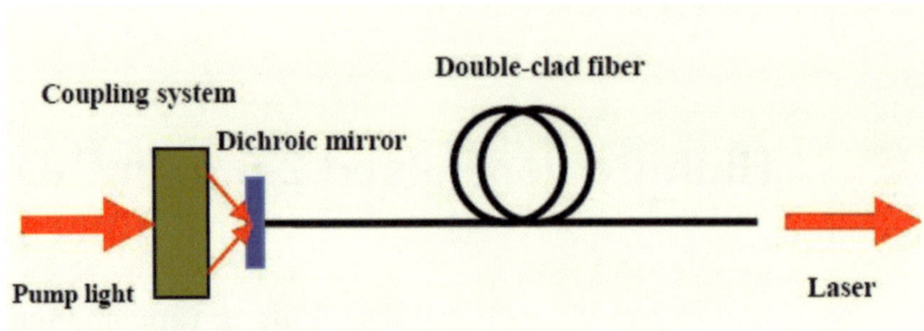

Fig. 1. The Fabry-Perot Tm^{3+}-doped fiber laser resonator.

The simplified energy-level diagram of Tm^{3+} ions is shown in Fig. 2. The pump light at ~790 nm excites Tm^{3+} ions from $^3H_6$ to $^3H_4$, which then nonradiatively decay to the upper laser level of $^3F_4$ with a fluorescence lifetime of 0.55 ms [6]. The transition from $^3F_4 \rightarrow ^3H_6$ will radiate photons at wavelength of ~2 µm. Due to large Stark splitting of the energy levels, the TDFL is a quasi-four-level system.

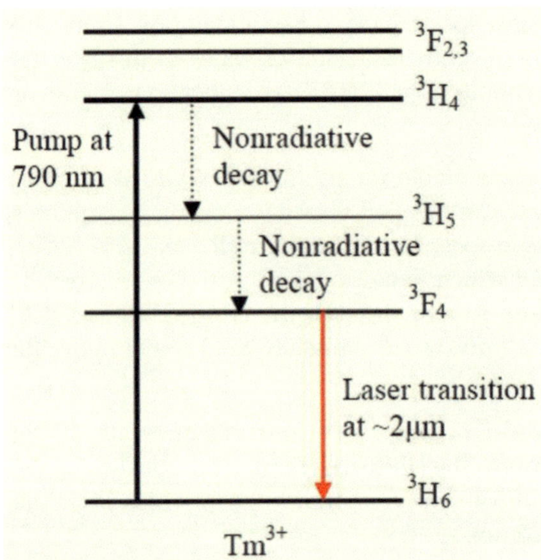

Fig. 2. The simplified energy-level diagram of Tm^{3+} ions.

The Tm^{3+} ion has fluent energy levels, providing complex energy-transfer processes. The **CR** process ($^3H_4+^3H_6 \rightarrow ^3F_4+^3F_4$), as shown in Fig 3, is very beneficial for improving slope efficiency of the TDFL. With the help of this process, high quantum-efficiency TDFLs have been realized [7-8], and the slope efficiency has surpassed 60% [9-10].

For high-power ~2µm TDFLs, 790-nm AlGaAs diode lasers are preferred pump sources. This pump wavelength is the only appropriate wavelength to stimulate the beneficial CR process for TDFLs. Aside from 790 nm, laser sources at other wavelengths can also be used as the pump source, including Nd:YAG lasers (1.064 and 1.319 µm), Yb-doped fiber lasers (1.09 µm), Er^{3+}-doped fiber lasers (1.57µm), and so on.

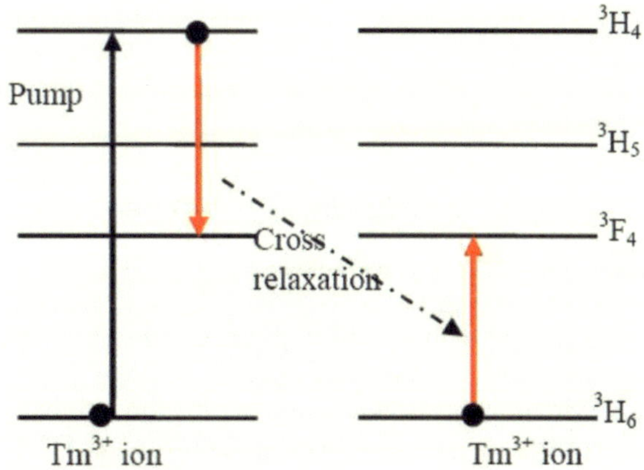

Fig. 3. Cross relaxation between Tm^{3+} ions.

1.2 Recent development of Tm^{3+}-doped fiber laser

The first ~2μm TDFL was achieved by Hanna et al. in 1988 with a 797-nm dye laser as the pump source [11]. With the advent of AlGaAs diode lasers as the pump source, the all-solid-state Tm^{3+}-doped silica [12] and fluoride [13] fiber lasers were both realized in 1990.

High-power TDFLs had not been achieved until the invention of double-clad pumping configuration by Snitzer et al. in 1989. The first high-power TDFL was reported by Jackson in 1998, with maximum output of 5.4 W and slope efficiency of 31% [14]. In 2000, Hanna et al. [15] improved the output power of the TDFL to 14 W with a slope efficiency of 46%. In 2002, the 2-μm TDFL has achieved an output power of 85 W by Jackson et al. [16], the slope efficiency being 56%. At nearly the same time, 75 W 2-μm laser output was also realized in an Yb-Tm codoped fiber laser [17], where Yb3+ ions were used as sensitizers to facilitate pumping with 976-nm diodes. However, the slope efficiency was just 32% due to the lack of the "cross relaxation" process. In 2007, G. Frith reported a TDFL with output power of 263 W and slope efficiency of 59% [18]. In 2009, the 2 μm output power from the TDFL was significantly enhanced to 885W with a slope efficiency of 49.2% [19], and exceeded one kilowatt [20] just one year later. At present, single frequency output has also been over 600 watts [21]. These significant developments in 2-μm TDFLs have made them a much attractive tool in many application areas.

Along with the power scaling of continuous-wave (CW) TDFLs, many works have also been carried out to realize high-peak-power or short-pulse-duration TDFLs. The first 2-μm Q-switched TDFL was carried out in 1993 with an acousto-optic (AO) modulator [22]. The pulse width was 130 ns, and the peak power was only 4 W. In 1995, 2-μm femtosecond (500 fs) laser pulse was first achieved in Tm fiber [23]. By also using AO modulator, a peak power of 4 kW (with pulse width of 150ns) TDFL was realized in 2003 [24], but the average power was just 60 mW. In 2005, by adopting an amplification configuration, the peak power of 2μm Tm fiber laser has reached 230 kW (108 fs) with improved average power to 3.1 W [25]. For achieving high pulse energy from TDFLs, the gain-switching technique is usually employed. In 2000, 2-μm TDFL has achieved 10.1-mJ pulse energy [26], which was further improved to 14.7 mJ in 2005 [27]. These gain-switching TDFLs were often accompanied by relaxation spikes, showing comparatively low system stability. Recently, special measures (such as employing high pump ratio or combined gain-switching and amplification) have been applied to scale the averaged 2-μm pulsed laser power to tens of and even hundred watt levels [28–32].

2. High power acousto-optic Q-switching Tm^{3+}-doped fiber laser

Nowadays, high-average-power 2-μm pulsed laser are eagerly required in pumping 3~4-μm mid-infrared lasers [33], environmental detecting [34], and medical surgery. For 2-μm TDFLs, pulsed operation can be realized through either Q-switching or mode-locking. Q-switching can usually provide a somewhat higher average power, but mode-locking operation can often provide much narrower pulse width thus higher pulse peak power. In this section, we mainly introduce AO Q-switching operation of TDFLs for achieving high-power 2-μm laser pulses in our laboratory.

2.1 Power scalability of pulsed Tm^{3+}-doped fiber laser

In our experiment, the double-clad Tm^{3+} silica fibers have a Tm^{3+} concentration of ~2wt.%. The fiber core has a diameter of ~30-μm and a numerical aperture (NA) of 0.09. The pure-silica inner cladding is D-shape, and has a 400-μm diameter and a NA of 0.46. The absorption coefficient of the fiber at the pump wavelength (~793 nm) is about 3 dB/m.

The experimental setup for the AO Q-switched TDFL is shown in Fig. 4. The pump source was a pig-tailed high-power laser diode (LD) operated at 793 nm with a total output power of ~110 W. The pig-tail fiber has a diameter of 200 μm and an NA of 0.22. Two aspheric lenses were used to couple the pump light into the fiber, with a coupling efficiency of ~90%. A dichroic mirror (T=97%@793nm&R=99.5%@2μm at 45° coating) was employed to transmit the pump light and extract the 2-μm laser beam. A high-reflection mirror (R=99.5%@2000nm) at the far end is used to provide the laser light feedback, which together with the perpendicularly cleaved pump-end fiber facet (~3.55% Fresnel reflection for laser oscillation) complete the laser cavity. The Tm^{3+}-doped gain fiber is wrapped on a copper drum or immerged in water for cooling. At both fiber ends, a short piece of gain fiber (with the polymer coating removed) is placed between a copper heat sink for efficient heat cooling. At the output end of the fiber, another dichroic (T=98%@2μm&R=99.8%@793nm) is used to filter the residual 793-nm pump power. The AO Q switch is inserted between the far fiber end and the high-reflection mirror. The far end fiber is angle-cleaved (~8°) to avoid the parasitic oscillation due to fiber-facet reflection. The laser output power was measured with a thermal power meter and the laser spectrum was tested with a spectrometer.

Figure 5 shows the average output power of the *Q*-switched TDFL with respect to the 793-nm launched pump power at the repetition rate (RR) of 50 kHz. The maximum laser power was ~32 W with a slope efficiency of 36%. At high power level, the AO switch could not switch off completely, leading to decreased signal-to-noise ratios. This is an important issue that should be addressed in high-power pulsed fiber lasers. The inset of the figure shows the laser spectrum that measured at the output power of 30 W. The emission wavelength was centered at 2017 nm with a FWHM (full width at half maximum) width of ~11 nm. This spectrum includes many peaks, showing that many longitudinal modes have oscillated in the cavity.

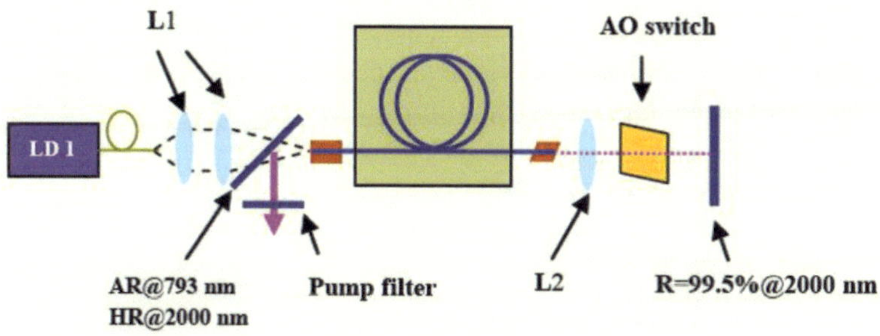

Fig. 4. Experimental setup for the AO Q-switched TDFL.

As shown in Fig. 5, the output power shows a linear dependence on the pump power at the high RR (50 kHz). However, at lower RRs, such as 10 kHz, the output power curve shows a severe deviation from linearity (not shown here), *i.e.*, there is a rollover of the output power curve. This rollover indicates that strong ASE was generated in the fiber and was emitted from both fiber ends. In order to achieve linear power dependence at lower RRs, such as 10 kHz, high diffraction-efficiency AO modulators are required. The appearance of ASE also put a limit on the maximum pulse energy that can be obtained in Q-switched fiber lasers. Further increasing the pump power just improves the ASE, and the Q-switched operation will become less effective.

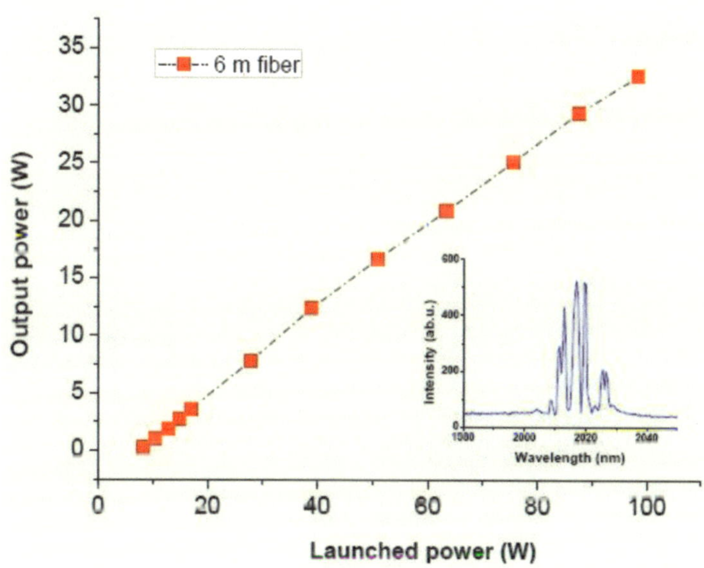

Fig. 5. Average output of the 6-m AO Q-switched TDFL.

Inset is the laser spectrum measured at the 30-W output level.

In pulsed fiber lasers, the RR usually has influence on the average output power when pump power is fixed. We measured the average power of the 6-m TDFL at a constant pump lever (50 W pump power) under different modulation frequencies, and the results are shown in Fig. 6. The maximum output power only shows little decrease when the RR decreases from 50 kHz to 20 kHz. However, when the RR further decreases to less than 20 kHz, the average power indicates a significant drop (over 20%). Therefore, the AO modulation frequency only has significant impact on the Q-switched TDFL when it is operated at low RRs. The pulse train recorded under the RR of 10 kHz is shown in Fig. 7, indicating somewhat low inter-pulse instability.

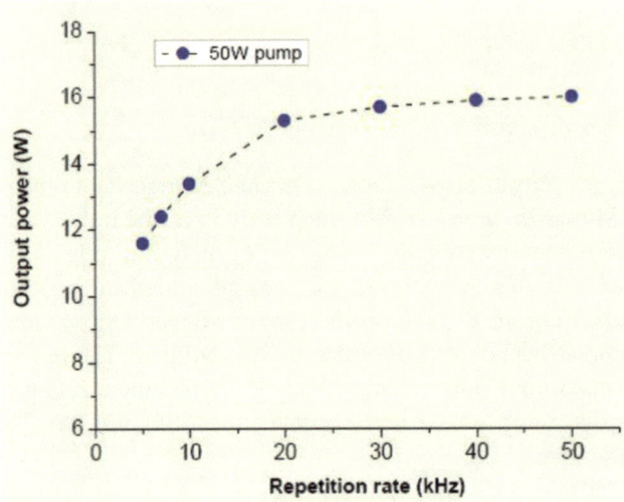

Fig. 6. Average output of the 6-m AO Q-switched TDFL under constant pump power but different RRs.

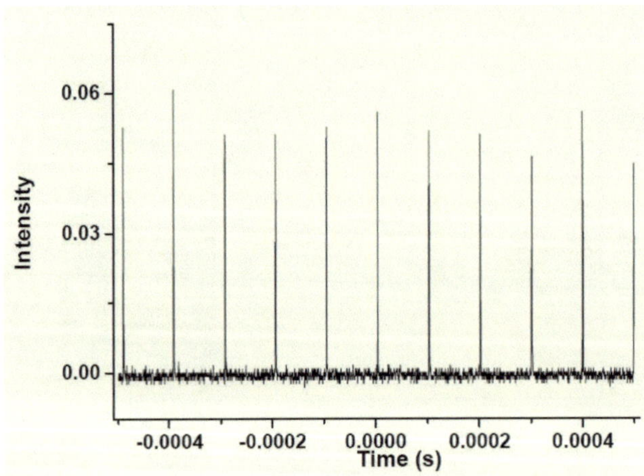

Fig. 7. Pulse train of the 6-m AO Q-switched TDFL at the RR of 10 kHz.

Operated under the RR of 10 kHz, the recorded pulse shape under different pump levels are shown in Fig. 8, where Pt denotes the threshold pump power. Just at the threshold, the laser pulse has a smooth shape with a pulse FWHM width of 240 ns. With increasing pump power, the pulse width narrows, and multiple peaks appears in the pulse. In a review description, Wang *et al.* proposed that the multi-peak pulse was initiated from the injection of ASE by quick switching of the AO modulator and the subsequent evolution of the switching-induced perturbation [35].

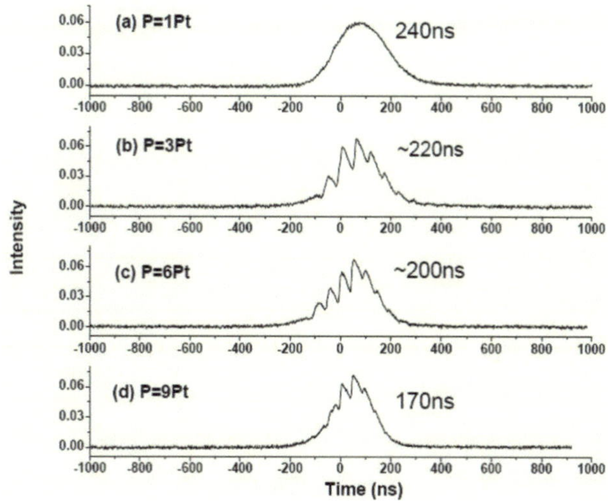

Fig. 8. Pulse evolution of the fiber laser at 10 kHz under different pump levels.

2.2 Pulse narrowing of Tm^{3+}-doped fiber lasers

For Q switched fiber lasers, the laser pulse width can be expressed as [36]

$$\tau_p = \frac{r\eta(r)}{r-1-\ln r} \times \frac{2L}{c\delta_0}, \tag{1}$$

where, δ_0 is the single-pass cavity loss, $\eta(r)$ is the energy extraction efficiency, r is the pump ratio (the ratio of the pump power to the threshold power), and L is the cavity optical length. Equation (1) implies that we can narrow the pulse duration through either increasing the pump strength r or shortening the fiber length. Simple calculations of the influence of the pump ratio on pulse narrowing are shown in Figs. 9 and 10, where $A = \frac{r}{r-1-\ln r}$ corresponding to the pulse width parameter. As shown in Fig. 9, the pulse width can be narrowed by a factor of ~6.3/1.5=4.2 when the pump ratio is increased from 2 to 10. When the pump ratio is over 10, further increasing the pump strength, such as from 10 to 100 (Fig. 10), can hardly shorten the pulse width to a appreciate level.

Based on the above analysis, we made an effort to obtain 2-μm laser pulse as narrow as possible from an AO Q-switched TDFL. For this aim, we shortened the fiber length to 0.4 m and increased the pump ratio as high as possible to carry out the Q-switching operation of the TDFL. With this fiber length, the maximum average output power was less than 2 W due to the fiber-length induced limited pump absorption. At a constant RR of 10 kHz, the pulse width narrowing characteristic with absorbed pump power is shown in Fig. 11. With increasing pump power, the pulse width reduces significantly. However, when the pump power was larger than 14 W, further increasing pump cannot lead to pulse narrowing. The shortest pulse width achieved with this 0.4-m TDFL was ~48 ns, as shown in the inset of Fig. 11.

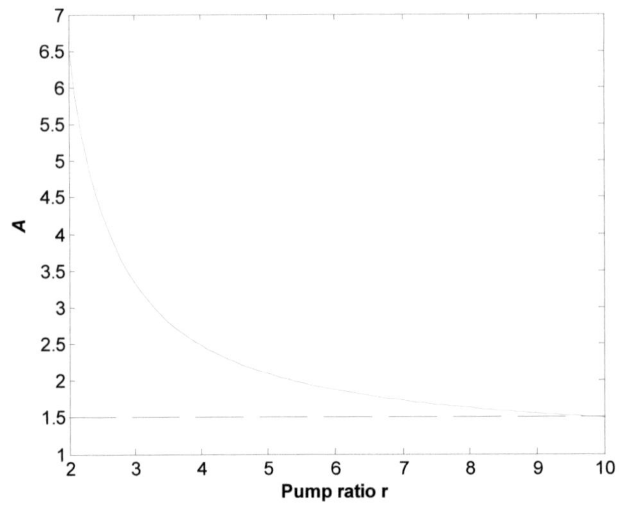

Fig. 9. Simulated evolution of the pulse width narrowing factor as a function of pump ratio from 2 to 10.

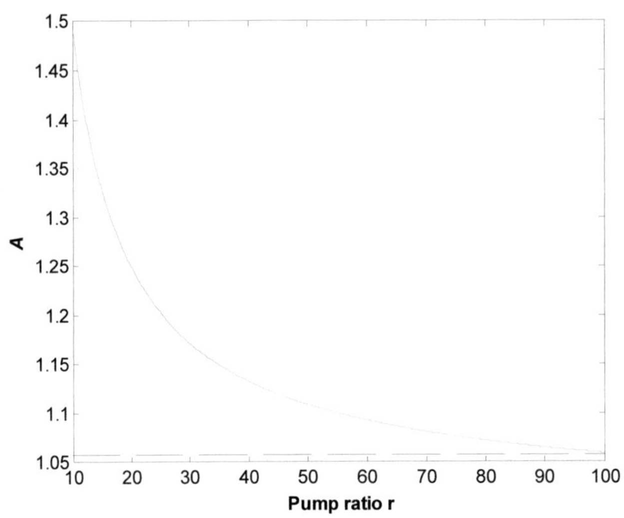

Fig. 10. Simulated evolution of the pulse width narrowing factor as a function of pump ratio from 10 to 100.

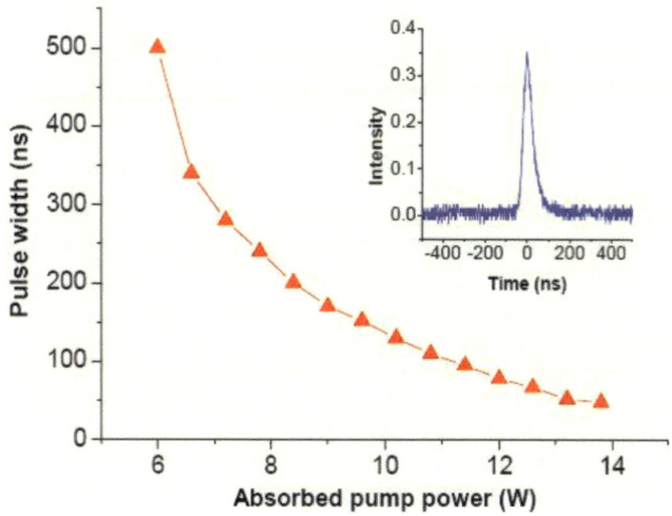

Fig. 11. Pulse width of the 0.4-m TDFL as a function of the absorbed pump.

Inset is the pulse shape with a width of ~48 ns.

Compared with bulk pulsed lasers, pulsed fiber lasers can provide higher average power due to their intrinsic high gain. However, it is still difficult to construct high-power 2-μm pulsed TDFLs at present. The difficulties mainly lie in the low damage threshold of the facets of fiber ends or other optical elements, and the onset of ASE. Therefore, how to improve the damage threshold of related optical components and efficiently suppress ASE will be the key problems for achieving high-power pulsed 2-μm TDFLs.

3. Gain-switching operation of Tm^{3+}-doped fiber laser

3.1 High-power gain-switching 2-μm fiber laser

With active Q-switching crystal lasers, stably controlled pulsed laser output can be achieved. With fiber amplifying systems, high gain can be easily obtained. Therefore, combining these two techniques is a potential way to scale 2-μm pulsed laser power to a new level. Besides, the advantages of high-doping concentration, efficient pump absorption, and long exited-state lifetime of Tm^{3+} fibers provide excellent prerequisites to achieve high-power pulsed 2-μm laser output.

In conventional fiber amplifiers, the seed source power is usually low. With such kind of seed source, a multi-stage fiber amplifier system must be constructed to achieve high average output power. The whole system is thus complicated and expensive. When the seed source has a narrow linewidth, the amplified output power of the fiber amplifier will be limited by the onset of stimulated Brillouin scattering (SBS). Ch. Ye *et al.* proposed using self-phase modulation (SPM) induced linewidth broadening to suppress SBS, and have obtained 50-W average power from a fiber amplifier [37]. For amplification of broad band ~2-μm laser pulses in Tm^{3+} fibers (~20 nm), SBS suppression is not a serious issue.

For obtaining 2-μm laser pulse, gain-switched fiber lasers can be excellent candidates owing to the well controlled pump pulse width and various available pump wavelengths. Gain-switched operation of TDFLs has achieved pulse energy of more than 10 mJ [38-40] and pulse width as narrow as 10 ns [41]. Compared with fiber amplifier systems, gain-switched devices can provide a more compact system configuration.

At present, actively Q-switched crystal lasers can provide moderate-power (several watts) 2-μm laser pulse with stably controlled characteristics. At the same time, TDFLs can provide high operation efficiency by taking advantage of the CR process [42-44]. Here, we will investigate how to improve the output from the pulsed TDFL through combining a high-power seed source with large-mode-area Tm^{3+} fiber amplification.

In experiment, we used a high-power Tm:YLF laser to gain switch the TDFL, which in turn was pumped by 793-nm LDs. This gain-switched system is different from conventional fiber amplifiers in that the high-power 'seed' laser acts just as a pulse switch, while the spectral characteristics are decided by the gain fiber. It is also different from conventional gain-switched lasers in that the gain and the switch are separated. We defined this laser system a combined gain-switched fiber laser (CGSFL) system. With this system, we will show how to achieve high-power 2-μm pulsed laser output through the combination of high-power switch laser, damage-threshold improvement of the fiber-end facts, and appropriate system configurations.

The experiment setup for the CGSFL system is shown in Fig. 12 [32]. The switch laser was an AO Q-switched Tm:YLF slab laser (pumped by a 793-nm LD) with the laser wavelength at 1914 nm. The Tm:YLF laser can provide a maximum average output power of ~4.4 W and RR from 500 Hz to 50 kHz. The pulse width can be varied between ~80 ns and ~1.2 μs by tuning the RR together with pump power. The 1914-nm laser pulse from the Tm:YLF laser was launched into the Tm fiber to switch the TDFL. The Tm^{3+}-doped silica fibers (both in the first-stage and the second-stage) had a ~25-μm diameter, 0.1-NA core doped with ~2wt.% Tm^{3+}. The octagonal pure-silica inner cladding had a 400-μm diameter and a NA of 0.46.

In the first-stage CGSFL, the pump source was a 120-W LD module at 793 nm. Two aspheric lenses (L1) were used to couple the 793-nm pump light into the gain fiber, with a coupling efficiency of ~80%. In this stage, a 6-m Tm^{3+} fiber was adopted with total pump absorption of ~18 dB. The amplified laser beam from the first stage was then launched into the second-stage Tm^{3+} gain fiber for further power scaling. In the second stage, the pump source was a ~230-W 793-nm LD module. The Tm^{3+} fibers were wrapped on 10-cm-diameter copper drums, which in turn were cooled by 18-°C circulating water.

In the first-stage CGSFL system, we launched ~1-W switch laser beam (with a pulse width of ~400 ns) into the fiber and kept the RR at 10 kHz. Under maximum pump power, this amplification stage can provide a maximum 2-μm pulsed laser of ~40 W with a slope efficiency of 50%. The maximum pulse energy was about 4 mJ, corresponding to a peak power of ~10 kW. At this level, no fiber facet damage was observed.

For amplification in the second stage (4-m Tm^{3+} fiber), the output of the first stage was kept at 15 W and the pulse RR was kept at 50 kHz. The laser output characteristics from the second stage Tm^{3+} fiber are shown in Fig. 13 [32]. The maximum output is ~105 W (pulse energy of 2.1 mJ) with a slope efficiency of 52.8%. When we decreased the RR to 40 kHz, the output power dropped down to ~100 W (corresponding to pulse energy of 2.5 mJ), and the fiber end facet was damaged. At this time, the pulse width was measured to be ~600 ns, leading to pulse peak power of ~4.2 kW.

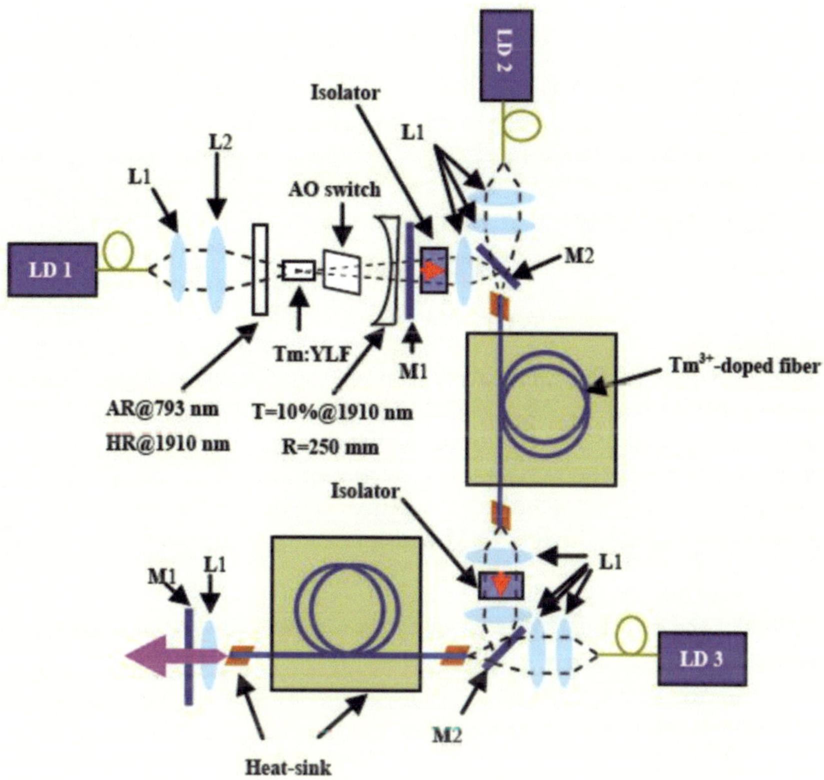

Fig. 12. Experimental setup of the combined gain-switched Tm^{3+}-fiber laser.
LD: laser diode; AR: anti-reflection; HR: high reflection; AO: acousto-optic;
L1: aspheric lens with f=11 mm; L2: convex lens with f=40 mm; M1& M2: dichroic mirrors.

In order to obtain higher output laser pulse energy, we fusion spliced a short piece (~2 mm) of passive silica fiber (1-mm diameter) to both ends of the active Tm fiber. At this time, only the first-stage CGSFL system (6-m fiber) was employed. We decreased the pulse RR to 500 Hz and kept the seed power at 200 mW. The maximum amplified ~2-μm output power was ~5.2 W, corresponding to a pulse energy of 10.4 mJ. The 1-mm-diameter endcaps greatly decreased the optical fluence at the output facet (<5 J/cm^2), which is less than the measured surface-damage fluence for nanosecond pulses in silica [45]. According to the empirical damage threshold for fused silica [46] of >$22t_p^{0.4}$ J/cm^2 (t_p is the pulse width in ns), the damage threshold of the 1-mm endcap facet would be hundreds of mJ. Therefore, this CGSFL system has the potential to scale the ~2-μm pulse energy even higher.

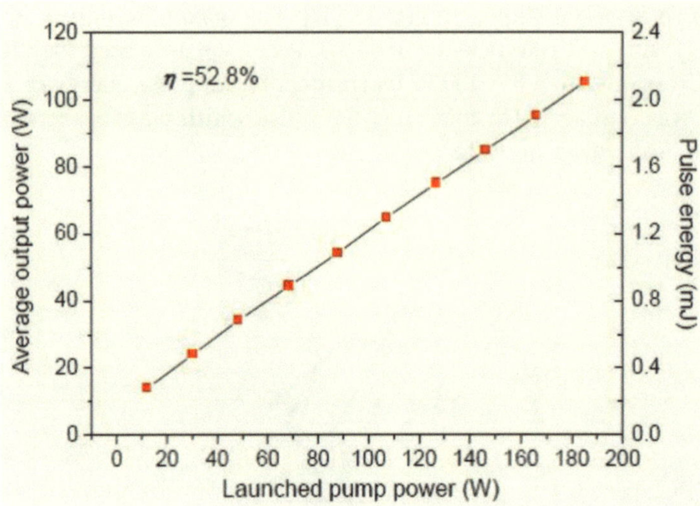

Fig. 13. Amplified output power from the two-stage CGSFL system at 50 kHz.

Under different RRs, the maximum average power and pulse energy achieved with this CGSFL system are shown in Fig. 14 [32]. The maximum output power increases near linearly with RR. Over 40 kHz, the roll-over of the average output was owing to the limited pump. On the other hand, the maximum pulse energy first decreases sharply with RR and then almost saturates due to limited stored energy. The inset is the pulse shape measured at the pulse energy of ~10 mJ with a FWHM width of 75 ns, corresponding to a peak power of ~138 kW.

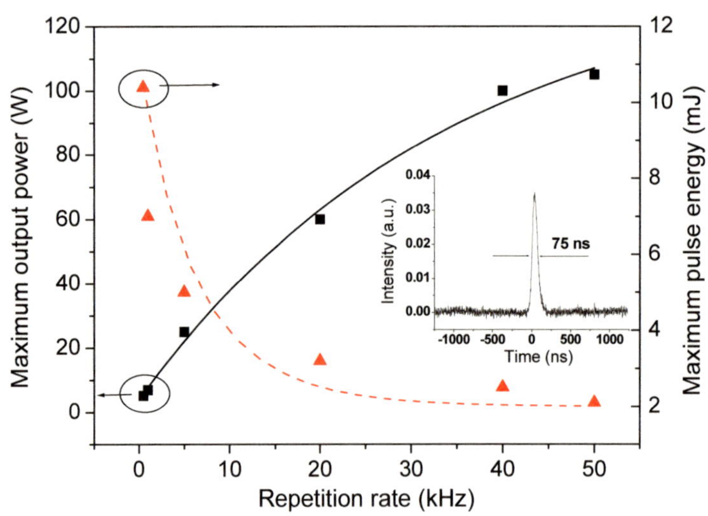

Fig. 14. Maximum average output power and pulse energy of the CGSFL system as a function of RR.

With the two-stage CGSFL system, the evolution of the 2-μm pulse shape (50 kHz) is shown in Fig. 15 [32]. The switch laser was kept at 1 W and had a pulse width of 900 ns. With increase of output power, the pulse width increased first and then narrowed, accompanied by steepening of the pulse leading edge. At 100-W power level, the pulse width was reduced to 750 ns, corresponding to pulse narrowing of ~17%. The pulse broadening at low power levels was probably originated from coexistence of the switch pulse and the ~2020-nm laser pulse (see Fig. 16). The pulse width narrowing and pulse steepening at high power levels was attributed to gain saturation and self-phase modulation [47-49].

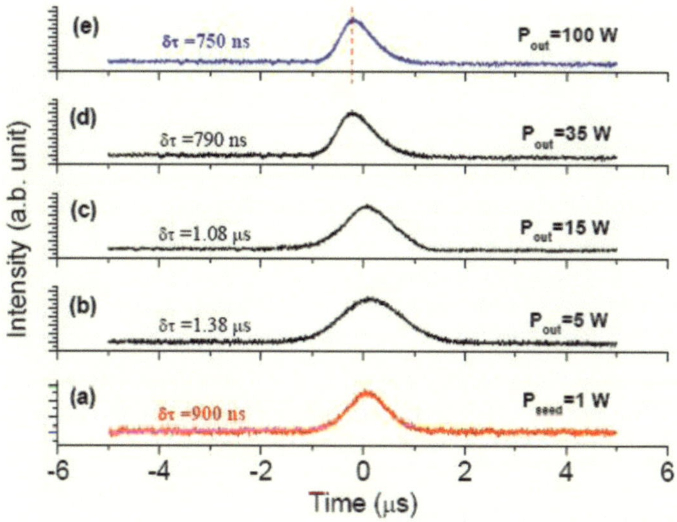

Fig. 15. Characteristics of the switch pulse and the fiber laser pulse at different power levels. (a)-(d) were measured after the first-stage CGSFL and (e) was measured after the second-stage CGSFL.

Under the same operating conditions (50-kHz RR and 1-W seed power), the spectrum of the seed and the output pulse at various power levels are indicated in Fig. 16 [32]. The fluorescence spectrum of the Tm^{3+} doped fiber is also shown in Fig. 16(a). It is clear that the 1914-nm seed laser was amplified at comparatively low power levels. When the 1914-nm laser was near 5 W, the ~2-μm laser pulse was stimulated, and thereafter more and more stored energy was extracted by the 2020-nm laser pulse. Finally, all energy was included in the 2020-nm laser beam. This is a unique characteristic of the CGSFL system, significantly different from fiber amplifiers and singly gain-switched devices. The detailed process can be described as follows. Owing to the broad fluorescence spectrum of our Tm fiber (1920-2040 nm), the switch pulse (1914 nm) lies in the wing of the gain spectrum. When the switch laser was launched into the fiber core, more than 90% of the laser was absorbed. At the same time, the Tm fiber accumulated population inversion through 793-nm pumping. The unabsorbed 1914-nm laser pulse will be amplified, while the absorbed 1914-nm laser pulse will modulate the gain of the system and act as a switch for the ~2-μm laser. Gain competition between the 1914-nm amplification and the stimulation of the ~2-μm laser emission leads to the evolution of the spectrum. Eventually, the 1914-nm laser was consumed completely and all stored energy was extracted by the ~2-μm laser emission. At 100-W level, the spectral width of the ~2-μm laser pulse was ~25 nm. We also observed the spectral evolution of a 2-m Tm^{3+} fiber laser with one-stage CGSFL configuration, and found that shorter fibers need stronger pump to switch on the CGSFL system.

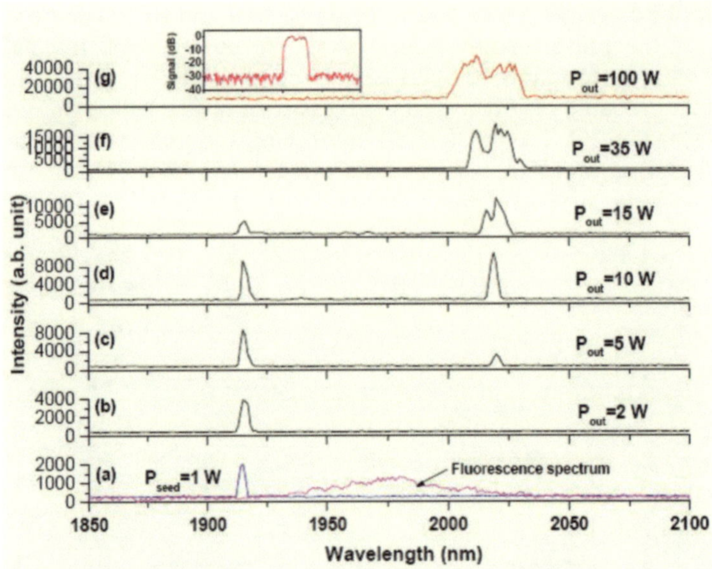

Fig. 16. Laser spectra (blue line) of the switch pulse (a), and that of the fiber laser pulse (b-f) at different power levels of the CGSFL system. (a) also shows the fluorescence spectrum of the Tm fiber.

3.2 Resonant-pumping 2-µm fiber laser

In high-power ~2-µm fiber lasers, the pump source is usually 790-nm LDs. Great difference between the pump wavelength and the laser wavelength causes a high quantum defect. This makes the optical-optical transfer efficiency of TDFLs be generally lower than that of high-power Yb fiber lasers. In order to improve the quantum efficiency thus the slope efficiency of high-power TDFLs, the pump wavelength must be elongated toward the laser wavelength. Such a pumping scheme with the pump wavelength approaching the laser wavelength is defined as resonant pumping. Resonant pumping TDFLs with ~1.6-µm wavelength has achieved slope efficiencies of ~80% with tens of miliwatts output [50].

In the pulsed 2-µm TDFLs achieved either by Q-switching [24] or by mode-locking [51], the laser power and pulse energy were usually limited. We have shown in the previous section that gain-switched TDFLs can produce high pulse energies, but their slope efficiencies were still low. Besides, the output pulses of these gain-switched TDFLs consist of a series of relaxation spikes, showing great chaotic temporal characteristics. In order to eliminate the chaotic spiking in gain-switched fiber laser, we must adopt highly-resonant pulsed pumping scheme, which is named as fast gain switching [52].

The experimental setup for the fast gain-switched TDFL is shown in Fig. 17 [53]. The pump source was an AO Q-switched Tm:YLF laser with 8.5-W maximum output at 1.914 µm. The M^2 beam quality of the pump beam was ~2. The double-clad Tm^{3+}-doped silica fiber had a ~30-µm diameter, 0.09 NA (numerical aperture) core doped with Tm^{3+} of ~2wt.% concentration. The pure-silica D-shape inner cladding had an average diameter of 400 µm and a NA of 0.46. One aspheric lens (*f* = 11 mm) was used to couple the pump light into the fiber core, with a coupling efficiency of ~90%. The absorption coefficient of the Tm^{3+} fiber at the pump wavelength (1914 nm) was measured with the cut-back method to be ~3 dB/m. At the output fiber end, a dichroic mirror (*R*=90%@1914nm & *T*=75%@1940nm, 45° coated) was used to filter the un-absorbed 1914-nm pump light. The 1940-nm laser output was calibrated by subtracting the un-filtered pump light and incorporating the filter-mirror-rejected laser light.

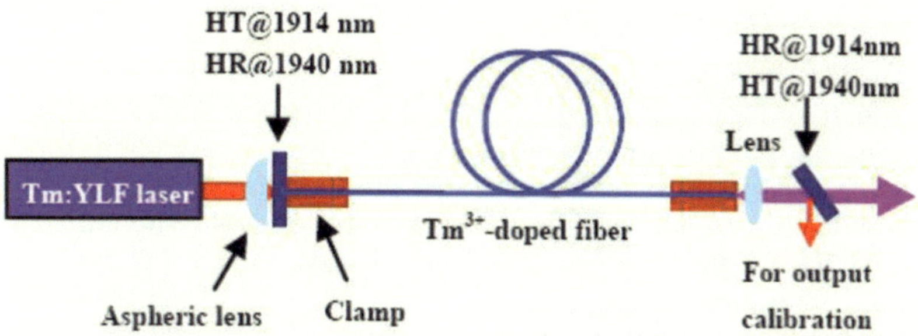

Fig. 17. Experimental setup of the resonantly gain-switched TDFL. HT: high transmission; HR: high reflection.

In experiment, three fiber lengths of 2, 4, and 6 m were used. Pumping at 1914 nm directly excites the Tm^{3+} ions from the ground state $^3H_6$ to the upper laser level $^3F_4$, which has a lifetime of ~340 μs [10]. The comparatively long lifetime of the upper laser level and the quasi-three-level nature of the laser transition guarantee efficient operation of the system through resonant pumping.

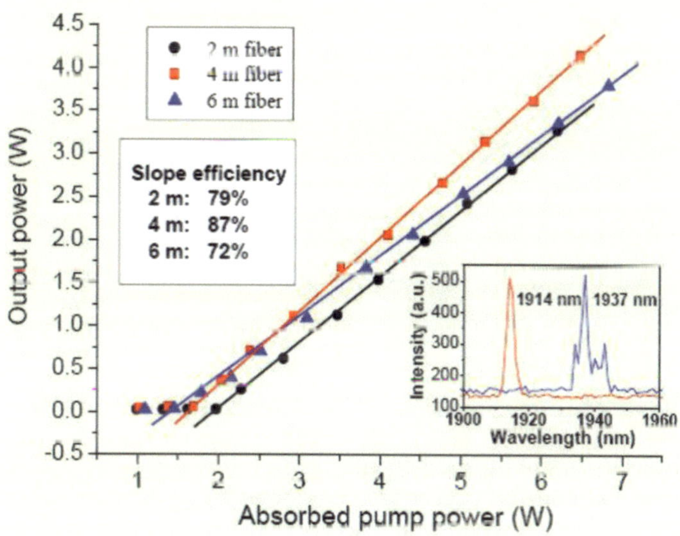

Fig. 18. CW Output performance of the resonantly-pumped TDFL.
Inset shows the laser spectra of the output laser beam and the pump light.

CW operation characteristics of the TDFLs (with the Q switch turned off) are shown in Fig. 18 [53]. The 4-m fiber provided a maximum output of 4.14 W and a slope efficiency of 87% with respect to absorbed pump power. Slope-efficiency dropping of the 6-m fiber can be attributed to laser re-absorption, while slope-efficiency reduction of the 2-m fiber mainly originated from the comparatively lower laser reflection coefficient of the pump-end dichroic mirror for the 2-m TDFL (1933 nm) than the 4-m TDFL (1937 nm). The figure inset shows the laser spectrum of the 4-m TDFL at 4-W output level, together with that of the 1914-nm pump source. The laser wavelength was center at 1937 nm, with a total FWHM width of ~7 nm and a main-transition-peak FWHM width of ~2.5 nm. Simple calculation shows that such a laser system only has a quantum defect of 1.4%.

When the Tm:YLF laser was Q-switched, the TDFL would be operated in the gain-switched mode. In order to achieve stable gain-switched 2-μm TDFLs, fast gain switching is required, i.e., the pump source must be able to realize a fast buildup of population inversion, which is depleted by only one signal pulse. Pumping the Tm^{3+} fiber with 790 nm, fast gain switching cannot be realized owing to the long relaxation time (>10 μs) from the pump level $^3H_4$ to the laser emission level $^3F_4$ [54], which will leads to relaxation spiking. On the other hand, if the TDFL is directly pumped to the laser emission band, population inversion can be built up instantaneously with the pump pulse, leading to the cavity gain being switched on and off in the same time scale as the pump pulse. Therefore, an AO Q-switched Tm:YLF laser at ~1.9 μm was chosen to pump the TDFL, and the pump pulse width was kept at ~120 ns. Output laser pulses of the 4-m gain-switched TDFL at RR of 10 and 20 kHz are shown in Fig. 19 [53]. It can be seen that single laser pulse is clean with no relaxation spiking present.

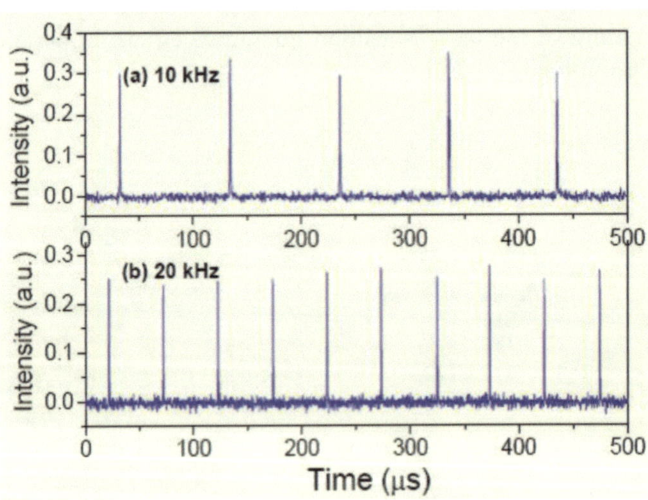

Fig. 19. Pulse trains of the 4-m gain-switched TDFL at the RR of (a) 10 kHz and (b) 20 kHz.

Laser output characteristics of the gain-switched TDFL under the RR of 20 kHz are shown in Fig. 20 [53]. The average output power was measured by adding up the output from both fiber ends, but the pulse energy was calculated only by dividing the far-fiber-end output by the RR. The optimum fiber length is also 4 m. The 4-m TDFL produced a maximum average power of 4 W and pulse energy of 0.155 mJ.

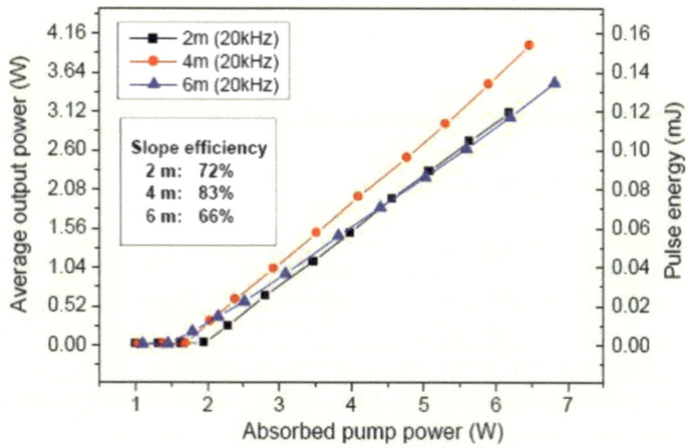

Fig. 20. Pulsed operation performance of the gain-switched TDFLs with different fiber lengths at the RR of 20 kHz.

In order to further scale pulse energy of the gain-switched TDFL, we kept the 4-m fiber laser near the maximum pump level and decreased the RR. Both average output power and pulse energy were measured only at the far fiber end, and the results are shown in Fig. 21 [53]. The average power tended to saturates at high RR. The maximum pulse energy reached 1.3 mJ at 2 kHz, and the pulse FWHM width was 61 ns, corresponding to peak power of ~21.3 kW.

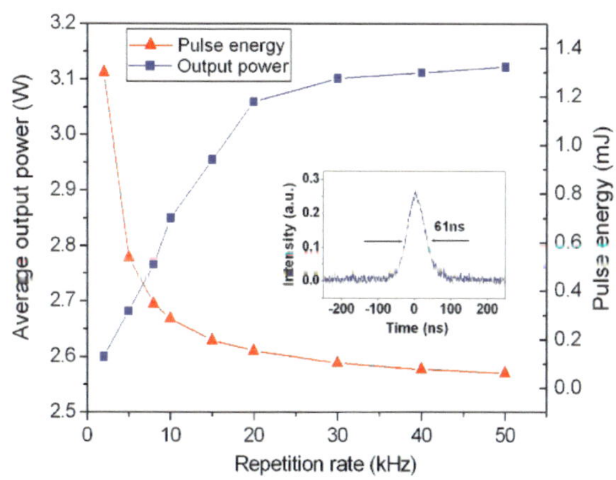

Fig. 21. Average power and pulse energy of the 4-m TDFL as a function of RR. Inset is the pulse shape measured near the maximum output power.

Due to high light density in fiber core, the peak power and pulse energy of TDFLs are still limited by optical damage and occurrence of nonlinear optical effects. Therefore, how to improve the pulse energy of 2-μm TDFLs is still a very important issue in the near future.

4. Self-pulsing operation of Tm^{3+}-doped fiber laser

When Tm^{3+} ions are doped into the medium such as glass, every energy manifold will be further split to many sub energy levels (Stark splitting). Therefore, TDFLs can produce fluent dynamical behaviors, including relaxation oscillation, self-pulsing, and self-mode-locking [55, 56]. Some of the dynamical behaviors can be advantages, but some others may be disadvantages. Among all kinds of the dynamical behaviors happened in TDFLs, self-pulsing phenomenon may be the most interesting one.

4.1 Self pulsing in Tm^{3+}-doped fiber lasers

It's well known that self-pulsing can be achieved in any laser with an adequate saturable absorber [57]. Self pulsing in TDFLs has been extentively observed, and was thought to result from saturable absorption due to ion-pair clusters, up-conversion and excited-state absorption (ESA) processes, or strong interactions among longitudinal modes [55,56,58–60]. In free running operation, the TDFL will operate successively in CW mode, self-pulsing mode and again CW mode with continuous increase of the pump power [61].

First, we construct a free-running laser configuration with a piece of Tm^{3+}-doped fiber to observe the formation and evolution of self-pulsing in TDFLs with increasing pump power. The experimental arrangement is shown in Fig. 22. The Tm^{3+}-doped fiber laser is pumped by a 40-W pig-tailed LD centered at ~792 nm. The Tm fiber has a 30-μm diameter, 0.22 NA core doped with Tm^{3+} of ~2 wt.% concentration. The inner cladding had a 410 μm diameter and a NA of 0.46. The fiber length used is 6 m. The dichroic mirror at the pump end and the far-end perpendicularly cleaved fiber facet (~3.55% Fresnel reflection) formed a Fabry-Perot cavity.

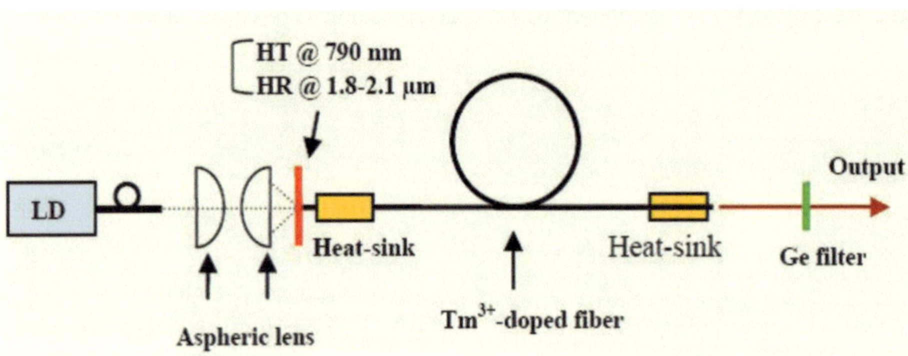

Fig. 22. Experimental arrangement of LD-pumped Tm^{3+}-doped fiber laser

The threshold pump power of the fiber laser is slightly less than 10 W. Increasing the pump power continuously, the fiber laser can operate at several stages. The evolution of the self-pulsing characteristics of the fiber laser is shown in Fig. 23. When the pump power is near the threshold (P=10 W), the laser delivers a regular train of pulses. When the pump power is increased to 12 W, the self-pulsing train is still high regular, and the pulse width narrows and the pulse frequency increases. Increasing the pump power over 15 W, the pulsing operation is randomized, and more like mode locking. When the pumping level is even higher (25 W), the laser output becomes quasi-CW.

The pulse width and frequency of the 6-m self-pulsing TDFL as a function of pump power are indicated in Fig. 24. At slightly over the pump threshold, the pulse width and pulse frequency, decreases and increases sharply with pump power respectively. When the pump power is over 14 W, the pulse width and pulse frequency show a near linear change with increasing pump power. At high pump levels, e.g. over 25 W, the pulse width begins saturating. Therefore, it is hard to further narrowing the pulse duration of self-pulsing TDFLs just through increasing pump power.

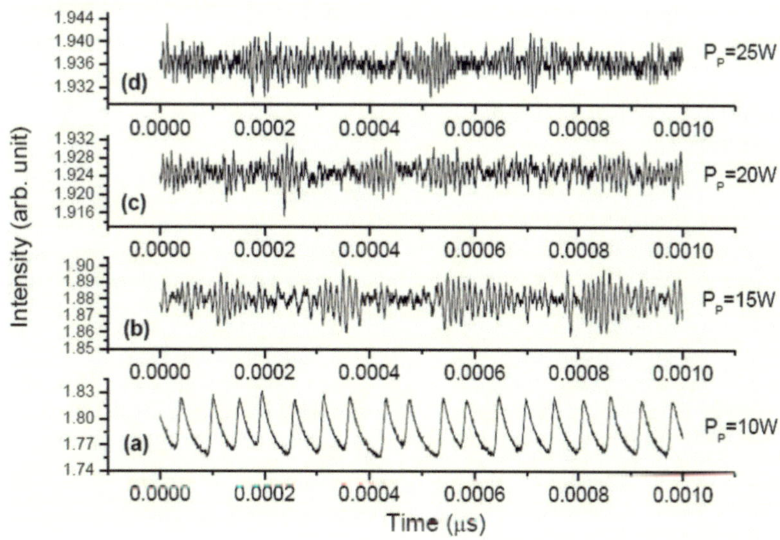

Fig. 23. Output intensity time trace of 6 m fiber laser with the end-facet output coupler for various pump powers.

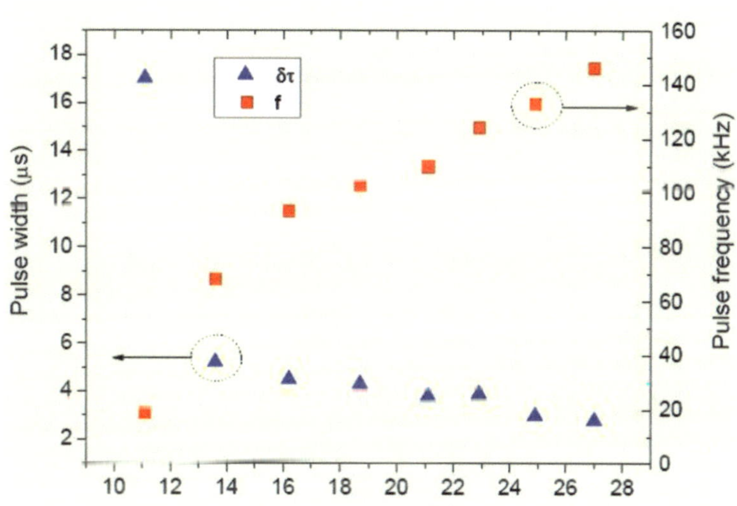

Fig. 24. Pulse width and pulse frequency versus pump power.

When the transmission of the output coupler is changed, the self-pulsing characteristics will be somewhat different. In order to make a efficient comparison of pulse width and frequency with different couplers, we define a intracavity laser power (equals to the output power divided by the coupling value). Three kinds of output couplers are used in the experiment:

T=5%, T=20%, and T=96% (the fiber facet). The dependence of the pulse width and frequency on the output coupler transmission (T) under different pump power is shown in Fig. 25.

It is clear that the pulse width decreases first sharply and then slowly with the intra-cavity power, which is in agreement with the description in the literature [62]. With increasing intra-cavity power level, the T=20% coupler shows much less change compared with the other two couplers, implying that self-pulsing operation can be sustained in a wider pump range. We suppose there is a optimum output coupling value that can support large-range self-pulsing operation.

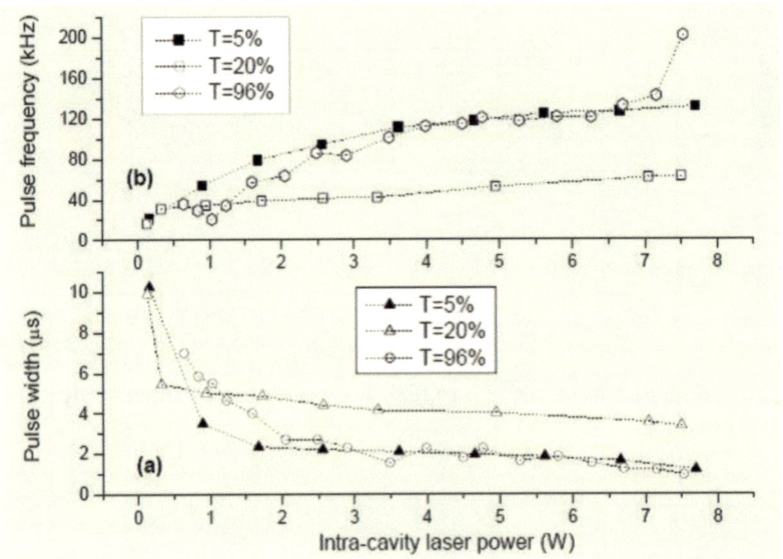

Fig. 25. Output coupling dependence of (a) pulse width and (b) pulse frequency under different intra-cavity power levels.

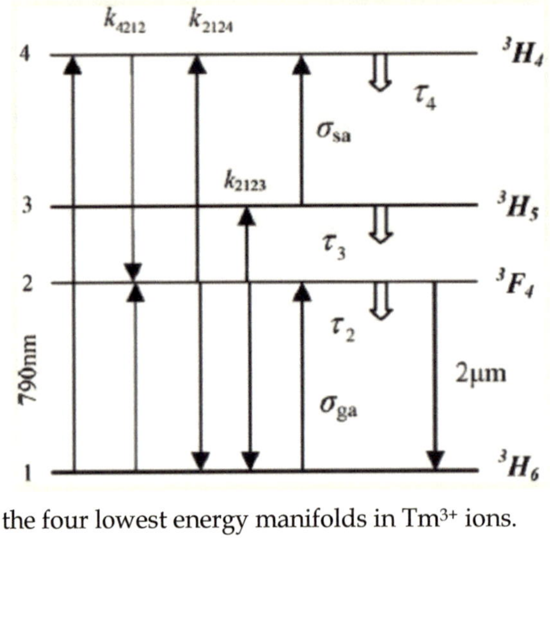

Fig. 26. Schematic of the four lowest energy manifolds in Tm^{3+} ions.

For Tm^{3+}-doped fiber lasers, as shown in Fig. 26 [63], the pump light at 790 nm excites the ions from $^3H_6$ state to $^3H_4$ state, which quickly relaxes to the upper laser level $^3F_4$. In heavily Tm^{3+}-doped fibers, the distance between Tm^{3+} ions decreases, strengthening the interaction between ions. This strong inter-ion interaction provides a loss mechanism, depopulating the 2-μm transition level $^3F_4$. This kind of loss mechanism acts as a saturable absorber, leading to the formation of self-pulsing. However, what is the real loss mechanism responsible for self-pulsing in the TDFL?

4.2 Origin of self-pulsing in Tm^{3+}-doped fiber lasers

Many mechanisms have been proposed to explain the origin of self-pulsing in TDFLs, but consistent agreement has not been reached yet. In this section, we will try to find out the real reason accounting for the self-pulsing formation in TDFLs through theoretical analysis and numerical simulation.

For Tm^{3+} ions, the energy-level structure diagram is sketched in Fig. 26, including several important transition and energy-transfer processes. The rate equations for the local population densities of these levels are as follows [64-66]

$$\frac{dN_4}{dt} = R(z,t)(N_1 - N_4) - k_{4212}N_1N_4 + k_{2124}N_2^2 - \frac{N_4}{\tau_4} + \sigma_{sa}c\phi(z,t)(N_3 - N_4), \tag{2}$$

$$\frac{dN_3}{dt} = k_{2123}N_2^2 + \beta_{43}\frac{N_4}{\tau_4} - \frac{N_3}{\tau_3} - \sigma_{sa}c\phi(z,t)(N_3 - N_4), \tag{3}$$

$$\frac{dN_2}{dt} = 2k_{4212}N_1N_4 - 2(k_{2124} + k_{2123})N_2^2 + \beta_{42}\frac{N_4}{\tau_4} + \beta_{32}\frac{N_3}{\tau_3} - \frac{N_2}{\tau_2} - c\phi(z,t)\sigma_e(N_2 - \frac{g_2}{g_1}N_1) + c\phi(z,t)\sigma_{ga}(N_1 - N_2), \tag{4}$$

$$\frac{d\phi}{dt} = c\phi(z,t)\sigma_e(N_2 - \frac{g_2}{g_1}N_1) + m\frac{N_2}{\tau_2} - \sigma_{ga}c\phi(z,t)(N_1 - N_2) - r_c\phi(z,t), \tag{5}$$

$$N_1 = N_{tot} - N_2 - N_3 - N_4, \tag{6}$$

$$R(z,t) = R(0,t)\cdot e^{-\alpha_p\cdot z}, \tag{7}$$

where N_i are the populations of four energy manifolds $^3H_6$, $^3F_4$, $^3F_5$, $^3H_4$, and N_{tot} is the total density of Tm^{3+} ions. R is the pump rate, and ϕ is the average photon density of the laser field. σ_e is the stimulated emission cross section of signal light, σ_{ga} and σ_{sa} are the absorption cross sections of ground state and excited state, respectively. Where g_1 and g_2 are the degeneracies of the upper and lower laser levels, τ_i is the level lifetimes of four manifolds, and r_c is the signal photon decay rate. β_{ij} are branch ratios from the i to j level, m is the ratio of laser modes to total spontaneous emission modes. The coefficients k_{ijkl} describe the energy transfer processes: k_{4212} and k_{3212} are the cross relaxation constants, and k_{2124} and k_{2123} are the up-conversion constants. The coefficient a_p is the pump absorption of the fiber, which is calculated by $\alpha_p = \sigma_{ap} \cdot N_{tot}$, where σ_{ap} is the pump absorption cross section. In the analysis and simulation, the phonon-assisted ESA process of $^3F_4$, $^3H_5 \rightarrow ^3H_6$, $^3H_4$ and ground-state absorption (GSA) through the $^3F_4$, $^3H_5 \rightarrow ^3H_6$, $^3F_3$ energy transfer process are considered. The corresponding parameters are listed in Table 1 [66-68].

Parameter	numerical value
k_{4212}	1.8×10^{-16} cm^{3}s^{-1}
k_{2123}	1.5×10^{-18} cm^{3}s^{-1}
k_{2124}	1.5×10^{-17} cm^{3}s^{-1}
τ_i	$\tau_4 = 14.2\mu s$ $\tau_3 = 0.007\mu s$ $\tau_2 = 340\mu s$
β_{ij}	$\beta_{43} = 0.57$ $\beta_{42} = 0.051$ $\beta_{32} \approx 1$
σ_e	2.5×10^{-21} cm^2
σ_{sa}	Variable (4×10^{-21} cm^2)
σ_{ga}	variable
m	8×10^{-7}
r_c	9.7×10^{6} s^{-1}
σ_{ap}	1×10^{-20} cm^2
N_{tot}	1.37×10^{20}cm^{-3}

Table 1. The parameters in the rate equations

Theoretical calculation

Provided the lifetime of level N_3 (0.007 μs) is much shorter than that of level N_2 (340 μs), we simplify the energy manifolds to three levels, i.e. we assume N_3~0 and let $N_{23} = N_2+N_3$. By adding Eq. (3) and (4), we get

$$\begin{aligned}\frac{dN_{23}}{dt} &= 2k_{4212}N_1N_4 - (2k_{2124} + k_{2123})N_{23}^2 + (\beta_{43} + \beta_{42})\frac{N_4}{\tau_4} - \frac{N_{23}}{\tau_2} \\ &-c\phi[\sigma_e(N_{23} - \frac{g_2}{g_1}N_1) + \sigma_{sa}(N_{23} - N_4) - \sigma_{ga}(N_1 - N_{23})]\end{aligned} \quad (8)$$

Then, the rate equations are simplified to a three-level system. Suppose the laser operating in the steady-state (or CW) regime, i.e. $\frac{d\phi}{dt}=0$ and $\frac{dN_i}{dt}=0$, we can solve ϕ, N_1 and N_{23} from the Eqs. (2), (5) and (8). Through solving the equations, we found that there was a certain pump range (defined as ΔR), where the steady-state solution for the rate equations could not be found. In this range, the laser would not be operated in the CW state. With increase or decrease of pump power out of the range ΔR, the operation of Tm^{3+}-doped fiber lasers would undergo phase transition (changed to CW operation). Such a case is in good agreement with the experimental observation in the self-pulsing TDFLs.

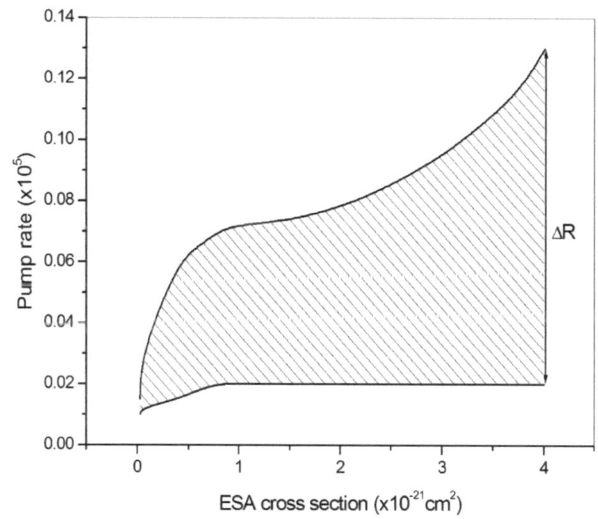

Fig. 27. The non-CW pump range ΔR as a function of the ESA cross section.

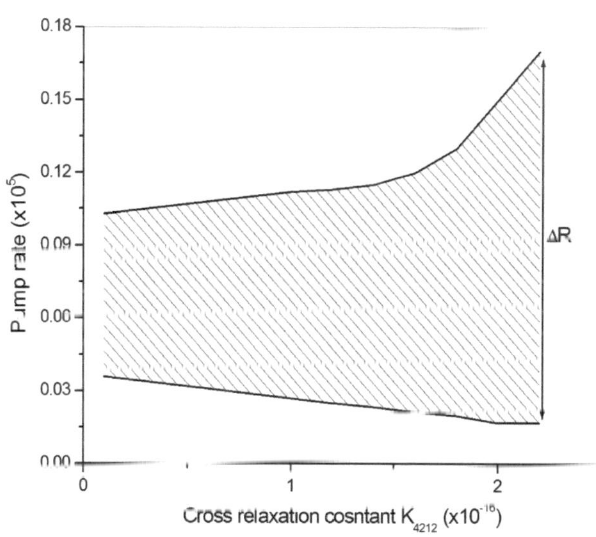

Fig. 28. The non-CW pump range ΔR as a function of the cross relaxation strength.

Then the non-CW range ΔR was calculated as varying the ESA cross section and the cross relaxation parameter k_{4212}. The variation of ΔR as a function of the ESA cross section is shown in Fig. 27 [63]. The non-CW range ΔR increases with the ESA cross section, especially increases exponentially when the ESA cross section is larger than 3×10^{-21} cm^2. When the ESA cross section is less than 1×10^{-21} cm^2, the range ΔR shrinks sharply, and goes to zero with a small value of ESA cross section. The CW operation of TDFLs can sustain for any pump rate when the ESA cross section is sufficiently small. On the other hand, with a larger ESA cross section, the CW operation will always be broken.

The non-CW range ΔR calculated as a function of the cross relaxation strength k_{4212} is shown in Fig. 28 [63]. Large values of k_{4212} will obviously enlarge the range ΔR . However, even when the cross relaxation k_{4212} is decreased to zero, the breaking of CW operation still preserves, implying that the cross relaxation process is not the key factor in the formation of self-pulsing in TDFLs.

Simulation results

The revolution of the photon density in TDFLs is investigated through numerical simulation based on the complete rate equations (2-7). In the simulation, four energy-transfer processes (cross relaxation, energy transfer up-conversion, GSA and ESA) are calculated separately to analyze their influence on the formation of self-pulsing.

The impact of the cross relaxation is evaluated by varying the value of the parameter k_{4212}. The simulation results are shown in Fig. 29 [63]. Stable CW laser operation preserves whatever value of k_{4212} is adopted. With the increase of k_{4212}, the laser intensity will be increased. Therefore, strong cross relaxation process may be helpful for improving the slope efficiency of TDFLs, but cannot account for the formation of self-pulsing.

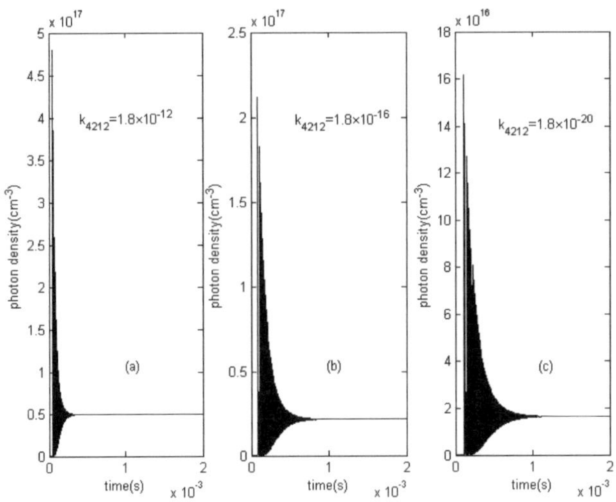

Fig. 29. Laser photon density dynamics characteristics with different cross-relaxation strength k_{4212}.

Then, we numerically observed the influence of the energy-transfer up-conversion process $^3F_4$, $^3F_4\rightarrow^3H_6$, $^3H_5$ (k_{2123}) and $^3F_4$, $^3F_4\rightarrow^3H_6$, $^3H_4$ (k_{2124}). The simulation results for k_{2124} are shown in Fig.

30 [63]. When the up-conversion is too strong, *i.e.*, k_{2124}>1.5×10^{-16} cm^{3}s^{-1}, the laser relaxation oscillation is suppressed, as shown in Fig. 30(a-c). The photon density is clamped in a very low level. When the parameter k_{2124}<1.5×10^{-17} cm^{3}s^{-1}, the laser relaxation oscillation occurs again. No matter which values of the parameters (from 1.5×10^{-6} cm^{3}s^{-1} to zero) are chosen, no self-pulsing phenomenon is observed. The behaviors of the parameters k_{2123} are very similar to that of k_{2124}. The up-conversion process does not directly connect to the self-pulsing operation in TDFLs. In the practical Tm^{3+}-doped system, the values of k_{2123} and k_{2124} are around 10^{-17} - 10^{-18} cm^{3}s^{-1}. The main influence of up-conversion is increasing the laser threshold.

The GSA is also called as the re-absorption in the Tm^{3+}-doped fiber lasers, which had been thought as a possible mechanism for the self-pulsing formation. The GSA process $^3H_6 \rightarrow ^3F_4$ can be thought as a reverse process of the laser transition $^3H_6 \rightarrow ^3F_4$. We kept the emission cross section σ_e constant and changed the value of g_2/g_1. The GSA strength increases with a larger value of g_2/g_1. When g_2/g_1=0, the fiber laser is a complete four-level system; when g_2/g_1=1, the fiber laser corresponds to a complete three-level system. In simulation, we changed the g_2/g_1 value from 0 to 1 to study the evolution of photon density, and found that stable CW operation always occurs. Therefore, the GSA process is definitely not the cause of the self-pulsing in TDFLs.

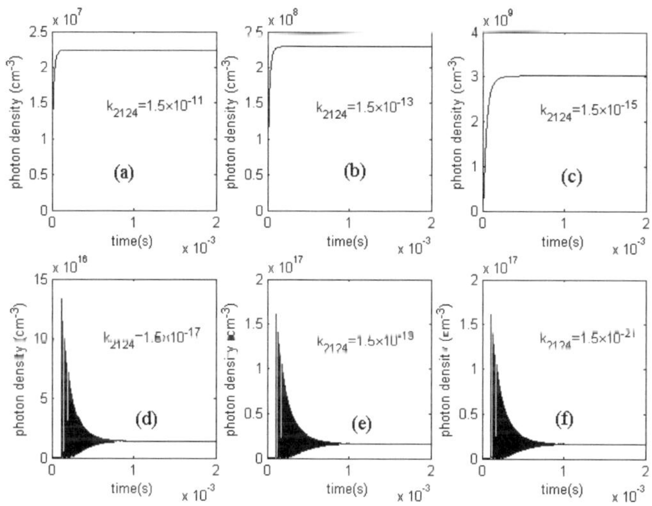

Fig. 30. Laser photon density dynamics characteristics with different energy-transfer up-conversion strength k_{2124}.

When only the ESA process $^3H_5 \rightarrow ^3H_4$ is taken into account, the revolution of photon densities for various ESA cross sections σ_{sa} are shown in Fig. 31 [63]. When the ESA cross section σ_{sa} is chosen in the range from 4×10^{-21} to 4×10^{-19} cm^2, it is clear to observe stable, regular self-pulsed trains. This verifies the theoretical predication that the ESA process is the key reason leading to the self-pulsing dynamics in TDFLs. When the ESA cross section σ_{sa} is much small, the ESA is too weak to hinder accumulation of the population in the level $^3H_5$ (N_3), and CW operation occurs after relaxation oscillation. With the increase of the ESA cross section σ_{sa}, the decay time of the relaxation oscillation becomes longer and longer, and finally, the relaxation oscillation evolves to a stable self-pulsed train.

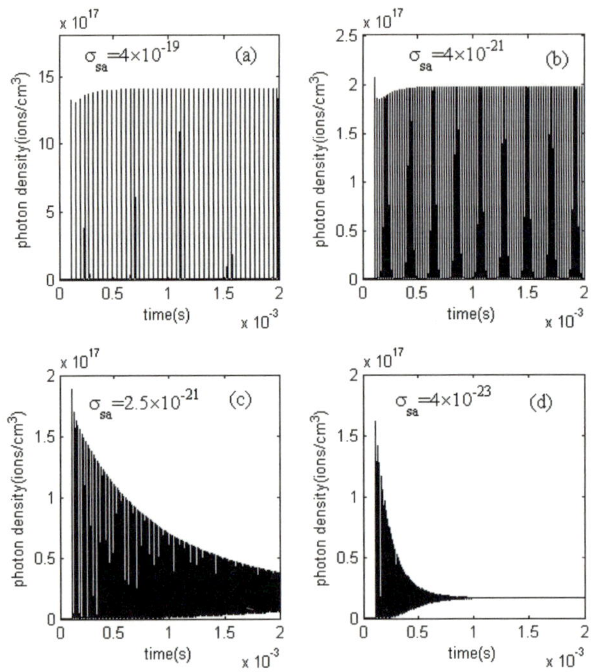

Fig. 31. Laser photon density dynamics characteristics with different ESA strength σ_{sa} (cm²).

4.3 Theoretical model and self-pulsing behaviors in Tm^{3+}-doped fiber lasers

Theoretical model

The schematic of our theoretical model to simulate the beam propagation in the Tm^{3+}-doped fiber is depicted in Fig. 32 [69]. The theoretical model is constructed based on the above mentioned rate equations (2) to (7), and the value of ESA cross section σ_{sa} is fitted from our experimental observations.

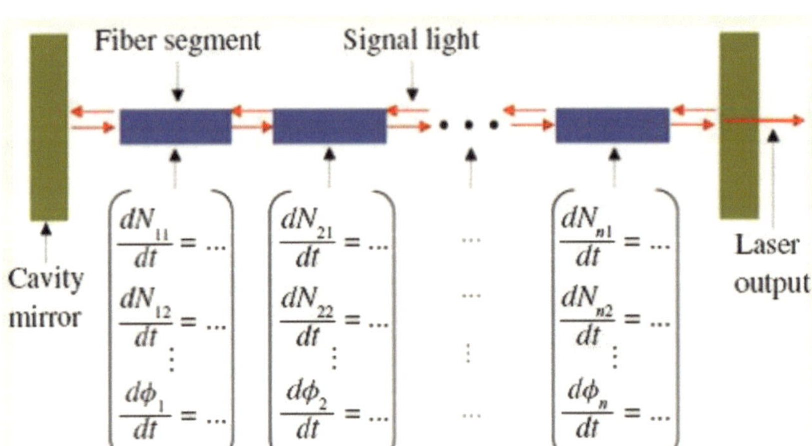

Fig. 32. Schematic of the simulation model.

The laser field S(z) propagating along the fiber can be expressed as follows:

$$\frac{dS_{f,b}(z)}{dz} = \pm S_{f,b}(z)[\sigma_e \Delta N(z) - \sigma_{sa} N_3(z) - \delta_s], \tag{9}$$

where the subscripts *f* and *b* denoting the forward and backward directions, respectively; δ_s is the intrinsic absorption of the host glass at the laser wavelength (set at 2.3×10^{-3} m^{-1}), and $\Delta N = N_2 - \frac{g_2}{g_1} N_1$.

The boundary conditions for the laser field are given by

$$S_b(L) = R_{S2} S_f(L)\,,\; S_f(0) = R_{S1} S_b(0)\,, \tag{10}$$

where R_{S1} and R_{S2} are reflectivities of the front and rear mirrors at the pump and laser wavelength, respectively; *L* is the length of the fiber.

We divide the fiber into many segments and construct separate rate equations for every segment. The absorption of pump power along the fiber is calculated, and the pump power is converted to the pump rate in the rate equations. We digitize the rate equations in every segment with the Runge–Kutta method. Based on the propagation equation (9), the laser field at segment (*t, z*) propagates to segment (*t* +Δt, *z*+Δz). The whole set of digitalized equations are solved through the iterative approach.

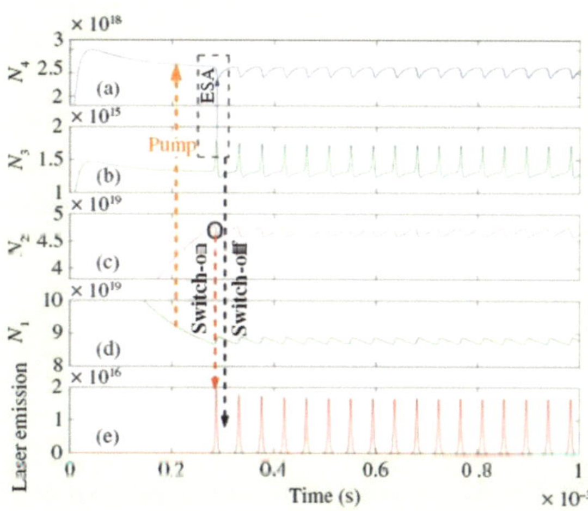

Fig. 33. Evolution of the populations of the four lowest energy manifolds of the Tm^{3+}-doped fiber laser and that of the corresponding laser emission.

Based on the theoretical model, the evolution of the populations of the four lowest energy manifolds of Tm^{3+} ions and the 2-μm signal photon density were simulated, and the results are shown in Fig. 33 [69]. The formation of self-pulsing can be explained as follows. When the fiber is pumped, the population N_1 of the ground-state level ($^3H_6$) is excited to the $^3H_4$ level (N_4). Through nonradiative transition and cross relaxation, the population jumps to the $^3H_5$ level (N_3) and $^3F_4$ level (N_2). After a time delay of ~0.28 ms, the photon density increases sharply, leading to the occurrence of laser emission, as shown by the first pulse in Fig. 33(e). At the same time, the population of N_3 increases and N_4 decreases simultaneously, leading to drop in the population difference N_4-N_3. This aggravates the ESA process [$^3H_5$ (N_3)$\rightarrow$$^3H_4$ (N_4)], hindering the accumulation of the population inversion ΔN ($N_2-N_1g_2/g_1$). Consequently, the laser emission is switched off by the ESA process ($^3H_5\rightarrow{}^3H_4$). After the laser emission is switched off, the population N_4 and the population inversion ΔN will be pumped up once more. The laser occurs again when the cavity loss is overcome by the gain for another time. This switching on and off process will operate repeatedly, accounting for the formation of self-pulsing.

Figure 34 shows the numerically calculated self-pulsing train of the TDFL near the threshold pump rate. This regular self-pulsing train has a pulse width and RR of 7.68 μs and 16.89 kHz, respectively, being in good agreement with experimental results.

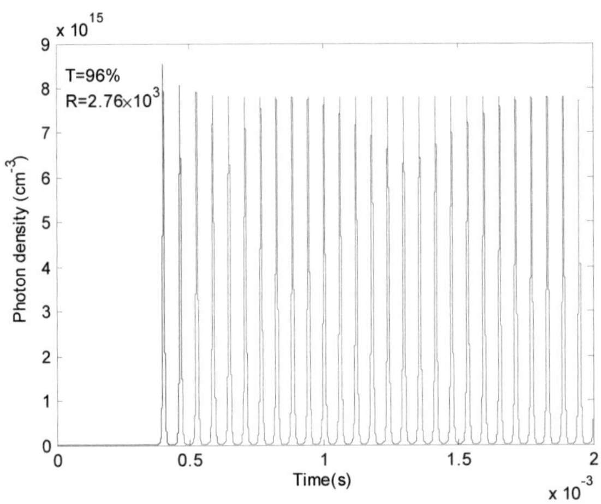

Fig. 34. Self-pulsing train at corresponding threshold pump rate with output coupling T=96%.

Influence of several parameters on the self-pulsing characteristics

With self-pulsing threshold pump strength, the laser pulse width and pulse frequency as a function of output coupling are shown in Fig. 35 [69]. The pulse frequency increases and the pulse width decreases nearly exponentially with the output coupling. Compared with the simulation results, the experimental observation shows slightly higher pulsing frequencies and narrower pulse widths. By using higher output coupling, higher pump level is needed to sustain self-pulsing. In such a case, accumulation of population inversion and switching of N_3 and N_4 populations by the ESA process are faster than that achieved by using lower output coupling, leading to higher pulse frequency and narrower pulse width.

To compare the simulation with the experimental results, we normalized the pump rate (power) to their corresponding threshold values, which is defined as the pump ratio $r=R/R_P$ (in simulation) or $r=P/P_{th}$ (in experiment). In the self-pulsing regime, the pulse width and pulse frequency as a function of r are depicted in Fig. 36 [69]. The pulse frequency and the pulse width display reverse tendencies with the increase of pump ratio. At high pump ratios, both pulsing frequency and pulse width tend to saturate. With *r*=3 in the experiment, the pulse width decreases to 0.8 μs, and the pulse frequency increases to 110 kHz, respectively. In the simulation, the maximum pulse frequency can be as high as 900 kHz with the minimum pulse width of about 200 ns (not shown here).

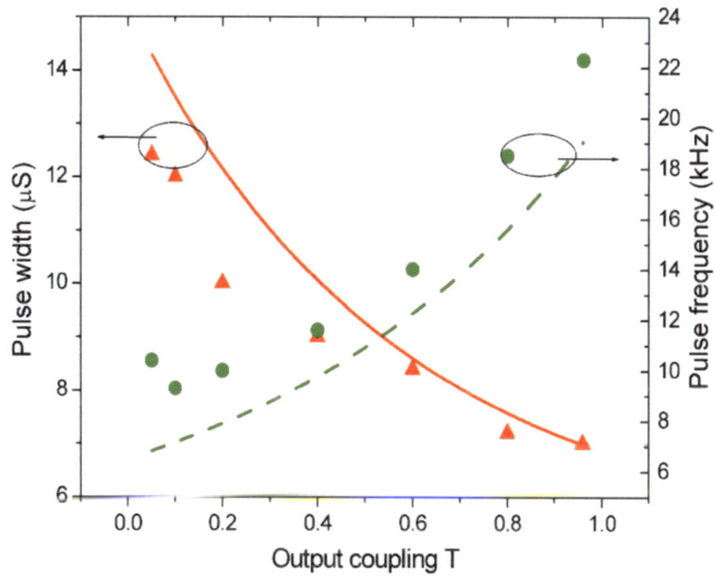

Fig. 35. Pulse width and frequency of the SP Tm^{3+}-doped fiber laser with various output couplings at the respective threshold pump levels. The solid and dotted lines show the simulation results, and the triangle and diamond dots indicate the experimental results.

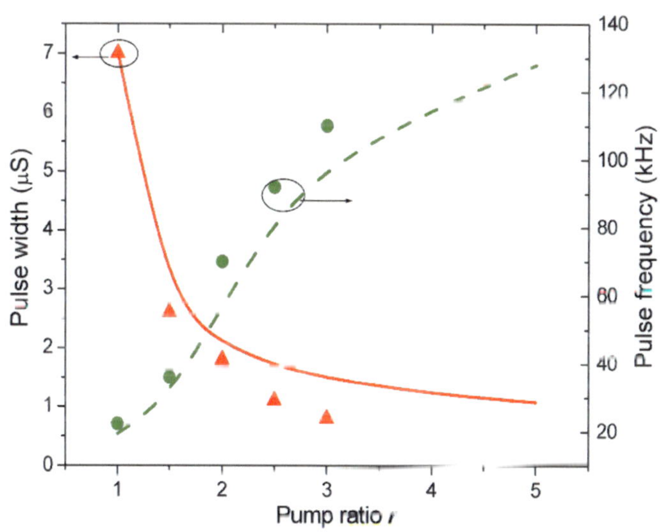

Fig. 36. Pulse width and pulse frequency of the Tm^{3+}-doped fiber laser with T~96% output coupler at various pump levels. The solid and dotted lines show the simulation results, and the triangle and circle dots indicate the experimental results.

In TDFLs, appropriately high Tm^{3+} doping concentration can strengthen the cross relaxation process, which significantly enhances the quantum efficiency of the laser.

Here, the SP operation is examined by changing the Tm^{3+} doping concentration in unit ppm (part per million). In experiment, Tm^{3+} doping concentration of ~1150 ppm [55] and ~2100 ppm were adopted. The pulse frequency and width as a function of Tm^{3+} doping concentration are shown in Fig. 37 [69]. As the Tm^{3+} concentration increases from ~1000 ppm to ~15000 ppm, the pulse frequency grows from 15 to ~110 kHz, and the pulse width narrows exponentially to ~350 ns and saturates. Higher doping concentration strengthens the CR, EUC, and re-absorption processes. The combination of these processes can speed up the recovery of population inversion after the pulse is switched off, thus improving the pulse frequency and narrowing the pulse width. Higher doping concentration is preferred to simultaneously achieve high pulse frequency and narrow pulse. However, very high doping concentration may form Tm^{3+}-ion clusters in the silica fibers and cause serious concentration quenching. Consequently, a tradeoff exists between obtaining a narrow pulse width and high laser efficiency.

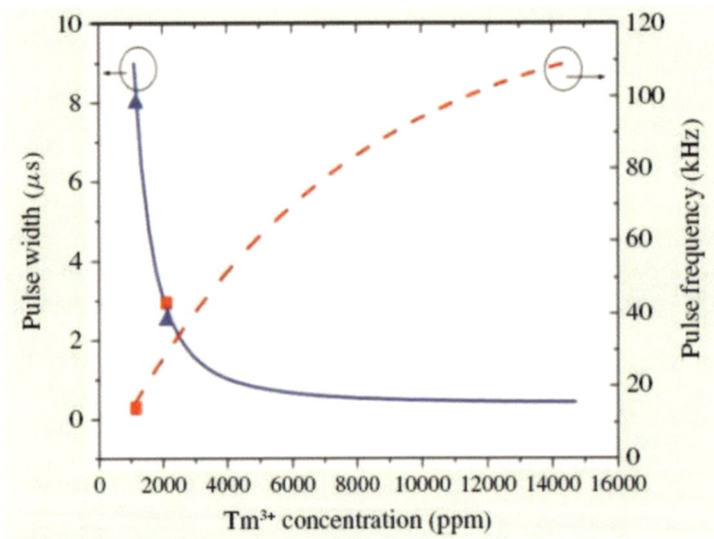

Fig. 37. Pulse width and frequency as a function of doping concentration at pump rate R=3.2×10^3 s^{-1}. The solid and dotted lines show the simulation results, and the triangle and rectangular dots indicate the experimental results.

Understanding the self-pulsing characteristics and methods as how to make use of it will help improving the performance and utility of 2-μm Tm^{3+}-doped fiber lasers.

5. Prospects of 2-μm Tm^{3+}-doped fiber lasers

Based on fast development of high-brightness laser diodes and optimizing of Tm^{3+} fiber fabrication technique, and further understanding about the spectral properties of Tm^{3+} ions, performance of TDFLs can be improved to a new stage. Due to its so many specific advantages, the Tm^{3+}-doped fiber laser has great potential in development toward high-power output, wide wavelength tunability, narrow pulse duration, and high peak power.

6. References

[1] P. Myslinski, X. Pan, C. Barnard, J. chrostowski, B. T. Sullivan, and J. F. Bayon, "Q-switched thulium-doped fiber laser," Opt. Eng. 32 (9), 2025-2030 (1993).

[2] L. Esterowitz, "Diode-pumped holmium, thulium, and erbium lasers between 2 and 3 μm operating CW at room-temperature," Opt. Eng., 29 (6): 676-680 (1990).

[3] R. C. Stoneman and L. Esterowitz, "Efficient, broadly tunable, laser-pumped Tm-YAG and Tm-YSGG CW lasers," Opt. Lett., 15 (9): 486-488 (1990).

[4] S. W. Henderson, P. J. M. Suni, C. P. Hale, S. M. Hannon, J. R. Magee, D. L. Bruns, and E. H. Yuen, "Coherent laser-radar at 2 μm using solid-state lasers," IEEE Trans. Geosci. Remote Sens., 31 (1): 4-15 (1993).

[5] I.T. Sorokina, K.L. Vodopyanov (Eds.): Solid-State Mid-Infrared Laser Sources, Topics Appl. Phys. 89, 219–255 (2003).

[6] H.W. Gandy, R.J. Ginther and J.F. weller, Stimulated emission of Tm3+ radiation in silicate glass, J. Appl. Phys. 38: 3030-3031 (1967).]

[7] R.G. Smart, J.N. Carter, A.C. Tropper and D.C. Hanna, "continuous-wave oscillation of Tm-doped fluorozirconate fibre laser at around 1.47m, 1.9 and 2.3 when pumped at 790nm", Opt. Commun. 82: 563-570 (1991).

[8] Jihong Geng, Jianfeng Wu, and Shibin Jiang, "Efficient operation of diode-pumped single-frequency thulium-doped fiber lasers near 2 μm," Opt. Lett., 32 (4): 355-357 (2007).

[9] N. Y. Voo, J. K. Sahu, and M. Ibsen, "345-mW 1836-nm Single-Frequency DFB Fiber Laser MOPA," IEEE Photon. Technol. Lett. 17, 2550 (2005).

[10] Jianqiu Xu, Mahendra Prabhu, Jianren Lu, Ken-ichi Ueda, and Da Xing, "Efficient double-clad thulium-doped fiber laser with a ring cavity," Applied Optics, 2001, 40(12): 1983-1988.

[11] D.C. Hanna, I.M. Jauncey, R.M. Percival, I. R. Perry, R.G. Smart, P. J. Suni, J. E. Townsend, A. C. Tropper: Continuous-wave oscillation of a monomode thulium-doped fibre laser, Electron. Lett. 24, 1222 (1988).

[12] W. L. Barnes, J.E. Townsend: Highly tunable and efficient diode pumped operation of Tm3+ doped fibre lasers, Electron. Lett. 26, 746 (1990).

[13] J. N. Carter, R.G. Smart, D.C. Hanna, A.C. Tropper: CW diode-pumped operation of 1.97 μm thulium-doped fluorozirconate fibre laser, Electron. Lett. 26, 599 (1990).

[14] Jackson, S.D., and King, T.A.: 'High-power diode-cladding-pumped Tm-doped silica fiber laser', Opt. Lett., 1998, 23, pp. 1462–1464.

[15] R.A. Hayward, W.A. Clarkson, P.W. Turner, J. Nilsson, A.B. Grudinin and D.C. Hanna, Efficient cladding-pumped Tm-doped silica fibre laser with high power singlemode output at 2μm, Electron. Lett., 36 (8): 711-712 (2000).

[16] G. Frith, D.C. Lancaster and S.D. Jackson, "85W Tm3+-doped silica fibre laser," Electron. Lett., 41, 687-688 (2005).

[17] Y. Jeong, P. Dupriez, J. K. Sahu, J. Nilsson, D. Shen, and W. A. Clarkson, "Thulium-ytterbium co-doped fiber laser with 75 W of output power at 2 μm," SPIE, 5620: 28-35 (2004).

[18] Evgueni Slobodtchikov, Peter F. Moulton, Gavin Frith, "Efficient, High-Power, Tm-doped Silica Fiber Laser," 2007ASSP-MF2.

[19] P. F. Moulton, G. A. Rines, E. V. Slobodtchikov, K. F. Wall, G. Frith, B. Samson, and A. L. G. Carter, "Tm-Doped Fiber Lasers: Fundamentals and Power Scaling," IEEE J. Sel. Topics in Quantum Electronics, 15 (1): 85-92 (2009).

[20] T. Ehrenreich, R. Leveille, I. Majid, and K. Tankala, "1 kW, all-glass Tm:fiber laser," in Proc. SPIE Photonics West 2010: LASE Fiber Lasers VII: Technology, System and Applicatons, San Francisco, CA, Jan. 2010, Paper 7580.

[21] G. D. Goodno, L. D. Book, and J. E. Rothenberg, "low-phase-noise single-frequency single mode 608 W thulium fiber amplifier," Opt. Lett. 34 (8): 1204-1206 (2009).

[22] P. Myslinski, X. Pan, Ch. Barnard, et al., "Q-switched thulium-doped fiber laser," Opt. Eng. 32, 2025-2030 (1993).

[23] L.E. Nelson, E.P. Ippen, and H.A. Haus, "Broadly tunable sub-500 fs pulses from an additive-pulse mode-locked thulium-doped fiber ring laser," Appl. Phys. Lett. 67, 19-21 (1995).

[24] A.F. El-Sherif and T.A. King, "High-peak-power operation of a Q-switched Tm3+-doped silica fiber laser operating near 2μm," Opt. Lett. 28, 22-24 (2003).

[25] G. Imeshev and M. E. Fermann, "230-kW peak power femtosecond pulses from a high power tunable source based on amplification in Tm-doped fiber," Opt. Express 13(19), 7424–7431 (2005).

[26] B. C. Dickinson, S. D. Jackson, and T. A. King, "10 mJ total output from a gain-switched Tm-doped fibre laser," Opt. Commun. 182(1-3), 199–203 (2000).

[27] Y. J. Zhang, B. Q. Yao, Y. L. Ju, and Y. Zh. Wang, "Gain-switched Tm3+-doped double-clad silica fiber laser" Opt. Express 13 (2005) 1085-1089.

[28] M. Eichhorn and S. D. Jackson, "Actively Q-switched Tm3+-doped and Tm3+,Ho3+-codoped Silica Fiber Lasers," in Conference on Lasers and Electro-Optics and Quantum Electronics and Laser Science Conference 2008, (San Jose, CA, 2008).

[29] M. Eichhorn and S. D. Jackson, "High-pulse-energy actively Q-switched Tm3+-doped silica 2 μm fiber laser pumped at 792 nm," Opt. Lett. 32(19), 2780–2782 (2007).

[30] M. Eichhorn and S. D. Jackson, "High-pulse-energy, actively Q-switched Tm3+,Ho3+ -codoped silica 2 μm fiber laser," Opt. Lett. 33(10), 1044–1046 (2008).

[31] D. Creeden, P. Budni, and et al., "High Power Pulse Amplification in Tm-Doped Fiber," in Conference on Lasers and Electro-Optics, OSA Technical Digest (CD), paper CFD1 (Washington DC, 2008).

[32] Yulong Tang, Lin Xu, Yi Yang, and Jianqiu Xu, "High-power gain-switched Tm3+-doped fiber laser," Opt. Express 18, 22964 (2010).

[33] D. Creeden, P. A. Ketteridge, and et al., "Mid-infrared ZnGeP2 parametric oscillator directly pumped by a pulsed 2 microm Tm-doped fiber laser," Opt. Lett. 33(4), 315–317 (2008).

[34] R. J. De Young and N. P. Barnes, "Profiling atmospheric water vapor using a fiber laser lidar system," Appl. Opt. 49(4), 562–567 (2010).

[35] Y. Wang and Ch-Q. Xu, "Actively Q-switched fiber lasers: Switching dynamics and nonlinear processes" Progress in Quant. Electron. 31, 131-216 (2007).

[36] A. E. Siegman, LASERS, 1986, P492, P1017, Miller/Scheier Associates, Palo Alto, CA.

[37] Ch. Ye, M. Gong, P. Yan, Q. Liu, and G. Chen, "Linearly-polarized single-transverse-mode high-energy multi-ten nanosecond fiber amplifier with 50W average power," Opt. Express 14, 7604-7609 (2004).

[38] B. C. Dickinson, S. D. Jackson, and T. A. King, "10 mJ total output from a gain-switched Tm-doped fibre laser," Opt. Commun. 182(1-3), 199–203 (2000).
[39] Y. J. Zhang, B. Q. Yao, Y. L. Ju, and Y. Zh. Wang, "Gain-switched Tm3+-doped double-clad silica fiber laser," Opt. Express 13(4), 1085–1089 (2005).
[40] S. D. Jackson and T. A. King, "Efficient Gain-Switched Operation of a Tm-Doped Silica Fiber Laser," IEEE J. Quantum Electron. 34(5), 779–789 (1998).
[41] M. Jiang and P. Tayebati, "Stable 10 ns, kilowatt peak-power pulse generation from a gain-switched Tm-doped fiber laser," Opt. Lett. 32(13), 1797–1799 (2007).
[42] Y. L. Tang and J. Q. Xu, "Effects of excited-state absorption on self-pulsing in Tm3+-doped fiber lasers," J. Opt. Soc. Am. B 27(2), 179–186 (2010).
[43] G. Frith, D. G. Lancaster, and S. D. Jackson, "85 W Tm3+-doped silica fibre laser," Electron. Lett. 41(12), 687–688 (2005).
[44] S. D. Jackson, "Cross relaxation and energy transfer upconversion processes relevant to the functioning of 2 μm Tm3+-doped silica fibre lasers," Opt. Commun. 230(1-3), 197–203 (2004).
[45] B. C. Stuart, M. D. Feit, S. Herman, A. M. Rubenchik, B. W. Shore, and M. D. Perry, "Nanosecond-to-femtosecond laser-induced breakdown in dielectrics," Phys. Rev. B 53(4), 1749-1761 (1996).
[46] W. Koechner, "Solid-state laser engineering", 5th ed. P685. Springer Verlag Berlin Heidelberg New York.
[47] E. C. Honea, R. J. Beach, S. B. Sutton, J. A. Speth, S. C. Mitchell, J. A. Skidmore, M. A. Emanuel, and S. A. Payne, "115-W Tm:YAG diode-pumped solid-state laser," IEEE J. Sel. Top. Quantum Electron. 33(9), 1592–1600 (1997).
[48] C. D. Brooks and F. D. Teodoro, "1-mJ energy, 1-MW peak-power, 10-W averagepower, spectrally narrow, diffraction-limited pulses from a photonic-crystal fiber amplifier," Opt. Express 13(22), 8999-9002 (2005).
[49] F. D. Teodoro and C. D. Brooks, "1.1 MW peak-power, 7 W average-power, high-spectral-brightness, diffraction-limited pulses from a photonic crystal fiber amplifier," Opt. Lett. 30(20), 2694-2696 (2005).
[50] R.M. Percival, D. Szebesta, C.P. Seltzer, S.D. Perrin, S.T. Davey, and M. Louka, "A 1.6 μm pumped 1.9-μm thulium-doped fluoride fiber laser and amplifier of very high efficiency," IEEE J. Quantum Electron., vol. 31, no. 3, pp. 489-493 (1995).
[51] R. C. Sharp, D. E. Spock, N. Pan, and J. Elliot, "190-fs passively mode-locked thulium fiber laser with a low threshold," Opt. Lett., vol. 21, no. 12, pp. 881-883 (1996).
[52] M. Jiang and P. Tayebati, "Stable 10 ns, kilowatt peak-power pulse generation from a gain-switched Tm-doped fiber laser," Opt. Lett., vol. 32, no. 13, pp. 1797 1799 (2007).
[53] Yulong Tang, F. Li, and J.Q. Xu, "High Peak-Power Gain-Switched Tm3+-doped Fiber Laser," IEEE Photon. Tech. Letts. 23, 893-895 (2011).
[54] S.D. Jackson and T.A. King, "Theoretical modeling of Tm-doped silica fiber lasers," J. Lightwave Technol., vol. 17, no. 5, pp. 948-956 (1999).
[55] Ashraf F. El-Sherif, Terence A. King, "Dynamics and self-pulsing effects in Tm3+-doped silica fibre lasers," Opt. Commun. 208, 381-389(2002).
[56] F.Z. Qamar and T.A. King, "Self-induced pulsations, Q-witching and mode-locking in Tm-silica fibre lasers", J. Mod. Opt. 52 (7), 1031-1043 (2005).

[57] D. Marcuse, "Pulsing behavior of a three-level laser with saturable absorber," IEEE J. Quantum Electron. 29, 2390–2396 (1993).

[58] S. D. Jackson and T. A. King, "Dynamics of the output of heavily Tm-doped double-clad silica fiber lasers," J. Opt. Soc. Am. B 16, 2178–2188 (1999).

[59] F. Z. Qamar and T. A. King, "Self-mode-locking effects in heavily doped single-clad Tm3+-doped silica fibre lasers," J. Mod. Opt. 52, 1053–1063 (2005).

[60] S. D. Jackson and T. A. King, "High-power diode-cladding pumped Tm-doped silica fiber laser," Opt. Lett. 23, 1462–1464 (1998).

[61] Y. Tang and J. Xu, "Self-induced pulsing in Tm3+-doped fiber lasers with different output couplings," Proc. SPIE 7276, 72760L–72760L-10 (2008).

[62] Michel J. F. Digonnet, "Rare-Earth-Doped Fiber Lasers and Amplifiers," P382, second ed. Marcel Dekker Inc., New York, Basel (2001).

[63] Yulong Tang, Jianqiu Xu, "Effects of Excited-state Absorption on Self-pulsing in Tm3+-doped Fiber Lasers," J. Opt. Soc. Am. B 27, 179-186 (2010).

[64] W. Koechner, "Solid-State Laser Engineering", Fifth Edition, springer-Verlag Berlin Heidelberg New York, pp.17–27, 1999.

[65] Gunnar Rustad and Knut Stenersen, "Modeling of Laser-Pumped Tm and Ho Lasers Accounting for Upconversion and Ground-State Depletion", IEEE Journal of Quantum Electronics, vol. 32, no. 9, pp. 1645–1656, September 1996.

[66] Igor Razdobreev and Alexander Shestakov, "Self-pulsing of a monolithic Tm-doped YAlO3 microlaser", Physical Review A, vol. 73, no. 5, pp. 053815 (1-5), 2006.

[67] S.D. Jackson and T.A. King, "Theoretical modeling of Tm-doped silica fiber lasers," J. Lightwave Tech. vol. 17, no. 5, 948-956, 1999.

[68] B. M. Walsh, N. P. Barnes, "Comparison of Tm:ZBLAN and Tm: silica fiber lasers; Spectroscopy and tunable pulsed laser operation around 1.9 µm," Appl. Phys. B, vol. 78, 325-333 (2004).

[69] Yulong Tang, Jianqiu Xu, "Model and characteristics of self-pulsing in Tm3+-doped silica fiber lasers," IEEE J. Quantum Electron. 47, 165-171 (2011).

17

Advances in High-Power Laser Diode Packaging

Teo Jin Wah Ronnie
Singapore Institute of Manufacturing Technology, Singapore

1. Introduction

Rapid evolution of semiconductor laser technology and its declining cost during the last decades have made the adoption of high-power laser diodes more readily affordable. The continuous pursue for higher lasing power calls for better thermal management capability in the packaging design to facilitate controlled operation. As these laser diodes generate large amount of heat fluxes that can adversely affect their performances and reliability, a thermally-effective packaging solution is required to remove the excessive heat generated in the laser diode to its surroundings as quickly and uniformly as possible.

For high-power applications, one needs to consider not only the thermal challenges, but also the mechanical integrity of the joint as well as the electrical coupling to the remaining components in the module. These poses significant packaging challenges as these factors complicate the effort to create an ideal packaging design. Firstly, laser diodes generate large heat fluxes, up to the range of $MWcm^{-2}$, which causes excessive temperature rise in the active region at high injection currents. The joint which secures the laser diode onto the assembly must be able to withstand the heat generated from the laser diode and capable of maintaining its structural integrity during the long service life of the device. Secondly, the parametric performance and the reliability of these laser diodes are sensitive to both temperature and stress. During the bonding process, excessively-induced bonding stress causes the parametric performances of the laser diode to change. Furthermore, laser diode packaging requires stringent alignment tolerance in order to achieve high optical fiber coupling. On top of the aforementioned, a cost-effective packaging solution is also important because packaging usually dominates the cost of the laser diode module. Hence, the development of laser diode packaging not only is a technological challenge for achieving better performances, but also a critical step for possible commercialization of the product.

2. Thermal management of high-power laser diodes

In a laser diode package, the heat generated in the laser diode is transferred to the ambient environment by attaching a heat sink or heat spreader onto the laser diode. The laser diode must be attached to the package optimally to ensure an efficient heat transfer through the thermal interface. A thin void-free bonding interface is desired to create an effective heat dissipation channel through the die attachment process. To understand the effectiveness of its heat dissipation capability, thermal resistance calculations are usually employed to evaluate the thermal design and performance. In interface engineering, the usual measure of the heat flow in a laser diode package can be expressed as:

$$R_{package} = R_{laser\ diode} + R_{interface\ material} + R_{contact} + R_{heat\ sink} \quad (1)$$

where $R = \dfrac{L}{kA}$, L is the thickness, k is the thermal conductivity of the material, and A is the area of heat transfer.

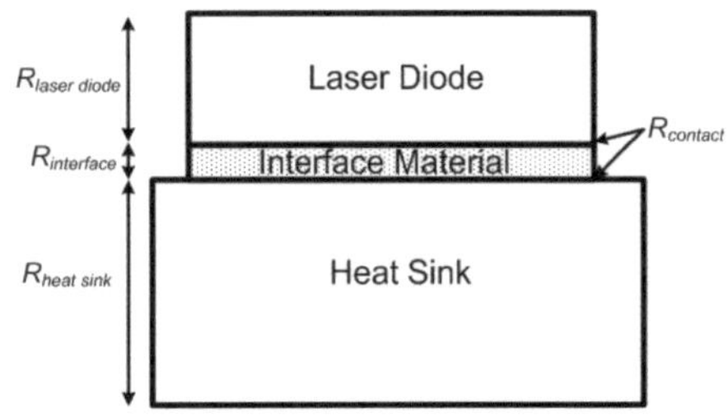

Fig. 1. Schematic diagram of the typical laser diode package and its associated thermal resistance.

Consequently, to improve the thermal design of the laser diode package, the thermal resistance should be minimized by:

- Bringing the heat source to the heat sink as close as possible,
- Making the interface as thin as possible,
- Increasing the thermal conductivity of the material,
- Providing intimate thermal contact between the laser diode and the heat sink.

For laser diode die attach, there are two bonding configurations; epi-side up and epi-side down (see Fig. 2). Eutectic die bonding processes for epi-side up bonding approaches have well been established by the semiconductor packaging industry (Qu, 2004; Larsson, 1990). The heat generated in the active region of the laser diode has to flow through the entire (GaAs/InP) substrate before reaching to the heat sink. The heat generated in the active region spreads laterally over the entire width of the laser as it flows to the heat sink, leading to a two-dimensional heat flow in the laser diode. This two-dimensional heat flow accounts for the logarithmical dependence of ridge width (Boudreau et al., 1993). Due to the low thermal conductivity of ternary alloys and multiple heterostructures (Capinski et al., 1999) of the laser diode, the heat generated in the active region cannot be dissipated onto the heat sink efficiently. Especially for single-mode (≤4 μm) ridge-waveguide laser diodes, the bulk of the heat generation is confined within the active region. This raises concerns as significant heat accumulation in the active region influences the spectral and spatial characteristics, with longitudinal modes broadening (Spagnolo et al., 2001). The optical output power will also be compromised since their performance characteristics are sensitive to the operating temperature of the laser diode.

As depicted in Fig. 2(b), epi-side down bonding is recommended for effective heat transfer since the proximity of the active region is only a few microns from the surface (Hayashi, 1992; Lee & Basavanhally, 1994; Katsura, 1990). The proximity of the active region to the top of the heat sink strongly influences the heat flow. Owing to the good thermal conductivity of the heat sink material, the heat produced in the active region is rapidly distributed to the heat sink. Substantial improvement can be achieved as the $R_{laser\ diode}$ is inversely proportional to the ridge width for epi-side down bonded approach (Martin et al., 1992). The $R_{laser\ diode}$ for a epi-side down bonded laser diodes was reported to be ~30% smaller than that of a epi-side up bonded laser diodes.

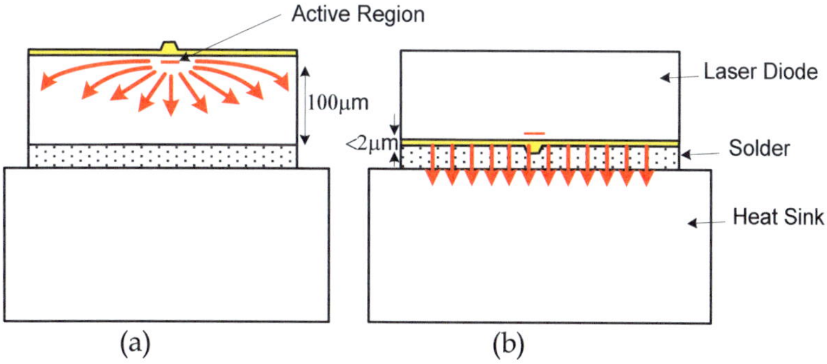

Fig. 2. Comparison of different bonding configurations of ridge-waveguide laser diodes. (a) For epi-side up bonded laser diode, the heat generated in the active region is ineffectively transferred through the substrate; (b) For epi-side down bonded laser diode, the heat flux is effectively reached the heat sink within several microns

Epi-side down bonding also eliminates the trade-off problem between high-frequency modulation (Delpiano et al., 1994) and temperature control. In a laser diode package, three forms of electrical parasitics were present; intrinsic diode, external chip and package parasitics. For high-frequency applications, the external parasitics of the laser diode chip and the package prevent the module from achieving higher frequency modulation. In order to drive the laser diode into higher frequency modulation, epi-side down bonding approach can help to reduce the external electrical parasitics significantly. However, there are physical constraints for epi-side down bonding. The stress associated with epi-side down bonding may cause physical distortion to the device due to the mismatch in the coefficient of thermal expansion (CTE) between the laser diode and the heat sink material. As shown in Fig. 3, the laser diode will experience stresses after the die-attachment process. When the package is cooled to room temperature, the laser diode will experience compressive stresses due to the large CTE mismatch between the laser diode and the heat sink material. The close proximity between the laser diode and the heat sink further promote the strain accumulation in the active region. Relieving this stress will, hence, improve the operating life of the device (Hayakawa et al., 1983).

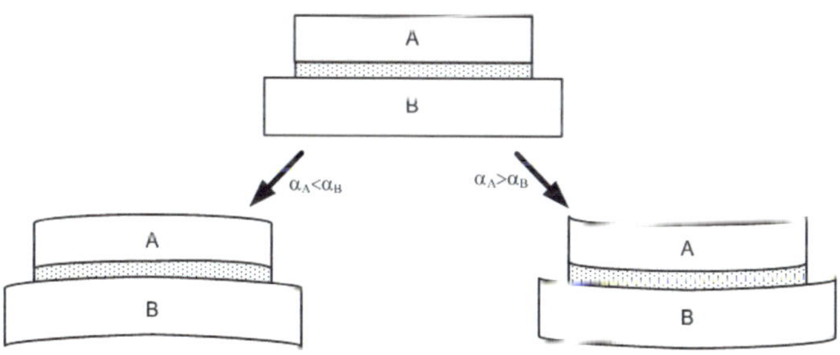

Fig. 3. Depending on the CTE properties of the two materials, compressive or tension stress can be observed after bonding.

2.1 Advanced Die-attachment techniques

After identifying that epi-side down bonding configuration as the preferred approach, the bonding process must also be able to realise the full potential of the improved thermal management capability. A major consideration in the bonding process is the formation of voids at the interface material. The primary concern with voids involves the loss of thermal conductivity. The void volumes add to the thickness to the joint volume, thus increasing the joint thickness, resulting in higher thermal resistance. Voids can cause hotspots by creating areas of poor heat dissipation paths. Voids in the thermal interface material not only limit the thermal dissipation capability but also deteriorate the electrical and mechanical properties of the joint. Continued product miniaturization and increased power density makes the importance of minimizing the void-fraction in the joint even more significant.

Hence, various die-attachment techniques have been introduced to tackle the plurality of complicating issues in epi-side down bonding approach (see Table 1). A flip-chip interconnection technique using small solder bumps (Hayashi, 1992) was introduced to resolve the alignment issues. When the solder bumps melt in the reflow process, the surface tension of the molten solder allows self-alignment and accomplishes precise chip positioning. However, thermal dissipation is compromised since heat can only be transferred through the solder bumps. Bridged die bonding (Boudreau et al., 1993) was introduced to exploit the advantage of epi-side down bonding while avoiding the stress issue by employing a solder pattern with a gap, preventing solder from having contact with the sensitive ridge. This bridged die bond geometry enhances the heat dissipation capability compared to epi-side up bonding configuration. However, this approach does not directly address the thermal issue as the heat flux generated in the active region has to be re-directed to the side of the laser diode before travelling towards the heat sink. To improve the thermal dissipation capability further, a vacuum-release process (Bascom & Bitner, 1976; Mizui & Tokuda, 1988) is recommended to produce a fluxless and virtually voidless solder bond. Epitaxial lift-off technique (Dohle et al., 1996) was also introduced to provide good bond quality between the semiconductor chip and the substrate. Ultrasonic bonding or bonding with scrubbing effect (Pittroff, 2001, 2002) was proposed to resolve any surface irregularities during bonding. However, special care must be taken as the solder material may bridge onto the facet easily, obscuring the emitted laser beam for edge-emitting laser diodes. The stress induced by the scrubbing process should be limited to avert any damage incurred on the laser diodes.

In general, $R_{interface\ material}$ can be reduced by applying a pressure to ensure good thermal contact between the laser diode and the solder material during the bonding (solder reflow) process. The pressure induces compressive stress onto the laser diode which may cause structural distortion to the device. Molten-state bonding (Tew et al., 2004) was proposed to alleviate the bonding stress induced. The solder material was pre-heated into molten state before a pressure is applied. Due to its molten state, the pressure applied onto the laser diode was minimal. At the same time, the bonding temperature can be lowered and the bonding time is reduced to mere seconds. This is potentially advantageous for high-volume production where a rapid bonding cycle is favoured. However, just like ultrasonic bonding or scrubbing approach mentioned earlier, the bonding parameters and conditions must be optimized (Teo et al., 2008).

Bonding Method	Bonding Parameters	Solder
Solder bump chip bonding (Hayashi 1990, 1992)	230 °C, 60 secs,	In-Pb
Vacuum-release process (Bascom & Bitner, 1976; Mizui & Tokuda, 1988)	133.3 Pa, 1 sec	Pb37Sn63
Composite solder structure (Lee & Wang, 1991, 1992)	320 °C, 13 mins 0.276 MPa	Au-Sn
Bridge die bonding (Boudreau et al., 1993)	230 °C, 1 min	In
Epitaxial lift-off (Dohle et al., 1996)	235 °C, 30 secs, 0.345 MPa	Au-Sn
Pulsed heated thermode (Pittroff, 2001, 2002)	370 °C, 5 secs, 0.375 MPa	Au80Sn20
Controlled solder interdiffusion (Merritt et al., 1997)	157-232 °C, 1 MPa	In-Sn
Molten-state bonding (Teo et al., 2004, 2008	290-310 °C, few secs	Au80Sn20

Table 1. Comparison of various bonding techniques proposed for laser diode packaging. The bonding parameters depend largely on the bonding technique and interface material used.

2.2 Thermal interface materials

As shown in Table 2, solders are utilized in every part of a laser diode assembly due to their electrical interconnect, mechanical support and heat dissipation capabilities. These solders can be commonly categorized into two types; hard solder and soft solder. The decision to use soft or hard solder is based on the optimization of a number of properties, including solder strength, solder migration, creep, fatigue, whisker formation, stress, thermal expansion, liquidus temperature, and thermal conductivity of each solder type. It is also dependent on the application as well as the hierarchy of the package.

In general, the solder material must satisfy the following requirements:

- Have the desired processing temperature to support high temperature operation
- Provide sufficient wetting characteristics to form metallurgical bond between the laser diode and heat sink
- Provide an efficient heat dissipating channel to the heat sink
- Reduce thermally induced stresses arise from the mismatch of thermal expansion between the laser diode and heat sink
- Exhibit no/low deformation during its long-term operation
- Exhibit low electrical resistivity to reduce Joules heating at high injection current

Table 2 shows a list of some common solder materials used in laser diode packaging. Soft solder, commonly containing large percentage of lead, tin and indium, has very low yield strength and incurs plastic deformation under stresses. Their capability to deform plastically helps to relieve the stress developed in the bonded structure. However, this makes soft solder subject to thermal fatigue and creep rupture, causing long-term reliability problems (Solomon, 1986; Lau & Rice, 1985). They are also attributed to solder instabilities like whisker growth, void formation at the bonding part, and diffusion growth (Mizuishi et al., 1983, 1984; Sabbag &McQueen, 1975). Whiskers growth may short the electrical connection

Solder	Physical Properties				
	Melting Point (°C)	Tensile Strength (MPa)	Creep Resistance	Thermal conductivity (W/m/°C)	Electrical Resistivity (μΩ.cm)
Hermetic Sealing: Eutectic die attach – fiber location – fluxless soldering					
Au88Ge12	356E	185	Excellent	44	-
Au80Sn20	280E	276	Excellent	57	16
Opto-package assembly step soldering – non amenable to fluxless soldering					
Sn63Pb37	183E	32	Moderate	50	15
96SnAgCu	215-225	~40	High	-	Na
Fiber ferrule joining					
Indium	156	2.5	Poor	80	8.8
In97Ag3	146E	5.5	Poor	73	7.5
In52Sn48	118E	11.9	Low	34	15
Component Anchoring					
Bi58Sn42	138E	55	Moderate	19	35
Bi40Sn60	138-170	-	Moderate	21	38

Table 2. Comparison of various solder materials used in butterfly laser diode package and their physical properties [32].

and obstruct its optical beam. As mentioned earlier, the laser diode package will experience elevated temperature during operation. When the laser diode package is subjected to temperature above 65 °C, the homologous temperature of soft solder material such as Indium is more than 0.8. The solder material will experience high creep deformation, which implies a reliability concern for these solder joints. Hence, bonding of laser diodes using soft solder will face reliability problems (Shi et al., 2000, 2002). The reliability of the joint is a critical issue for the practical design and fabrication of a mechanically stable and reliable assembly.

Hard solder, on the other hand, has very high yield strength and thus incurs elastic rather than plastic deformation under stresses. Eutectic Au80Sn20 alloy are usually adopted for high-power laser diode applications to overcome the reliability issues (Fujiwara, 1982). Accordingly, it has good thermal conductivity and is free from thermal fatigue and creep movement phenomena (Matijasevic et al., 1993). Unfortunately, hard solder does not help to release the stresses developed during the bonding cycle because of low plastic deformation in the solder material.

2.3 Next generation heat sink materials

The standard heat sink material in nearly all commercially available laser diode packages is copper owing to its excellent thermal conductivity, its good mechanical machining properties, and its comparably low price. However, with the global demands for increasing output power, heat sink material with even higher thermal conductivity is desired. In response to these needs, an increasing number of ultra-high thermal conductivity materials have been and are being developed that offer significant improvements that may be suitable for high-power laser diode applications (Zweben, 2005). Not only do these materials possess very high thermal conductivity compared to traditional packaging materials, they also offer low CTE properties to reduce thermal stresses that can affect the performance and reliability of the package. While some of these materials are still at its infancy, they offer an alternative perception of how they can contribute to future thermal management problems, especially in the area of high-power applications.

2.4 Cooling approaches for high-power applications

In most thermal management problems, removing the heat flux generated from the heat source (in this case, laser diode) to the heat sink by means of conduction alone is insufficient. To maintain its high efficiency and long lifetime, the operating temperature of the laser diode package is kept as low as possible, typically below 60 °C. Depending on the thermal power density, these heat sink materials are further cooled by means of passive- or active-cooling approaches. The design and analysis of heat sinks is one of the most extensive research areas in electronics cooling (Rodgers, 2005). The heat sinks function by extending the surface area of heat dissipating surfaces through its use of fins.

Air-cooling is traditionally associated with the use of heat sinks. The use of fan technology faces scrutiny as the acoustic noise generated and fan bearing may in some way or another affect the functionality and reliability of the application. Furthermore, the limits of air-cooling capabilities force the migration of air- to liquid or thermoelectric cooling. Liquid cooling solutions have proven to be able to manage large transient heat loads, especially in areas where design/space constraints limit the use of forced air-cooling approach. Micro-channel heat sinks are the state-of-the-art solution for maximum cooling performance (Leers, 2007, 2008). However, the integration of liquid-cooling technology raises reliability, cost and weight issues. Alternatively, thermoelectric cooler (TEC) can also be used to cool and regulate the operating temperature of the assembly. In fact, thermoelectric coolers have been widely used in a pump laser package to cool the laser diode and achieve wavelength and power stability. The TEC provide an effective negative thermal resistance to regulate the temperature in the laser diode package. Currently, TECs have very low efficiency (more commonly called coefficient of performance) and efforts to develop new TE materials with superior figure of merit (ZT) are ongoing. While each cooling approach has its advantages and limitations, the decision to employ which type of cooling approach depends largely on the thermal power density and construction of the system architecture.

2.5 Microstructure evolution of solder joint

During the bonding (solder reflow) process, a metallurgical bond is formed between the laser diode and the heat sink through the formation of intermetallic compound (IMC) at their respective interfaces. The initial formation of IMCs ensures a good thermal contact $R_{contact}$ at their interfaces. However, these IMCs continues to grow, though much more slowly, during storage and service. The growth of the interfacial IMCs depends on a number of factors, such as temperature/time, volume of solder, property of the solder alloy and the metallizations on the laser diode and heat sink. The IMC growth rate in terms of temperature and time are usually represented according to the empirical relationship,

$$X(t) = X_o + At^n \exp\left(\frac{-Q}{RT}\right) \tag{1}$$

where $X(t)$ is the layer thickness at aging time t; X_o is the initial thickness; A is a numerical constant; Q is the apparent activation energy; T is the aging temperature, and R is the gas constant.

Continuous interfacial reaction may compromise on the integrity and reliability of the solder joint as the IMCs makes the solder joint less ductile and less capable of releasing stresses through plastic strain. Increased interfacial IMCs in the solder material also increases the thermal and electrical resistances of the bonded structure during storage and aging (Kressel, 1976; Fujiwara et al., 1979). This undesirable diffusion growth is of particular technological concern since it is always ongoing and may cause cracks and delaminations, especially in the presence of residual stress. In electronics packaging, the thickness of their solder bumps is typically several hundreds of microns while the thickness of the interfacial IMCs is a few microns. However, for laser diode die-attach, the solder joint has a thickness of only several microns. The influence of the IMC volume in the solder material will have a significant impact on the mechanical strength and reliability of the solder joint (Wong et al., 2005). In this section, three different solder materials commonly found in laser diode packaging will be reviewed.

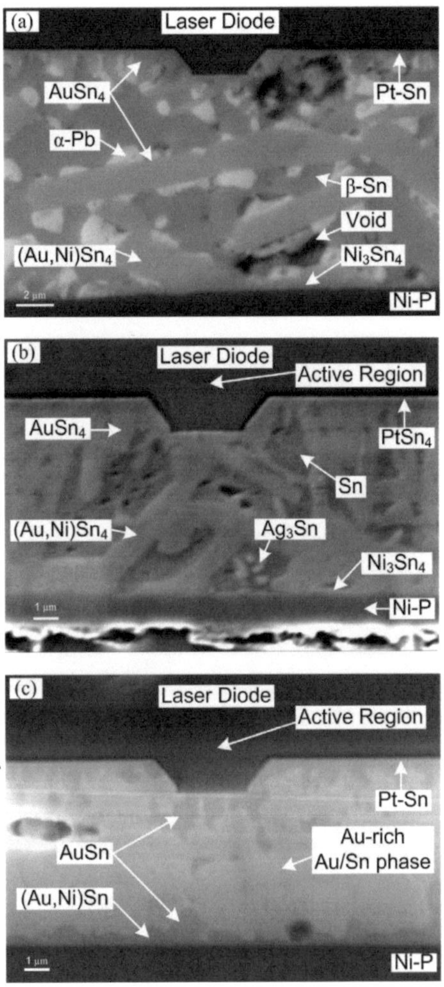

Fig. 4. Typical microstructure behaviour of as-bonded laser diode package using (a) 63Pb37Sn, (b) 3.5Ag96.5Sn, and (c) 80Au20Sn solder systems.

Fig. 4(a)-(c) shows the interfacial reaction of the as-bonded laser diode package using 63Pb37Sn, (b) 3.5Ag96.5Sn, and (c) 80Au20Sn solder systems. During reflow, PtSn and $PtSn_4$ IMCs were observed at the laser diode/solder interface. At the solder/heat sink interface, diffusion of Ni from the heat sink into the solder joint could be detected within 2-3 μm from the solder/heat sink interface. For both the 63Pb37Sn and 3.5Ag96.5Sn solder systems, a layer of Ni_3Sn_4 IMC was formed, followed by $(Au,Ni)Sn_4$ IMCs. Due to the thin solder joint, the $AuSn_4$ and $(Au,Ni)Sn_4$ IMCs could be found in the solder as well as at the interfaces. The $AuSn_4$ and $(Au,Ni)Sn_4$ IMC precipitates were randomly dispersed into the solder joint. However, for the 80Au20Sn solder system, only a thin layer of (Au,Ni)Sn IMCs was observed. The solder joint consists of three Au-Sn phases; δ (AuSn), ζ′ (Au_5Sn) and β ($Au_{10}Sn$) phases (Teo J.W.R et al., 2008). As shown in Fig. 4(c), the δ phase was observed to coalesce to the interfaces while the Au-rich ζ′ and β phases remained at the center of the solder.

During operation, the heat generated in the laser diode will caused the package to experience thermal loading. Metallurgical interaction in the solder joint will continue to occur by means of solid-state processes. Consequently, the composition, microstructure and physical properties of the solder joint changes with the device life. Fig. 5 shows the microstructure evolution of 3.5Ag96.5Sn solder joint during aging at 150 °C. Although the laser diode package may not experience such high temperature during operation, the accelerated temperature aging was often adopted to screen if the solder material and the joint is capable of meeting the desired operating lifespan of the applications.

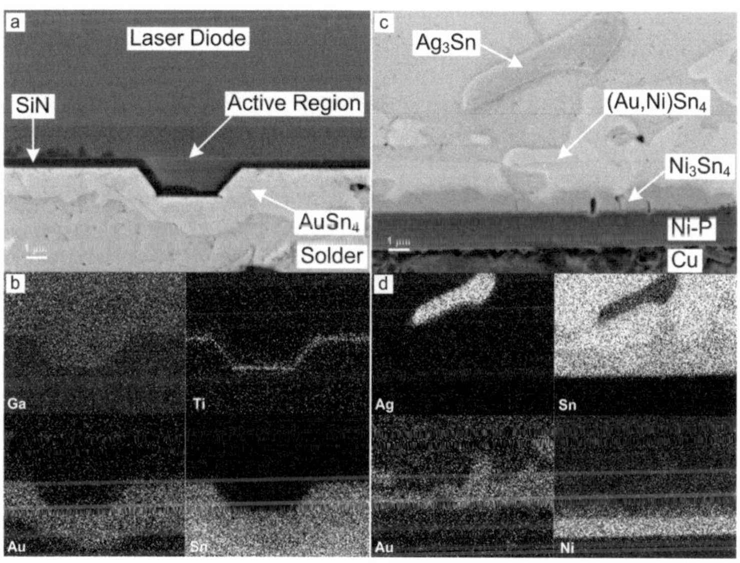

Fig. 5. Typical SEM micrographs and EDX mapping showing the development of intermetallic compound layers in a 3.5Ag96.5Sn soldered laser diode package as a result of solid-state aging at 150 °C at the (a) laser diode/solder and (c) solder/heatsink interface.

During aging, the interfacial IMC thicknesses were found to increase with aging time. For the 63Pb37Sn and 3.5Ag96.5Sn solders, the $AuSn_4$ and $(Au,Ni)Sn_4$ IMCs first settled to the interfaces and then grew with time. As depicted in Fig. 5, the $(Au,Ni)Sn_4$ grains separated from the Ni(P) layer at the roots of the grains in the process of breaking off and a Ni_3Sn_4 IMC layer was formed in between the $(Au,Ni)Sn_4$ IMC and the Ni(P) layer. These interfacial IMC layers, which formed a large portion of the solder joint, could grow up to a thickness of 8 µm. A thick IMC layer at the interface pose reliability concern as stress is usually concentrated around a thick IMC layer (Lee & Duh, 1999). To understand the growth kinetic, the total IMC thickness at the interfaces for the three solder systems was shown in Fig. 6. The IMC thickness for 63Pb37Sn and 3.5Ag96.5Sn solder systems was significantly large while 80Au20Sn solder joint exhibited limited Ni solubility into the solder. The IMC growth rate for 63Pb37Sn solder was initially much faster than 3.5Ag96.5Sn solder, followed by 80Au20Sn solder. When the bonded laser diodes were subjected to thermal aging at 150 °C, the homologous temperature for 63Pb37Sn, 3.5Ag96.5Sn, and 80Au20Sn solders were 0.928, 0.856, and 0.765, respectively. The diffusion activation energy for 63Pb37Sn solder was lower than 3.5Ag96.5Sn and 80Au20Sn solders. Hence, the IMC thickness for the 63Pb37Sn solder joint was initially observed to be the largest. As the aging duration increased, the IMC growth rate for the 63Pb37Sn solder joint reduced while the IMC thickness for 3.5Ag96.5Sn solder joint continued to increase. During aging, the participation of IMC formation at the interfaces reduced the overall Sn content in the solder joint. The reduced Sn composition lowered the Sn activities at the interfaces and hence, kirkendall voids were introduced in the 63Pb37Sn solder joint. On the other hand, the main constituent of 3.5Ag96.5Sn solder is Sn. Even after 49 days of aging, the solder joint still have significant Sn content in the solder joint for further Ni-Sn and Au-Sn interdiffusion. Hence, with increased aging duration, the IMC thickness for 3.5Ag96.5Sn solder continued to grow and surpass the IMC thickness for 63Pb37Sn solder joint. It is important to reinstated that the interfacial IMCs such as $AuSn_4$ and $(Au,Ni)Sn_4$ forms a large part of the $R_{contact}$ highlighted in Eq. 1.

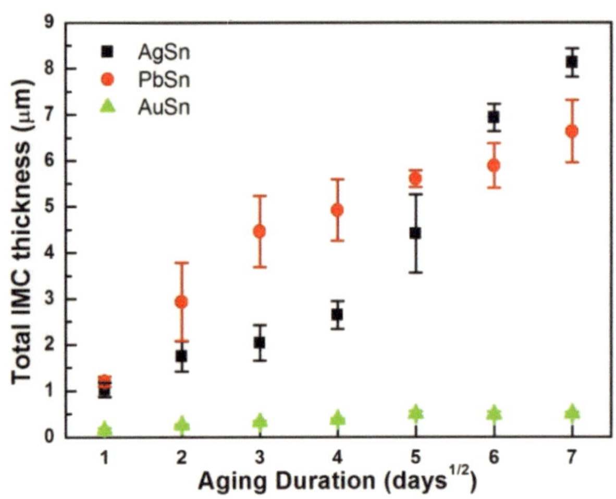

Fig. 6. Interfacial IMC growth of the three solder systems with Ni(P) metallization in solid-state reaction at 150°C.

In the 80Au20Sn system, the solder microstructure did not change significantly from the as-reflowed state (Fig. 5 (c)), even after 49 days of thermal aging. Only a thin layer of $(Ni,Au)_3Sn_2$ IMC was introduced between the AuSn(Ni) IMC and Ni(P) layer. The slow interfacial IMC growth was due to the low Sn content in the solder joint. Furthermore, diffusion of Sn to the interfaces was limited as the Au-rich ζ′ and β phases at the center of the solder joint essentially behaved as a diffusion barrier by preventing more Sn from diffusing into the IMC layers. Hence, the introduction of $(Ni,Au)_3Sn_2$ IMC rather than the growth of the (Au,Ni)Sn layer. The microstructure details of the three solder systems were summarized into Table 3.

Solder	Bulk	Interface ($R_{contact}$)	
		LD	Heatsink
80Au20Sn	$Au_{10}Sn$ (β), Au_5Sn (ζ′)	Pt-Sn, AuSn	AuSn (δ),(Au,Ni)Sn, $(Ni,Au)_3Sn_2$
63Pb37Sn	α-Pb, β-Sn	Pt-Sn, $AuSn_4$	Ni_3Sn_4, $(Au,Ni)Sn_4$
3.5Ag96.5Sn	Sn, Ag_3Sn	Pt-Sn, $AuSn_4$	Ni_3Sn_4, $(Au,Ni)Sn_4$

Table 3. Microstructure summary of the three solder systems.

2.6 Structural Integrity of solder joint

Microstructure and failure mode are closely related in solder joints as the composition, microstructure and physical properties of the joint changes during aging. To assess the mechanical integrity of the solder joint, the laser diode package was subjected to shear testing, a mechanical overloading condition, to determine the weakest interface or material. Good bonding integrity with brittle fracture occurring within the laser diode was observed for all the three solder systems after bonding. The bonded laser diode exhibited a complete fracture after several microns. As shown in Fig. 7, the brittle fracture consisted of wallner lines at the GaAs material and interfacial delamination at the GaAs/SiN passivation interface of the laser diode.

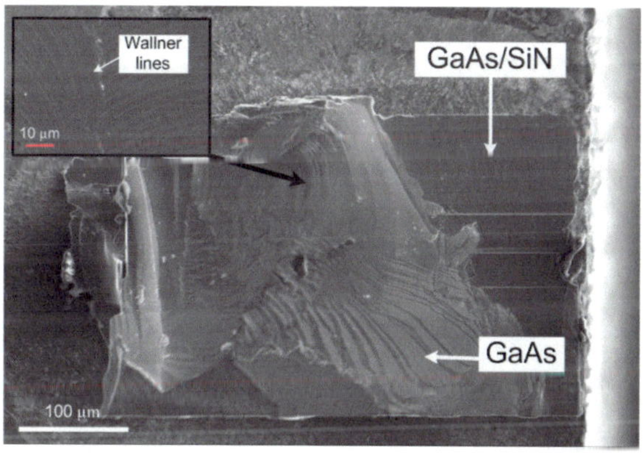

Fig. 7. Typical fracture surface examination of as-bonded LD package for the three solder systems.

During aging, the fracture mode for both 63Pb37Sn and 3.5Ag96.5Sn systems changed to ductile solder fracture as shown in Fig. 9(a)-(b). The peak shear load reduced after aging and the bonded laser diodes were completely removed after shearing off a distance of more than half the length of the laser diode (see Fig. 8). During aging, the $AuSn_4$ and $(Au,Ni)Sn_4$ IMCs settled to the interfaces and grew. These IMCs grew into thick planar morphologies and gross defects were formed in the solder joint. These planar IMC layers reduced the

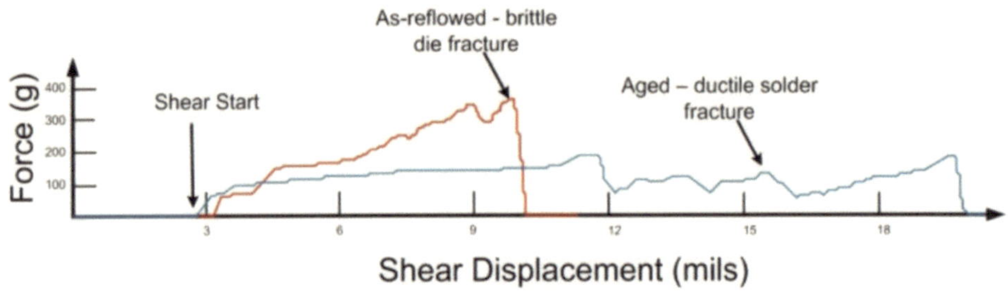

Fig. 8. Typical shear strength profile of the bonded laser diode. The mechanical strength of 63Pb37Sn and 3.5Ag96.5Sn soldered laser diode package reduced with aging.

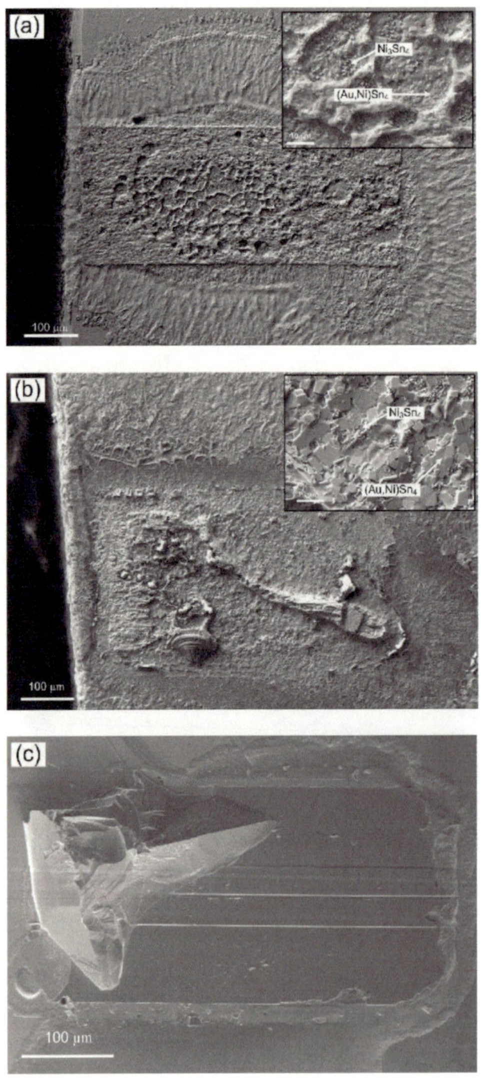

Fig. 9. Fracture surface examination of aged laser diode package for (a) 63Pb37Sn, (b) 3.5Ag96.5Sn, and (c) 80Au20Sn solder.

interfacial adhesion strength, weakening the mechanical integrity of the solder joint. This tally with the fracture surface examination that the failure mode for the aged samples has becomes more ductile. A large amount of plastic deformation was observed to occur inside the solder joint and interfacial solder/(Au,Ni)Sn_4 and (Au,Ni)Sn_4/Ni_3Sn_4 fracture started to take place. As shown in Fig. 10, cracks were observed at the (Au,Ni)Sn_4/Ni_3Sn_4 interface. However, in the 80Au20Sn system, the failure mode did not change even after 49 days of thermal aging (see Fig. 9(c)).

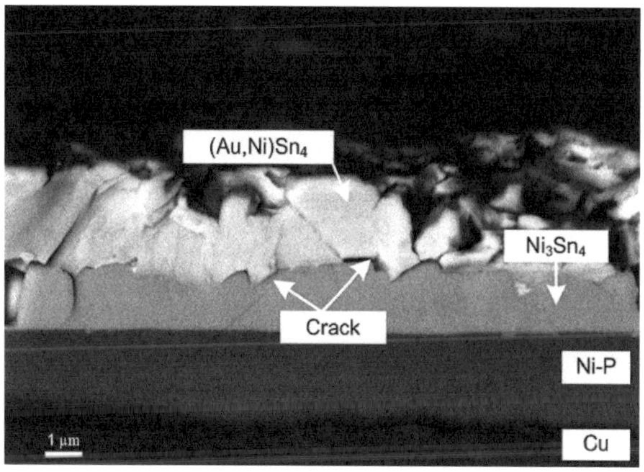

Fig. 10. Cross-sectional examination of the fracture surface at the heat sink surface after 49 days of thermal aging. The thick IMC layers have the tendency to generate structural defects.

3. Thermal behavior of laser diodes

In the previous section, we have highlighted on the importance of devising an efficient thermal management capability in the packaging design to meet the global demand for high-power applications. The heat dissipation capabilities not only depend on the selection of material and means of external cooling but also on the bonding configuration of the laser diode. Since the parametric performance of the laser diodes is strongly influenced by the heat dissipation capabilities, in this section, the thermal behaviour of the laser diodes under different bonding configurations will be evaluated and compared.

3.1 Heating response of laser diodes

Uncoated 980nm single-mode ridge-waveguide laser diodes with a cavity length of 600nm were used for this comparison study. The laser diodes were bonded onto a copper heat sink using the molten-state bonding technique (Teo et al., 2004, 2008). As shown in Fig. 11, the laser diode performance improved with higher optical power achieved after bonding. The typical power achieved for epi-side up bonded laser diodes was ~1.3 times higher whereas in epi-side down bonding approach, the optical power further improved to ~1.5 times before catastrophic damage. This shows that the lasing optical output of the laser diodes is strongly influenced by the heat dissipation capability through the die-attachment interface. Hence, understanding the heating response of the laser diodes is important to the thermal design and optimization.

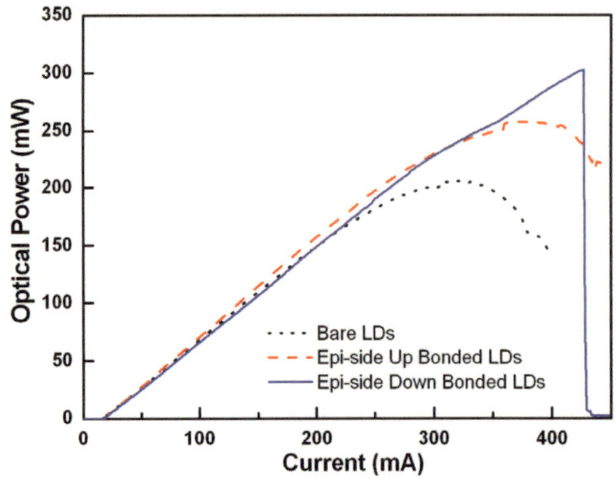

Fig. 11. Influence of bonding on the electrical-optical characteristics of the laser diode.

To understand the heat flow in the diode, it is important to study the transient behaviour of the LDs (Teo et al., 2009). Fig. 2 shows the transient heating response of the LDs at different pulse durations and duty cycles. The emission wavelength did not increase within the first 1 µs of operation. As the pulse width increased, transient heating (emission wavelength) could be observed. The temperature increased at a rate of 2.84 °C/ms and saturated within several milliseconds. Likewise, the temperature in the active region was also observed to vary with duty cycle. When the frequency of the pulse repetition rate increased above 10%, the temperature distribution across the LD was non-uniform, and the temperature in the active region increased exponentially until CW operation. At high pulse repetition rate, the temperature rise in the active region might lead to performance deterioration. From the analogy of heat conduction, the time for the excess heat energy to be transported to the GaAs substrate and reached thermal equilibrium depends on the device and its surrounding medium.

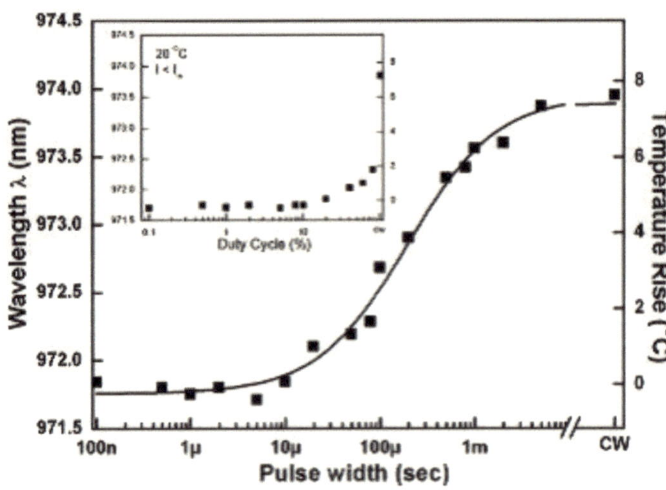

Fig. 12. Emission spectra of LDs as a function of pulse width and duty cycle. Transient heating response was observed from the spectrally resolved emission measurements.

Using Paoli method (Paoli, 1975), the temperature rise in the active region of the laser diode was estimated. Fig. 13 shows the effects of different bonding configuration on the temperature rise in the active region under pulse and continuous-wave operating conditions. In pulse operation, the temperature in the active region did not increase significantly even at a high injection current of 300 mA. This suggested that, at pulse operation, the heat sink did not have a significant influence on the thermal behaviour of the laser diodes and that the heat generated was localized within the laser diode.

However, in continuous-wave operation, Joule heating was evidently shown (see Fig. 13). For the unbonded samples, measurements were conducted until 220 mA before catastrophic damage occurred at its emitting facets. A large temperature rise of more than 100 °C could be observed in the laser diode. For epi-side up and epi-side down bonding, the heat removal means from the laser diode to the heat sink reduced the temperature in the active region at 220 mA to an average of ~70 °C and ~40 °C, respectively. Hence, higher electrical-optical measurements were permissible for epi-side up and epi-side down bonding approach. Two other characteristics were observed. First, at low injection current, the temperature rise for epi-side up bonded laser diodes and unbonded laser diodes were higher than epi-side down bonded laser diodes. The heat generated in the active region could not be removed effectively in unbonded and epi-side up bonded samples and, hence the temperature in the active region was larger than the epi-side down bonded laser diodes. Second, as the injection current increased, an exponential increment of device heating could be observed. At high injection current, additional heating source due to the series resistance of the laser diodes becomes apparent. This behaviour suggested that Joules heating was the dominant heating mechanism at high continuous-wave operating conditions.

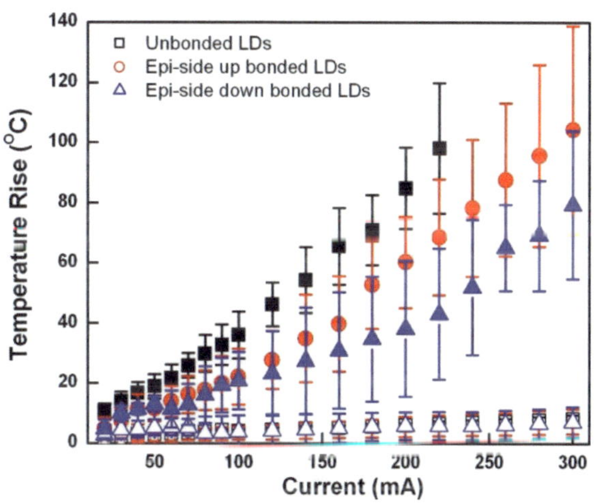

Fig. 13. Heating response of unbonded and bonded laser diodes at high pulse and continuous-wave operation. Joules heating could be observed at high continuous-wave injection current.

3.2 Thermal resistance calculation for laser diodes

Similar to interface engineering, thermal resistance calculation is a common practice to evaluate the thermal behaviour of semiconductor lasers. Typically, thermal resistance is defined as the ratio of the temperature rise in the device to the input power

$$R_{th} = \frac{\Delta T}{\Delta P} \tag{2}$$

where ΔT is the average temperature rise in the active region for a given injected power.

However, the calculation of $R_{laser\ diode}$ is not as straightforward. Equation 3 is valid only for heat generated well below the threshold current since most of the electrical input is converted into heat energy. As the current increases nearer to its threshold current, photon emission becomes more apparent (see Fig. 14). The electrical incremental input is now converted into both heat energy and coherent radiation, which can be extracted from the laser. The exponential temperature dependence of threshold current and the relative importance of Ohmic heating at high injection current further complicate in the thermal analysis as the temperature rise in the active region may not be proportional to the input power.

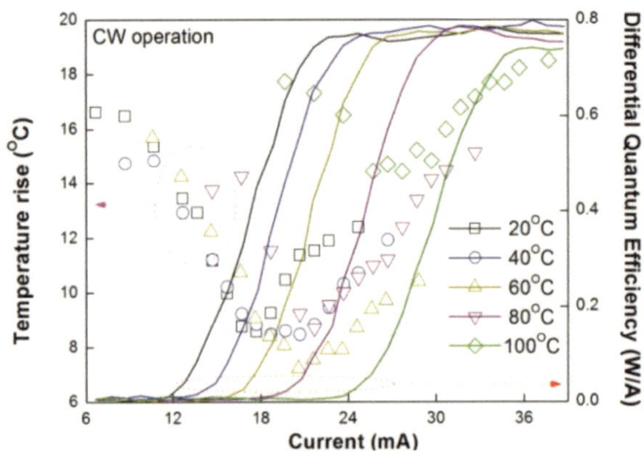

Fig. 14. Temperature rise and quantum efficiency of laser diode at different operating temperatures.

Hence, when measurements are conducted near the lasing threshold, the emitted optical power must be corrected to obtain the correct heat generation rate. Consideration of the heat generation in the active region alone is insufficient to deduce the $R_{laser\ diode}$ for high injection current. Other sources of heating element may also surface; radiative absorption of free carriers and series resistance of the diode. Firstly, the rate of photon absorption differs at different current densities. At high injection current, thermal rollover exists due to an increased of photon absorption. The heat generated in the active region is significantly large and, the effective heat generation rate for ΔP is therefore

$$\Delta P = P - \eta \bullet P = (1-\eta)P \tag{3}$$

where η is the differential quantum efficiency of the diode. η is extracted from the LI curve. Following Eq. (2) and (3), to account for the optical absorption, the thermal resistance is change to

$$R_{laser\ diode} = \frac{\Delta T}{(1-\eta)P} \tag{4}$$

In addition to the heat generated at the junction, Joules heating due to the series resistance R_S may also be present. At low injection current, R_S can be neglected as $IV \rangle\rangle I^2R$. However, as the injection current increases i.e. $IV \langle\langle I^2R$, Joules heating becomes apparent as it increases to the square of current. Hence, the thermal resistance of the laser diode is expressed as

$$R_{laser\ diode} = \frac{\Delta T}{P + I^2R} \frac{1}{(1-\eta)} \tag{5}$$

As shown in Fig. 15, the thermal resistance of the laser diode differs at below and above the lasing threshold. A large $R_{laser\ diode}$ of as much as 1000 °C/W could be observed below the threshold current and it dropped abruptly to 200 °C/W as it approached towards its lasing threshold value. Below the lasing threshold, the thermal resistance followed a linear regression of 130-150 °C/W per mA for all heat sink temperatures, and it remained relatively constant thereafter. The stabilized thermal resistance is terms as the 'effective' thermal resistance. The change of thermal resistance was induced by the transfer of non-radiative energy (non-stimulated emission) into radiative emission of free carriers as discussed earlier. This shows that the dominant cause for the temperature rise, at low injection current, is dominated by the non-radiative recombination.

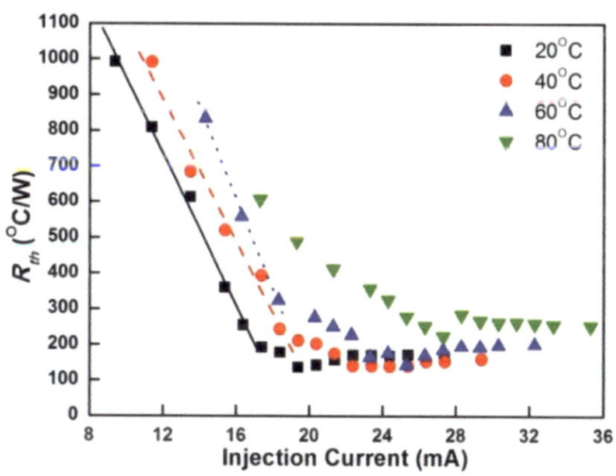

Fig. 15. Thermal resistance of laser diode at a function of current. The effective thermal resistance of the diode varies with current.

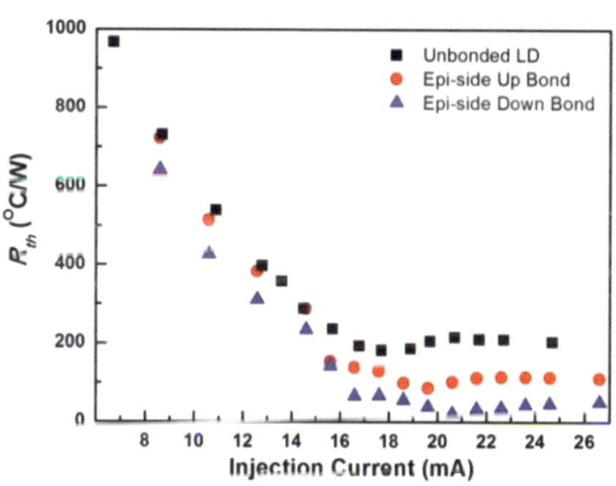

Fig. 16. Comparison of bonding configuration on the thermal resistance of laser diode package. The associated thermal resistance of the package reduced after bonding.

As shown in Fig. 16, the effective thermal resistance of the laser diode package is reduced after bonding, lowest achieved in epi-side down bonding approach. From an initial 'effective' thermal resistance of ~200 °C/W for the unbonded laser diodes, the 'effective' thermal resistance has now dropped to ~100 °C/W and ~40 °C/W after epi-side up and epi-side down bonding, respectively. The reduction in the thermal resistance led to improved lasing performance shown in Fig. 11.

4. Conclusion

In this chapter, the challenges in high-power laser diode packaging were identified. Material-oriented problems concerning electrical, mechanical and thermal issues must be resolved, while design-oriented issues strive for ease of manufacture, rework and inspection. The attributes of various die-attachment techniques using different kinds of interface materials to overcome the thermal management issues in the packaging design were discussed. A well-controlled void-free bonding interface is required to enable an effective heat dissipation path through the die-attachment process. The heat dissipation capabilities has a strong relationship to the parametric performance of the laser diode. Epi-side down bonding approach has lower thermal resistance, resulting in lower temperature rise in the active region and hence, permitting higher optical output power compared to epi-side up bonding approach. A preview of some of the state-of-the-art heat sink materials and various cooling methods were also discussed to expand the possibility of providing design flexibility for future demand of high-power applications.

As applications continue to demand for higher optical output power and longer lifetime, thermo-mechanical stresses on the die-attachment interface pose a challenge in the laser diode package. To quantify the reliability of the laser diode package, one needs to consider not only the parametric performances of the laser diode device, but also the integrity of the joint. Knowledge on the physical changes at the interface is crucial to the understanding on the device performance and reliability. Three different solder systems - 63Pb37Sn, 3.5Ag96.5Sn and 80Au20Sn - were compared. Metallurgical bond is form between the laser diode and the heat sink through the formation of IMCs at the interfaces during bonding. For the three solder systems, the total chemical driving force arises from the dissolution of Ni from the heat sink in molten solder and from the interfacial reaction in forming IMCs. 63Pb37Sn and 3.5Ag96.5Sn solder exhibit large IMCs growth in the solder joint and the integrity of the solder joint degrades during aging. The mechanical strength of the solder joint weakens significantly and large amount of plastic deformation was observed in the solder joint during shear test. Only 80Au20Sn solder has exhibited a stable microstructure with minimal interdiffusion at the interfaces and the structural integrity of the joint was excellent. Hence, for a reliable assembly, 80Au20Sn solder is the preferred interface material to support high-power laser diode applications.

5. References

Bascom, W. D. & Bitner J. L. (1976). *Microelectron. Reliability,* Vol. 15, No. 1, pp. 37-39, ISSN 0026-2714

Boudreau, R.; Tabasky, M.; Armiento, C.; Bellows, A.; Cataldo, V.; Morrison, R.; Urban, M.; Sargent, R.; Negri, A. & Haugsjaa, P. (1993) Fluxless die bonding for optoelectronics, *Proceedings of 43rd Electronic Components and Technology Conference,* pp. 485 - 490, ISBN 1-4244-0985-3

Capinski, W.S.; Maris, H.G.; Ruf, T.; Cardona, M.; Ploog, K. & Katzer, D.S. (1999) *Phys. Rev. B*, Vol. 59, No. 12, pp. 8105-8113, ISSN 1098-0121

Delpiano, F.; Paoletti, R.; Audagnotto, P. & Puleo, M. (1994) *IEEE Trans Comp., packag., & Manufact. Technol. B*, Vol.17, No. 3, pp. 412-417, ISSN 1070-9894

Dohle, G.R.; Callahan, J.J.; Drabik, T.J. & Martin, K.P. (1996) A new cost effective packaging technique for optoelectronic devices *Proceedings of 46th Electronic Components and Technology Conference*, pp. 1301 – 1307, ISBN: 0-7803-3286-5

Dohle, G.R.; Drabik, T.J.; Callahan, J.J. & Martin, K.P. (1996) *IEEE Trans. Comp., Packag., & Manufact. Technol. B*, Vol. 19, No. 3, pp. 575-580, ISSN 1070-9894

Fujiwara, K.; Fujiwara, T.; Hori, K. & Takusagawa M. (1979) *Appl. Phys. Lett.*, Vol. 34, No. 10, pp. 668-670, ISSN 0003-6951

Fujiwara, K.; Imai, H.; Fujiwara, T.; Hori, K. & Takusagawa M. (1979) *Appl. Phys. Lett.*, Vol. 35, No. 35, pp. 861-863, ISSN 0003-6951

Hayakawa, T.; Miyauchi, N.; Yamamoto, S.; Hayashi, H.; Yano, S. & Hijikata T. (1983) *Appl. Phys. Lett.*, Vol. 42, No. 1, pp. 23-24, ISSN 0003-6951

Hayashi, T. (1992). *IEEE Trans. Comp., Hybrids, & Manufact. Technol.*, Vol. 15, No. 2, pp. 225 - 230, ISSN 0148-6411

Katsura, K.; Hayashi, T.; Ohira, F.; Hata, S.; Iwashita, K.; (1990) *J. Lightwave Technol.*, Vol. 8, No. 9, pp. 1323-1327, ISSN: 0733-8724

Kressel, H. (1976). *Characterization of Epitaxial Semiconductor Films*, Elsevier, Amsterdam.

Larsson, A.; Forouhar, S.; Cody, J. & Lang, R.J. (1990) *IEEE Photon. Technol. Lett.*, Vol. 2, No. 5, pp. 307-309, ISSN 1041-1135

Lau, J. H. & Rice, D. W. (1985) *Microelectron. Reliability*, Vol. 26, No. 6, pp. 1189, ISSN 0026-2714

Lee, C. C. & Wang, C. Y. (1992) *Thin Solid Films*, Vol. 208, No. 2, pp. 202-209, ISSN 0040-6090

Lee, C.C.; Wang, C.Y. & Matijasevic, G.S. (1991) *IEEE Trans. Comp., Hybrids, & Manufact. Technol.*, Vol. 14, No. 2, pp. 407-412, ISSN 0148-6411

Lee, Y. C. & Basavanhally N. (1994) *J. Minerals, Metals Mater. Society*, Vol. 46, No. 6, pp. 46-50, ISSN 1047-4838

Lee, Y. G. & Duh J. G. (1999) *J. Materials Sci: Materials in electronics*, Vol. 10, No. 1, pp. 33-43, ISSN 0957-4522

Leers, M.; Scholz, C.; Boucke, K. & Oudart, M. (2007). Next Generation Heat Sinks For High-Power Diode Laser Bars. *Proceedings of 23rd IEEE SEMI-THERM Symposium*, pp. 105-111, ISBN 1-4244-09589-4

Leers, M. & Boucke, K. (2008) Cooling Approaches for High Power Diode Laser Bars. *Proceedings of 58th IEEE ECTC*, pp. 1011-1016, ISBN 978-1-4244-2230-2

Martin, O.J.F., Bona, G.-L. & Wolf, P. (1992) *IEEE J. Quant. Electron.*, Vol. 28, No. 11, pp. 2582-2588, ISSN 0018-9197

Matijasevic, G. S.; Lee, C. C.; Wang C.Y. (1993). *Thin Solid Films*, Vol. 223, No. 2, pp. 276-287, ISSN 0040-6090

Merritt, S.A.; Heim, P.J.S.; Cho, S.H.; Dagenais, M. (1997). *IEEE Trans. Comp., Packag., & Manufact. Technol. B*, Vol. 20, No.2, pp. 141-145, ISSN 1070-9894

Mizuishi, K. (1984). *J. Appl. Phys.*, Vol. 55, No. 2, pp. 289-295, ISSN 0021-8979

Mizuishi, K.; Sawai, M.; Todoroki, S.; Tsuji, S.; Hirao, M. & Nakamura, M. (1983) *IEEE J. Quantum Electronic*, Vol. 19, No. 8, pp. 1294-1301, ISSN 0018-9197

Mizuishi, K. & Tokuda, M. (1988). *IEEE Trans. Comp., Hybrids, & Manufact. Technol.*, Vol. 11, No. 4, pp. 447-451, ISSN 0148-6411

Nishiguchi, M. (1991) "Highly reliable Au-Sn eutectic bonding with background GaAs LSI chips." *IEEE Trans. Comp., Hybrids, & Manufact. Technol.* Vol. 14, No. 3, pp. 523-528, ISSN 0148-6411

Rodgers, P.; Eveloy, V. & Pecht, M.G. (2005). Limits of Air-Cooling: Status and Challenges. *Proceedings of 21st IEEE SEMI-THERM Symposium*, pp. 116 – 124, ISBN 0-7803-8985-9

Paoli, T. L. (1975) *IEEE J. Quant. Electron.* Vol. 11, No. 7, pp. 498-503, ISSN 0018-9197

Pittroff, W.; Erbert, G.; Beister, G.; Bugge, F.; Klein, A.; Knauer, A.; Maege, J.; Ressel, P.; Sebastian, J.; Staske, R. & Traenkle, G. (2001) *IEEE Trans. Adv. Packag.* Vol. 24, No.4, pp. 434-441, ISSN 1521-3323

Pittroff, W.; Erbert, G.; Klein, A.; Staske, R.; Sumpf, B. & Traenkle, G. (2002). Mounting of laser bars on copper heat sinks using Au/Sn solder and CuW submounts. *Proceeding of 52nd Electronic Components and Technology Conference*, pp. 276 – 281, ISBN 0-7803-7430-4

Qu, Y.; Yuan S.; Liu, C.Y.; Bo, B.X.; Liu, G.J. & Jiang,H.L. (2004) *IEEE Photon. Technol. Lett.* Vol. 16, No. 2, pp. 389-391, ISSN 1041-1135

Sabbag, N. A. J. & McQuuen, H. J. (1975) *Metal Finishing*, pp. 27-31

Shi, X. Q.; Pang, H. L. J.; Wei, Z. & Wang, Z.P. (2000) *Int. J. of Fatigue*, Vol. 22, No. 3, pp. 217-228, ISSN 01421123

Shi, X.Q.; Wang, Z.P.; Zhou, W.; Pang, H.L.J. & Yang, Q.J. (2002) *ASME J. Electron. Packag.*, Vol. 124, No. 2, pp. 85-90, ISSN 1043-7398

Solomon, H. D. (1986) *IEEE Trans. Comp., Hybrids, & Manufact. Technol.* Vol. 9, No. 4, pp. 423-432, ISSN 0148-6411

Spagnolo, V.; Troccoli, M.; Scamarcio, G.; Becker, C.; Glastre,G. & Sirtori, C. (2001) *Appl. Phys. Lett.* 78(9): 1177-1179, ISSN 0003-6951

Teo, J.W.R.; Goi, L.S.K.; Xiao, L.H.; Lim, W.C.; Wang, Z.F. & Li, G.Y. (2009) *IEEE Trans. Adv. Packag.*, Vol. 32, No. 1, pp. 130-135, ISSN 1521-3323

Teo, J.W.R.; Shi, X.Q.; Yuan, S.; Li, G.Y. & Wang, Z.F. (2008) *IEEE Trans. Comp. and Packag. Technol.*, Vol. 31, No. 2, pp. 159-167, ISSN 1521-334X

Teo, J.W.R.; Ng, F.L.; Goi, L.S.K.; Sun, Y.F.; Wang, Z.F.; Shi, X.Q.; Wei, J. & Li, G.Y. (2008) *Microelectron. Eng.* Vol. 85, No. 3, pp. 512–517 ISSN 0167-9317

Tew, J. W. R.; Shi, X. Q. & Shu, Y. (2004). *Materials Lett.*, Vol. 58, No. 21, pp. 2695-2699, ISSN 0167577X

Tew, J.W.R.; Wang, Z.F.; Shi, X.Q. & Li, G.Y. (2004). An Optimized Face-Down Bonding Process for Laser Diode Packages. *Proceeding of IEEE 6th EPTC Electronics Packaging Technology Conference*, pp. 390 – 395, ISBN 0-7803-8821-6

Wong, C.K.; Pang, J.H.L.; Sun, Y.F.; Ng, F.L.; Tew, J.W. & Fan, W.(2005) Influence of Solder Volume on Interfacial Reaction between Sn-Ag-Cu Solder and TiW/Cu/Ni UBM, *Proceeding of IEEE 7th EPTC Electronics Packaging Technology Conference*, pp. 390 – 395, ISBN 0-7803-9578-6

Zweben, C. (2005) Ultrahigh-Thermal-Conductivity Packaging Materials, *Proceeding of 21st IEEE SEMI-THERM Symposium*, pp. 168 – 174, ISBN 0-7803-8985-9

18

Laser Diode Gas Spectroscopy

Pablo Pineda Vadillo
University of Dublin, Trinity College Dublin
Ireland

1. Introduction

The use of laser sources for optical spectroscopy experimentation is a well-established, cross-disciplinary field with an enormous number of applications in research areas as broad as Physics, Chemistry or Biology (Demtröder, 1996). The high optical intensity achieved with these types of sources, along with the extraordinary monochromaticity of their optical output, has a huge impact in spectroscopy research. In particular, the application to gas sensing in weak absorption spectral regions significantly benefits from the specific properties of laser. Gas species exhibit well-defined, complex spectral absorption structures that require the utilisation of tuneable, high resolution and spectrally narrow sources to be resolved. Such requirements are fulfilled by several types of Tuneable Diode Lasers (TDL's) readily available from the optical telecommunication industry, where much work was done during the last decades to develop and improve the characteristics of these laser devices. Further technology development and intensive research has lead to TDL's of high optical power and narrow emission linewidth to be routinely available for research groups across the world, thus opening the possibility for investigation of gas absorption processes of increasing complexity and very narrow spectral scales. In addition, advances in manufacturing technology have enabled the design and fabrication of new photonic microstructures in parallel to the development of laser sources. The combination of both factors has thus multiplied the possibilities for challenging gas-sensing research, which in turn reveals new fundamental characteristics of gaseous systems whose study requires more and more stable, narrow linewidth and powerful laser sources.

This chapter contains a summary of the main properties and operational principles of laser diodes in connection to gas spectroscopy. Analysis is presented related to the main laws and principles applying to light absorption by gases, though many concepts can be extended to other types of absorbing media. Description of the main detection techniques is included along with selected experimental results to complement the theoretical discussions. Finally, a selection of references gives the reader the option to get a deeper insight into this rich and multidisciplinary research area.

2. Light absorption and gas spectra

2.1 Light absorption by a medium

Light of angular frequency $w=2\pi\upsilon$ travelling through a material of refractive index n can experience a number of different processes. In general, part of the incoming radiation is absorbed by the medium, while the rest is free to propagate until it eventually escapes, retaining its original propagation direction –transmitted light- or presenting some deviation - scattered light-. *Absorption* processes are accompanied by *dispersion* of the light, the latter defined by the change in phase velocity of light from its value in vacuum *c*. These properties of the interaction of electromagnetic radiation with an absorbing medium are classically described by an oscillator model for the atomic electrons, leading to the definition of a complex refractive index to explain both effects.

The basic rules governing linear absorption of light by a medium of refractive index *n* are summarized in the so-called Beer-Lambert law. Consider a light beam travelling along the Z direction of an absorbing medium of number density *C* and transverse area *A*. The total number of illuminated molecules for a material thickness *dz* is *CAdz*, though the effective area presented by this ensemble is given by $\sigma CAdz$, σ being the cross-section for absorption/scattering in this case. Thus, the total probability for a molecule of being either absorbed or scattered out of the beam will be (Garrett, 2006)

$$\frac{dI_Z}{I_Z} = \frac{\sigma CAdz}{A} \tag{1}$$

which upon integration gives the so-called Beer-Lambert law for linear attenuation

$$I = I_0 e^{-\sigma Cz} = I_0 e^{-\mu z} \tag{2}$$

If scattering is neglected, then the *linear attenuation coefficient* $\mu=\sigma C$ is equal to the *linear absorption coefficient* α and the Beer-Lambert relation can be rewritten in the form

$$I(\lambda) = I_0(\lambda) e^{-\alpha(\lambda)z} \tag{3}$$

where the dependence of the absorption coefficient on the wavelength λ of the incoming light is explicitly indicated. In Optical Spectroscopy, the Beer-Lambert law is usually expressed in terms of the incoming plane wave angular frequency $\omega = 2\pi c/\lambda = 2\pi\upsilon$, and the medium, frequency-dependent, complex refractive index $n(\omega)$. The treatment presented by Measures (Measures, 1988 - special emphasis on the use of tuneable laser diodes in laser absorption measurements) is followed here, hence it can be written

$$n(\omega) = \eta(\omega) + \frac{1}{2} ix(\omega) \tag{4}$$

where the real part of the refractive index $\eta(\omega)$ accounts for the dispersion of the light as it travels through the medium and the imaginary term $\chi(\omega)$ is related to its absorption properties. The oscillator model for an electromagnetic wave travelling through a medium of complex refractive index (4) leads to the well-known *Kramers-Kronig* dispersion relations relating these two terms, which apply in the proximity of an atomic transition frequency ω_0 and for the case of the imaginary term reads (Measures, 1988)

$$x(\omega) = \frac{Ne^2 f_0 \gamma}{4 \epsilon_0 m_e \omega \left[(\omega - \omega_0)^2 + (\gamma / 2)^2 \right]} \tag{5}$$

where N is the number density of oscillators, e and m_e the charge and mass of the electron, respectively, and γ the damping constant for the oscillator model. The absorption coefficient introduced in (3) is related to the imaginary term of the refractive index so that

$$\alpha(\omega) = \frac{\omega x(\omega)}{c} \tag{6}$$

where c is the speed of light in vacuum. It is thus convenient to refer to the frequency-dependent form for the Beer-Lambert law for consistency,

$$I(\omega) = I_0(\omega)e^{-\alpha(\omega)z} \tag{7}$$

2.2 Gas molecular structures and absorbing transitions

In this chapter absorption will be studied regarding gas species. Optical absorption by a gas occurs when the energy carried by an incoming photon is resonant with the difference between two molecular energy states. The distribution of energy states for a gas is determined by three types of motion, namely *rotational* (molecules rotating about its centre of mass, CM), *vibrational* (vibrations of individual/sets of atoms within a molecule with a static CM) and *electronic* (electrons moving within the molecule). Depending on the particular symmetry of the targeted molecule, the energy levels are distributed in a specific manner. In addition, these three types of motion interact with each other according to established selection/combination rules, giving rise to very complex absorption spectra in some cases.

In section 5 of this chapter some experimental results are presented regarding acetylene gas (C_2H_2), targeting absorbing transitions which belong to the *rotational-vibrational combination band* $\upsilon_1+\upsilon_3$. The main properties of such bands can be described as follows. First of all, we are working within a *vibrational* band characterized by its corresponding vibration frequency υ. Secondly, the band is not centred at a *fundamental* vibration frequency υ_i, but at a *combination* value $\upsilon_i + \upsilon_j$ (υ_1 and υ_3 correspond to two of the total five normal vibration modes available for acetylene (Steinfeld, 2005)). Thirdly, the absorbing transitions occur between *rotational* energy levels whose distribution is embedded on top of the vibrational modes. This type of rotational-vibrational energy map is illustrated in figure 1, corresponding to some of the relevant energy levels and transitions in the simple case of a diatomic molecule (this figure is adapted from the very comprehensive book on molecular spectroscopy by Banwell (Banwell & McCash, 1994)). Note how the absorption spectrum schematically shown at the foot of the figure consists of several absorption lines with varying intensities, due to the fact that not all rotational J levels are populated to the same degree in thermal equilibrium. The $\upsilon_1+\upsilon_3$ band for acetylene is centred at a wavelength of approximately 1525.8 nm within the Near Infra Red (NIR) region, where the intensity of absorption lines is much smaller than in the IR fundamental absorption region. This is a very important fact to take into account, because these weak transitions are difficult to detect unless very sensitive detection schemes are implemented or the gas concentration is high. A more extended, good comprehensive summary of main molecular concepts relevant to spectroscopy can be found in Weldon (Weldon, 2005), while an extremely detailed treatment on Infrared and Raman spectra of polyatomic molecules was published by Herzberg and can be consulted for advanced topics (Herzberg, 1945).

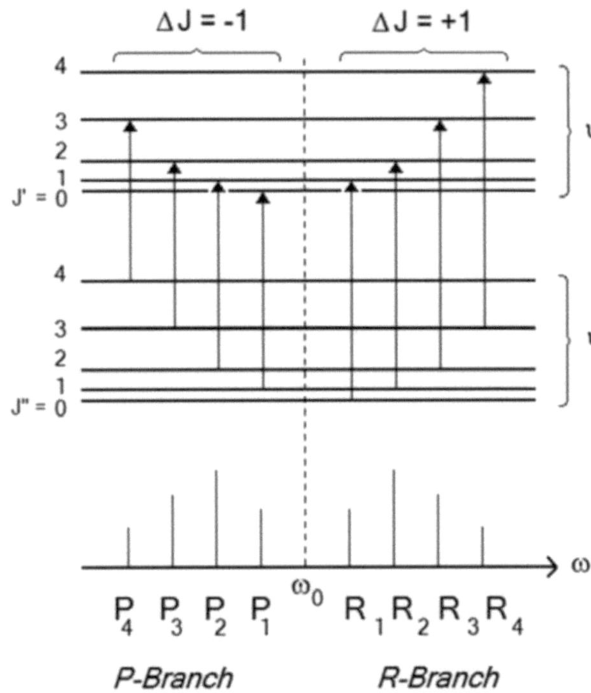

Fig. 1. Schematic diagram for the rotational (J) – vibrational (υ) energy level distribution of a diatomic molecule (figure adapted from (Banwell & McCash, 1994)).

2.3 Broadening processes and absorption lineshapes

In spectroscopic applications, the Beer-Lambert law (7) can be written as

$$I(\tilde{\upsilon},z)=I_0 e^{-\kappa(\tilde{\upsilon})z_D}=I_0 e^{-\kappa(\tilde{\upsilon})Cz} \tag{8}$$

where $\kappa(\tilde{\upsilon})$ is the absorption coefficient at the frequency $\tilde{\upsilon}$ when the latter is expressed in inverse length units according to the relations

$$\tilde{\upsilon}\left(cm^{-1}\right)\equiv\frac{1}{\lambda(cm)}\equiv\frac{\upsilon(Hz)}{c\left(cm\ s^{-1}\right)}=\frac{\omega(Hz)}{2\pi c\left(cm\ s^{-1}\right)} \tag{9}$$

and z_D = Cz is the so-called *optical density* with generic units of concentration times length (using the notation introduced in equation 2). The absorption $\kappa(\tilde{\upsilon})$ thus has units of reciprocal concentration times length, and is normally expressed in 1/(molecules cm^{-3} cm) or (cm^2/molecule). We will simplify the notation from now on and use υ when referring to $\tilde{\upsilon}(cm^{-1})$ for more clarity. Because the shape of the absorption profile is generally of interest, the absorption coefficient can be rewritten in a more useful form according to

$$\kappa(\upsilon) = Sg(\upsilon - \upsilon_0) \tag{10}$$

where the so-called *linestrength* S is essentially the integrated absorption coefficient and includes information about the absorption profile shape,

$$S = \int_{-\infty}^{\infty} \kappa(\upsilon)d\upsilon \tag{11}$$

while the *lineshape function* $g(\upsilon - \upsilon_o)$ is normalised according to

$$g\left(\upsilon-\upsilon_0\right)=C\phi\left(\upsilon-\upsilon_0\right),\ \int_{-\infty}^{\infty} g\left(\upsilon-\upsilon_0\right)d\upsilon=1 \tag{12}$$

and allows intensity profiles described by $\phi(\upsilon\text{-}\upsilon_0)$ to be compared directly. A common choice is that the concentration C in (8) is expressed in molecules/cm^3 since these units are independent of temperature (if the absorbing medium is a gas and concentration is expressed in terms of pressure units, then its equation of state needs to be specified). In that case S has units of cm/molecule and the lineshape function $g(\upsilon\text{-}\upsilon_0)$ is given in cm according to (10).

Real absorption profiles for gaseous systems are not δ-like functions but exhibit some finite linewidth due to several broadening effects. The convolution of these effects determines the final lineshape function for the absorbing transition, empirically characterized by its Full Width at Half Maximum (FWHM) as illustrated in figure 2. Several physical processes affect the absorption profiles during typical experimental work, and can be classified/described according to different criteria.

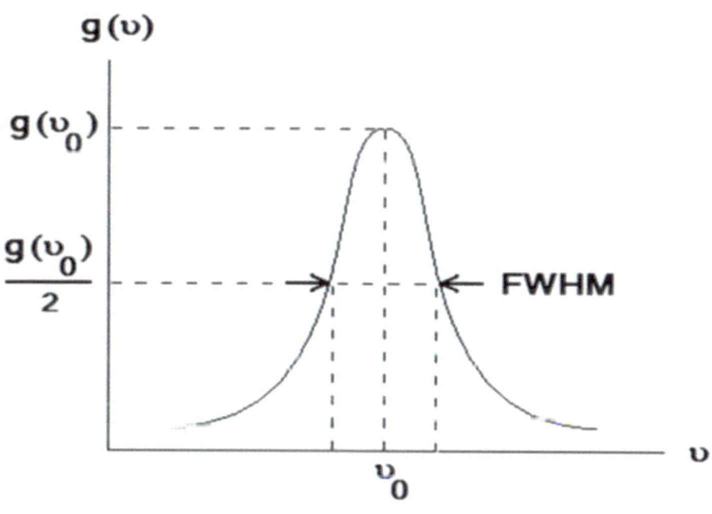

Fig. 2. Lineshape function $g(\upsilon\text{-}\upsilon_o)$ characterized by its Full Width at Half Maximum (FWHM).

Three main broadening mechanisms contribute to the experimental measurements presented at the end of the chapter. The most straightforward mechanism is that related to the well-known **Doppler** effect, where a molecule with a non-zero velocity component in the propagation direction of the radiation field absorbs a photon at some frequency shifted from the resonant transition frequency *at rest*. The distribution of molecular speeds across the whole ensemble of active molecules is given by the Maxwell-Boltzmann distribution, which tells us which molecules fall in which velocity class. The final result is a spectral absorption profile given, to a first approximation commonly used, by a **Gaussian** lineshape function of the form (Measures, 1988)

$$g_D(\upsilon-\upsilon_0)=\left(\frac{2}{\Delta\upsilon_D}\right)\sqrt{ln2/\pi}\,\exp\left[-\frac{2(\upsilon-\upsilon_0)^2\,ln2}{\Delta\upsilon_D}\right] \tag{13}$$

with a FWHM of

$$\Delta\upsilon_D=\frac{2\upsilon_0}{c}\sqrt{2kTln2/m} \tag{14}$$

Note how the magnitude of the Doppler broadening increases with temperature as $T^{1/2}$ and resonant frequency υ_0. During the experimental results presented in section 5 both the gas temperature and the operating frequency (within some small tuning range) are kept constant, and so the associated Doppler width is essentially constant.

The second effect to take into account is the spectral broadening introduced by molecular **collision** processes. Several models exist to explain such an effect, but we will restrict the discussion to the simple Lorentz model followed in (Measures, 1988). This model assumes that the electron oscillation described by the classical absorption model halts on collision and then restarts with a completely random phase, totally unrelated to that prior to the collision event. This incoherence eventually leads to molecular absorption at a frequency different than the resonant value, and it can be demonstrated that the collisionally broadened lineshape function is given by a **Lorentzian** curve

$$g_L(\upsilon-\upsilon_0)=\left(\frac{\Delta\upsilon_L}{2\pi}\right)\frac{1}{\left[(\upsilon-\upsilon_0)^2+\Delta\upsilon_L^2/4\right]} \tag{15}$$

A rather cumbersome expression for the FWHM ($\Delta\upsilon_L$) of this normalised lineshape is found in the treatment by Measures that we are following, so for explanation purposes we can refer to Svelto (Svelto, 1998) where the FWHM for a collisionally-broadened transition is given in a different approach by

$$\Delta\upsilon_L=\frac{1}{\pi\tau_C}=\Delta\upsilon_L^0\left(\frac{P}{P_0}\right)\left(\frac{T_0}{T}\right)^{1/2} \tag{16}$$

where τ_C represents the mean time between collisions and the pressure and temperature dependences for the linewidth parameter have been taken from (Measures, 1988) again ($\Delta\upsilon^0_L$ is the linewidth at standard pressure and temperature conditions). The most important fact illustrated by relation (16) for experimental purposes is that, at a fixed gas temperature, the collisional broadening increases linearly with pressure. Note that this type of broadening can be introduced not only by intermolecular collisions, but also by collisions of the active molecules with the **walls** of the gas hosting structure. The latter effect is small when gas cells of the order of tens of centimetres of height/widths are used, but in the specific case of the Hollow-Core PBF used in section 5 the wall collisions play a crucial role (the diameter of the utilised fibre was approximately $d \approx 11\ \mu m$). However this fibre-related broadening is in practice almost indistinguishable from the transit-time effect explained next, with treatment in the literature using one of the two possible approaches.

The last of the three significant broadening mechanisms to take into account during experiments in PBF fibres is that related to the **transit-time** effect: under certain physical conditions the transit time of gas molecules *moving* across the laser beam becomes smaller than their excited state lifetime. This *effective* shorter lifetime eventually gives rise to a spread of the absorbing frequencies around the resonant value, as time and optical frequency can be related via the Heisenberg uncertainty relations. It can be demonstrated (Demtröder, 1996) that the spectral intensity profile for atoms perpendicularly traversing a Gaussian laser beam of 1/e diameter $2w$ with a velocity v is given by a **Gaussian** curve

$$I(\omega) = I_0 \exp\left[-(\omega - \omega_0)^2 \frac{w^2}{2v^2}\right] \tag{17}$$

with a transit-time limited FWHM

$$\Delta \upsilon_{TT} \approx 0.4v / w \tag{18}$$

Figure 3 further illustrates the previous discussion. Note that this effect depends on the geometry of the fibre, but also on the *gas temperature*: the value for the mean velocity of the gas molecules is given by the kinetic theory of gases and ultimately depends on the temperature T for a fixed molecular mass M (Nave, 2009).

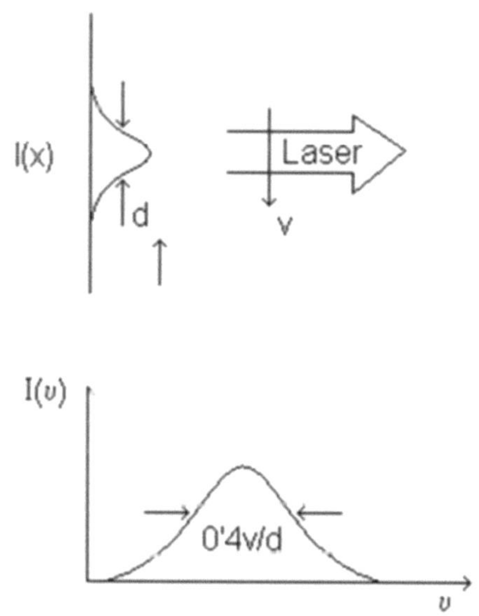

Fig. 3. Transit-time broadening of an absorbing transition (figure adapted from (Demtröder, 1996)).

Thus, the main broadening processes affecting an absorbing transition during the experimental work presented at the end of the chapter have been both qualitatively and mathematically described. Note that, in a real situation, the final absorption profile is the result of more than one broadening factor at the same time and cannot be described by the *pure* lineshapes described above. In such case the so-called *Voigt* profile must be used, a convolution of a Lorentzian and Gaussian profiles that must be evaluated numerically (Measures, 1988). A graphical comparison between the normalised Lorentzian, Gaussian and Voigt profiles with the same area under the curve $A = 1$ is included in figure 4, where it can be observed that the Voigt profile essentially follows a Lorentzian profile near its wings and a Gaussian curve around the line centre. Further analysis of absorption lineshapes in gases can be found in (Lepère, 2004; Varghese & Hanson, 1984).

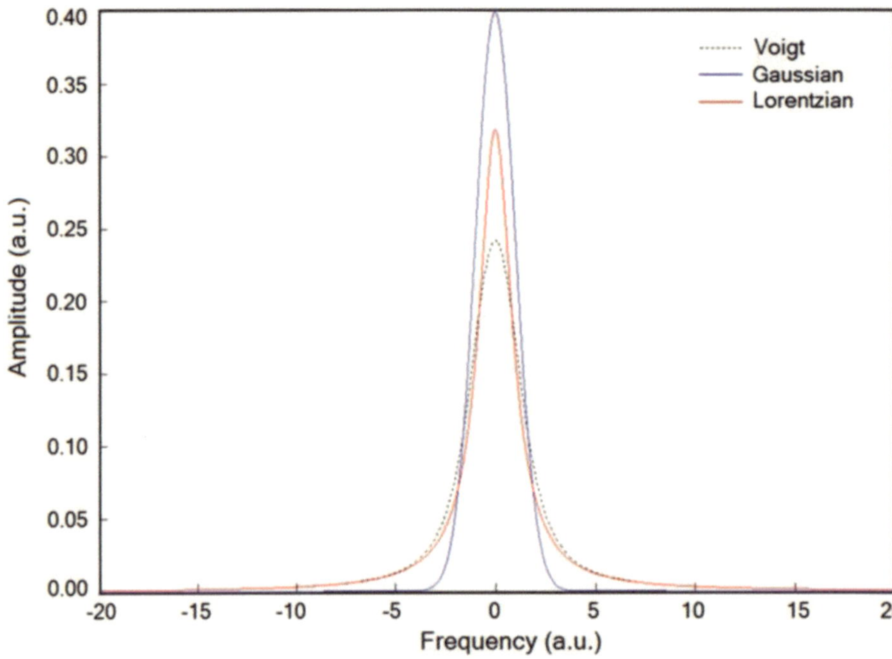

Fig. 4. Simulation results for normalised Gaussian, Lorentzian and Voigt lineshape profiles of the same area under the curve A = 1.

It is useful to estimate the order of magnitude of the broadening processes affecting measurements under typical experimental conditions. It helps with optimizing the experimental set-up when required, as well as helping to select the proper relation when performing some theoretical fit of acquired data. According to the previous discussion in this section, there are basically two experimental parameters that determine the final broadening contributions: the gas temperature and pressure. Typical conditions during the experimental work presented later are gas temperature of $T_{GAS} \approx 22$ °C and pressure $P_{GAS} \approx 1$ mbar. For these values the approximate order of magnitude for the natural, collisional, transit time and Doppler broadenings is respectively 100 Hz, 10 MHz, 50MHz and 500 MHz. The laser linewitdh was typically $\Delta\upsilon_{LAS} \approx$ 1-5 MHz. As it can be observed the broadening is dominated by the Doppler effect at low pressure values. As pressure increases so does the collisional term, eventually dominating the total absorption linewidth at pressures above tens of millibars. The natural linewidth is negligible in this scenario. Note how the estimated values for these broadening processes -except the negligible natural linewidth term- are significantly higher than the typical emission frequency linewidth of the diode lasers utilised during experimental work.

3. Laser diodes in spectroscopy

3.1 Important parameters in laser spectroscopy

The absorption signals of gas species are not ideal in most real scenarios and acquiring good quality data is not a straightforward task. Low gas pressure or concentrations, rapidly changing temperature gradients within the targeted volume (ex. combustion jets in airplanes) or presence of multiple gases and overlap of gas absorption lines are examples of issues encountered in many real field experiments. For laser spectroscopy applications to be feasible, some parameters of the laser source become crucial and must be met:

- *Spectral purity (laser linewidth)*: the output from a laser source is broadened by several mechanisms, such as random phase changes due to spontaneous emission within the laser cavity or electronic noise from the laser current or temperature controllers. In most cases, a rule-of-thumb for successful gas absorption spectroscopy is that the laser linewidth should be smaller than 5% of the gas absorption linewidth.
- *Tuneability*: the output of the laser source must be ideally tuneable by using temperature or current control. A typical tuning range of 5 nm is required in multigas sensing applications (with typical temperature/current tuning rates of 0.1 nm/°C and 0.01 nm/mA respectively).
- *Modulation*: in order to apply high-sensitivity detection techniques such as WMS and FMS (see section 4.2), a high bandwidth modulation of the laser source is both desirable and necessary.

Additional parameters such as reduced size for portability, fiberized laser output for easier system integration, or cost-effective fabrication processes, are also beneficial for the design and implementation of effective gas detection schemes.

3.2 Widely tunable telecommunication laser diodes

Laser diodes (LD's) developed by the optical communications industry in the last decades are ideal candidates for spectroscopy studies, especially in the Infrared region where the absorption strength of some gases is significantly weak. An optical telecommunication window was opened in 1977 centred at 1550 nm (the *C band*), so much effort was made to improve the performance of laser sources working in this Near Infrared (NIR) region of interest. These affordable LD's fulfil the linewidth characteristics required for spectroscopic applications, their emission frequency can be tuned across a wide range of values and they exhibit a great flexibility when modulation techniques need to be applied. Many different types of tuneable LD's (TLD's) exist that fulfil usual spectroscopic requirements, based on different laser structures and tuning mechanisms. Some characteristics of two of these laser diodes, namely *SG-DBR (Sampled Grating Distributed Bragg Reflector)* and *SGC-DFB (Strongly Gain Coupled Distributed Feed Back)*, are summarized in table 1 as an example.

	SG-DBR (Sampled Grating Distributed Bragg Reflector)	SGC-DFB (Strongly Gain Coupled Distributed Feed Back)
WAVELENGTH RANGE (NM)	1535-1570	1537.5-1539.5
TUNING MECHANISM	Current	Current/Temperature

Table 1. Wavelength and tuning characteristics of two sample NIR tuneable laser diodes

A very detailed treatment on tuneable laser sources by Buus (Buus et al, 2005) can be consulted for further design and working principles applying to these types ofLD's.

4. Typical detection techniques

4.1 Direct absorption

The most simple and straightforward detection technique is that where the light is collected *directly* after it interacts with the absorbing gas or medium. The unmodulated emission wavelength of the laser is scanned across the spectral region of interest and the transmitted signal recorded using a photodiode/photodetector. A typical gas absorption line using this scheme is depicted in figure 5. As it can be observed, the absorption peak sits on the top of a sloped background. This slope is a result of the optical power of the laser being proportional to the scanning laser current/temperature. This is an undesired effect as it reduces the Signal-to-Noise Ration (SNR) of the collected data, and thus limits the maximum sensitivity achievable with this detection technique.

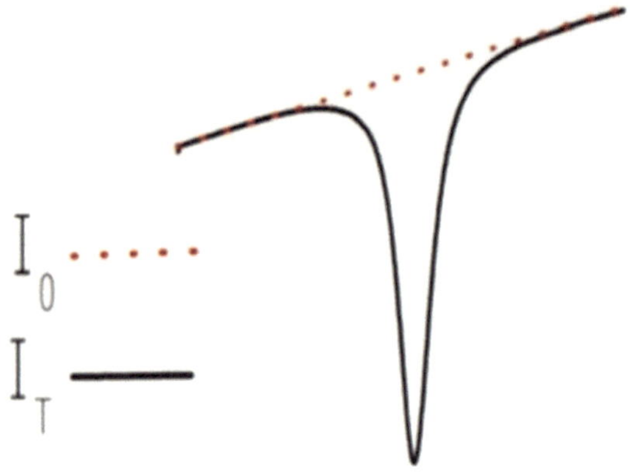

Fig. 5. Gas absorption signal utilizing a direct absorption technique.

4.2 Modulation techniques

In order to enhance the detection of weak absorption signals some high sensitivity techniques can be applied, based in the modulation of the emission wavelength of the laser diodes utilised. Two main techniques are commonly used, namely Wavelength Modulation Spectroscopy (WMS) and Frequency Modulation Spectroscopy (FMS).

In the case of WMS, the emission wavelength of the laser source is modulated at frequencies smaller than the FWHM of the absorption profile (≈ kHz/MHz). Modulation is typically achieved using an external sinusoidal signal applied to the modulation facility of the current controller driving the laser. Following the treatment by Reid and Labrie (Reid & Labrie, 1981), the emission frequency therefore may be written as

$$\upsilon(t) = \bar{\upsilon} + a\cos\phi_m t \tag{19}$$

Where $\bar{\upsilon}$ is the mean emission frequency, ϕ_m the angular modulation frequency and a the modulation amplitude. Because absorption is weak for the targeted gas transitions, the Beer-Lambert law presented in (8) can be approximated by the expression

$$I(\upsilon) = I_0(\upsilon)\left[1 - \kappa(\upsilon)z_D\right] = I_0(\upsilon)\left[1 - \kappa(\bar{\upsilon} + a\cos\phi_m t)z_D\right] \tag{20}$$

The time dependent term, periodic and even function, can be expanded in a cosine Fourier series

$$\kappa(\bar{\upsilon} + a\cos\phi_m t) = \sum_{n=0}^{\infty} H_n(\bar{\upsilon})\cos(n\phi_m t) \tag{21}$$

Where $H_n(\bar{\upsilon})$ is the nth Fourier component of the modulated absorption coefficient. The mean frequency is considered constant over a modulation period and the laser power assumed to remain constant while modulation occurs. Equation (21) essentially summarises the key aspect of this technique: demodulating the detected signal using a *Lock-In Amplifier* (LIA, (Bentham Electronics, 2009; John, 1994)) referenced at some multiple nf of the modulation frequency $f = \phi_m/2\pi$ makes the LIA output proportional to the nth *Fourier component* according to $I_0(\upsilon)H_n(\bar{\upsilon})z_D$. A detailed theoretical analysis shows that such components are dependent on the absorption lineshape or its derivatives, so the above discussion means that, in practice, the output of a demodulator referenced at the nth multiple of the laser modulation frequency is proportional to the nth derivative of the absorption profile. A simple, more intuitive way of understanding this important result is presented by Suplee et al (Supplee, J. M. et al, 1994). If the small absorption coefficient $\kappa(\upsilon) << 1$ in (20) is expanded into a Taylor series around the mean frequency $\bar{\upsilon}$ one gets

$$I(\upsilon) = I_0(\upsilon)\left[1 - z_D\left(\kappa(\bar{\upsilon}) + \left.\frac{d\kappa}{d\upsilon}\right|_{\bar{\upsilon}}(\upsilon - \bar{\upsilon}) + \frac{1}{2!}\left.\frac{d^2\kappa}{d\upsilon^2}\right|_{\bar{\upsilon}}(\upsilon - \bar{\upsilon})^2 \ldots\right)\right] \tag{22}$$

which using (19) and applying trigonometric identities can be rewritten as

$$I(\upsilon)=I_0(\upsilon)\left[1-z_D\left(\kappa(\bar{\upsilon})+\left.\frac{d\kappa}{d\upsilon}\right|_{\bar{\upsilon}}a\cos(\phi_m t)++\frac{1}{2!}\left.\frac{d^2\kappa}{d\upsilon^2}\right|_{\bar{\upsilon}}a^2\frac{1}{2}\cos(2\phi_m t)...\right)\right] \tag{23}$$

Only terms up to second order have been considered, as higher terms for the Taylor expansion rapidly decay according to the weighting amplitude factor a^n (the modulation amplitude a in WMS experiments is usually small compared to the FWHM of the targeted absorption feature). Thus, equation (23) explicitly reveals the proportionality between the first and second derivatives of the absorption coefficient κ and the components of the transmitted intensity demodulated at frequencies ϕ_m and $2\phi_m$, respectively. Note that the unique Fourier transform of the absorption coefficient in (21) is essentially the same as the Taylor expansion, because trigonometric identities allows for terms in consecutive frequencies $n\phi_m$ to be grouped together. Figure 6 represents a typical schematic set-up when using this Lock-in detection technique.

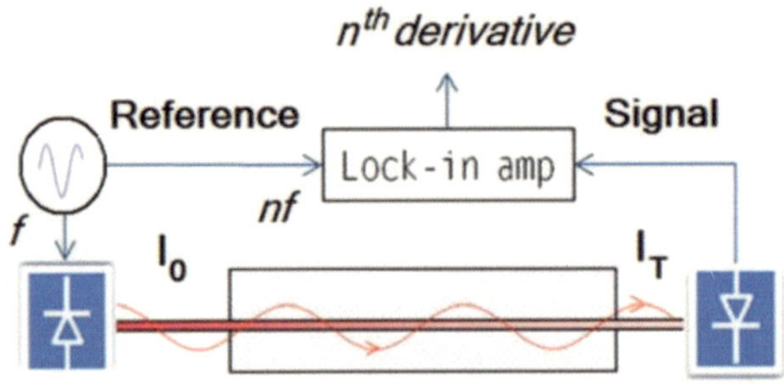

Fig. 6. Schematic diagram for a WMS lock-in detection technique.

It can be demonstrated that the amplitude of the detected n^{th} harmonic signal depends on the amplitude a of the modulation signal (*modulation depth)*, which in practice allows for optimization of the LIA output amplitude. A particular advantage of this WMS detection technique is that any residual sloped background detected in direct transmission measurements is effectively suppressed, significantly improving the Signal-to-Noise Ratio (SNR) of the acquired data.

The second case, FMS, is essentially based in the same principle as WMS, but only differs in the order of magnitude of the applied modulation frequency: In the FMS technique, the modulation frequency is much higher than the FWHM of the absorption line. The main effect of this high-frequency modulation is the appearance of several sidebands in the optical power spectrum of the laser source, separated from the carrier emission frequency υ_0 by an amount exactly equal to the modulation frequency *f*. The number of sidebands can be controlled by adjusting the modulation amplitude applied to the laser, with only the first sidebands typically considered. Several beat effects can occur between the sidebands and carrier peaks that may affect the detected signal, and must be assessed if some detailed information needs to be extracted. On the other hand, the main advantage of FMS detection is shifting the detection frequency to higher values where the low-frequency *1/f* noise affecting the laser is greatly reduced. However, the lock-in amplifier must be substituted by discrete high-frequency components which are more expensive and sensitive to external noise coupling. Formal analysis of both WMS and FMS techniques can be found in references (Silver, 1992; Suplee et al, 1998).

4.3 Feedback techniques

It is common to find dynamic physical systems where one or more of the outputs drift with time and need to be stabilised to a reference value. These kind of systems are widely spread both at macroscopic and microscopic scales with examples including temperature regulation in buildings, microscopes where vibrations need to be minimised and effectively suppressed or cruise speed control associated with different means of transport such as cars or ships. In order to stabilise the desired output a certain signal is chosen which contains the relevant information to assess the stability, and then processed before being fed back to the system *S* implementing a loop structure. A schematic diagram corresponding to a simple case is presented in figure 7, where only one time-dependent output parameter needs to be stabilised and its magnitude *y(t)* become exactly equal to that of a chosen reference value *r(t)* at all times.

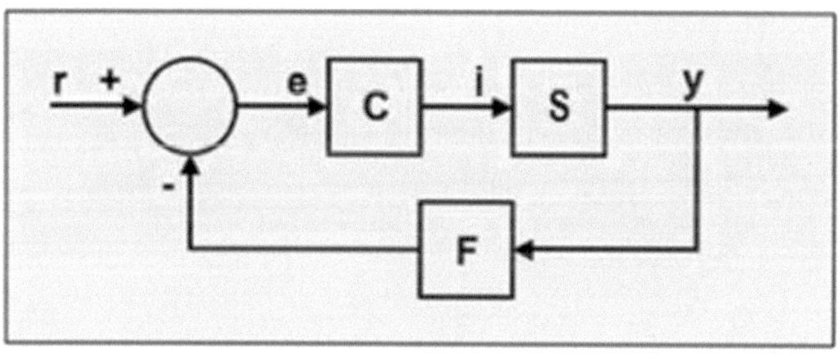

Fig. 7. Schematic diagram for a feedback control loop to stabilise a dynamic system S.

In the ideal case where the output is stable in time the condition $y(t) = r(t)$ holds. However, under normal operation conditions the magnitude of the output of interest drifts with time from the desired value. In that case the output signal is collected by a sensor scheme *F* and subtracted from the reference signal, defining a finite error signal $e(t) = r(t) - y(t)$. This signal is a measure of the departure of the system from its stable state and its magnitude relative to the output is typically small. A controller C receives this error signal and generates the input value *i(t)* to be finally fed back to the system *S*. In the ideal case of successful implementation of a stabilization loop this input value compensates the undesired system drift. The error value asymptotically approaches zero within some stabilizing time interval, which will be characteristic of the physical process involved in the system and can be defined as the system time constant τ_{SYS}. It is customary to define such an arrangement as an *active feedback loop*. Small signal analysis is usually applied in this type of loops as the magnitude of the error signal is much smaller than the system output, and compensation for the undesired drift thus carried out in a linear regime. Transient behaviour becomes important when the loop is initially closed, and conditions must be chosen to assure that the magnitude of the error signal does not exceed a certain limit during the associated transient time scale. On the other hand the frequency response of the loop is determined by its effective time response τ_{SYS}, and the active loop will be only able to compensate system drifts whose associated time scale τ_{DRIFT} is longer than that effective response.

Active feedback loops can be applied to stabilization of a laser output frequency. The emission frequency of a laser source exhibits some finite drift in time under normal operation conditions, mainly due to 1/f noise in the laser controllers. This becomes a serious drawback in fields where frequency stability plays a crucial role (such as optical metrology or precise frequency referencing), so active feedback loops can be applied to minimise such noise effects. Figure 8 includes the experimental scheme to implement an active feedback loop in order to lock the optical frequency $\upsilon(t)$ of a laser source, making use of a spectral feature of gas in an absorption cell. The laser optical frequency is now the output of the system to be stabilised $y(t) = \upsilon(t)$.

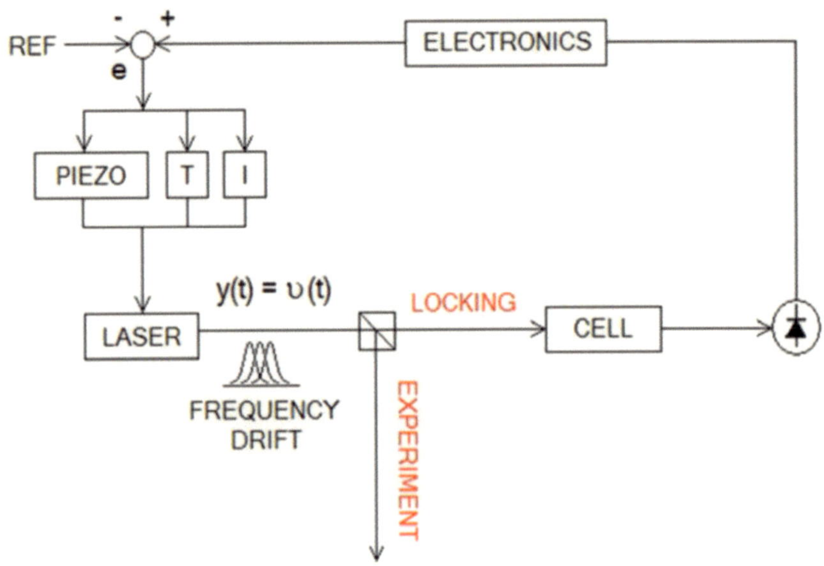

Fig. 8. Active feedback loop to lock the operating frequency of a laser, based in a gas absorption cell.

As observed in the figure, a usual arrangement is that where the main beam is divided in two with the help of a suitable splitting element. A small fraction is used for locking purposes, while the remaining light is sent towards the main experiment to be carried out.

5. An example: Gas sensing in hollow-core photonic bandgap fibres

Some experimental results are presented in this section to illustrate the various concepts covered so far, pointing out the main characteristics regarding the associated experimental scheme.

5.1 Hollow-core photonic bandgap fibres

Standard optical fibres work on the basis of the Total Internal Reflection (TIR) happening between the fibre core and cladding materials due to the step in the refractive index profile. However, a completely different guiding mechanism can be utilised based on the so-called *Photonic Crystal* (PC) structures (reference (Joannopoulos, 2008)). PC's are periodical structures in which materials of different refractive index alternate according to the geometry imposed by a certain lattice configuration. By means of multiple interference effects, based on a detailed analysis of the Maxwell equations, either allowed or forbidden frequency states for photons arise within the PC structure. The total frequency interval across which light is not able to propagate defines the so-called *Photonic Band Gap* (PBF), analogous to semiconductors and *electronic* bandgaps. Thus, when a photonic bandgap structure is embedded within a standard silica fibre the new light guiding mechanism applies, giving rise to a *Photonic Bandgap Fibre* (PBF). In an ideal PBF the bandgap structure is designed so that only one fundamental mode is allowed to propagate along the fibre core length. Secondary modes rapidly decay across the surrounding fibre region due to the photonic band gap effect. A particular type of PBF is the so-called *Hollow-Core Photonic Bangap Fibre* (HC-PBF) (Benabid et al, 2005;Konorov et al, 2005; Ritari et al, 2005), characterized by a hollow core along with a surrounding array of air holes distributed according to a PC structure (figure 9). This type of fibre can easily be filled with either gaseous or liquid materials by standard pumping techniques or just by using capillary or diffusion effects.

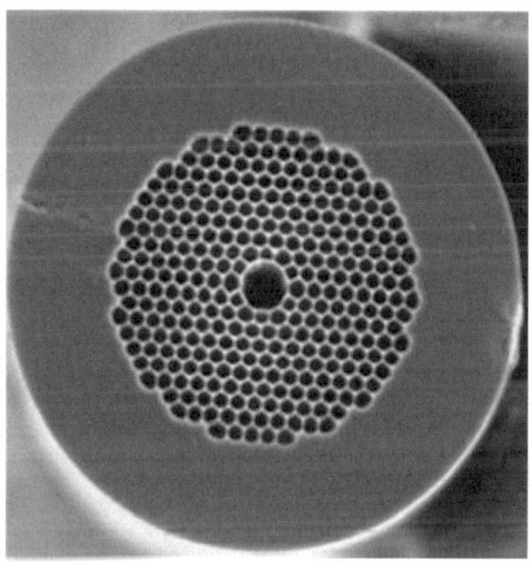

Fig. 9. Cross-section of a HC-PBF. Taken from (NKT Photonics, 2009).

5.2 Linear gas sensing in HC-PBF

Results are presented here regarding linear acetylene gas $^{12}C_2H_2$ hosted within a HC-PBF of core size ≈ 11 um. A SG-DBR laser diode operating at a wavelength λ ≈ 1530 nm is coupled into the fibre using high precision translational stages and lenses. The fibre sits within a gas chamber at a gas pressure p = 5mbar and temperature T = 22 °C. The diffusion time of acetylene inside the hollow fibre structure is t_{DIFF} ≈ 10 min. The emission wavelength of the LD is scanned using temperature and current controllers, and the transmitted signal of the gas collected using an InGaAs photodiode. High sensitivity WMS is achieved by means of current modulation of the LD, and the operating LIA referenced to the second harmonic (*2f*) of the modulation frequency. Results for several acetylene absorption lines are presented in figure 10, including signals from weakly absorbing lines that are enhanced due to the WMS technique. The particular shape of the 2f harmonics for selected gas lines around 1530 nm can be observed in the inset.

The results can be extended to the non-linear absorption regime by utilising a high-power laser diode. In such regime the absorbing transition gets saturated and the transmitted (absorbed) light is no longer directly proportional to the input optical power. The Beer-Lambert law given by equation 7 is no longer valid and a deeper theoretical analysis must be carried out in order to find out the analytical expression for the lineshape function $g(\nu-\nu_0)$ (a detailed analysis can be found in (Demtröder, 1996)). Non-linear absorption experiments are commonly based on the so-called *pump-probe* scheme, where an additional weak beam is used in conjunction with the high-power source in order to detect the nonlinear signal. This experimental arrangement is of particular complexity but can lead to many useful applications, such as stabilisation of the laser diode emission wavelength (Pineda Vadillo et al, 2009) by implementing an active feedback loop, as explained in section 4.3. Nonlinear absorption signals normally possess a very narrow spectral intensity profile so it is necessary to utilise laser diodes with ultranarrow emission linewidths, in order to prevent laser-induced broadening of the gas absorption lines.

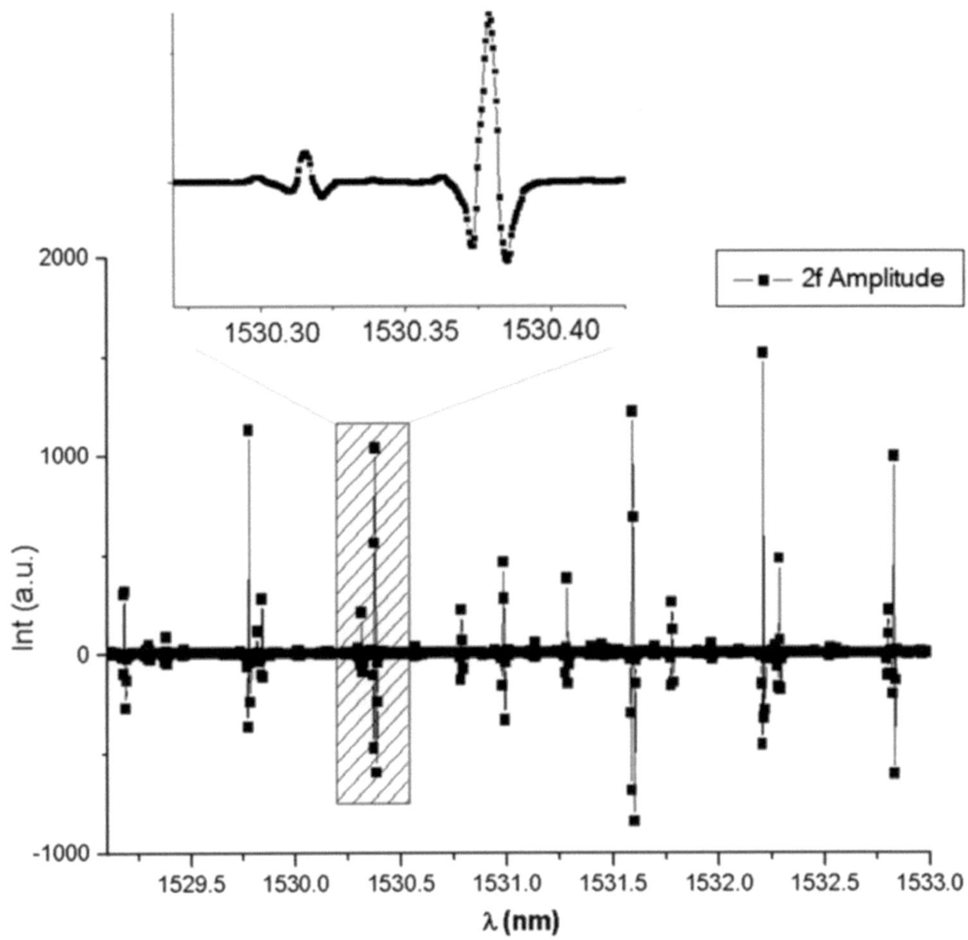

Fig. 10. Results for the successful WMS detection of the second harmonic of multiple acetylene absorption lines within a HC-PBF microstructured fibre.

6. Conclusion

Laser Diodes are a fundamental tool in spectroscopy research. Some of their fundamental physical properties, along with an affordable cost, make them nowadays an ideal light source for high precision and sensitivity spectroscopic measurement techniques. Their operational versatility makes them well-suited for both laboratory and field-based applications, ranging from metrology or laser stabilization techniques to monitoring of environmental important gases or safety gas detectors in industrial processes. In addition, laser diodes can be utilized in conjunction with novel types of photonic microstructures in order to further enhance their spectroscopic capabilities. Current research including ultra-narrow linewidth laser diodes (see for example the chapter by Dr. R. Phelan within this book) is even pushing the limits of spectroscopic applications further, making the near-future a promising time for this type of laser sources.

7. References

Banwell, C. N. & McCash, E. M. (1994). *Fundamentals of Molecular Spectroscopy*, McGraw-Hill.

Benabid, F. et al. (2005). Compact, stable and efficient all-fibre gas cells using hollow-core photonic crystal fibres, *Nature* 434, pp. 488-491.

Bentham Electronics. (2009). Lock-In Amplifier, www.bentham.co.uk/pdf/F225.pdf

Buus, J et al (2005). *Tunable Laser Diodes and Related Optical Sources*, Wiley-IEEE Press.

Demtröder, W. (1996). *Laser Spectroscopy: Basic Concepts and Instrumentation* (2nd Edition), Springer.

Garrett, P. P. (2006). Absorption and Transmission of light and the Beer-Lambert Law, www.physics.uoguelph.ca/~pgarrett/teaching/PHY-1070/lecture-21.pdf.

Herzberg, G. (1945). *Infrared and Raman Spectra of Polyatomic Molecules*, D. Van Nostrand Company, Inc.

Joannopoulos, J. D. (2008). *Photonic crystals : molding the flow of light*, Princeton University Press, Princeton.

John, H. S. (1994). Frequency-domain description of a lock-in amplifier, *American Journal of Physics* 62, pp. 129-133.

Konorov, S et al. (2005). Photonic-crystal fiber as a multifunctional optical sensor and sample collector, *Opt. Express* 13, pp. 3454-3459.

Lepère, M. (2004). Line profile study with tunable diode laser spectrometers, *Spectrochimica Acta Part A: Molecular and Biomolecular Spectroscopy* 60, pp. 3249-3258.

Measures, R. M. (1988). *Laser Remote Chemical Analysis*, John Wiley & Sons.

Nave, C. R. (2005). Kinetic Theory, Department of Physics and Astronomy, Georgia State University, hyperphysics.phy-astr.gsu.edu/hbase/Kinetic/kinthe.html, 2009.

NKT Photonics, "HC-1550-02 fibre datasheet, www.nktphotonics.com/side5334.html, 2009.

Pineda Vadillo, P. et al (2009). Non-resonant wavelength modulation saturation spectroscopy in acetylene-filled hollow-core photonic bandgap fibres applied to modulation-free laser diode stabilisation, *Opt. Express* 17, pp. 23309-23315.

Reid, J. & Labrie, D. (1981). Second-harmonic detection with tunable diode lasers – Comparison of experiment and theory, *Applied Physics B: Lasers and Optics* 26, pp. 203-210.

Ritari, H.et al (2005). Photonic Bandgap Fibers in Gas Detection, *Spectroscopy* 20, pp. 30-34.

Ritari, T. et al (2004). Gas sensing using air-guiding photonic bandgap fibers, *Opt. Express* 12, pp. 4080-4087.

Silver, J. A. (1992). Frequency-modulation spectroscopy for trace species detection: theory and comparison among experimental methods, *Appl. Opt.* 31, pp. 707-717.

Steinfeld, J. I. (2005). *Molecules and Radiation: An Introduction to Modern Molecular Spectroscopy* (Dover Reprint Edition).

Supplee, J. M. et al (1994). Theoretical description of frequency modulation and wavelength modulation spectroscopy, *Appl. Opt.* 33, pp. 6294-6302.

Svelto,O. (1998) *Principles of Lasers* (4th Edition), Springer.

Varghese, P.L. & Hanson, R. K. (1984). Collisional narrowing effects on spectral line shapes measured at high resolution, *Appl. Opt.* 23, pp. 2376-2385.

Weldon, V. (2005). Spectroscopic Based Gas Sensing Using Tuneable Diode Lasers, *Encyclopedia of Sensors*, C. A. Grimes, E. C. Dickey, and M. V. Pishko, eds., American Scientific Publishers.

19

CW THz Wave Generation System with Diode Laser Pumping

Srinivasa Ragam
Lehigh University, Department of ECE, Bethlehem, USA

1. Introduction

This chapter contains the work of continuous-wave (CW) terahertz (THz) radiation; the THz waveform generation, propagation and detection. THz technology attracts increasing interest due to its versatile application possibilities in medical imaging, spectroscopy, THz communication, nondestructive screening, and identification of many chemical elements. For such applications, it is necessary to realize compact and cost-efficient THz sources, which emit a broadband spectrum on the one hand, and can be tuned in frequency with a narrow linewidth on the other hand. Thus, the realization of a compact, cost-efficient and frequency-tunable CW THz radiation source was presented in this chapter. The principle of THz wave generation is by exciting the phonon- polariton in bulk GaP crystal on the basis of nonlinear optical method. The theoretical concept of difference frequency generation (DFG) on the basis of non-linear optics is well known. The generation of widely tunable CW single-frequency THz waves from GaP based on laser diode (LD) pumping is described in this chapter. DFG method is a parametric process based on second-order nonlinearities has recently been proved to be one of the most promising techniques for the efficient generation of widely tunable, monochromatic, high-power and coherent THz wave with a suitable combination of light sources and nonlinear optical (NLO) crystals. The region of the electromagnetic spectrum from 0.1 THz to 10 THz is very attractive because THz waves have a variety of applications in several fields. In 1963, Nishizawa [1], [2] proposed the generation of THz waves via resonance of phonons and molecular vibrations in compound semiconductors, following the realization of a GaP semiconductor laser [3], [4]. An electromagnetic wave with a frequency of 12.1 THz was generated from GaAs pumped by a GaP Raman laser at a power of 3W [5] Our group succeeded in generating wide frequency-tunable high power THz wave signals from GaP with Q-switched pulse pumping [6]. Also, time domain THz sources have been developed based on femto-second pulsed lasers [7-10]. Compared to pulsed THz sources, continuous-wave (CW) THz sources provide a THz spectrum with a narrower line width and a higher spectral THz power [11, 12]. The potential applications of CW THz waves include high resolution THz spectroscopy, multi-channel telecommunications and imaging technologies [13-19]. Several CW THz wave sources such as, Gunn diodes, TUNNETT diodes, backward wave oscillators, CO_2-laser pumped gas lasers, sources based on nonlinear optical difference frequency generation(DFG) [20], optical parametric oscillators [21], free-electron lasers, quantum cascade lasers [13,14] and photo mixers [15-16] have been developed.

Broadband tunability and power levels of at least tens of microwatt are highly desirable for CW THz spectroscopy. In addition, it is also desirable that the THz source is compact, cost-efficient and operating at room temperature. Some CW THz sources listed above can deliver sufficient power. However, either they are bulky, expensive and/or have a limited frequency tunability. CW THz sources that are compact and tunable throughout the frequency range from 100 GHz to 10 THz with tens of microwatt power levels still remain a technological challenge. A simple configuration of THz sources even is desired for easy maintenance of the devices. Moreover, the stability of THz radiation is a requirement for reliable measurements. Pumping with narrow linewidth diode lasers in nonlinear optical method of DFG will enable the generation of narrow linewidth CW THz waves of a MHz order. Semiconductor lasers are stable at light intensities, and spectrometers using CW THz waves do not need prolonged measurements with much repetition. With this motivation, we had successfully generated tunable CW THz waves from GaP crystal with diode laser pumping in the frequency range of 0.69 - 2.74 THz [22]. However the output power was limited to in the order of pW in that configuration. By using a two fiber amplifier system for amplifying the optical power of input diode lasers, we have reported that the output power of THz waves was improved nW order with employing a 20 mm long GaP [23]. In addition, the frequency of THz wave was tuned up to 4.42 THz in that configuration.

2. CW THz wave generation emission system with diode laser pumping using GaP crystal

Semiconductor NLO crystals with birefringence, such as $ZnGeP_2$ and GaSe, have been widely used as DFG crystals for THz generation due to their exotic nonlinear optical properties in the THz region [24, 25]. When the process of difference frequency mixing is phase matched using birefringence, a coherent and powerful THz wave will be obtained. Zinc blende semiconductor crystals, such as GaP is an excellent candidate for the generation of the THz radiation via difference frequency mixing. These crystals have relatively high damage threshold compared with organic crystals and can be grown with high purity and large size. Especially, their second-order nonlinear susceptibilities, $\chi^{(2)}$, are one or two orders of magnitude larger than those of typical second-order nonlinear materials such as $LiNbO_3$ and KTP [26, 27]. Furthermore, their absorption coefficient in the THz region is several times smaller than that of $LiNbO_3$ [28]. It is well known that, in the anisotropic birefringent crystals, the input pump and signal waves, λ_p and λ_s, for difference frequency mixing, and the generated THz wave, λ_T, are generally all in the same transmission window of the nonlinear crystal. In other words, their corresponding refractive indices all belong to the same dispersion curve. The dispersion is negative for three interactive waves, that is,

$$\frac{n_T - n_{p,s}}{\lambda_T - \lambda_{p,s}} < 0 \; ; \; (\lambda_p < \lambda_s < \lambda_T) \tag{1}$$

Where n_k is the refractive index at the wavelength λ_k (k = p, s, T). It is easy to satisfy the phase-matching conditions for DFG based on the normal dispersion property of nonlinear crystals and birefringence phase-matching technique. As for optically isotropic materials, such as GaP doesn't possess any birefringence. Fig. 1 shows the dispersion relation for the GaP crystal, which are generated according to its Sellmeier equation [29]. For the case of

THz-wave generation using DFG technique, the interacting pump and signal sources are generally in the NIR transmission window of the nonlinear crystal (curve with blue color as shown in Fig. 1), while the generated THz wave is in the FIR transmission window (curve with red color as shown in Fig. 1), on the other side of the crystal's Reststrahlen band, corresponding to the larger refractive indices. In the process of THz-wave DFG, the energy conservation condition is well known and is given by

$$\frac{1}{\lambda_p}-\frac{1}{\lambda_s}=\frac{1}{\lambda_T} \tag{2}$$

and the momentum conservation condition is given as

$$k_p-k_s=k_T, \quad \text{or} \quad \frac{n_p}{\lambda_p}-\frac{n_s}{\lambda_s}=\frac{n_T}{\lambda_T} \tag{3}$$

On the basis of DFG, continuous-wave (CW) THz wave generation is possible from GaP by pumping with laser diodes. We demonstrated the generation of a narrow line width, higher power, and wide frequency tunable THz waves using various GaP crystals and NIR beam spot sizes. It is known that the frequency resolution and the power stability of the THz waves are determined by the input lasers line width and power stability.

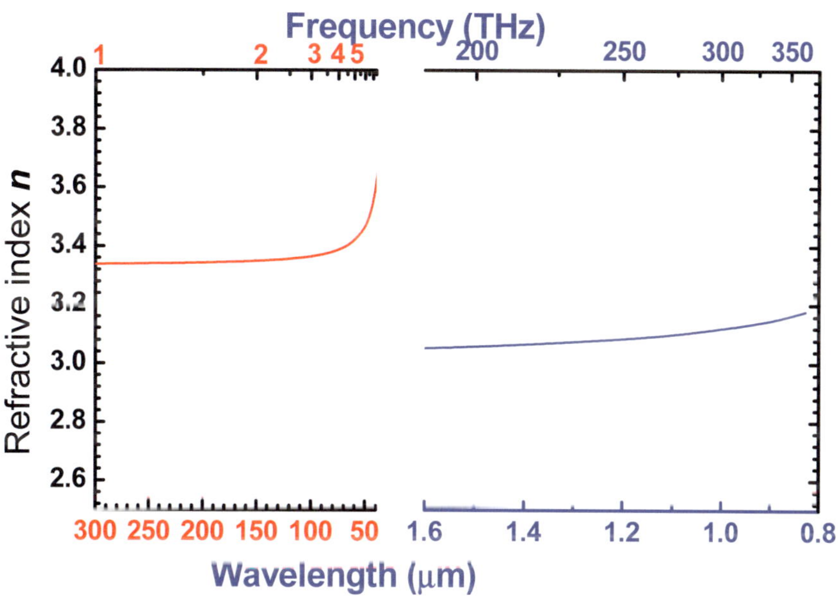

Fig. 1. Dispersion relation of GaP.

The advantages of laser diodes are: narrow line width of a few MHz, wide frequency tunability with stable output power and hence we chosen a distributed feedback (DFB) laser and an external cavity laser diode (ECLD) as pump and signal lasers. It is well known that the output power of the generated THz waves in difference frequency generation (DFG) is related to the input laser powers as follows [30]

$$P_{THz} = A\left(\frac{P_1 P_2}{S}\right) L_{eff}^{\ 2} \tag{4}$$

where P_1, P_2 are the effective powers of the pump and signal beams. L is the effective interaction length of the two near-IR beams inside the crystal (GaP), and S is the cross-sectional area of the pump and signal beams defined by the $1/e^2$ radius of the Gaussian beam, and A is a material constant for generating THz waves. From above equation (4), it is known that THz power scaling can be possible by (i) increasing the pump and signal powers P_1, P_2. (ii) proportional to the square of crystal length (iii) inverse proportional to the pump beam cross-sectional area, S. Hence we used two Yb doped fiber amplifiers (FAs) to amplify the output power of each laser diode as shown in the fig. 2(a).We studied the THz output power relation with various lengths of GaP crystals over a wide frequency range and also the THz power dependence with beam spot size of pump and signal lasers

3. Experimental method

The experimental set up for CW THz wave generation is shown in the Fig. 2(a). The pump and signal lasers are DFB laser and an ECLD respectively.

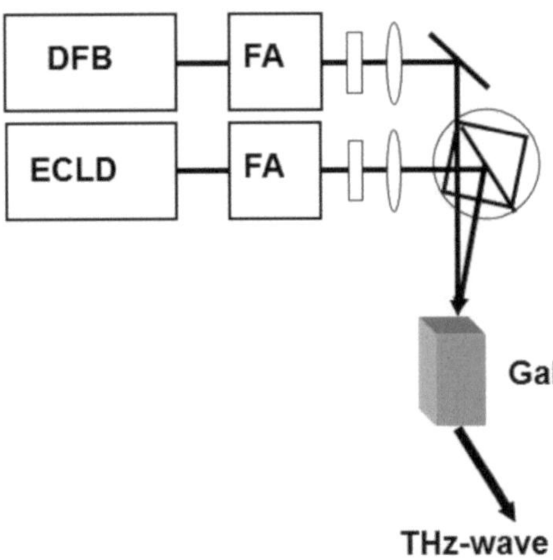

Fig. 2. (a) Schematic of the experimental set up used for generation of CW THz waves in GaP crystal.

The wavelength of DFB laser is automatically tuned from 1058 -1061 nm with a line width of 2 MHz by changing the temperature of the diode. The wavelength of ECLD can be manually tuned from 1020 nm to 1080 nm with a line width of 2MHz. Polarization maintained (PM) ytterbium-doped two Fiber amplifiers (FAs) are used to separately amplify the output power of each laser diode up to 2 W. The amplified laser beam from each FA is collimated to a diameter of 3 mm with a circular shape by using micro lenses of focal length 8.5 mm. A very small angle between the two beams is needed for DFG to satisfy non-collinear phase matching condition. So finally the collimated beams from each FA are combined on a cubic polarizer, which is mounted on a linear and rotating stage platform. An efficient spatial overlap of NIR laser beams is automatically realized on the input surface of a GaP crystal at any angle of the beams by adjusting the translation movement of the beam combiner. To investigate the THz wave power dependence on GaP crystal length, we used 2.5mm, 5 mm, 10mm, 15 mm, 20 mm long and 3 mm thick undoped semi-insulating GaP crystals in the experiment. The GaP is cut into a rectangular shape of length in <110> direction and thickness in <001> direction.

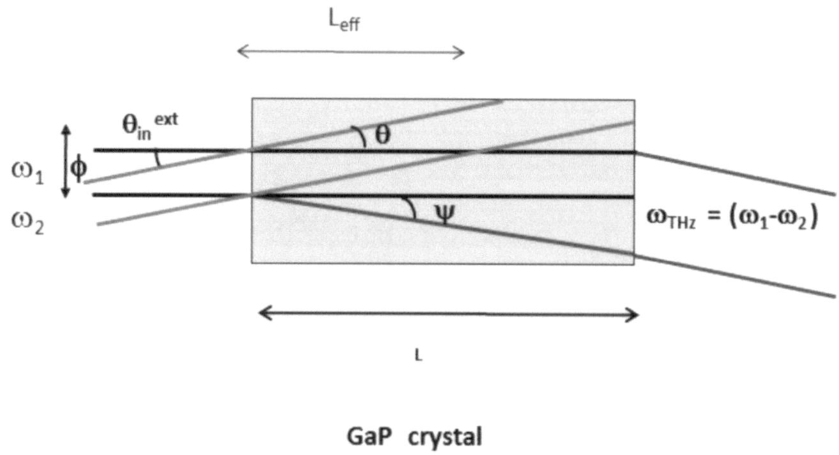

Fig. 2. (b) Typical GaP crystal cut for non-collinear phase-matched DFG.

Both the input and output faces of crystal are coated with Al_2O_3 using EB- evaporation (JEOL) after mechanical polishing and chemical etching. The polarizations of the pump and signal beams are adjusted in <001> and <110> directions respectively. The line-width of each laser is measured to be 2 MHz using a fabry-perot etalon. The generated CW THz waves are either collected by a polyethylene (PE) lens or with parabolic mirrors and detected with a Si Bolometer, cooled at 4.2 K. The Si bolometer is placed in the direction of the generated THz waves, which leave the GaP crystal output face at 40°- 45° to the surface normal. Black polyethylene film is used to filter out near IR radiation. The THz signal is measured with lock-in-amplifier.

4. Results and discussion

Figure 2(b) shows typical GaP crystal cut for non-collinear phase matched DFG at $\omega_{THz} = \omega_1 - \omega_2$ where ω_1, ω_2 are frequencies of ECLD and DFB lasers respectively. The ω_1 beam incident along normal to the input face and ω_2 beam is incident at an external angle θ_{in}^{ext} so that the angle between two input beams inside the GaP crystal is

$$\theta = sin^{-1}\left(\frac{1}{n} sin\theta_{in}^{ext}\right) \tag{5}$$

Where n = 3.105 is the refractive index at the frequency, ω_2. Under phase matched conditions, the ω_{THz} propagates at an angle (ψ) relative to the ω_1 beam as shown in fig. 2(b). The critical angle for total internal reflection ~ 17. 4° in GaP. We generated THz waves over a wide frequency range from 0.69 – 4.42 THz as θ_{in}^{ext} is varied from 8′ to 15′. Figure 3 shows the phase-matching angle (θ_{in}^{ext}) dependence of the THz wave output power at various THz frequencies, where θ_{in}^{ext} is angle between the two lasers outside the GaP crystal. THz waves are generated over a wide range from 1.4 THz - 4.42 THz as θ_{in}^{ext} is varied from 8′ to 45′. In this measurement, the GaP crystal is 10 mm long and the pump beam is incident nearly normal to the crystal surface. The linewidth of THz wave is estimated to be 4 MHz. The combined beam spot size of lasers on the GaP surface is 700μm and polyethylene lens is used to collect the emitted THz waves.

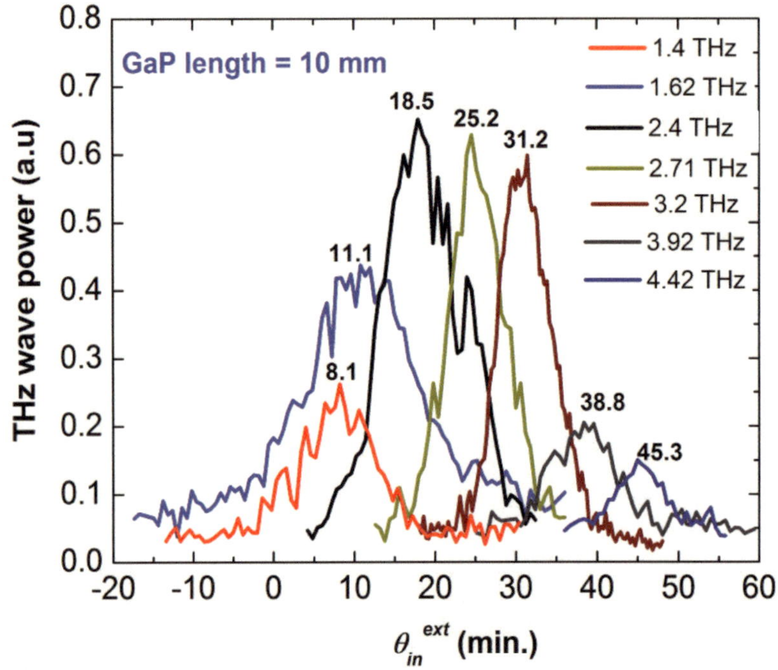

Fig. 3. The phase-matching angle $\left(\theta_{in}^{ext}\right)$ dependence of the THz wave output power at various THz frequencies, where θ_{in}^{ext} is angle between the two lasers outside the GaP crystal.

Fig. 4 shows the phase matching angle relation with THz wave frequency. The slope of the experimental data is 0.154/ THz = 2.67 mrad/ THz while slope of the solid curve is calculated as 2.91 mrad/THz. The experimental data agrees with the theoretical value, which is the θ_{in}^{ext} curve as a function of THz wave frequency for phonon-polariton branch of GaP. Increasing θ_{in}^{ext} shifted the generated THz wave frequencies slightly higher, similar to the result of pulse pumping. A continuously frequency tunable from 1.4- 4.42 THz can be generated in GaP crystal by sweeping the wavelength of the pump laser. Fig. 5 shows the frequency dependence of THz wave maximum power. It is seen that the THz wave power increases rapidly from 1.4 - 2.1 THz as the θ_{in}^{ext} is varying from 8-16.2′. The THz wave maximum power is remained over a frequency range 2.1 - 3.2 THz.

In the lower frequency (below 2 THz) the free carrier absorption is dominant. The carrier concentration of GaP is $n \approx 10^{12} cm^{-3}$. The THz wave power in the 3.2 - 4.42 THz frequency region decreases rapidly. For frequencies above 4.5THz, the deviation increases considerably because of the dispersion curve of the polariton branch of GaP. The length of the GaP crystal (L) is 10 mm and the beam spot size (ϕ) of NIR lasers on the GaP input surface is 700μm. It is estimated that for the frequency 4.5 THz, the generated THz wave propagates with an angle ψ ~ 17.5°, which is same as the critical angle for total internal reflection in GaP. To generate THz waves above 4.5 THz, we need to rotate the GaP crystal relative to the input lasers to prevent the total internal reflection.

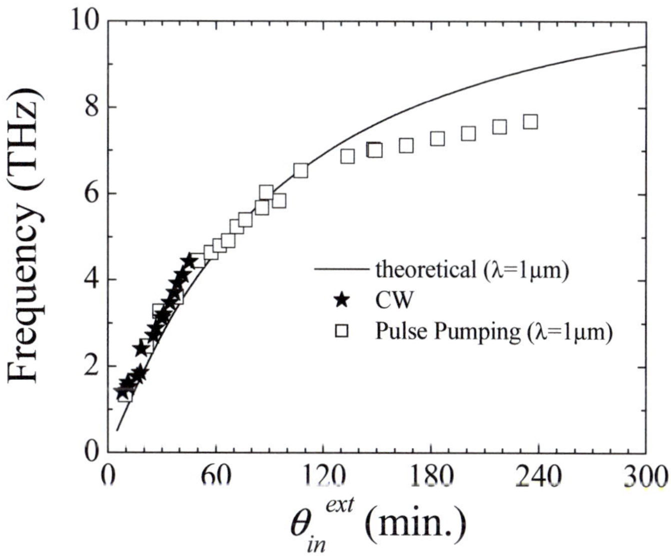

Fig. 4. The phase matching angle relation with THz wave frequency.

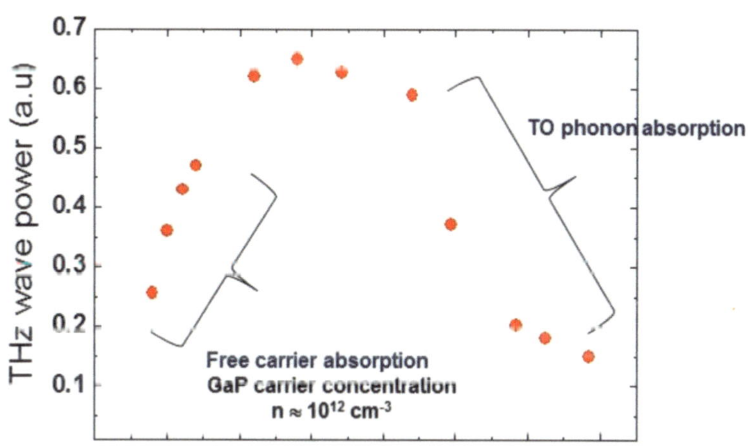

Fig. 5. The frequency dependence of THz wave maximum output power.

5. Power scaling of the CW THz source

Assume two Gaussian pump beams with the same beam radius and no diffraction effect. The output power of the generated THz waves in DFG can be calculated by using the following formula [30]

$$P_{THz} = \frac{4\ P_1 P_2 \pi^2 d^2_{effL_{eff^2}}}{c\epsilon_0 n_1 n_2 n_3 \lambda_1^2 S} \tag{6}$$

where P_1, P_2 are the effective powers of the pump and signal beams. L_{eff} is the effective interaction length of the two near –IR beams inside the crystal (GaP), d_{eff} is the effective nonlinear coefficient, λ_1 is the output wavelength of THz wave, n_1, n_2, and n_3 are the indices of refraction for pump, signal and THz beams respectively, c is the speed of light in vacuum, ε_0 is the permittivity constant and S is the cross-sectional area of the pump and signal beams defined by the $1/e^2$ radius of the Gaussian beam. The power density of the two laser diodes are enhance up to 2 W each by employing two ytterbium doped polarization maintained fiber amplifiers (FAs). When the pump power is limited the THz wave power can be improved by (i) employing longer GaP crystal and (ii) decreasing the beam spot size of the near-IR lasers.

5.1 THz wave power dependence with GaP crystal length

We have investigated the THz wave output power dependence on GaP crystal length. THz waves are generated using 2.5, 5, 10, 15 and 20 mm long GaP crystals over the frequency range from 1.5 – 3 THz as $\theta_{in}{}^{ext}$ is varied from 9′ to 23.6 ′. The bandwidth for half the maximum power was about 600 GHz when the $\theta_{in}{}^{ext}$ is fixed. Generated THz waves are detected with a calibrated Si bolometer. The beam spot size (ϕ) of the pump and signal beams are 700 μm. Fig. 6(b), shows the theoretically predicted THz wave output power behavior against the GaP crystal length at frequencies 1.5, 2 and 3 THz respectively. In estimating the THz wave power, we taken into account of the effective interaction length (L_{eff}) of the two NIR beams in non-collinear configuration and also the THz wave absorption effects in GaP crystal at each frequency. The experimental data of THz wave power with GaP crystal length for frequencies 1.5,2 and 3 THz is shown in fig. 6(a), and it is seen that for frequencies 1.5 and 2 THz, the THz power is increased with 2.5 - 20 mm long crystals as expected from the theory. The THz wave power is not saturated even for 20 mm long GaP crystal. But for frequency 3 THz, the THz wave power increased for 2.5- 10 mm long GaP crystals and decreased with 15 and 20 mm long crystals. This behavior is clearly agrees with the theoretical prediction as shown in the fig. 6(b).

The reasons may be the power losses due to absorption effect in GaP are not dominant even for 20 mm long crystal at 1.5 and 2 THz in the lower the frequency region, however at higher frequencies (above 2.5 THz) the THz absorption effects in longer crystals may not be negligible in GaP crystal [31]. The THz power is expected with proportional to the square of the effective crystal length L_{eff}. Here L_{eff} represents the effective interaction length of the NIR beams in non-collinear phase-matched mixing and is given by

$$L_{eff} = \left(\frac{L\phi}{\sin\psi} \right)^{1/2} \tag{7}$$

Where L is length of GaP crystal, ϕ is beam spot size of the input laser beam, ψ is angle of generated THz wave inside GaP relative to the pump beam as shown in fig. 2(b).

The fig. 7 indicates that the advantage of using a longer crystal is not applicable at higher frequencies (above 2.5 THz) due to the THz wave absorption effects in longer GaP crystals.

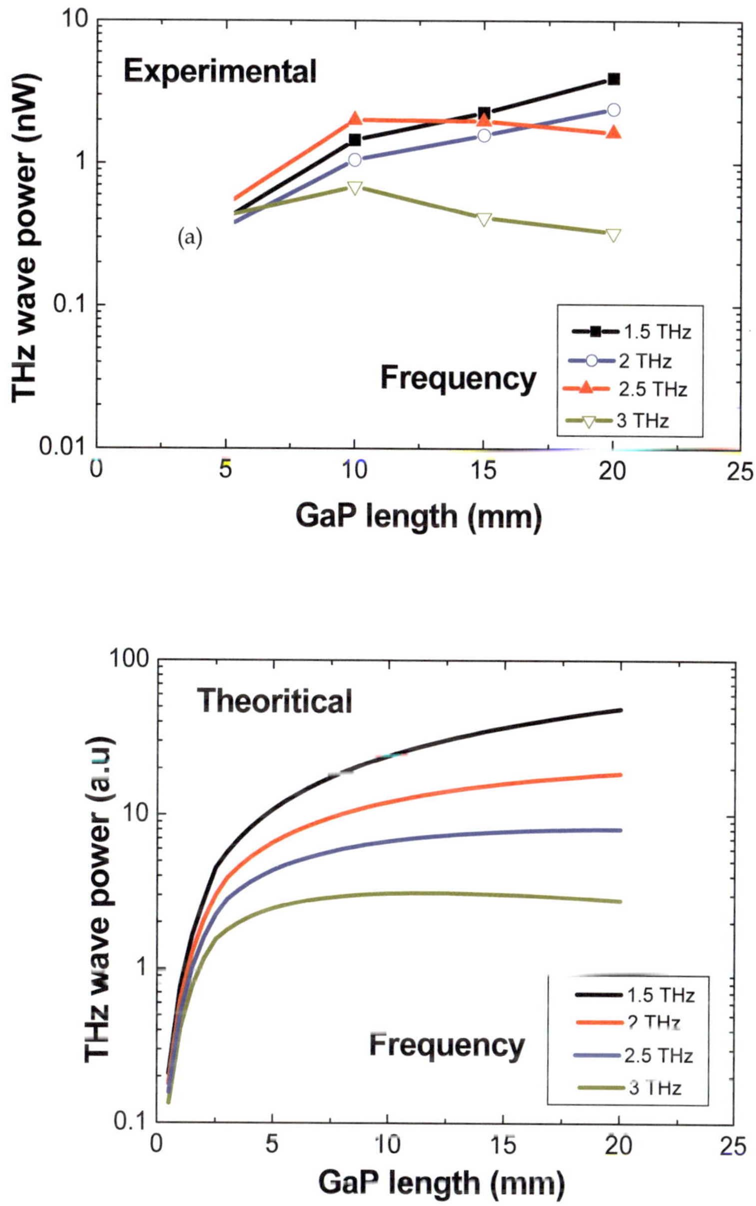

Fig. 6. (a) THz wave power dependence on GaP crystal length at 1.5-3 THz.
(b) The theoretical behavior.

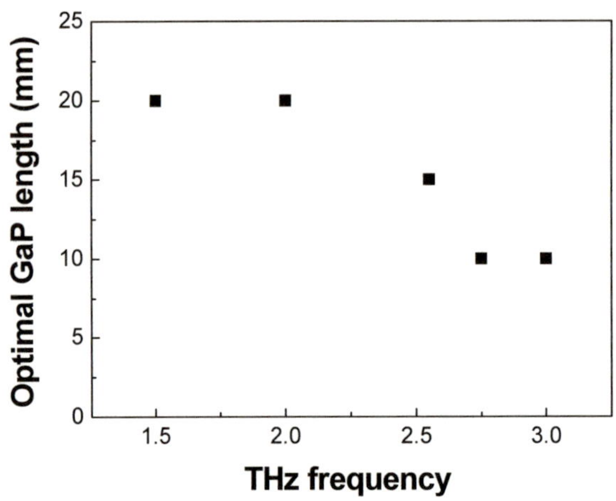

Fig. 7. The optimal GaP crystal length for THz wave frequency to obtain the maximum power.

5.2 THz wave power dependence with beam spot size of pump and signal lasers

The THz wave power is inversely proportional to the spatial overlap of cross-sectional areas of pump and signal lasers, and hence the enhancement of output power is possible by decreasing the beam spot size of the pump and signal lasers. We have investigated the THz wave power dependence for various beam spot sizes (1.2 mm-300 μm) of input near-IR lasers at a frequency of 1.62 THz using a 10 mm long GaP crystal. Focal lengths of 600 - 150 mm were used to achieve the beam spot sizes of 1.2 mm - 300 μm. A polyethylene lens with a diameter 2.5 cm was used to collect the THz waves in this configuration. It is seen from fig. 8(a) that THz wave power sharply increased as the beam spot size was reduced from 1.2 mm–500 μm.

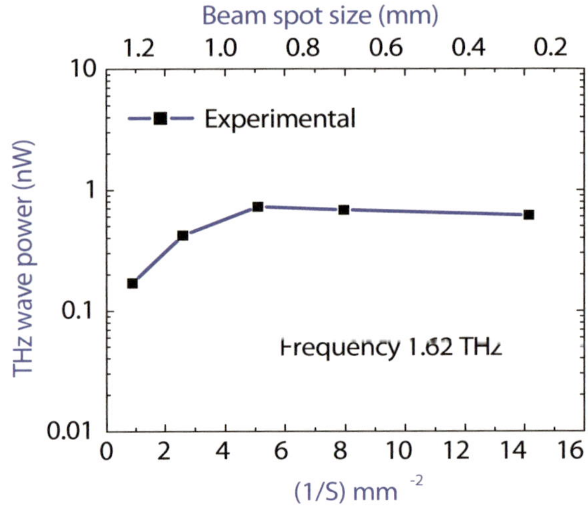

Fig. 8. (a) The THz wave power dependence on the inverse of spatial overlapping of cross-sectional areas of pump and signal lasers at 1.62 THz.

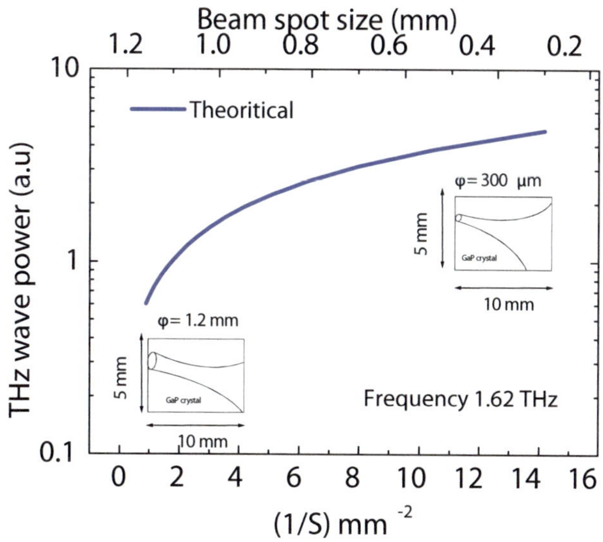

Fig. 8. (b) The theoretical behavior.

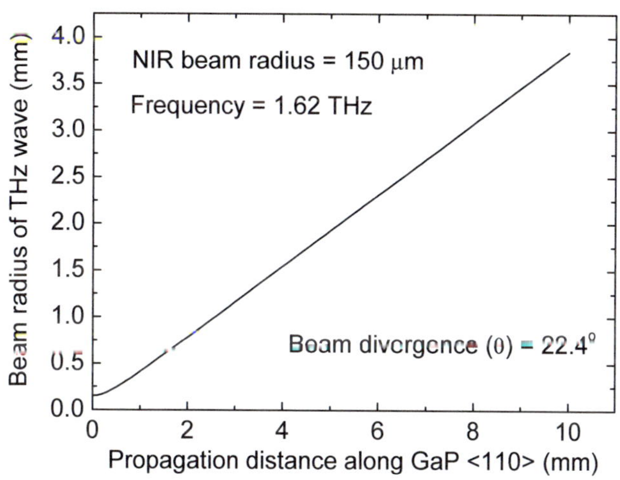

Fig. 9. The THz wave divergence inside the GaP crystal along <110> direction for the near-IR beam spot size 300 μm.

This behavior agrees with the theoretical calculation as shown in the Fig. 8(b). However, for beams spot sizes of 400 μm and 300 μm, the THz power was decreased. The output power discrepancy between the experimental data and theory in this region may have been caused by the divergence effect of the THz waves. Thus, in order to under-stand the reason for power decrement, the THz wave divergence effect at beam spot sizes of 1.2 mm and 300 μm was considered.The beam divergence of THz wave was estimated to be ~ 5.6° and ~ 22. 4° for near-IR beam spot sizes of 1.2 mm and 300 μm, respectively. Fig. 9 shows the THz wave divergence inside the GaP crystal along <110> direction for the near-IR beam spot size 300 μm. This shows that the beam divergence of THz wave was very dominant for near-IR beam spot sizes of 300 μm, as the beam spot size is of the same order as that of the wavelength of the generated THz waves. Another reason for reduction of the THz power in this region (500- 300 μm) may be that the effective interaction length L_{eff} is increases as square root of beam spot size for a given value of crystal length. Fig. 10. shows the dependence of L_{eff} on beam spot size of input lasers. It is seen that for the near-IR beam spot sizes below 400 μm, the effective interaction length is less than 4 mm and THz beam divergence is very significant. So in order to achieve high-power, one has to consider effects of beam divergence, the effective interaction length of the NIR beams, and also the absorption in the crystal. It is known that the GaP lattice absorption is dominant in the high frequency region (above 2.5 THz), and also effective interaction length is increases with beam spot size of NIR lasers. Therefore it is very important to select these parameters (frequency, crystal length, beam spot size) in order to improve the THz wave power.

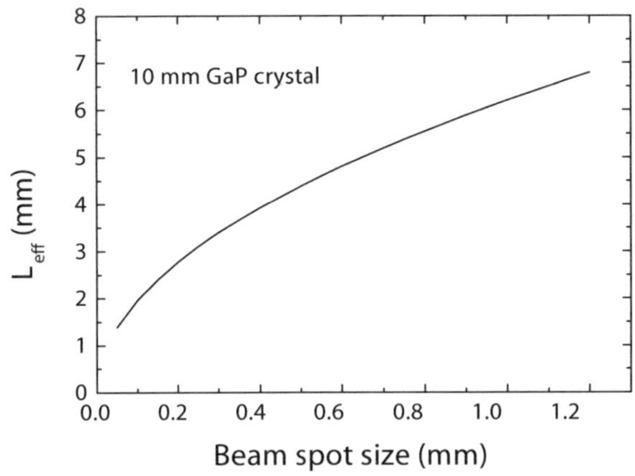

Fig. 10. The dependence of L_{eff} on beam spot size of input lasers.

6. Applications

We constructed a high resolution terahertz (THz) spectroscopic system with an automatic scanning control using our CW THz wave source 11(a). The wavelength of DFB laser was automatically tuned from 1058-1061 nm with a line width of 3 MHz by changing the temperature of the laser.

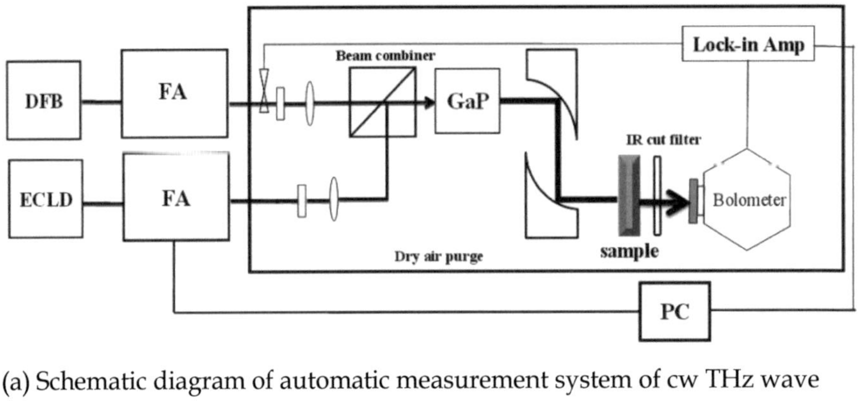

Fig. 11. (a) Schematic diagram of automatic measurement system of cw THz wave spectrometer using GaP.

The wavelength of the ECLD was manually tuned from 1020 nm - 1080 nm with a line width of 4 MHz. Two polarization maintained (PM) ytterbium doped FAs were used to separately amplify the output power of each laser diode up to 2 W. Using micro lenses with focal lengths of 8.5 mm, the amplified laser beam from each FA was collimated to a diameter of 3 mm with a circular shapeA very small angle between the two near-IR beams was required for the DFG to satisfy the non-collinear phase matching condition [22]. The phase-matching condition is fulfilled automatically by rotating and linearly translating the cubic polarizer with a PC controller via GPIB interface using visual basic software. Thus, an efficient spatial overlap of near-IR laser beams was automatically realized on the input surface of a GaP crystal at any beam angle. The near-IR beam spot size is focused to 700μm on the GaP input surface by using lenses with focal lengths of 500 mm. GaP crystals of 10 mm and 15 mm long were cut into a rectangular shape, its length being in the <110> direction and its thickness being in the <001> direction. These parameters such as beam spot size and crystal length are very important to obtain the improved THz wave power [23].

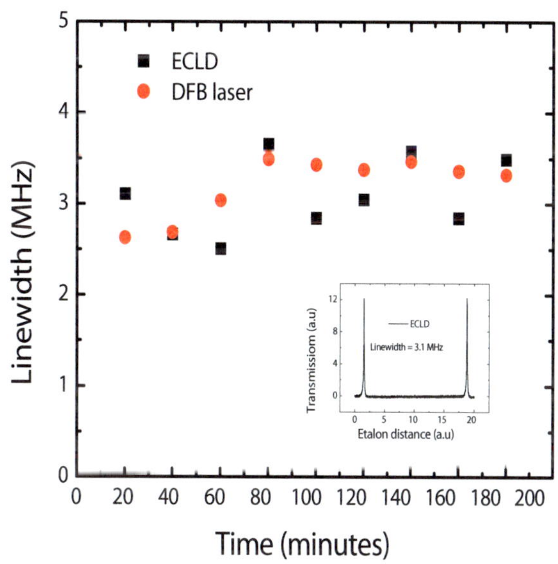

Fig. 11. (b) Frequency stability of DFB and ECLD lasers.

Both the input and output faces of crystal were coated with Al_2O_3, using EB-evaporation after mechanical polishing and chemical etching. The polarizations of the pump and signal beams were adjusted in the <001> and <110> directions, respectively. The frequency stability of diode lasers was measured with Fabry-Perot interferometer in confocal configuration for duration of 3 hours at room temperature (298 K). Figure 11(b) shows the frequency stability of DFB and ECLD lasers and the inset shows interference pattern of the ECLD laser. The linewidth of the diode lasers was calculated by the following formula:

$$Linewidth = \left(\frac{\text{FWHM of the interference wave}}{\text{Distance between the Interference peaks}} \right) \times free\,spectral\,range \quad (8)$$

The free spectral range (FSR) of the etalon was 2 GHz. The calculated linewidth of ECLD for full width at half maximum (FWHM) was ranging from 4 ~ 8 MHz and that of DFB laser was 4 ~ 7 MHz, respectively for the duration of 3 hours. Note that the linewidth is sensitive and depends on the conditions such as acoustic noise and thermal vibrations. That the spectral distribution of the diode lasers was well described with Gaussian function. Then, in order to know the frequency resolution of THz spectrometer, we can well estimate the linewidth of diode laser by full width at quarter maximum (FWQM). The FWQM linewidth of the ECLD was 2.5 ~ 4 MHz and that of DFB laser was 2.5 ~ 3.5 MHz, respectively. Thus it is concluded that the estimated linewidth of the THz wave for FWQM was < 8 MHz. The centre frequency fluctuation of the diode lasers were within 4 MHz and thus the frequency drift of generated THz waves were estimated to be within 8 MHz. The generated CW THz waves were collected by a pair of off-axis parabolic reflectors and detected with a Si bolometer, cooled to 4.2 K. The Si-bolometer was calibrated with a black body source at 350° C. Black polyethylene film was used to filter out near-IR radiation. The THz wave signal was measured with lock-in amplifier. It is known that the THz wave power stability is dependant on the input diode lasers power stability and overlapping efficiency of the two near-IR beams. A transmission spectrum of water vapor from 1.5- 3 THz was measured. Water vapor is one of the main atmospheric constituents, especially near the earth surface. The THz wave power spectrum was measured and the resonance frequencies of the water vapor were observed in the frequency range from 2 THz to 2.5 THz under room temperature (298 K) and atmospheric pressure conditions by automatically tuning the DFB laser temperature at a fixed phase-matching angle. The wavelength of the DFB laser was tuned from 1058.246–10560.043 nm by changing the temperature of DFB laser diode from 10 - 39.6 C deg. Figure 12(a) shows the water vapor transmission spectrum of THz wave. It is seen that several water vapor absorption lines occur at THz resonance frequencies with FWHM of a few GHz order and every single line of water vapor absorption can be clearly resolved. The thick lines indicate the data from NASA data base [32]. Determining the THz frequency by measuring the wavelength of the pump and signal beams is not sufficiently accurate. The THz frequency was calibrated by using the resonance frequencies of water vapor absorption lines.

We also measured the absorption spectrum of white polyethylene (PE) in the frequency range from 1.97-2.45 THz under room temperature (298 K) and atmospheric pressure conditions. Dry air is purged to eliminate the water vapor absorption in the path of the THz wave. The spectral peak at 70 cm^{-1} (2.1 THz) is known as B_{1u} translation lattice vibration mode of the orthorhombic polyethylene crystal [33]. Such low frequency lattice vibration arises from weak H··H atom interactions forming a PE crystal lattice. The PE sample has dimensions of 5 mm thickness, 10 mm width 15 mm long. The Si- bolometer window contains an edge type white polyethylene. The dimensions of the wedge on the poly window are, the thick side is 2.032 mm and the thin side is about 0.762 mm. So in order to accurately obtain the absorption coefficient of PE sample, we considered the absorption of the wedge type polyethylene window inside the Si-bolometer. We measured the absorption spectrum of wedge on poly of the bolometer by using a pulsed THz wave spectrometer system based on pumping with Cr: forsterite lasers.

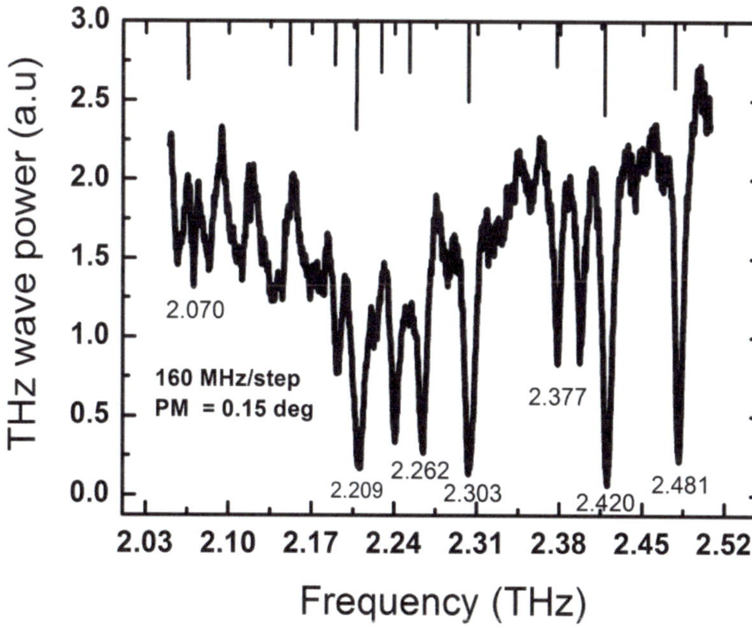

Fig. 12. (a) The CWTHz wave power spectrum with water vapor absorption lines.

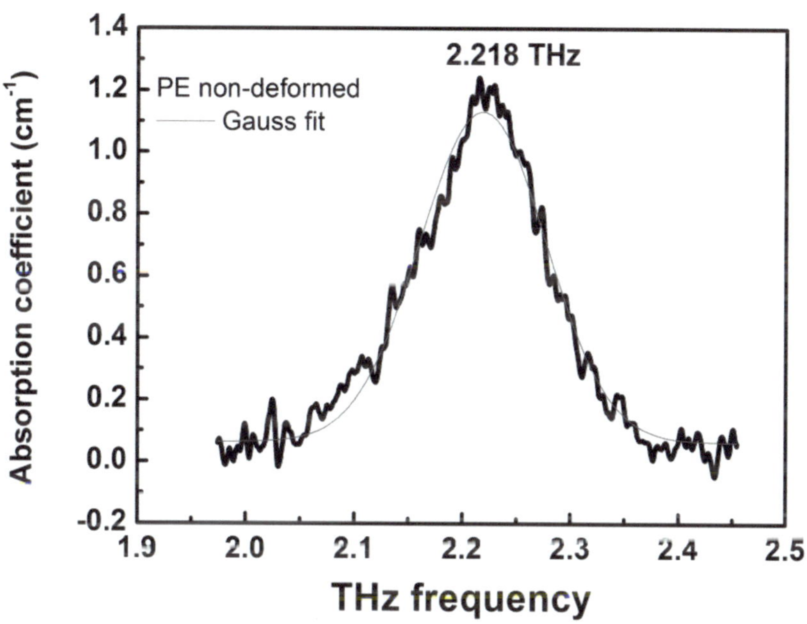

Fig. 12. (b) The absorption spectrum of white PE in the frequency of 1.97- 2.45 THz.

Figure 12(b) shows the absorption spectrum of white polyethylene in the frequency from 1.97-2.45 THz. The data analysis was done by cancelling the absorption effect of the wedge type polyethylene window inside the Si-bolometer. The B_{1u} translation vibration mode of white PE is resolved to be 2.203 THz. Note that the linewidth of the THz wave source estimated to be < 8 MHz. The B_{1u} vibration mode of the PE is very sensitive to the deformation direction. It is shown that the polarization THz measurements are helpful for non-destructive diagnosis of polymer chains in deformed ultra-high molecular weight polyethylene (UHMWPE) [34]. This CW THz spectrometer has potential application to precisely resolve the vibration mode of the PE crystals with respect to the crystal deformation. Also, this narrow linewidth frequency tunable THz wave source has potential applications in high resolution spectroscopy for investigating resonance properties of various materials.

7. Conclusion

Continuous-wave (CW) and frequency tunable coherent THz radiation was successfully generated at room temperature based on laser diode pumping via excitation of phonon-polariton in GaP crystal. The pump and signal laser sources were an ECLD and a DFG lasers. A two fiber amplifier system was implemented to separately amplify the optical power of the pump and signal lasers respectively. Two methods to improve the generated THz wave output power was proposed and investigated. In order to achieve improved output power, the effects of beam divergence, the effective interaction length in non-collinear mixing and also the absorption in the nonlinear optical crystal must be considered. A high resolution terahertz (THz) spectroscopic system with an automatic scanning control using the CW THz wave generator was constructed. The frequency of THz waves was tuned automatically by changing the temperature of the DFB laser using a system control. The power and frequency stability of the diode lasers were studied. The water vapor transmission characteristics of the THz wave and also absorption spectrum of white polyethylene were investigated. The advantage of our CW THz wave system is its very wide frequency tunability with narrow line width and it has potential applications in high resolution spectroscopy.

8. Acknowledgements

I appreciate Prof. Yutaka Oyama and Dr Tadao Tanabe of Tohoku University for their discussion and suggestions. This work was also supported in part by Global COE program "Materials Integration (International Centre of Education and Research), Tohoku University," MEXT, Japan.

9. References

[1] Nishizawa.J (1963), History and characteristics of semiconductor laser, *Denshi Kagaku,* vol.14, pp.17-31, 1963.
[2] Nishizawa.J (1965), Esaki diode and long wavelength laser, *Denshi Gijutu,* vol. 7, pp.–106, 1965.
[3] Nishizawa.J (1957), Semiconductor Maser, *Japan patent 273217,* Apr.1957.
[4] Nishizawa.J & Suto.K (1980), Semiconductor Raman laser, *J.Appl.Phys,* vol. 51, pp.2429-2431, 1980.
[5] Suto.K, Nishizawa.J (1983), Low threshold semiconductor Raman laser, *IEEE J.Quantum Electron,* vol.19, pp. 1251-1254, 1983.

[6] Tanabe. T., Suto.K, Nishizawa. J, Saito. K & Kimura. T (2003),Tunable terahertz wave generation in the 3- to 7-THz region from GaP, *Appl.Phys.Lett.*, vol. 83, pp. 237- 239, 2003.

[7] Auston. H, Cheung. K.P. & Smith. P.R. (1984), Picosecond photoconducting Hertzian dipoles, *Appl.Phys.Lett.*, vol. 45, pp. 284-286, 1984.

[8] Fattinger. Ch. & Grischkowsky. Ch. (1988), Point source terahertz optics, *Appl. Phys.Lett.*, vol.53, pp.1480-1482, 1988.

[9] Hu. B. B., Darrow. J. T, Zhang. X-C, Auston. D. H & Smith. P. R (1990), Optically steerable photoconducting antennas, *Appl.Phys.Lett.*, vol. 56, pp. 886-888,+ 1990.

[10] Xu. L., Zhang. X-C & Auston. D. H. (1992), Terahertz beam generation by femto optical pulses in electro optic materials, *Appl.Phys.Lett.*, vol. 61, pp. 1784-1786,1992.

[11] Karpowicz. N, Zhong. H., Jingzhou. X, Lin. K-I., Hwang. J-S & Zhang. X-C. (2005), Comparison between pulsed terahertz time-domain imaging and continuous terahertz imaging, *Semicond.Sci.-Technol.",* vol. 20, pp. S293-S299, 2005.

[12] Siebert. K.J, Quast. H., Leonhardt. R, Loffler. T., Thomson. M, Bauer. T, Roskos H. Gand Czasch. S (2002), Continuous-wave all optoelectronic terahertz Imaging, *Appl.Phys.Lett.*, vol.8, 3003-3005, 2002.

[13] Kohler. R., Tredicucci. A, Beltram. A, Beere. H. E., Linfield. E.H, Ritchie. A., Iotti. R. C., and. Rossi. F (2002), THz semiconductor hetero structure laser, *Nature*, vol. 417, pp. 156-159, 2002.

[14] Yu. J. S, Slivken. S, Evans. A, Darvish. S. R, Nguyen. J and Razeghi. M (2006), High power λ ~ 9.5 μm quantum- cascade lasers operating above room temperature in continuous-wave mode, *Appl.Phys.Lett.*, vol. 88, pp. 091113, 1-3, 2006.

[15] McIntosh. K.A, Brown. E. R., Nichols. K. B, McMahon. O. B, DiNatale. W. F and Lyszczarz. T. M (1995), Terahertz photomixing with diode lasers in low-Temperature grown GaAs," *Appl.Phys.Lett.*, vol. 67, pp. 3844-3846, 1995.

[16] Kleine-Ostmann. T, Knobloch. P, Koch. M., Hoffmann. S., Breede. M, Hofmann. M, Hein. G, Pierz. K, Sperling. M and Donhuijsen. K (2001), Continuous THz imaging, *Electron.Lett.*, vol. 37, pp.1461-1462, 2001.

[17] Sasaki. Y, Yokoyama. H, and Ito. H (2005), Surface-emitted continuous-wave terahertz using periodically poled lithium niobate, *Electron.Lett.*, vol. 41, pp. 712- 713, 2005.

[18] Baker. C, Gregory. I, Evans. M.,.Tribe. W, Linfield. E. and Missous. M (2005), All optoelectronic terahertz system using low-temperature grown InGaAs photo mixers," *Opt.Express*, vol. 13, pp. 9639- 9644, 2005.

[19] Stone. M. R, Naftaly. N, Miles. R.E, Mayorga I.C, Malcoci. A and Mikulics. M (2005), Generation of continuous-wave terahertz radiation using two-mode sapphire laser containing an intracavity Fabry-Perot etalon,"*J.Appl.Phys.*, vol. 97, pp.103108.1-4, 2005.

[20] Ding. Y.J and Shi. W (2003), Widely tunable, monochromatic, and high power terahertz and their applications", *Journal of Nonlinear Optical Physics & Materials*, vol. 12, no. 4, pp- 557-585, 2003.

[21] Kawase. K, Shikata. J.-I, Imai. K and Ito. H (2001), *Appl. Phys. Lett.* 78, 2819,2001.

[22] Nishizawa. J, Tanabe. T, Suto. K, Watanabe.Y, .Sasaki. T and Oyama. Y (2006), Continuous-wave frequency tunable Terahertz-wave generation from GaP," *IEEE Photon.Technol. Lett.*, vol.18, no. 19, pp. 2008, 2010, 2006.

[23] Ragam.S, Tanabe.T,. Saito.K, Oyama. Y and Nishizawa. J (2009), Enhancement of CW THz wave power under non-collinear phase-matching conditions in DFG, *Journal of Light wave Technology,* vol.27, no. 15, pp 3057-3061, 2009.

[24] Shi. W and Ding. Y. J(2003), *Appl. Phys. Lett.*, 83, 848, 2003..

[25] Shi. W and Ding. Y. J (2002), *Opt. Lett.*, 27 ,1454, 2002.

[26] Kondo. T and Shoji.I (2005), *IPAP Books*, 2 , 151, 2005.

[27] http://www.clevelandcrystals.com/Default.htm

[28] Tomita. I, Suzuki. H, Ito. H, Takenouchi. H, Ajito. K, Tungsawang. T and Ueno. Y(2006), *Appl. Phys. Lett.*, 88, 071118, 2006.

[29] Bahoura. M., Herman. G. S., Barnes. N. P, Bonner. C. E, and Higgins. P. T (2000), *Proceeding of SPIE*, 3928 132, 2000.

[30] Shen. Y,R (1977), *Nonlinear infrared generation,* page 62, 1977.

[31] Saito. K, Tanabe. T, Oyama. Y, Suto. K, Kimura. K, and Nishizawa. J (2008), Terahertz wave absorption in GaP crystals with different carrier densities," *Journal of Physics and Chemistry of Solids*, 69, pp. 597- 600, 2008,.

[32] http://www.jpl.nasa.gov/

[33] Tasumi M and Shimanouchi T (1965), crystal vibrations and intermolecular forces of polymethylene crystals *J.chem.phys.* vol.43, 1245, 1965.

[34] Ragam.S, Tanabe.T,. Oyama. Y and Watanabe. K (2010), Comparison of CW THz spectrometer developed with laser diode excitation and pulsed THz wave spectrometer with Cr: Forsterite source, *JInfrared, MilliTerahz Waves, 31(10),1164-1170, 2010.*